Birkhäuser

Alan Gluchoff

# From Fréchet Differentials to Firing Tables

## The Scope and Sources of Gilbert Ames Bliss' Contributions to World War I Era Ballistics

Birkhäuser

Alan Gluchoff
Mathematics and Statistics
Villanova University
Villanova, PA, USA

ISBN 978-3-031-68266-7 ISBN 978-3-031-68267-4 (eBook)
https://doi.org/10.1007/978-3-031-68267-4



This book is published under the imprint Birkhäuser, www.birkhauser-science.com by the registered company Springer Nature Switzerland AG
The registered company address is: Gewerbestrasse 11, 6330 Cham, Switzerland

# Preface: The Art and Science of Gunnery

The student of mathematics unfamiliar with the realities of artillery fire, yet wanting to understand the role of theoretical ballistics in the gunnery practices of a certain nation in a certain era, quickly faces several obstacles. To begin with, there are usually several military services from which to choose: a navy, land army, and coastal defense option must be selected, each with its own, probably differing, set of artillery problems. In a given service there are many types of guns in use, with new ones being introduced constantly in more modern times, a given kind of gun having its own peculiarities which must be mastered. Will the theory be applicable to any weapon, or will some feature of the weapon, say muzzle velocity, change some aspect of the theory? (Examples: The effect of the atmosphere on the trajectory of a shell became more important as muzzle velocities became so large that violent resistance of the air was met. It was found that the simple parabolic theory of Galileo was adequate for some mortar fire, but not other weapons.) The ballistics of high-angle and low-angle fire for a given weapon may be different.

At the level of activity known as preparation of fire, the actions surrounding the setting up of a gun for battle or artillery practice, there is a collection of instruments, each serving a different function: cannon sights, gunner's quadrants and levels, anemometers, thermometers, hygrometers, barometers, signal and telegraph services, plotting boards, azimuth instruments—these form only a partial list of the devices a gunner may have to master.[1] A ballistic theory will come up against the precision of these instruments for measuring relevant quantities such as atmospheric density; why take a certain quantity into account if it can't be accurately measured? The order of their use in a firing sequence can change, and the chain of command must be established; artillery officers, gunners, cannoneers, each relaying or executing orders. Who, if anyone, is to be responsible for the use of the ballistic theory? In what form is the theory to be made useful? Will calculations using the theory be done on the spot at the time of engagement? (Several advocates of this

[1] These devices as well as the guns and ammunition themselves were collectively referred to as the *Artillery Material* in [5, p. 48].

procedure will be encountered in what follows.) How much of the theory do the participants in a military action need to understand? (Opinions on this issue varied widely in the era under consideration.) The description of the parts of a given weapon and their relation to the trajectory of a shell is daunting; the accidental slant of an axle can throw off a range calculation; how does one account for this accident? A purely mathematical theory of the trajectory may be so sophisticated that its "solution" involves expressions which cannot be easily reduced to calculable quantities, rendering it of interest to theoreticians only.

It is important to emphasize that the range firing table, the embodiment of the mathematical ballistics theory chosen by a branch of the service (if in fact its entries are based on theory and not a compilation of firing data), is but one of the tools necessary for the preparation of fire. The table itself can be a product of several methods of construction available at the time of its making, even if the theory remains unchanged. Indeed, caution is urged by most in believing its values to be sacrosanct; they are often treated as guidelines. The theoretical basis for a chosen model of the ballistic trajectory is virtually never of interest to a gunner, and its academic instruction to officers was only accomplished, at least in the time and place considered in this book, in fits and starts, subject to an ever-growing body of competing military subjects.

Consequently, the place of ballistic theories in the practice of gunnery needs to be related in light of the practices and controversies of the era of their use, lest we succumb to the temptation to think that mathematics determines all. An attempt to do this is presented in this preface, after which mathematics will take center stage. Often in reading the documents issued by the mathematical theorists of the World War I era, one encounters suggestions, such as the need for a certain calculation to be performed during a military action, or the need for a data measurement for a new formula to be available at a certain moment in the chain of fire preparation. These suggestions can be quite at odds with the realities of fire, and must be viewed with this in mind.

The story told here is largely one concerning the literature of mathematical ballistics as it developed during the period from about 1880 through 1930, focusing mostly on the World War I era, and most closely associated with the United States Coast Artillery during that time. The literature consists of books devoted to mathematical ballistics and gunnery, articles in newly created professional military journals, articles in established mathematical journals, technical reports of new investigations taking place at Aberdeen Proving Ground, government-issued circulars, and range firing tables. So it is natural to select a piece of this literature in order to gain some insight into the interface of theory and reality. We consider three articles, dating from the 1890s, by Captain James Chester, an American Civil War combat veteran whose academic posts included a university position in Military Science and an instructorship at the US Artillery School at Fort Monroe, Virginia. Also discussed is a 1924 article by General Johnson Hagood, a World War I combat veteran whose military service continued after the War, which initiated an intense debate in the 1920s regarding the roles of preparation for and adjustment of fire. A final paper from 1929 by Philip Schwartz, a young independent-minded mathematically educated officer, is also mentioned. None of these presents all the relevant issues, but they are nonetheless instructive of the conflicts involved. The authors did not see themselves as representative of any larger group or movement; if anything, their pieces provoked such discussion as to reveal the range of opinion held by members of the military, a range surprising to one unfamiliar with the possibilities inherent in gunnery.

In this preface, extensive quotes of the writings on military issues will be given. It is thought best to allow these men to speak for themselves, both to give a sense of their style and to acknowledge that they themselves are the best spokesmen for their opinions and reasoning.

In the 1880s,[2] there began to appear publications by Major Francesco Siacci, an Italian artilleryman, who proposed an approximate solution to the differential equations of motion of a projectile. His method became the standard used by all European nations at that time due to its simplicity of calculation; indeed, it remained in use in one form or another through World War II. Its adoption in the United States occurred at the hands of several military authors, but James Ingalls, another American Civil War veteran, published in 1885 the volume [6] which had the most influence in our story. This volume had as one of its goals the production of range firing tables, the tables used by artillerymen to connect the range of a trajectory to all relevant data: the angle of elevation of the gun, the initial velocity and weight of the projectile, its shape, and atmospheric condition, for example, were all included. Many such tables were constructed using the theory as presented by Ingalls. James Chester became familiar with this work, admired it greatly, and used it as the basis of a somewhat naive but well-intentioned 1892 article [1] in the first volume of the new *Journal of the United States Artillery*, itself a product of progressive[3] thinking: the journal was intended as a forum for the discussion and dissemination of new information regarding developments in artillery material and practice.

Chester began with a classification of artillery troops into three types: cannoneers, who were to be able to "execute the manual" [1, p. 171] of any given piece of artillery, and see to its mechanical operation; gunners, who were to be able to instruct the cannoneers and also to be expert in the instruments of fire and use of ballistic formulae to determine quantities of interest such as ranges; and finally the artillerists, expert gunners who could test the accuracy of instruments, have a general knowledge of history and science of artillery, and be able to organize and plan artillery defenses. It was to the instruction of gunners that his article was

[2] Details as to exact year and nature of some publications, biographical information about authors, and exposition of Siacci's methods will be given in "Introduction".

[3] The word "progressive," used by the present author as well as those cited in this book, is likely to be associated with the "Progressive Era" in America. That era is generally regarded as having ended by the close of World War I; its beginning is less universally agreed upon, but it can safely be included in the years following 1880. The term as used by the present author carries no association with this era unless specifically stated, but that connection by others in this story is hard to judge. Like "efficiency," which became a cornerstone of some engineering management techniques, the words "progress" and "progressive" were in the air.

addressed, with special attention paid to the theoretical aspect, in light of his reading of Ingalls' book. His first point is that artillery recruits must meet certain special qualifications, especially with respect to "mental capacity" [1, p. 172], which he felt were not sufficiently recognized by the service. "The artillery will never be able to extricate itself from the slough of inefficiency in which it has been wallowing for the last quarter of a century, until some ... remedy is applied" [1, p. 173]. Speaking of artillery instruction in general, he claims–

> The artillery has been cursed with a conservatism which is bred in the bone of the majority of its officers. An artillery garrison is looked upon as an infantry battalion and treated accordingly. Artillery instruction is like a side show at a circus, permitted rather than prescribed. It is never allowed to interfere with infantry work. The effect is deplorable. Artillery instruction should be the chief occupation of artillery troops ... The two months artillery instruction at Fort Monroe is a perfect god-send to the Third Artillery. [1, p. 181]

Turning to the main topic of the paper, instruction of gunners in ballistics, Chester admits that there are some differences of opinion as to the usefulness of this topic: "Some artillery captains hold that ballistics in any form are not essential to the gunner ..." [1, p. 180], an opinion which he does not share. To make this point, he presents a battle scenario as he envisions it, aided by the presence of properly instructed gunners:

> Imagine an action about to open, and, for the sake of simplicity let it be between some hostile ships and a four gun battery. ... [All personnel are at their stations.] Get me the distance of such and such a ship says the captain, and the sergeant gets it—that is, if he and the other non-commissioned officers [who might have to take the place of the sergeant] have been properly trained. What is the normal elevation for that range? The sergeant calculates or consults a table previously calculated [e.g., by using Ingalls' method], and gives the desired information. How many inches of iron can we penetrate at that distance? And the well-instructed sergeant is able to answer in a very short time. And so on.
>
> We maintain therefore, that efficiency demands that the non-commissioned officers of a battery should know how to use ballistic formulas ... And not only should they know how to use them, but they should be expert in their use. ... They should know arithmetic and the use of logarithms sufficiently well for a gunner's purposes, and be able to solve simple problems in algebra. ... thanks to Captain Ingalls, the practical use of ballistic formulas has been brought well within the capacity of any man qualified to enter the gunner's class. [1, pp. 183-184]

There follow several pages describing how Ingalls' formulae may be taught, then a key demand is made "Turning therefore to the computations in ballistics which every gunner ought to able to make we find them ... to be *range*, *penetration*, *energy*, *angle of fall*, and *remaining velocity*. The number of formulas to be committed to memory, therefore, is five" [Emphasis in the original] [1, p. 185]. As for the mathematics behind these symbols, "The instructor should then explain that the symbol [example] represents a value which is the result of a series of abstruse mathematical calculations, which fortunately it is unnecessary to go into. Captain Ingalls has already made these calculations for rifle projectiles..." [1, p. 186]. The article continues with more suggestions on instruction, and concludes with a plea for uniformity in education: "[If these suggestions are followed] The grade of gunner would then mean something definite, and the meaning would be the same in New York and California" [1, p. 191].[4] Of equal or even greater importance than Chester's arguments are the responses, included after his presentation, of nine artillery officers. These provide an opportunity to survey a sample of opinions and concerns about military realities of the time relevant to the absorption and use of ballistic knowledge. Taken as a whole, they fall into natural categories.

Regarding in general the instruction of gunners in ballistics, interest was expressed at the time taken up by teaching new material. The limited time for schooling and the large amount of material of all sorts to be learned forced the consideration of what was worth teaching, and whether adding anything new was justified. One also needed to face the fact that if too much instruction was required, a recruit might rebel and seek service elsewhere. The "mental" quality of the recruits was bluntly assessed[5]and found wanting. Chester's methods of instruction, which favored large amounts of instructor involvement, were, however, largely praised, though all agreed that the use of a textbook by the gunners was doomed to failure.[6] Such obstacles would indeed be relevant to the educational efforts in this setting in the coming decades, a topic of later sections of this book. An opinion was expressed about the instruction of artillery officers, with a caveat: "The study and application of scientific principles in artillery affairs, is an occupation advantageously and successfully pursued by artillery officers, as such ... A dangerous tendency of our prevailing instruction is the apparent assumption that artillery officers are to be invariably absent [comment made with reference to artillery exercise], and that their duties are to devolve entirely on enlisted men. Absenteeism is the bane of our artillery service of to-day; has been so in the past—disastrously in time of war—and will be so in the future if methods be adopted that encourage it" [1, p. 204].

The crux of most comments, as one might expect, involved the irrelevance of learning the formulae for any reason, be it for general knowledge or use in battle. "It will be sufficient, in addition to what we already require, if the men are taught the use of the tables *deduced from the formulas* [emphasis in the original]..." [1, p. 193] runs a typical comment. "I can conceive of no duty properly pertaining to the grade of gunner, using the designation in its largest sense, which will require a knowledge of such formulae, or their computation under such circumstances [in battle]. If a gunner is properly and fully instructed, he will be familiar with the range table of his gun, and in action, on the announcement of the object and its distance, it will be necessary for him to glance at a printed copy of it, only to make sure of the elevation, deviation and time of flight; to apply these, and correctly and promptly

[4] The desire for uniformity is reminiscent of the issue of standardization within engineering circles, also a topic of interest at this time.

[5] "There are not six men in my battery who can take out a logarithm, and what could they do at taking out interpolated values of $V$, $I(v)$, $S(v)$, etc., ..." [variables found in Siacci's theory] declared one captain [1, p. 197]. Another opined: "How can you make a man learn what a sine of an angle is, if you cannot make him thoroughly understand what is meant by an angle?" [1, p. 199].

[6] "The mental training of the average citizen who seeks to honor himself by service to our sea-coast forts has been such that the sight of a text-book with its discussion of circular functions, theory of logarithms, etc., produced mental nausea..." [1, p. 195].

lay the gun, especially on a moving target, will entirely absorb all his intelligence" [1, p. 197].

A compromise of sorts between the need for tables and the chaotic conditions under which they might have to be used was reached by the embrace of many of the respondents of *Whistler's Graphical Tables*. This collection was a presentation of the necessary range firing information in a neat, systematic form, including a graph, from which the needed data could be obtained "mechanically" [1, p. 194], to cite the word used by Whistler, who was one of the commentators on Chester's paper. More on these tables will be given below; in particular, Chester himself wrote a favorable review of them.[7]

Respect for the work involved in computing the tables themselves was shown: "...those tables should be the result of the most careful, deliberate and accurate computation; they should be based on and compiled from the annual practice reports. These reports should cover with scientific exactitude every condition of atmosphere, powder, gun, density of loading, etc., which in any way will affect the range" [1, p. 198]. Such a comment, taken at face value, commends the effort of range table construction, but its description avoids any mention of the role played in their construction by ballistic theory, which had established a beachhead in the 1880s in America.

One final matter, of significance to the larger evolution of gunnery and its increasing incorporation of scientific and mathematical material, which emerges is the qualified comparison of sea-coast gunnery with field gunnery expressed by a first lieutenant in the artillery:

> The infantry and cavalry sharpshooter...does not have to take a book of logarithms and of tables and compute the value of $C$ [another Siacci variable] for any variation of barometer or thermometer, or other change in any form of its factors. He learns by *practical experience* [emphasis in the original] that, under certain conditions of temperature, light, position, or what not, he must vary his elevation from day to day to make good shooting. In the course of a time of greater or less length he learns to almost instinctively give somewhere near the proper allowance for the first or trial shot, and to make the proper alteration to correct any inaccuracies observed in this trial shot, and once on the target he can stay there indefinitely. Admitting at once that the heavy gun practice does not agree exactly with small arm firing, particularly in the cost of ammunition, and that for the latter reason if no other every shot should be made to tell if possible, the question arises as to how to make that happen? [1, p. 200]

This comparison opens the door to the role of "practical experience" in gunnery, a way of conducting fire that relies on methods not translatable in scientific terms. It also brings up "adjustment of fire," an activity with its own set of problems. The lieutenant later remarks:

> I object to the scheme of theoretical instruction on another ground, and that is the spirit of pedantry and pedagogy that it would be bound to foster. We are scientific enough now in

[7] It is not entirely clear at what point Chester became aware of the tables, as the review cited appeared in the same year as the article under consideration. Whistler prefaced his remarks cited above with the phrase "as I soon hope to show."

> certain ways, and need to become somewhat more practical instead of more scientific. … The science … should be developed and worked out while we are in garrison at peace … We then can learn to shoot somewhat in the same manner as the sharpshooter with his hand rifle, or the field-battery with its field-guns, with the advantage over both that science has foreseen the difficulties to be met, and has provided the corrections therefore in advance, as aide and assistants to experience and memory that are now the main reliance of the sharpshooter and field-artillery. [1, p. 202]

The lieutenant's comments emphasize the need for a more cautious and scientific style in the employment of sea-coast guns. But they also point to the inevitable balance between the scientific treatment of an activity, which involves the language of mathematics, and the reservoir of experience filled by those who have engaged in it. Numerous similar instances of this conflict in developing technologies at this time could be given; aircraft design is a slightly later example. Born of this duality is the phrase "the art and science" of a subject, which one often finds in ballistic literature of the era.[8] This tension was to increase with the coming of the contributions of mathematicians to the ballistics of the World War I era.

The dichotomy is present in the opening lines of [2], Chester's 1892 review of *Whistler's Graphical Tables*, which he praises for their bringing to the gunner in easy-to-read form all the firing table data necessary for gunnery. "The precision and power of modern guns make it possible to operate upon an enemy at distances beyond the limits of distinct vision, but to do so effectively and with due regard to economy, the gunner must have scientific assistance. 'Finding the range' is an art which never can be thoroughly acquired without some knowledge of the science of ballistics" [2, p. 1028].

Since the ultimate subject matter of this book is the variations from a "normal" trajectory that will occur under differing conditions of fire, it will be instructive to quote Chester's description of a typical use of the *Tables* as he envisions it in action. The reader can then obtain a rough idea of the activity involved.[9]

> No. 1 is a table of three elements, viz: Initial Velocity, Angle of Departure, and Range of Shot; any two of which being known, the third can be read from the table. … the gunner can determine the practical initial velocity for himself, and then read the table, with that initial velocity, as part of the velocity argument … We should like to call it the gunners initial velocity, which is the true one. With the Reduced Initial Velocity and the Angle of Departure as arguments, Table 1 becomes a table of ranges for the day.
>
> While the gunners of the past, especially in active service, have been content with close approximations in ballistic calculations, and have drawn largely upon their experience and judgment for the slight corrections demanded by insignificant variations in ballistic data, modern gunners must be more exact. Rule of thumb[10] corrections are inadmissable. In Table

[8] As might be expected, this phrase became subject to alteration, with references to the "art of gunnery" meant to convey the entire subject to be mastered. For example, the opening sentence of [p. 38]: "The art of applied exterior ballistics has flourished from earliest times."

[9] A more comprehensive account involving more instruments and procedures can be found in [5, Chapter 9].

[10] This phrase was repeated used in a perjorative sense in Forest Ray Moulton's critiques of the ballistics of this era.

No. 2 he can read the variations in range due to variations in weight of projectile...This may be called the first correction. Having found the range from Table No. 1 he corrects it for the variation of the shot...

The second correction is that due to direction and velocity of the wind. The wind vane and anemometer give him the needful data, which he resolves into components, and entering No. 3 with the range reads the increase or decrease due to wind of one mile velocity. [Another operation gives the desired second range correction.]

No. 4 is a double table intended to supply the correction for variation in atmospheric conditions. The elements of the right-hand table are Barometric reading, Thermometric reading, and $\frac{\delta_1}{\delta}$ (the ballistic value of the density of air). [Directions involving the table and multiplication applied to this data are then given to determine the third correction.]

Table No. 5 deals with ... "Difference of Level [the height of the gun above the water]." The elements of the table are the initial velocity, and the variations due to a difference in level of one foot. [Again, directions follow on how to get the correction for this element.]

Table No. 6 ... gives the jump of the 8" M. L. R. [magnum long rifle] at elevations ranging from 0 to 11 degrees. ... The jump according to Ingalls or Whistler can be read directly from it. [The jump is defined as the difference between the angle of elevation and angle of departure, the latter caused by firing the gun.]

[Tables No. 7, 8, and 9 give respectively the time of flight, angle of fall, and remaining velocity.]

A final word of praise for these tables is sounded by the last respondent to Chester's paper: "It effectually quiets many conflicting opinions regarding the theoretical instruction of gunners, and as such, is a boon that requires only to be utilized to be appreciated" [1, p. 204].

Four years later, perhaps after having more experience with coast artillery target practice, Chester published a second article on instruction [3] which struck some of the same themes as his first, and sounded some new warnings. Here the art and science of gunnery are explicitly connected at the outset to practical and scientific perspectives, respectively. Furthermore, the distinction has become associated with the gunner himself, the contrast being between a "practical" and a "scientific" gunner. The claim to be justified is that the teaching and training of the practical gunner has been ignored in the coastal service. Immediately dealt with is the irrelevance of ballistic formulae to such a gunner, based on their uselessness in the follow-up activity to the preliminary gun-laying as outlined in the steps above, namely the correction of observed deviations. This last becomes the focus of a large portion of the article.

"He [the gunner] has no time to analyze a deviation in action, and if he had, what would be gained by resolving it into its elements? He already has their algebraic sum before him, and that is the thing to be corrected. ... The practical gunner is indifferent as to the contributing causes. He has a fact before him, which contains, and is the resultant of all known and deviating causes, and his art tells him how to correct it. Why should he appeal to ballistics?" Even if he were a scientific gunner as well as a practical one, he would have no time to do so in action.... Observation and judgment should be his guide" [3, p. 209]. Whistler's charts are praised, but with the qualification that "... if his observation and judgment have been untrained, he would be helpless with it in battle. Neither time nor the turmoil of a bombardment would permit the calm consideration of even such a ready reckoner. If the gunner cannot correct observed deviations without appealing to chart or formula, he is untrained" [3, p. 210].

It appears that in the converse problem of adjustment of fire, Chester has gone to the opposite extreme on the spectrum of judgment and calculation. And this conclusion is borne out. Criticizing the existing *Course of Instruction for Artillery Gunners*, he finds it "scientific" [3, p. 211], suitable for use in "any first class college," but unsuitable for the practical gunner. Chester even goes to the extent of criticizing an overdependence on the ballistic firing tables:

> However [after learning that the chart contains angle of departure, muzzle velocity, and range], as the muzzle velocity, the angle of departure, and the range hang together so beautifully on the chart, the gunner gradually gets the idea that gunnery is an exact science, and that no reason exists why he should not be able to hit anything visible within range. The reaction that follows target practice may be imagined. ... The gunner is fascinated with it [the table] in proportion to his ignorance of the laws on which it is constructed. His faith in it is sublime until he comes to test its accuracy on the practice ground. Then the instructor has to explain the difference between theoretical and practical results as best he may. [3, p. 211]

The mode of instruction for the practical gunner is of course target practice, and the remainder of the paper deals with suggestions to be employed in that domain. The specifics of equipment to be mastered are not relevant, nor are different protocols for distinct scenarios (target within sight or not, for example), but general skills and facts emerge. Focusing on the problem of correcting observed deviations, Chester repeats the fact that the gunner is presented with "a single fact, which he knows to be the algebraic sum of many misty elements. To him the fact is everything, the misty elements, nothing" [3, p. 212]. It is asserted that the gunner must learn to recognize the deviations by sight, at least within certain distances, and quickly correct them; officers and gunners must be trained observers. Beyond this demand, his remaining advice can be summarized in the phrase, not of Chester's coinage, "Know your gun." He points out that no two guns shoot alike, and that the gunner must know the peculiarities of his weapons, and, in particular, what the typical dispersion of fire is for a given gun. How much of a deviation from "normal" can be attributed to, for example, weather factors, and how much is due to the characteristic dispersion of the piece? These are, of course, very difficult questions, but surely a knowledge of the particular gun is needed to begin to answer it, lest an undue amount of deviation be attributed to quantities for which there are abstruse formulae to analyze. Disaster in battle will follow. The production of a deviation rectangle, within which all shots landing are considered to be on target, is discussed. An economic motive, to be repeatedly invoked, is also brought to the fore: "There is no surer way to waste ammunition than continual adjustment of the cannon sight for errors within the normal deviation of the gun" [3, p. 2213].

This concludes our account of Chester's efforts at gunnery instruction. The third paper in particular is meant as a corrective to overly scientific thinking, while acknowledging the role of science. When reading the literature of theoretical ballistics, one is tempted to forget such simple realities as emphasized by this article, though an attempt was made to mathematically account for dispersion of fire by the use of probability theory. This effort, extending well into the 1920s, was hampered by a lack of understanding of statistical variance, in the guise of a quantity called "probable error." Reading the accounts of this idea led some more mathematically inclined officers to attempt expositions of their own.[11] In a similar vein, where in any theory of the variation of the ballistic trajectory is there mentioned the fact that, as related by Chester, the most unsatisfactory part of laying guns for fire beyond the limit of vision is the "traversing" involved: "The cannoneers have no guide but the gunner's signals or commands, and as it requires the combined strength of two or more men to move the gun, they are sure to traverse it beyond the desired point, several times perhaps, before the gunner is satisfied. During the excitement of actual battle, this difficulty would doubtless be increased" [3, p. 215].

The years between the two collections of articles here presented—from 1896 through 1924—saw many changes in artillery material, military practice, and the use of ballistics in all branches of the American military, the latter forming the topic of this book. In particular, the twin poles of preparation of fire and adjustment of fire (correction for errors) became a source of controversy in the Coast Artillery. Usually considered as tandem activities, arguments emerged which pitted the effectiveness of one over the other in firing against moving targets. This resulted in a series of articles in the *Coast Artillery Journal*, the successor to the *Journal of the United States Artillery*. This collection stretched over a six-year period beginning in 1924. The pieces are of interest for many reasons, but for our purposes, the salient features have to do with the value of preparations of fire, in all its increased complexity, and in the perspectives it brought to the range firing table.

The first article, [4], appearing in 1924, was by General Johnson Hagood, a veteran of both the Spanish-American War and World War I and recipient of several medals for his service. It is worth quoting from extensively, and and appreciating its bombastic tone:

> Eighty five percent of the heavy artillery fire by the American forces (155-mm. G. P. F. and above) was conducted without the aid of observation, either aerial of terrestrial. If we had not had methods of fire independent of observation we might as well have dumped the bulk of our ammunition into the sea. The accuracy and effectiveness of eighty five percent of our fire depended upon a system that did not require us to know the effect of our fire—a system that on the other hand did require us to make, in advance, the most accurate mathematical and ballistic computations, the most precise adjustment of guns and instruments, to make the most careful calibration and registration of our pieces; and then to shoot, for effect, upon an invisible target, not knowing until days or perhaps weeks later whether the shots fell true or missed.
>
> This was the old system in the Coast Artillery, when correction of fire for observed error was forbidden... But today there is an ambition upon the part of many Coast Artillery officers, especially in the progressive bloc, to discard the old-fashioned and tedious methods of careful training, laborious computations, painstaking adjustments, etc., and to depend entirely upon correcting errors as they are observed during the shoot.
>
> These officers might be described as Imitation Field Artillerymen; that is, they want to imitate the methods of the Field Artillery; some without having any practical knowledge or

[11] See [7–9].

> experience in these methods; and some hoping against hope that a scheme will be devised by which the methods used by them to adjust upon a fixed land target can be used to adjust upon a moving water target. [4, pp. 87-88]

Thus opens the first article. The importance of preparation of fire in the recent World War, under conditions in which adjustment was literally impossible, is emphasized, and a respect for any necessary ballistic calculation for successful fire is shown. The United States Artillery Corps had split into the Coast Artillery and Field Artillery in 1908, and the issue at hand is whether the methods of the latter were applicable to some problems of the former (recall the discussion of Chester's first paper). The use of the airplane for observation in the World War added a new dimension to this discussion. Hagood goes to the extent of declaring that in the Coast Artillery of old, adjustment of fire was forbidden, though Chester's first paper considers it. His reasons for coming out against the adoption of adjustment under the given circumstances shows a preference for reliance on ballistics: "General Weaver [James Weaver, Civil War veteran], whose name will go down in history as one of the great pioneers in the technique of modern Coast Defense, said that the Coast Artillery was divided into two camps; those that believed in ballistics and those that believed in guesses" [4, pp. 88-89].

Hagood relates that this issue was of such importance that it was once debated on the national level:

> The question as to whether "Spotting" or "Correction of Fire by Observed Error" is or is not a desirable thing for Coast Artillery was fought out once between General Arthur Murray[12]—on one side, and the father of modern naval gunnery—Admiral Sims[13]—on the other. Theodore Roosevelt decided between them and sided with General Murray on the negative. [4, p89]

From the point of view of our topic, the conclusions seem to be paradoxical: the activity of preparation of fire, often criticized as too time-consuming in battle, is here promoted, despite its "painstaking" nature, and adjustment of fire, sometimes involving elaborate observation, is seen as "impracticable" in combat. The point of citing this quote is not to highlight the information it contains but to illustrate the complexity Schwartz experienced as he dealt with the effect of any ballistic variable on another. One also notes the variation in opinions on such a matter among military services of a single country as well as those of other countries. There is no place for it in the Standard System of Fire Control [4, p. 92]. It would appear that easy dichotomies, tempting though they may be, are not appropriate in this context.

Hagood ends with hopes for the future of the Coast Artillery, which "lies with its young men" [4, p. 92]. He gives a list of captains and lieutenants, which begins with Ingalls and Whistler, representing men who "brought us out of black powder and muzzle loading guns thirty years ago," and says that similar men will emerge

---

[12] Murray was the first chief of the newly created Coast Artillery, serving in this capacity from 1908 to 1911.

[13] Sims was the commander of all naval forces in Europe during the First World War; he served twice as the president of the Naval War College.

who will get the service out of its present difficulties, but only if they do so with "careful, painstaking and scientific development; not with a divining rod" [4, p. 92].

Absent in Hagood's discussion is the above noted attempt to apply mathematics to one aspect of adjustment of fire, the analysis of the inevitable dispersion of shots by probable error. If a shot fails to hit the target, should it be counted as a miss, or is it a result of inevitable dispersion which can somehow be accounted for statistically? (Again, recall the "rectangle of dispersion" from the discussion of Chester's paper.) How do such "misses" effect the decision to adjust? The series of articles debating the issue raised by Hagood came to an end in 1930 with a few articles devoted to probable error, one by Philip Schwartz [8], a newcomer to the military whose work will be considered in some depth in Chap. 1 of this book. Suffice it to say here that during his first decade in the military he wrote many articles on ballistics topics, some of them involving the use of mathematics, in which he earned a B.S. prior to joining the Coast Artillery. Schwartz had written an article on this topic independent of the debate [7], and in 1929s. He returned to this topic in 1929, when he published in the *Field Artillery Journal*. The article is of interest to us not primarily for its treatment of probable error, but rather its concluding overall assessment, which occurs at the end of the piece, of the accuracy and use of the ballistic tables. The assessment was made in light of the fact that the value of the probable error for each range value was included in the tables; Schwartz was interested in the accuracy of these values as well as others.

Schwartz's study of probable error was extensive, including the perusal of articles in the *Field Artillery Journal*, the U.S. Naval Institute Proceedings, and the Coast Artillery Journal, and his writing gives an attempt to account for these differing points of view. Pages of definitions, calculations, and graphs follow, leading to the conclusion that the probable error may vary in certain ways depending on the conditions of fire, thus calling the single table value into question, but that in some sense the tables give an adequate average value. The final two pages of his paper deal with the accuracy of other entries in the table, such as the effect of powder temperature on muzzle velocity. The discussion at times becomes quite technical, but shows a range of writings consulted for comparison with Army tables:

> It is undoubtedly true that increasing the temperature of the powder increases the muzzle velocity. However, the exact magnitude of this increase is considerably in doubt. The formula on which this correction is based gives values which differ from all other similar values. [An example then follows.] It is understood that the Navy uses a formula which gives approximately 2 f.s. change per degree Fahrenheit change in powder temperature, and that the French use .1 per cent change in velocity per degree Centigrade, or .06 per cent per degree Fahrenheit. Some such simple formula would be just as accurate as the present one. [8, p. 608]

This quote is cited mainly to illustrate how complex matters relating to the effect of any ballistic variable on any other became at this time. It also shows how greatly opinions varied within this nation's military services as well as those of other countries. When reading the following chapters, one needs to be mindful of these differences. Any purely mathematical model of, say, the effect of density of atmosphere on range could ultimately be subject to such scrutiny.

Schwartz concludes his paper with some statements concerning the accuracy of the Army firing tables:

> In spite of these criticisms, the U. S. Army firing tables may be considered to be as accurate as any firing tables published either by the Navy or any foreign countries. The inaccuracy in the powder temperature effect, air temperature effect, and weight of projectile effect, may all be attributed to lack of information which is being sought at the proving ground and elsewhere. No foreign country has better information on these subjects than we have, nor is any other country making any greater efforts to secure this information than we are. [8, p.608]

Also included are his opinions on the use of the tables:

> It must be realized that the firing tables constitute a tool which is prepared by the Ordnance Department and furnished to the using branches to help in firing. Artillery can be fired like shotgun shells and results will be obtained, but the results will, in the long run, not compare with those obtained with the help of firing tables. To use this tool properly it is necessary that the user have some notion of its limits of accuracy. No one would think of using a common celluloid protractor in setting the elevation of a 155 mm howitzer; a gunner's quadrant, accurate to the mil or tenth of mil would undoubtedly be used. ... Similarly, one cannot blindly apply firing table values to gun fire and hope to get the best results. It is necessary to have some idea of the limits of accuracy of every quantity in the tables in order to make use of it properly. [8, p.609]

In conclusion, it is hoped that this preface has given the reader some grounding in the realities surrounding the range and use of firing tables and their use, before so much effort is expended in explaining the mathematics used to derive the values they contain. Their relative importance in target practice and battle, their place in the protocols of preparation of fire, the importance of the ease of their use, the range of opinion as to the importance of understanding the theories behind them, the issues inherent in attempts to teach their construction, the debates of the relative merits of preparation of fire and adjustment of fire, the importance of knowing their limits of accuracy—all these will likely fall into the background as the narratives of Forest Ray Moulton and Gilbert Ames Bliss starts. Let a reminder of the risk of doing so be considered as given!

Villanova, PA, USA Alan Gluchoff

# Introduction

This book tells the story of the introduction, by American mathematicians in the latter years of World War I, of new mathematical methods to address a problem in the study of the ballistic trajectory. The problem was that of the variability of that trajectory due to changes in factors such as wind, atmospheric conditions, different types of shells, alterations in initial velocity, and so on. The two key figures in this introduction, Forest Ray Moulton and Gilbert Ames Bliss, were, respectively, professors of astronomy and mathematics at the University of Chicago. Their involvement in ballistic research was part of a larger program, sponsored by Oswald Veblen of Princeton University, to recruit academic mathematicians for the war effort at the newly established Aberdeen Proving Grounds. Moulton headed a branch of the theoretical work in Washington, D.C. and made numerous contributions to the revision of ballistic theory by analyzing a number of problems and administering a huge variety of initiatives. He also introduced a method to deal with the variation problem, a method that owed much to his knowledge of astronomy and in particular to the work of Henri Poincaré, for whom he had an often expressed admiration. In his hands, the problem of perturbations of the trajectory became analogous to the "secondary forces" which altered the orbits of planets; this connection was repeatedly used by those seeking brief descriptions of his work. He was also responsible for emphasizing the importance of mathematics and its sophisticated use in addressing ballistic problems in America, a symbol of the heavy dose of mathematical energy provided by new American military institutions. Such use of contemporary mathematics was common in European nations long before.

In treating the variability problem, each man dipped into the mathematical resources with which he was familiar–the calculus of variations and ongoing line of abstraction, which culminated in functional analysis in Bliss' case, and the current writings on celestial mechanics in Moulton's–to produce methods of solution. Moulton was the initiator in American ballistics of the idea of treating the variations as a separate entity for analysis by differential equations, whose solution he was able to effect; Bliss introduced what became known as the "adjoint method" for their solution. Both methods were used in the construction of range firing tables, but Bliss' approach became a touchstone achievement at the time not only for its efficiency in dealing with the problem but also for its illustration of the point, beginning to be stressed by contemporary mathematicians, that the development of abstract concepts can have unexpected practical application. It was often cited in this manner by spokesmen for the American mathematical community in the decades after the war.

One might well ask why the solution to such a specific problem deserves a book-length treatment. In this Introduction, an attempt is made to justify this effort.

When one thinks of mathematical contributions arising during war-time activities in the past century, the examples that come to mind are those most closely associated with the Second World War-the physicists' or mathematicians' war: the code-breaking activities of Alan Turing, Norbert Wiener's cybernetics, the mathematics of radar and atomic physics, the beginnings of operations research. The First World War, perhaps due to its remoteness, yields no obvious counterparts-the introduction of numerical methods for the solution to differential equations is the most frequently cited example, if indeed it is even mentioned. Moulton was responsible for this introduction, and his familiarity with analogous methods used in astronomical work was a staple of contemporary accounts. Moulton himself had reservations about being awarded the honor of adding this innovation to the mathematical mix, and his expositions involved factors not often stressed: rigorous proofs of convergence of the methods, for example. Delving further into the variability problem-an overlooked special case of mathematics in the World War-one encounters several persistent general themes, which can be captured by familiar labels: professionalization, the ascension of the expert, the emergence of the scientist as the exemplar of progress, replacing the "wizards" of inventive achievement (e.g., Edison), efficiency, and standardization. The supercession of the old rule-of-thumb methods in ballistics-what Norbert Wiener described as the "rough, old, formal ballistics"-by rigorous mathematical treatments, executed by mathematicians, needs to treated as an example to be added to this list. In this respect, these war-related developments were part of a movement in American engineering, the move toward a more scientifically based approach in which mathematics played a key role, as reliance on accumulated hands-on knowledge only began to be questioned. These themes were a hallmark, in particular, of Forest Ray Moulton's writings. The work of the engineers and managers associated with Scientific Management, of which Frederick Taylor was the most prominent figure, in the study of metal-cutting and related efficiency issues can be cited as a similar contemporary example in the mechanical engineering world. Taylor's efforts had a mathematical base, albeit of a different nature: studies of labor by fitting curves to data and the incorporation of complex formulae onto slide rules in order to solve systems of equations using these formulae constituted the innovations here. Moulton's and Bliss' work will be discussed from this perspective, as well as several others, in the story that follows.

The adoption of the new ballistic ideas, as mathematically appealing as they are, in the wake of the first World War is a phenomenon seldom addressed. It is an interesting feature of treatment of the variability problem by the two mathematicians that not only did it draw its methods from and add to contemporary cutting edge pure mathematics, but it found its way into the practical matters of firing table production and preparation of fire in battle. It is then natural to inquire about the reception of these ideas by the military. The story relates this reception by focusing on certain key figures, their roles, and attitudes toward the new material. Considerable space is devoted to a description of the state of ballistic discourse prior to the advent of the First World War to highlight the transition from the old to the new methods. Attention is paid to efforts to teach these ideas to officers and the approaches and reinterpretations of the mathematicians' papers that were effected to make the ideas accessible. Critical appraisals by junior military officers and professional "ballisticians" working outside the mathematical community but associated with the military are assessed. The reception of these ideas by the US Navy is addressed, and comparisons between French and English methods are made. The roughly half-dozen post-War books on ballistics, authored by Americans of different mathematical backgrounds, mentioned to varying degrees the variability problem; their contents are reviewed. Opinions of physicists are examined. Articles in newly formed artillery journals are reviewed. In sum, as comprehensive an effort as possible is made to view these new ideas from all relevant perspectives, an effort to avoid the easy conclusions of acceptance, which often follow from sentence-long descriptions of innovations. The hands and minds through which Moulton's and Bliss' ideas passed included pure mathematicians, editors and readers of military journals, human computers who performed the calculations necessary for translating variation formulae into range table entries, skeptical and critical artillery officers, gunners, instructors at artillery schools, and spokesmen for the American mathematical community. This work tries to do credit to all of these parties.

The work of Forest Ray Moulton on the variability problem was just one of many ballistic problems dealt with by him. His supervisory activities were enormous: reports on wind tunnel experiments, alteration of forms of projectiles, statistical analyses of range firings, study of the law of aerodynamic resistance of the air-these are just some of the topics he juggled and assessed in his reports. This book does not draw a complete picture of the relation between his astronomical theories and the variability problem; those wanting more information should consult [11] and [12], where a detailed account of some of related work can be found. The present volume, by contrast, focuses on a parallel account of Bliss' contributions.

Bliss' work, though typically mentioned in its military context, is part of a story in a different world, the development of mathematical analysis from the post-Weierstrassian era, beginning roughly around 1880, through the establishment of Banach's functional analysis in the 1930s. In this view of the story, one encounters the over-arching idea of rigor in mathematics. This was also a key concern of Moulton's, who used it as a rallying point in his reworking of ballistic theory, as well as Bliss, whose early publications addressed Weierstrass' attempts to provide a sound foundation for theorems in the calculus of variations. The third chapter of the present book is devoted to this aspect of the adjoint method-its surprisingly natural emergence from issues of the smoothness of solutions to systems of ordinary differential equations, its connections with implicit function theory, the calculus of variations, Volterra's functions of a line, Hadamard's functional calculus, new developments in integral equations, and to a lesser degree Hadamard's work in partial differential equations. These disciplines, though often considered today as unrelated subjects, emerged as a tightly connected package at the time. For example, one can find in Hadamard's volume on calculus of variations material on functions of a line, applications to potential theory, statements of implicit function theorems, and references to Darboux's work on theory of surfaces. Through Bliss' interest in different notions of the differential of a functional, a natural outgrowth of his concern with calculus of variations, the evolution of the analysis of functionals can be viewed, an evolution including the work of Volterra, Hadamard, Fréchet, and Banach. It will be shown that some of this development took place in what might be termed a "school" of mathematicians centered at the University of Chicago in the first half of the twentieth century and involving Bliss and his students, some of whom made significant contributions to the understanding of the new discipline. The more abstract notions of the differential and in particular the general implicit function theorems in which they occur, though a natural outgrowth of Bliss' line of inquiry, are perhaps not as familiar to modern students of functional analysis as they were at the time of their development.

The mix of firing tables, astronomy, and functionals mentioned above points to the range of worlds touched upon by Moulton's and Bliss' efforts. There is an element of simplification here. Although Moulton's work, when covered in general terms in military and scientific journals, was nearly always accompanied by descriptions of him as an astronomer, he was more accurately classed as a celestial mechanician, a mathematical theorist. Bliss' material, when similarly covered, either was linked directly to variations in trajectory, referred to as the Bliss adjoint system, or simply called "additional contributions." He himself pointed out the connections between the method and the calculus of variations and the functions of a line but the complete story is more involved. The word "functional" was not generally used in these accounts, though Norbert Wiener used it in his autobiography.

Though it may be thought that the mathematical ideas developed in America during the First World War had little lasting impact, one surprising aspect of events related in this narrative is the evolution of these ideas through the Second World War and beyond. Bliss' adjoint material was refashioned to deal with control issues in rocketry in the early 1950s, and the roots of the general notions of implicit function theory and the differential in functional analysis owe at least some of its history in the Chicago school to Bliss' work. These developments are described in the third and fourth chapters, as are some modest attempts of the American mathematical community to promote their war-time contributions in military journals and to the general public as well. The last chapter of the book attempts to address more general historical issues encountered in this story. The developments in mechanical engineering in America during the era of these events, the ongoing contrast between empirical and rational approaches to formula development, the place of this story in the larger context of mathematics and war, are all subjects worthy of longer study; an attempt to address their relevance is made here.

This last matter has been explored in recent times, for example, in the work of R. Siegmund-Schultze.[1] The First World War produced many vivid technical illustrations of the new theories of the ballistic trajectory. For example, one has the blueprinted depictions of full trajectories of a projectile at different angles of elevation accompanied by cutoff arcs for the timing of explosion of the shells; one such is depicted in this book. These artifacts are so striking that they can lead one to overlook the simple fact that they are diagrammatic representations of destruction.[2] It is not surprising that this aspect of military work was not stressed in post-war accounts by involved parties in America, though one does not have to look far to encounter it in critiques of the ascendancy of technology, let alone in the literature, of the time. It was however present in the mind of Forest Ray Moulton, who concluded his lengthy report of the activities of his branch of mathematical researchers with these remarks:

> It will not be profitable to make further recommendations which must be carried out by others if they are carried out at all. Plenty of others spring to mind. They all have elements of scientific interest. The sad thing is that whether we regard these as having for their purpose the destruction of our fellow human beings, or adopt the more comforting theory that they are being solved to deter men from engaging in war, they are nevertheless important only because of unworthy human ambitions and the wickedness of the human heart.

---

[1] Specific references to his work will be made in the story that follows.

[2] This juxtaposition is present in the straightforward introduction to the chapter on "Gunnery" contained in [10], a 1908 book on ballistics: "Gunnery in its widest sense embraces a knowledge of Exterior and Interior Ballistic [the trajectory of the shell and mechanics of the gun, respectively] and the use of *Artillery Material* [emphasis in the original] the object of the art and science involved being the destruction or of maximum damage to the object at which the artillery first is directed" [10, p. 48].

# Contents

# Chapter 1
# The Practical Contributions of Forest Ray Moulton and Gilbert Ames Bliss to Exterior Ballistics

## 1.1 Introduction

The enormous changes in science and technology, which occurred in the United States in the last decades of the nineteenth and first decades of the twentieth centuries, were reflected in the developments in its military branches. The replacement of the horse by new Ford automobiles and trucks, the introduction of the airplane, the use of the telephone and radio, and the harnessing of electrical power all took place in the armed forces as well as in civilian life. Steel was used to build elevated train lines as well as becoming a component of artillery shells. Technological developments made possible new and more effective weapons, required changes in tactics and battlefield strategy, and brought about new methods of manufacturing and testing ordnance. "The material of war has undergone greater changes in the past thirty years than in the previous hundreds of years since the introduction of gunpowder," declared the author of a 1907 textbook on ordnance and gunnery. "The weapons of attack and defense have become more numerous, more complicated, and vastly more efficient....The science of gunnery constantly requires of the officer greater knowledge and higher attainments..." [226, preface]. A more scientific attitude toward accumulation and interpretation of firing data as well as improvements in related measuring devices, such as the chronograph, resulted in the evolution of the new Aberdeen Proving Grounds into a scientific laboratory, as the present author has noted elsewhere [134, p. 331]. New educational institutions and journals sprung up to help propagate and discuss these new developments. So numerous were the changes that a post-World War editorial in the *Journal of the US Artillery*, a professional journal that began publication in 1892, warned against change for the sake of change. It contrasted change with "progress," the catch phrase for the "progressive era": "It is still a fair question as to the extent that new weapons have predicated progress in addition to change. In viewing the possibilities of a new weapon it is altogether easy for any of us to become over sanguine as to its importance and power, and thereby overthrow the balance of our judgment" [16].



A. Gluchoff, *From Fréchet Differentials to Firing Tables*,
https://doi.org/10.1007/978-3-031-68267-4_1

Among these changes, but not nearly as visible, was a revamping of the mathematical basis for exterior ballistics, the science of the trajectory of a projectile. This revision took place roughly in two waves. The first was initiated in 1885 by the appearance of the first book by an American author devoted exclusively to this subject, *Exterior Ballistics in the Plane of Fire*, by James Ingalls (1837–1927) [199]. It was continued by other writers such as Alston Hamilton (1871–1937) through the turn of the twentieth century. These men were officers in the Army, and their work was largely an adaptation of methods developed in Europe. The second wave was brought about by the first World War and was overseen by Oswald Veblen and Forest Ray Moulton, the first professor of mathematics at Princeton University and the second professor who taught astronomy and mathematics at the University of Chicago. The cumulative effect of these two waves was to lift the study of ballistics from chapter-length treatments in standard ordnance and gunnery textbooks to a higher-level study using some of the most advanced mathematical analysis of the day. The second round was sometimes referred to as the "New Ballistics" in various contemporary journals [184].

Our initial goal is to focus on the second wave of this transition by choosing a particular problem in exterior ballistics, the variability of the ballistic trajectory. Work on this problem reflects a new level of mathematical sophistication brought to an old problem. It also reflects the difficulties in making the transition to this level experienced by instructors and users of the material. This story can be seen as a study of the increasing mathematization of a technological problem occurring in the "third period" of the evolution of mathematics in America, as delineated by Parshall and Rowe [269] covering the period from 1900 to 1933. It can also serve as a study of the stratification of mathematical discourse during this era, in the spirit of David Roberts' study of Albert Harry Wheeler's geometric work [284]. It is this context that we introduce the practical aspect of the work of Gilbert Ames Bliss, as the final stage in the evolution of the treatment of the variability problem by earlier authors.

Before beginning our story proper, we will give some background on the history of mathematical ballistics, its development in the United States, and the variability problem.

## 1.2 General Ballistics History and Its Instruction in the United States

In a paper read before the International Congress of Mathematicians in Toronto in 1924, Prosper Charbonnier (1862–1936), Engineer General of Naval Artillery in France, made the following remarks:

> Exterior Ballistic Theory... is often called the terrestrial sister of celestial mechanics; it is an older sister, since Galileo created it in its modern form well before Newton, who generalized its fundamental laws to the world of the planets. The relationship between the two sciences, at their source, is so close that Newton's second book of The Principles of Natural Philosophy is actually a treatise of pure ballistics. The two sciences later split: the

> one, which occupied itself with unchanging objects, eternal and perfect, developed itself in the pure domain of mathematics and absolute laws; the other, which participates more in terrestrial imperfection, and is used for very few speculative ends, goes very slowly, step by step, hesitates, turns back ... ; as a science of application, it must follow the requirements of the fashionable ideas of the moment. ... The history of exterior ballistics is firstly closely linked to the history of rational mechanics and the discovery of the fundamental laws of dynamics [author's translation] [93, p. 317].

This passage gives a concise summary of the contemporary view of the relationship between mathematics, ballistics, celestial mechanics, and mechanics from the point of view of a career ballistician of the day.[1] Indeed, a 1953 account of the history of ballistics [242], written from the point of view of a mathematician, reads like a course in mathematical physics in which, as David Alan Grier has recently remarked [155], names not usually associated with military applications abound. In this history discoveries in subjects such as astronomy, fluid dynamics, and aerodynamics, though not necessarily leading to easily solvable problems, translate rapidly into results usable in ballistics. The following paragraphs offer a brief summary of exterior ballistic history prior to the first World War.[2]

"The early modern development of ordnance engineering in Italy began with the work of Leonardo da Vinci?" [242, p. 751]; he made studies in mechanics and stated laws of motion based on his work.[3] Niccolo Fontana (Tartaglia) (1500–1557) and Girolamo Cardano (1501–1576), known for their work on the solution of the general cubic equation, consulted with arsenals of Italian principalities; Tartaglia investigated the ballistic trajectory at the behest of the chief of ordnance of Verona. Galileo (1564–1642) held a consulting position with the Venetian Arsenal; his experiments and conclusions are well known. His parabolic path of a projectile in a vacuum was still used to direct mortar fire in the era of the World War.[4] The fourth chapter of his *Two New Sciences* [130] contains tables of characteristics of the parabolic paths corresponding to different angles of elevation of fire. Isaac Newton (1642–1727) explicated his law of gravitation by considering a projectile fired horizontally from the top of a mountain in a vacuum, culminating in a description of the moon's orbit around the earth as arising in a similar manner, with sufficiently high initial velocity. In his *Principia*, Newton also considered the motion of a particle through a medium offering resistance proportional to the square of the velocity, a study of obvious ballistic application.

---

[1] The now-antiquated term "ballistician" refers to one whose career entirely or in part involved the study of ballistics.

[2] "Exterior ballistics" is the second of three aspects of the subject as defined in the modern era, the other two being interior ballistics, or study of the means of propulsion of the projectile, and terminal ballistics, the study of the effect of the impact of the projectile.

[3] In [242, pp. 751 ff.] one can find more details plus remarks concerning his dealings with political figures of the day.

[4] The phrase "World War" in what follows refers to the first World War, as it was called in the era between the wars.

John Bernoulli (1667–1748) found the analytic solution for the motion of a projectile subject to gravitational and drag force, the latter proportional to the square of the velocity. His solution required numerical integration, one technique for which was developed by Thomas Simpson (1710–1761), a professor of mathematics at the Woolwich Arsenal in England. The close association of numerical approximation techniques with astronomical and ballistic problems was exemplified by the work of Leonhard Euler (1707–1783), who set up a system of differential equations to describe the ballistic trajectory, used a numerical technique to solve it, and also investigated planetary motion by using Taylor expansions. His translation of Benjamin Robins 1742 work *New Principles of Gunnery* and his additional contributions, which constituted a "ballistics revolution" according to Brett Steele [312], were undertaken at the request of Frederick the Great to improve the gunnery knowledge of his officers. Jean le Rond D'Alembert (1717–1783) solved the equations of motion of an object through air offering a resistance whose force was a specific polynomial function of velocity; Legendre (1752–1833) solved a similar problem, which involved an additional air density function. Pierre Laplace (1749–1827) devoted a portion of his *Méechanique Céleste* to the problem of the effect of Coriolis acceleration on the trajectory of a projectile; he became examiner of the Royal Artillery in 1784. Francis Bashforth (1819–1912), known for his contribution to the Adams/Bashforth method of numerical integration, was also an experimentalist who developed methods for finding the drag on projectiles.

Bret Steele has argued that "The ballistics revolution [credited by him to the work of Leonhard Euler and Benjamin Robins] generated new theories that offered a rational understanding of gunnery, the technology of controlling gunfire. This made the teaching of calculus and mechanics to artillery and engineering officers profitable for Western governments during the second half of the 18th century" [312, p. 350]. In particular, "The ballistics revolution offers one explanation of why the École Polytechnique provided such advanced scientific and mathematical education - to prepare cadets with sufficient knowledge to comprehend advanced ballistics textbooks such as the fourth volume of Étienne Bézout's *Cours de mathématiques, a l'usage du corps de l'artillerie...*" [312, pp. 369–371]. He goes on to describe how these ideas came to the newly emerging United States, and the role they played in the establishment of the military academy for engineering and artillery officers at West Point, New York, in 1802. *A Course in Mathematics* (1798) by Charles Hutton (1737–1823), a ballistic experimenter and instructor at the Royal Military Academy at Woolwich, England, could have been used as preparation for understanding Euler's work, Steele notes. The former volume, edited much later by Robert Adrian of Columbia University, had, according to an American account of the history of ballistics given in 1943, "a wide influence on mathematical education in the United States" [126]. In what follows, we will focus on the educational institutions of the Army and their effect on developments.

It was from West Point that textbook treatments of exterior ballistics came forth in a series of volumes for courses of instruction in ordnance and gunnery. These volumes continued to be issued well into the twentieth century. Typical were the books of James G. Benton, first appearing in 1859 and revised in editions [40] as late as 1889, which were held in high regard. In Benton's books, ballistics was but a single chapter among, for example, 13 in the 500 plus pages of the third edition. But in the following decades, a single textbook could no longer encompass all that needed to be known. In an 1892 review of a new West Point textbook appearing the previous year, the reviewer writes "When Benton's ordnance and gunnery was first published, its excellence as a text-book was at once established. For many years the growth of artillery science had been relatively slow... Under these conditions it was possible for a single text-book to present both fully and accurately, that state of military science... Today, however, it is for obvious reasons, a task of no ordinary difficulty to write a textbook of ordnance. ... We cannot reasonably expect today any textbook to fill the place once occupied by Benton" [14]. The book under consideration by the reviewer failed in his estimation to do the job. He noted in particular that "The instruction in ballistics proper, as a distinct element of the general subject, was almost nil."

The increasing complexity of military science in these years was mirrored in the founding of a series of schools relevant to our story, namely, those culminating in the Coast Artillery School of 1907. All were located at Fort Monroe, Virginia. It was in the journal associated with this institution that articles about the "New Ballistics" appeared. An article from the journal, appearing in 1927, gives the history of the school itself [21]. We shall mention some highlights, since this institution has not received the attention it warrants in connection with military education in the United States. The author refers to military academy at West Point as having become a "preparatory school" in the years following its founding, a school in need of additional support. Such support was forthcoming, and in 1824, there came into being the Artillery School of Practice. It was envisioned as a school "of every branch of military science....There are now no means of obtaining that knowledge, nor can the means be furnished in any other way than by establishing an artillery school of practice, into which shall be admitted all the officers of the army, without distinction of corps...There would also be a competent number of professors of the sciences connected with the artillery..." The orders from the Adjutant General's Office regarding the founding of the school called for the formation of a board to determine the courses of instruction; on this board were to be the field officers, the director of artillery, and the instructor of mathematics. In its second incarnation as The Artillery School in 1858, it included among its personnel "One instructor of Mathematics, to give such instruction therein as may be directed by the commandant; to attend experiments in practical gunnery; and to make and keep a record of the calculation for the same." In 1867, after the American Civil War, the school was reorganized into The United States Artillery School, where the mathematics courses used editions of Davies' *Elementary Algebra* and *Elementary Geometry and Trigonometry*. These were later replaced by Davies' *Legendre*. For some subject matter, it was also realized that existing literature was inadequate for the school's advanced goals, and in the years, some instructors wrote monographs of their own, among them being Ingalls's text on exterior ballistics of 1885. Captain Ingalls had been put in charge of the newly created Department of Ballistics at the school in 1883. By the time of this incarnation's closing in 1898, the course of study was considered by the author of the article as post graduate to that at West Point.

Shortly after reopening at the end of the Spanish-American War, a decision was made to limit the instruction to coastal defense matters only, the three departments being coast artillery, ballistics and seacoast engineering, and electricity. The name Coast Artillery School was adopted in 1907. It would appear that the school considered itself on the vanguard of progress in the military science in the United States both because of its focus on artillery and its attention to the defense of the coast. In the first issue of the *Journal of the United States Artillery*, in 1892, the Announcement prefacing the articles states that "In no branch of the military service is progress so rapid, development so unexpected, as in the Artillery," and ... one great and essential difference exists between the war-conditions that this country may have to satisfy, as compared with those of other first-rate powers. So far as these are concerned, it is almost demonstrably certain that the theater of any armed struggle into which circumstances may force us will be our maritime borders. And this means that our artillery must be prepared for the duties that will then devolve upon it...[13].[5] Illustrative of the new level of sophistication sought by the school were articles appearing in its journal, which tried to advance the discussion of ballistics. The first issue contained one such on a method for adjusting artillery fire in a wind, part of the variability problem on which we will focus. Later issues contained book reviews, including those for such mathematical works as Bateman's *Differential Equations* (1918), Dickson's *History of the Theory of Numbers* as well his *Plane Trigonometry with Practical Applications* (1922). A review of Greenhill's book on elliptic functions (1892) by Thomas Craig, professor of mathematics at Johns Hopkins University, appeared in the second volume [98].[6] Works on chess and history were also reviewed. Articles by Charbonnier on the history of ballistics and activities at French proving grounds were translated [91, 92]. New artillery instruments were highlighted, and the new techniques of nomography for illustrating physical laws were introduced in a summary article [212]. Field ballistics problems modeling battlefield situations were posed and solutions published, in the "hope[s] that [they] may be of assistance in illustrating the practical value of ballistics in the field." The nature and content of instruction for gunners was hotly debated.

The establishment and development of the Coast Artillery School can be seen as part of a general trend in the US military, which, of the increased professionalization of the American Army officers, itself a reflection of forces at work in American

[5] The *Journal* changed its title to the *Coast Artillery Journal* in 1922. Since this series plays an important role in our story, we will from now on refer to it as JUSA/CAJ.

[6] Alfred George Greenhill was professor of mathematics at the Royal Military Academy at Woolwich, London, at the time the book was published. One of the applications indicated in the title was deriving information on the trajectory of a projectile using a cubic law of resistance of the projectile to atmospheric forces.

society from 1880 to 1917.[7] In the late nineteenth century, American society had come to regard professionalization as involving more than simply gaining experience through spending a career in a certain line of work; education by knowledgeable instructors who upheld standards for both entrance and advancement became a requirement. This trend in the Army was consistent with the expectations of the Progressive Era in the early twentieth century, with its emphasis on the supposed ability of rational, educated managers to solve the problems of society. America's emergence on the world stage after the Spanish-American War also impelled the American military to operate like, act like, and even look like the professional European armies. The emphasis on continuing education became increasingly important in the post-Civil War period with the establishment of Reserve Officer Training Corps [ROTC] military education in land-grant colleges to supplement the pool of West Point officers, and the creation of the General Staff College in 1881 and the Army War College in 1901 is considered major milestones in the professionalization of the US Army.[8] Instruction in these institutions would presumably provide opportunity for exposure to mathematics.

## 1.3 Range Firing Tables, The Mathematics of Exterior Ballistics, and the First Wave of Reform

As we come closer to the mathematics of exterior ballistics as used at this time, it will be helpful to distinguish three levels of discussion. The first is the theoretical level, at which mathematical models of the trajectory are formulated, using whatever scientific laws are deemed necessary, and solved by whatever methods seem appropriate, without regard to use. The model is usually a system of differential equations incorporating analytic expressions for such variables as air resistance, shape of shell, and density of atmosphere and solved either by exact techniques or numerical approximation. In this view, the contributions of Galileo, d'Alembert and Legendre, for example, occurred at the theoretical level. A contemporary example was the application by Sir George Greenhill of elliptic functions to solve the equations of motion for a certain form of the drag law. This achievement, in Charbonnier's view (expressed in an article translated for JUSA/CAJ in 1908), elevated the status of these functions so much that they began to be taught in courses of analysis in France [92, p. 62].

The second level we will call the firing range level: here, ballistics is studied under scientifically controlled conditions, where target practices are held, data is collected, theories are tested, and firings for range tables conducted. The Coast

[7] For the information in this paragraph I am indebted to Richard J. Sommers, PhD, Senior Historian, US Army Heritage and Education Center at the US Army War College, Carlisle, Pennsylvania, whose remarks to the author are closely paraphrased here.

[8] General references for these topics are [264] and [29].

Artillery School had its firing ranges, as did many arsenals; Aberdeen Proving Grounds was the preeminent firing range in America during the World War era. We include textbook presentations of ballistics and journal articles at this level, since results of firing range tests are reported and explained in these media. The third level is on the battlefield, where tactical and strategic factors enter into the use of ballistics, and decisions are made on imperfect information. These levels clearly overlap: any theory can be tested on the firing range, and the information gathered about the performance of a gun is needed for battle.

This information was compiled in a range firing table, the material product of exterior ballistic work. The tables included such basic "elements" as range and time of flight for a specific type of projectile and gun fired at a given angle of elevation, though many more elements came to be included over time. The first such tables in modern times were those presented to the Royal Society by Benjamin Robins in 1746. Robins followed this in 1750 by a comparison of the tables, which were based on Euler's solution of a version of the equations of motion, with experimental firings; he found the agreement satisfactory. In 1764, more comprehensive tables, again based on Euler's analysis, were provided by Henning Friedrich, Graf von Graevenitz, a German Infantry officer.[9] Details of the subsequent historical development of range firing tables can be found in the Historical Appendix of [242]. We merely note here that it was not until the first World War that it was considered necessary to compute the entire trajectory of a projectile; prior to this the range, maximum ordinate and time of flight alone were considered sufficient.

The relationship between mathematical ballistic theories, actual firings, and construction of range firing tables is more complex than is usually discussed. One can imagine two extremes: a table based only on computations following from a ballistic theory and involving no firing and a table based only on the results of firing and collecting data. Thus, Galileo's parabolic arcs could be considered an early example of the former, with Robins' and Graevenitz's tables as instances of the latter. Additionally, though, the accuracy of any theory needs to be tested by actual firing; this was a routine activity at firing ranges. The matter is complicated further by the fact that, as more elaborate theories developed, firing data was required to provide accurate numerical values for constants occurring in the theory itself. An example of this is the introduction of the "coefficient of form," denoted by $i$ (less politely referred to in some American texts as the "coefficient of ignorance"). This constant appearing in the differential equations of motion was an attempt to take into account the shape of the projectile. (See Fig. 1.1.) Without going into details [91, p. 255], a preliminary value of $i$ was determined from the ogival, or nose angle of the projectile, and two successive firings were then necessary, with data collected from each one, to arrive at a satisfactory value. Other similar constants were used in the theories. The most prominent was the "ballistic coefficient," a constant that incorporated $i$ as well as the diameter and weight of the projectile,

[9] An example of the Graevenitz's tables can be found in [312, p. 370].

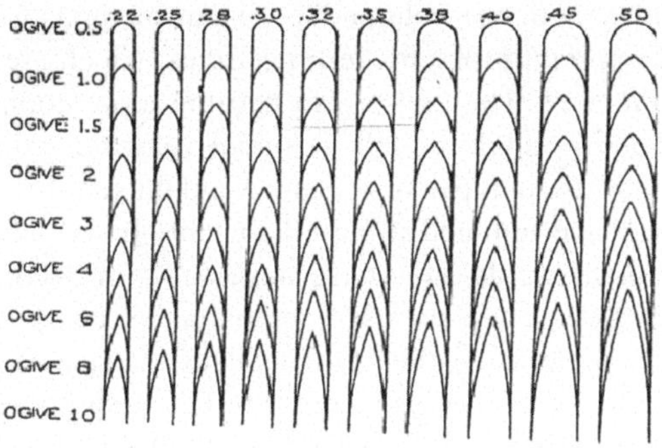


**Fig. 1.1** Various shapes of ogives used in the calculation of $i$, the coefficient of form, for projectiles. Reproduced from [96], *Exterior Ballistics Charts*

among other factors, and was roughly described as measuring the ability of a projectile to penetrate the atmosphere. Ignoring all these difficulties, one could compile a table on the basis of a few firings at different elevations followed by linear interpolation for the remaining trajectories. This involved mathematics only at the arithmetic level. Computation of a few key trajectories followed by interpolation was another option. One of the methods of compiling a range firing table devised by the second wave of ballistics researchers will be described in detail in Chap. 1 when we consider the work of Roger Sherman Hoar.

In the United States, during our era, the tables show evidence of the two extremes previously mentioned. In the *Confederate Field Manual* of 1862 , used by the Southern states in the American Civil War, ranges for different kinds of guns were given as "the mean results of such trials of the ranges of our ordnance as have been made from time to time by the ordnance department" [95, p. 110]. The information for the 6 pound field gun is comprised of amount of powder used (1.25 pounds in all cases), type of ball (shot or spherical case shot), and six elevations with their corresponding ranges. Benton's book contains tables resulting both from theory-based calculation (including the same 6 pound field gun) as well as those determined "by practice." The table for the six pound gun takes about a half a page. As the century progressed, the two extremes were explicitly discussed: an 1877 Navy pamphlet [325, pp. 3–4] describes–

> two ways of constructing such a table. One depending upon the application of formulas which have been deduced partly from practice and partly from abstract laws ... [With regard to this method] As yet, however, it has been impossible to form any law of resistance [of air] that will apply to all the circumstances attending the flight of a projectile; and, even were such laws known, their application to the known formulas would be extremely difficult...The second means of arriving at the required results is by making observations of the effects of the firing of any gun as actually carried on....This second method may be called the practical one in contradistinction to the first, which is purely theoretical. It is simpler than the first, and, in the hands of others than the most skillful mathematicians, is less liable to error.

The author described the second method, which was used at a naval experimental battery for a "3-inch B.L. rifle." The artillerymen of the first wave of reform, beginning around 1880, presented new theories as specifically tied to range table construction: Ingalls, in the introduction to his book, states "The aim has been to present in one volume the various methods for calculating range tables and solving important problems in trajectories, which are in vogue at the present day ..." [199, Preface]. The tables also became more complex, introducing additional elements such as angle of fall, terminal velocity, ordinate, and height of the summit. The increase in types of ordnance and the rapidity of change in their performance seems to have initiated a hope that table-based calculations could provide information, which would otherwise have to come from costly firings. *Artillery Circular M* (1900), a widely used set of tables compiled by Ingalls, contains a suggestive example using the United States magazine rifle, caliber 0.3 inches. "The muzzle velocity of the magazine rifle has been considerably increased since 1898, and is now in the neighborhood of 2200 f.s. [feet per second], with a promise of still further increase. We will therefore make similar computations for other muzzle velocities, forming the following table, which gives important information, more accurate than could be obtained by experiment, and which costs nothing save a little labor" [204, p. iv]. The table extends the elements of the trajectory up to a muzzle velocity of 2500 feet per second. This example clearly states an economic motive for the preference of mathematics over firing.

This economic motive became an explicit concern in the second wave of reformers. Forest Ray Moulton, the first of the second-wave mathematician-reformers, whose work we describe below, wrote the following in 1919:

> At present the performance of a gun at all ranges can only be determined by trial at the proving grounds. The range firing of a large gun costs tens of thousands of dollars besides wearing out the gun itself. This expense in time and money is only an expression of our ignorance of the fundamentals of the problem. The flight of a projectile is as much subject to the laws of physics as the performance of an engine. There is no doubt that at some time a few shots will be sufficient for the accurate prediction of what would happen under any conditions [255, p. 90].

The conditions referred to by this time had been extended to include larger angles of elevation, several types of shells for the same gun, and a host of weather-related factors such as wind, density of air as a function of altitude, and temperature. (See Fig. 1.2) Each of these factors had to be considered in the firing of each type of weapon. Moulton spoke here with characteristic imperious optimism about the power of the New Ballistics to handle all these elements with a minimum of actual firing. He also directed his ire at the work of his predecessors: "... the range tables were based on large numbers of experimental firings at various angles of elevation and were largely independent of any mathematical theory. The theory was adjusted to the firings by choosing such a ballistic coefficient at each elevation that the ballistic tables and practice would be in harmony." This is an instance of his piercing criticism of the use of the ballistic coefficient, also noted by Grier [154, p. 928]. Moulton characterized the tables as "not much more than interpolation formulas for supplying results in the intervals" [255, p. 10].

The tables had several uses. Some of these occurred at what we have called the firing range level of discussion. Ingalls' work as well as that of his successor Alston

| Range | Elevation | | Change in Elevation for 100 Yd. Change in Range | Change in Range for 1 Mil Change in Elevation | Maximum Ordinate | Terminal Velocity | Angle of Fall | Slope of Fall | Ballistic Coefficient | Fork (Change in Elevation for 4 P. E. Change in Range) | Probable Error in | | Time of Flight | Range Setting | Range |
|---|---|---|---|---|---|---|---|---|---|---|---|---|---|---|---|
| | | | | | | | | | | | Range | Deflection | | | |
| (1) | (2) | (3) | (4) | (5) | (6) | (7) | (8) | (9) | (10) | (11) | (12) | (13) | (14) | (15) | (1) |
| Yd. | Mils | ° ′ | Mils | Yd. | Ft. | F/s | Mils | 1 on— | | Mils | Yd. | Yd. | Sec. | M. | Yd. |
| 0 | −0.2 | −0 01 | 1.4 | 71.4 | 0 | 1505 | 0 | | 1.44 | 0.7 | 13 | 0 | 0.0 | 85 | 0 |
| 100 | 1.2 | 0 04 | 1.5 | 67.8 | 1 | | 2 | 623 | | | | | 0.2 | 180 | 100 |
| 200 | 2.8 | 0 10 | 1.6 | 64.5 | 2 | | 3 | 312 | | | | | 0.3 | 275 | 200 |
| 300 | 4.4 | 0 15 | 1.7 | 61.5 | 2 | | 5 | 202 | | | | | 0.5 | 365 | 300 |
| 400 | 6.2 | 0 21 | 1.7 | 58.7 | 3 | | 6 | 149 | | | | | 0.7 | 445 | 400 |
| 500 | 8.0 | 0 27 | 1.8 | 56.1 | 4 | 1565 | 8 | 117 | 1.44 | 0.9 | 13 | 0 | 0.9 | 525 | 500 |
| 600 | 9.8 | 0 33 | 1.9 | 53.6 | 6 | | 10 | 94 | | | | | 1.1 | 605 | 600 |
| 700 | 11.8 | 0 40 | 2.0 | 51.2 | 8 | | 13 | 76 | | | | | 1.3 | 690 | 700 |
| 800 | 13.8 | 0 47 | 2.0 | 49.0 | 10 | | 16 | 63 | | | | | 1.5 | 775 | 800 |
| 900 | 16.0 | 0 54 | 2.1 | 47.0 | 13 | | 19 | 53 | | | | | 1.7 | 865 | 900 |
| 1000 | 18.0 | 1 01 | 2.2 | 45.2 | 16 | 1377 | 22 | 46 | 1.44 | 1.2 | 14 | 0 | 1.9 | 955 | 1000 |
| 1100 | 20.4 | 1 09 | 2.3 | 42.8 | 20 | | 25 | 40 | | | | | 2.1 | 1045 | 1100 |
| 1200 | 22.8 | 1 17 | 2.4 | 41.9 | 24 | | 29 | 35 | | | | | 2.3 | 1135 | 1200 |
| 1300 | 25.2 | 1 25 | 2.4 | 40.3 | 29 | | 32 | 31 | | | | | 2.6 | 1225 | 1300 |
| 1400 | 27.6 | 1 33 | 2.5 | 38.8 | 34 | | 36 | 28 | | | | | 2.8 | 1315 | 1400 |
| 1500 | 30.2 | 1 42 | 2.6 | 37.4 | 39 | 1225 | 40 | 25 | 1.44 | 1.6 | 15 | 1 | 3.1 | 1410 | 1500 |
| 1600 | 32.8 | 1 51 | 2.7 | 36.1 | 45 | | 44 | 23 | | | | | 3.3 | 1500 | 1600 |
| 1700 | 35.6 | 2 00 | 2.8 | 34.9 | 52 | | 48 | 21 | | | | | 3.6 | 1595 | 1700 |
| 1800 | 38.6 | 2 10 | 2.9 | 33.8 | 60 | | 53 | 19 | | | | | 3.8 | 1690 | 1800 |
| 1900 | 41.6 | 2 20 | 3.0 | 32.8 | 68 | | 58 | 18 | | | | | 4.1 | 1790 | 1900 |
| 2000 | 44.8 | 2 31 | 3.1 | 31.8 | 77 | 1106 | 63 | 16 | 1.44 | 2.0 | 16 | 1 | 4.4 | 1890 | 2000 |
| 2100 | 48.0 | 2 42 | 3.2 | 30.9 | 86 | | 68 | 15 | | | | | 4.6 | 1990 | 2100 |
| 2200 | 51.2 | 2 53 | 3.3 | 30.0 | 95 | | 73 | 14 | | | | | 4.9 | 2090 | 2200 |
| 2300 | 54.6 | 3 04 | 3.4 | 29.2 | 108 | | 78 | 13 | | | | | 5.2 | 2190 | 2300 |
| 2400 | 58.0 | 3 16 | 3.5 | 28.4 | 121 | | 84 | 12 | | | | | 5.5 | 2290 | 2400 |
| 2500 | 61.6 | 3 28 | 3.6 | 27.6 | 134 | 1016 | 90 | 11 | 1.44 | 2.5 | 17 | 1 | 5.8 | 2390 | 2500 |
| 2600 | 65.2 | 3 40 | 3.7 | 26.9 | 148 | | 96 | 11 | | | | | 6.1 | 2495 | 2600 |
| 2700 | 69.0 | 3 53 | 3.8 | 26.2 | 163 | | 102 | 9.9 | | | | | 6.4 | 2600 | 2700 |
| 2800 | 73.0 | 4 06 | 3.9 | 25.5 | 179 | | 108 | 9.3 | | | | | 6.7 | 2705 | 2800 |
| 2900 | 77.0 | 4 20 | 4.0 | 24.8 | 197 | | 115 | 8.8 | | | | | 7.0 | 2815 | 2900 |
| 3000 | 81.2 | 4 34 | 4.1 | 24.2 | 215 | 949 | 122 | 8.3 | 1.44 | 3.1 | 19 | 2 | 7.3 | 2920 | 3000 |

**Fig. 1.2** A page from *Range Tables for 75 mm. Model 1916 American and Model 1917 Guns*

Hamilton[10] involved designing classroom exercises using their tables in problem books. Ingalls' 1885 volume promoted their use "for determining in advance the ballistic efficiency of those [guns] which may be proposed in the future" [199, Preface], that is, as a planning tool. Similarly, of mortar tables for harbor defense published in the first issue of the JUSA/CAJ, he wrote "These tables show that to provide for the various ranges which it may be necessary to employ ... several different muzzle velocities are required; and the selection of the most suitable of these muzzle velocities and the preparation in advance of the proper charges to secure them are matters of the first importance" [201].

But it is also clear that the tables were at least envisioned as being used in battle. Moulton, describing some of the new ballistic problems brought about by the World War, cited "the barrage, and especially the moving barrage, [which] made it more necessary than before to direct artillery fire with great accuracy by theory alone. Under such conditions with hundreds, and even thousands, of guns in action,

[10] Ingalls, an American Civil War veteran, was an 1872 graduate of the Artillery School who in turn taught many subjects there. He was put in charge of the Department of Ballistics in 1883. Hamilton was a career Army man, a graduate of West Point, and served in the Spanish-American War. For references on the life of Hamilton, see *General Hamilton, Ballistics Expert, Retired Army Officer Past 40 Years in the Artillery, Dies in Maryland at 66*, New York Times, December 21, 1937, p. 23; Hamilton, A., entry in *Who Was Who In America, Volume 1*, nnn, Marquis Who's Who, Chicago, 1968, p. 510. For Ingalls, see *Colonel J. M. Ingalls, Veteran of Two Wars and Expert on Ballistics, Dies at 90*, New York Times, 1927, May 2, p. 32; Ingalls, J., entry in *Who Was Who In America, Volume 1*, nnn, Marquis Who's Who, Chicago, 1968, p. 617. Authorship is absent for these references.

it would be essentially impossible to correct imperfect firing, even if there were time" [255, p. 11]. Thus, he is speaking of the need to supplant adjustment by sight with accurate tables providing information for one-time firing. This sentiment is supported by a quote from MacFarland's *Ordnance and Gunnery* of 1929, the first of the West Point Ordnance and Gunnery volumes to include the New Ballistics: "The Firing Tables are the means by which the commander of an artillery unit determines the range and deflection settings that will enable him to place his shells on the target with a minimum of adjustment after firing is open...Familiarity with these tables and facility in their use are required of all officers"[11] [238, p. 428]. Such familiarity by this time would have been no small task. The volume of information to be digested had risen from the Civil War total of one-half page to 60 pages for the *Provisional Range Tables* (Fig. 1.2) used with the British 75-mm. Gun, Model 1917, issued under Moulton's direction.[12] The 1921 edition of this same volume added detailed instructions for the use of the "meteorological message" to account for weather factors among many others in adjusting the elevation of the gun. An example in the book giving the order of actions involved in incorporating all the information for a single firing runs about two and one-half pages [278, pp. v-vii].

And what of the mathematics used to construct these tables? The United States had no native tradition of theoretical exterior ballistics comparable to those developing in European countries in the eighteenth and nineteenth centuries. D'Alembert's 1744 solution of the equations of motion of a projectile under a certain retardation law occurred more than 30 years prior to America's declaration of independence from England. Such a tradition was impossible in a country whose mathematical community only had an institution of higher learning to grant advanced degrees in the subject in 1876. The textbooks issued from West Point were dependent upon the theories extant in Europe, and such theories were reported. This state of affairs continued into the period prior to the first World War, a fact frankly stated by Alston Hamilton in the introduction to his 1908 text: "On account of the finished classical methods current at the present time little remained for the writer to do beyond selecting the best for use in this school [the Coast Artillery School]" [171]. The United States did not have the parallel of theoretical and practical ballistics as for example occurred in France at the Gâvre, where the two approaches melded in the 1880s, as Charbonnier relates in [91, p. 242]. The closest the United States came to a theoretical ballistician in the classical Euorpean sense was T.H. Gronwall, whose post-World War contributions are discussed in [134] and whose role in the development of Bliss' ideas is discussed later in this book.

The first wave of innovation came with the publication in *Revue d'Artillerie* in October of 1880 of an article by Major Francesco Siacci[13] (1839–1909) of the

---

[11] MacFarland's book will be mentioned again in the next chapter, which his step-by-step procedure for using corrections is cited.

[12] The latter volume was issued on May 24, 1918, as a handy 4-inch by seven-inch volume.

[13] Siacci was an Italian ballistician and army officer, also a professor of mechanics at University of Milan.

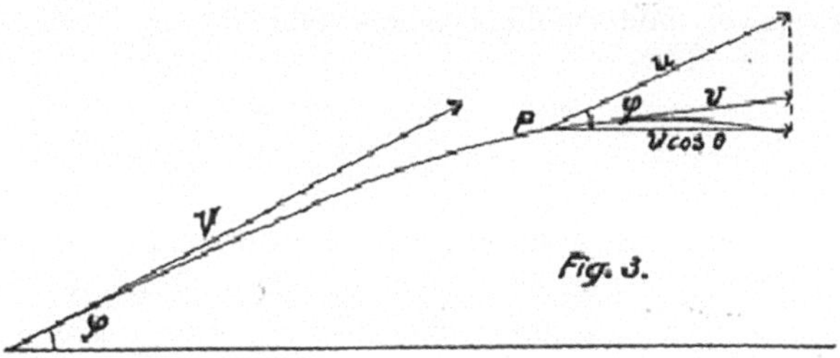


**Fig. 1.3** Reproduced from *Exterior Ballistics* by Alston Hamilton, [170]

Italian Artillery. This article described of a method of approximate solution of the equations of motion giving the ballistic trajectory. The article was translated in the United States in 1881 and included in that year in a report to the Chief of Ordnance of the Army. Siacci's method was so successful that it eventually became used by all the world's major military forces. This was the theory first explicated in the United States by Captain James M. Ingalls in his 1885 book published by the Coast Artillery School. The method was officially adopted at West Point in 1891 and replaced all other textbook approaches by order of the Academic Board. We will give an idea of the theory, combining the treatments given by the two advocates of the first wave of ballistic reform, James Ingalls and Alston Hamilton. A complete treatment here is of course impossible. The reader is referred to the books for details and cautioned that what follows is a mere fraction of the information contained in these books.

Suppose that a projectile is fired with initial velocity $V$ at an angle of inclination $\phi$. Letting $x$ and $y$ be the Cartesian co-ordinates of the position of the projectile at time $t$, taking the origin to be the location of the gun and $x$ and $y$ postive, let $\theta$ be the angle tangential to the curve of travel at time $t$, and $v$ the velocity at time $t$. (See Fig. 1.3.)

The projectile is considered as being acted on by two forces, gravity, denoted by $g$, and the force of retardation due to the atmosphere, which is considered as acting opposite to the tangential direction of the projectile with magnitude $F(v)$. Resolution of the forces along horizontal and vertical lines through the projectile easily gives the so-called Principal Equation

$$\frac{d(v\cos(\theta))}{d\theta} = \frac{vF(v)}{gC}. \tag{1.1}$$

where $C$ is the notorious ballistic coefficient. The function $F(v)$ was to be experimentally determined for a standard projectile under standard conditions of fire, and $C$ was an attempt to account for any variation from those conditions, so that $\frac{F(v)}{C}$ was effectively an adjusted atmospheric force function. At its simplest, $C$ is defined as $C = \frac{w}{id^2}$, where $w$ is the weight of the standard projectile, $d$ its diameter, and $i$ the coefficient of form. This definition had a theoretical justification involving the manner in which diameter and weight could affect the force of retardation.

However, the formula became swollen to

$$C = \frac{\delta_1}{\delta} f_a f_w \frac{w}{i\gamma d^2} \tag{1.2}$$

where the other constants are $f_w$ for a wind, $f_a$, an altitude factor, $(\delta_1/\delta)$, a ratio of density factors to account for the atmospheric density, and $\gamma$ a "constant of curvature" to account for the curvature of the trajectory. Some of these factors were omitted in exposition, especially in problem exercises.[14] This complication could have been adduced by Moulton for his assertion that "For a quantity that was treated as a constant in the mathematical theory the ballistic coefficient was made to carry heavy burdens" [255, p. 10]. One of Moulton's contributions, to be explained later, was his attempt to remove the variable sources from $C$ and account for them in a functional way in the differential equations of motion.

Historically, the function $F(v)$ for a standard projectile was first taken to be proportional to a monomial in $v$; Newton argued on theoretical grounds for $v^2$. A vast literature devoted to the form of this law eventually developed. The experimental investigation was begun by Robins, and his results were reported in his 1742 book. It continued into the nineteenth century, during which test firings with different projectiles were undertaken at various European firing ranges. Ingalls devotes the first chapters of his book to a summary of the results of these firings. At the time Siacci formulated his method for the solution of the Principal Equation, several such results were available. Ingalls and Hamilton used those made by the Krupp firings at Meppen[15] in 1881, analyzed by General Mayevski in an article in *Revue d'Artillerie* in April 1883 and extended by Colonel Zaboudski to include higher velocity shells.[16] The mathematical form of $F(v)$ is a continuous function of piecewise monomials in $v$, including some fractional powers; see [154] for the specific function.[17]

The difficult part of the analysis comes in the attempt to integrate the principal equation. ("De la non-integrabilité, en general, de cette equation, decoulent toutes les misères et tous les travaux des balisticiens," as Charbonnier poetically put it in his 1924 Toronto address.[18] [93, pp. 317–318]) Many strategies for the integration

[14] An additional complicating factor was the practice of allowing $C$ to vary with the angle of elevation of the trajectory, though remaining constant throughout flight. Tables showing this variation were used by the second wave reformers even after their other innovations were in place.

[15] Frederich Krupp AG was the well-known steel concern and weapons manufacturer in Germany, provider of armaments for that country in the World War. In 1877, the company set up a firing range in the town of Meppen in Lower Saxony to test long-range naval and army artillery.

[16] Mayevski and Zaboudski were Russian artillery officers.

[17] A trace of an older mathematical tradition is evident when Hamilton refers to this law as "these discontinuous functions."

[18] "From the general non-integrability of this equation follow all the difficulties [literally misery] and work of the ballisticians." [author's translation] The attitude conveyed is an acknowledgment of the complexity of the problem followed by a determined attitude toward its solution.

problem were given by mathematicians throughout the years for $F(v)$ of various forms. Ingalls summarized some of these techniques. Siacci's major contribution consisted in introducing an approximation which made this integration possible for all laws simultaneously. He defined a quantity $u$ by $u\cos(\phi) = v\cos(\theta)$. This quantity has a physical interpretation as a velocity parallel to the muzzle velocity and having the same horizontal component as the actual velocity $v$; see Fig. 1.3. Siacci named $u$ the pseudo-velocity. He then introduced a function $\beta$ defined by the equation

$$F(v) = \beta F(u)\cos^2(\phi)\sec(\theta) \tag{1.3}$$

"with the practical certainty that, for direct fire, the value of $\beta$ will never differ greatly from unity," as Hamilton states [171, p. 18].[19] The constant $\beta$ had to be determined for any particular law and, since it was generally a function of $\theta$ and $\phi$, replaced by an appropriate constant approximation, thus making the principal equation integrable. These definitions and approximations, when substituted into the principal equation, yielded $\dfrac{du}{uF(u)} = \cos^2(\phi)\dfrac{d(\tan\theta)}{gC}$, where the $C$ now was the original $C$ divided by $\beta$. Integrating between $\theta$ and $\phi$, equivalently from $V$ to $u$, gives

$$\tan(\theta) - \tan(\phi) = \frac{-C}{2\cos^2(\phi)}\int_V^u \frac{-2g}{uF(u)}du = \frac{-C}{2\cos^2(\phi)}(I(u) - I(V)) \tag{1.4}$$

where

$$I(u) = \int_0^u \frac{-2g}{uF(u)}du. \tag{1.5}$$

Thus, the tangent of angle of inclination, and thus the angle itself, is given as an integral function of the pseudo-velocity $u$. The differentials $dt$, $dx$, and $dy$ can also be written in terms of $du$, yielding, upon integration, formulae for $t$, $x$, and $y$. (In this, treatment time is not considered an independent variable with all others depending on it.) These three variables are expressed in terms of integrals of rational functions of $u$ and $F(u)$, similar to $I(u)$, called $S(u)$, $A(u)$, and $T(u)$. Thus, all the elements of the trajectory are simply expressible in terms of the four functions $S$, $A$, $I$, and $T$, each of which incorporated the retardation law $F$. The law itself did not need explicit expression at this point. The formulae for the elements in terms of $S$, $A$, $I$, and $T$ can be found in [154, p. 925]. For a specific $F$ such as that given by the Krupp firings, these functions were computed and tabulated; in fact, Siacci used the term "ballistics table" to refer to these tabulated values. A determination of a constant to approximate the function $\beta$ was also necessary to incorporate into the ballistic coefficient.

[19] Direct fire was defined as occurring when $\phi < 15$ degrees.

Ingalls' book contains many sample problems using these four functions and various initial or terminal data; the ballistic coefficient was given as part of the data. The solutions are given and involve only table lookup and knowledge of logarithms and trigonometry. A miniature range table for the 8-inch rifle was developed giving angle of elevation, angle of fall, range, striking velocity, and time of flight. Adaptations of the method by Hamilton addressed higher angle fire as well, invoking a host of new variables.

In spite of mass of mathematics present, it would be misleading to say that this formed the whole of exterior ballistics as presented in these books either in content or milieu. A striking feature of the work is the number of approximations necessary to carry out the solutions. This was not a new feature peculiar to Siacci's method or the Americans' application of it. Ingalls reports the approximations made by earlier ballisticians such as Francis Bashforth, for example, who assumed a cubic law of resistance but needed to replace certain expressions by their mean values over intervals. Approximations presented by Ingalls and Hamilton include, among others, the assumption that the time from the point of projection to the summit is one-half the time of flight, the use of a mean constant resistance at low velocities, or a square law throughout for mortar fire. Methods of numerical interpolation such as the rule of "Double Position" [p. 112] were needed to work with the tables for $S$ and $A$ jointly, since some elements of the trajectory were defined using both variables simultaneously. For the calculation of a mean value for $\beta$ as used in Siacci's theory, several techniques were given in Hamilton's book. For example, assuming a mean square law of resistance throughout, $\beta$ is in fact just $\sec(\theta)$; a mean value for this is calculated by the formula

$$\frac{1}{\beta} = \int_{-\omega}^{\phi} \left(\frac{1}{\sec\theta}\right) \frac{\sec^3\theta}{(\phi) + (\omega)} d\theta = \frac{\tan\phi + \tan\omega}{(\phi) + (\omega)} \tag{1.6}$$

where

$$(\phi) = \int_0^{\phi} \sec^3(t) dt \tag{1.7}$$

where $\phi$ is the angle of fire and $\omega$ is the terminal angle. Several other options are given. These instances are over and above the averaging used in interpreting firing data on which some constants of the theory depend.

These techniques of approximation, though anathema to a pure mathematician, were part of ballistics at our first and second levels of discussion as practiced at this time. This style was characterized as the "rough old formal ballistics" by Norbert Wiener, a participant in the activities at the Aberdeen Proving Grounds during the first World War [328]. It required familiarity with the many types of ordnance and the judgment to choose the right technique for a given problem, leading to the large number and variety of examples in the books. This experience would presumably be a prerequisite for intelligent study of these problems by the officers. The style included rules of thumb such as the Principle of the Rigidity of the Trajectory:

> If the angle of departure necessary to reach a certain point at a horizontal range, $x$, from the gun and on the same level, is known, it will only be necessary, in order to reach another point $h$ feet below the former and at the same horizontal range, $x$, to subtract from the first angle of departure, the angle $\epsilon \ldots$ given by the equation $\tan(\epsilon) = h/x$.

Practical translation:

> The term "rigidity" in this connection refers to the supposed rigid shape of the trajectory and its chord (drawn from the muzzle to the target or point of impact), and the assumption practically involves the hypothesis, that the figure whose outline is composed of the trajectory and its chord, behaves as if it were cut out of cardboard and rotated up or down with the muzzle of the gun as a center [170, pp. 4–5].

This was routinely assumed in direct fire, though its effectiveness was debated in JUSA/CAJ articles. Wiener stated that this manner of considering ballistic problems was giving way in his time to the "point by point solution of differential equations," that is, to the numerical solution of the differential equations of motion ushered in by Moulton. This method made some of the approximations unnecessary. These new techniques would ideally be accompanied by a less *ad hoc* treatment of the problems of exterior ballistics. It would not be clear that newer methods were more effective, however. In fact, the Siacci theory was retained by later ballisticians for use in some low-angle fire cases.

The result of the first wave of reform in exterior ballistics in the United States was the proliferation of techniques based on the Siacci theory and illustrated by problem books and range firing tables constructed by Ingalls and Hamilton. The latter's work introduced algorithms and variables to stretch the application of the Siacci theory to all types of ordnance, resulting in a book that has the appearance of complication but whose level of pure mathematics goes no deeper than calculus and separable differential equations. There was no approaching, for example, the mathematical level of Siacci himself in his 1892 French edition of *Balistique Extérieure* [304]: the author's preface has him noting his development of the quantity $\beta$ in inverse powers of the ballistic coefficient. Such a development and its calculated values could, he suggests, theoretically be used to solve the complete problem of the ballistic trajectory; he ponders the question of the convergence of his series. Such considerations were alien to American exterior ballistics at this time. But the results of Ingalls' and Hamilton's work appeared sufficiently complex in the eyes of some military men to be off-putting. In a document prepared after a post-armistice tour of Europe, Oswald Veblen opined that "The theoretical methods which cluster about these tables are objectionable because the mathematics employed is at once too recondite and clumsy. It is largely because of this mathematical defect that exterior ballistics has established for itself a reputation in the American Army of being an extremely difficult subject" [323, p. 1]. It is hard to gauge the truth of this view, and the degree to which the antipathy described could be attributed to run-of-the-mill distaste for mathematics itself.

## 1.4 The Variability of the Ballistic Trajectory

All of the preceding considerations assume that the weapons used and discussed were sufficiently accurate that roughly the same trajectory could be expected on each firing if all conditions could be kept the same. This state of affairs was in fact only achieved during the nineteenth century as a result of advances in engineering. A consequence of this was that one could with confidence begin to speak of the "differential effects," as they came to be known, of phenomena that could slightly vary the "normal" trajectory. These included such factors as wind, density of the atmosphere, temperature, a slight change in the angle of fire, muzzle velocity, and later a change in the type of shell used in a given gun. These forces were sometimes called the "secondary effects," as opposed to appreciable changes in, for example, angle of fire. The astronomical analogy of the "secondary forces" of distant planets in determining the orbit of a given planet around the sun was also cited. Some of these factors could be anticipated, and others had to be considered accidental. It was of course understood that, regardless of how similar conditions were from one firing to the next, different trajectories were inevitable, due partly to undetected variations and the realities of gun performance. (It is also important to note that technological advances in instrumentation allowed for more accurate measurement of these forces: new chronometers to measure muzzle velocity and improved weather-related devices were profiled in journal articles.) This leads us finally to the statement of the problem under consideration, the problem of the variability of the trajectory: loosely speaking, how can one provide a mathematical account of all the factors that vary a "normal" trajectory and a description of their effects? This is the problem that came to occupy Forest Ray Moulton and Gilbert Ames Bliss, the protagonists of our story.

This question is relevant at all levels of ballistic discussion. From a modern point of view, mathematical tools that provide measures of variation run from simple subtraction and differentials of elementary calculus through differentials of multivariable calculus and Banach spaces, the calculus of variations, and perturbations theory in differential equations, when deterministic theoretical models, are used. At the firing range level, the problem manifests itself in the ever-changing conditions of fire. Once it is admitted that weather factors, for example, have a measurable if small effect on the range of a fired shell, then some set of weather conditions needs to be accepted as standard for range table data. The results of firing under nonstandard conditions have to be adjusted to standard by taking into account the effect that the differences in the factors would have. This required having acceptable results describing these effects available in some combination of empirical laws and theoretical constructs. On the battlefield, the battery commander needs to have constant updates on meteorological data for accurate fire control.

The changing attitude toward factors altering an accepted trajectory can be traced in the textbook sources we have discussed previously. Benton devotes about ten pages to these causes, noting variations in the weights of powder and projectile, temperature of the piece, manner of loading, wind, and atmosphere among others.

Tables showing ranges and corresponding drifts are given, based on experiment. Much of the material involves the effect of different loading methods for various types of projectiles. Advice on wind: "It is difficult to calculate the effect of the wind in any particular case; in making allowances for it, therefore, the gunner should be guided by experience and judgment. For the same projectile, velocity, and wind, the deviation [from the vertical plane of fire] varies nearly as the square of the range" [40, p. 433]. No justification is given for this rule.

By the first wave, the situation had evolved. Ingalls, for example, discusses the effect of the density of air by describing the computation of density factors by which to multiply the ballistic coefficient. If $\delta_1$ is the weight of a cubic foot of air at 62°F and 30 inches of mercury and $\delta$ its weight at any given barometric pressure $b$ and temperature $t$, then $\delta = 20.212b/(1 + .002178t)$. The ballistic coefficient is multiplied by $\delta_1/\delta$ to incorporate this difference at sea level. For higher altitudes of flight, this $\delta$ was replaced by $\delta' = \delta e^{-y/\lambda}$, where $\lambda$ is an experimentally determined constant and $y$ is the mean height of the trajectory of the projectile.[20]

Hamilton offered two views of the problem. In Part I of his Ballistics book, he devoted two chapters to methods of calculating range and deviation corrections. These largely involve table lookup, using underived formulae from interior and exterior ballistics, and interpolation by differences. He also treats the altering of the ballistic coefficient as did Ingalls but incorporating more factors as previously discussed. Empirical formulae for these factors are again given. In Part II, however, he uses the relations among $X$, $C$, and $V$ developed in the Siacci theory and derives changes in these quantities by calculating their differentials. One such relation, for example, is given by

$$\frac{dX}{X} = (2 - \frac{nN}{2N+1})\frac{dV}{V} + \frac{N+1}{2N+1}\frac{d(\sin(2\phi))}{\sin(2\phi)} + \frac{N}{2N+1}\frac{dC}{C} \tag{1.8}$$

where $N = V^2\dfrac{\sin(2\phi)}{gX} - 1$, and $n$ is the coefficient of $v$ used in $F(v)$. Hamilton provided a few examples of use of this relation. If, for instance, one wishes to find the change to be made in muzzle velocity $V$ when the gun and projectile are changed but the range and angle of elevation are not, the left-hand side of the equation and the second term in the right are zero, and the change in projectile is accounted for by $dC$, the change in the ballistic coefficient. A value of $n$ is chosen as either 2 for curved fire (higher elevation than 15 degrees) or 3.5 for field artillery. These methods were still in use after the methods of the second wave of reformers were introduced: an article on charts for calculating, using Hamilton's theory, the effects of small changes in the elements of fire appeared in the *Coast Artillery Journal* at the same time as those of the Moulton school.

[20] This latter formula was attributed to Chauvenet's *Practical Astronomy*.

## 1.5 Forest Ray Moulton

The World War brought about new phenomena, which required extensions and revisions of ballistic theory in all countries.[21] In battle, guns had to be fired for greater ranges, which required information on firing at higher elevations, not just at the "direct fire" level of 15 degrees or less, which had previously sufficed. The older methods, when applied to these situations, did not provide adequate results. We have already cited the moving barrage, which required many guns to be fired with accuracy "by theory alone," without possibility of correction by sight. The airplane as an offensive weapon made it necessary to be able to fire at all elevations and at a moving target; once again, the time for correcting fire by sight was not available. A knowledge of the entire trajectory, not just the elements, of a projectile was needed to deal with the airplane, since fuzes on the projectiles had to be timed to explode in the air.[22] Another new weapon with which to contend was the long-range gun used in the German offensive on Paris in the Spring of 1918. The long-range gun (colloquially known as "Big Bertha" but more commonly referred to as the "Paris cannon" or "75 mile gun" in the American artillery) presented a frightening new prospect-with a range of 75 miles several such guns sieged Paris for weeks, well out of sight. Their angle of fire was 50 degrees, and the projectiles achieved an altitude of 24 miles. The prospect of these guns required more of ballistic theory-a knowledge of the effect upper atmosphere at least was needed.[23] This gun was apparently still on the minds of the American ballisticians in 1922, as a hand-drawn trajectory labeled "German 75-mile gun," computed by J.J. Arnaud, exists as a document in the files of the group for the Ballistics Branch [NARA Washington].[24]

Forest Ray Moulton (1872–1952) was the man responsible for organizing and overseeing the second wave of reform in exterior ballistic theory during and after the World War. He was raised on a farm in southern Michigan and did his first teaching at age 16 in a rural school near his home. He received a BA at Albion College in 1894, having taught astronomy there. After working his way through the instructor ranks at University of Chicago, Moulton received his PhD in Mathematics and Astronomy in 1899; his thesis was entitled *Periodic Oscillating Satellites*.[25] He became full professor in 1912. He headed the Department of Astronomy at Chicago

[21] For a general treatment of mathematics and its use in the first World War, see [28], which includes articles on the technical and sociological issues dealt with by several of the involved countries.

[22] One method to derive the entire trajectory, based on the Siacci theory, was introduced by Alston Hamilton in a document issued from the Sandy Hook Proving Ground in August 1917, but this had no influence on subsequent events. It appears to be his sole mathematical contribution to the War.

[23] Interestingly, it was a contemporary astronomer, Charles Nordmann, who first theorized the characteristics that a gun would have to possess in order to achieve what the Paris cannons had done, in the time before it was even known how the Paris bombardment was being effected.

[24] Arnaud was listed as a "Master Computer" at Aberdeen Proving Grounds and later worked under Moulton.

[25] See p. 127 of [313] for more information about this thesis. See also [314].

until 1926. From 1898 to 1904, he worked with Thomas Crowder Chamberlain, the chair of the geology department, on a theory of solar system formation. He published the first edition of his *Introduction To Celestial Mechanics* in 1902 [253] and authored several other books on astronomy before the War.[26]

Mathematics formed an important part of his treatment of astronomy. In a 1911 expository paper [251], he described the influence of astronomy on mathematics, stating among other claims that "Astronomy not only turned the attention of mathematicians to analysis, but it often determined the precise form their theories should take" [pp. 362–363] and continuing by naming five methods for solving differential equations, "...all of which were devised under the pressure of astronomical problems..." [p. 363]. A recent biography [105] of a contemporary astronomer, Princeton University's Henry Norris Russell, refers to Moulton as a "mathematical theorist" and a "celestial mechanician" [p. 4].

At the beginning of the American involvement in the War, Moulton was posted as Fort Sill, Oklahoma, as a major in the Army ordnance. He was appointed head of the newly organized Ballistics Branch of the Artillery Ammunition Section, Engineering Division of the Ordnance Depart, to give it its full title, in April 1918. Prior to his involvement, ballistics matters were handled by other personnel in the Artillery Ammunition Section, including the Princeton topologist J.W. Alexander II. According to Moulton, "During 1917 and the first three months of 1918 the demand for ballistic results and researches had increased to such an extent that the Ballistics Branch [as a separate unit] was organized. The list of personnel which follows and the account of the work carried out by it are a measure of the growth of ballistic activities during one year under the pressure of war conditions" [255, p. 2]. The parallel activities of Oswald Veblen at the Sandy Hook and Aberdeen Proving Grounds should also be kept in mind as part of the total American ballistics effort, which involved for the first time the call for academic mathematicians to participate in a war effort.

We begin by entering into Moulton's role in the War effort, his mathematical predecessors, and his investigation of the variability problem (and its sources) to form an idea of the origin of the problem itself in American ballistics at this time. One can then contrast Moulton's solution, which was the only existing method upon Bliss' arrival, with Bliss'. Thus, one can ultimately compare the two methods.

The magnitude of the efforts coordinated by Moulton can be grasped by reading the 98-page document [255], the most comprehensive documentation of his contributions. One can readily agree with the assessment in [119, p. 31] (a preliminary version of [26]) that "Moulton appears to have taken it on himself to wholly recast ballistics." Not including his own considerable mathematical papers, he devised instrumental aids and nomograms for computing trajectories, authorized tests to determine maximum muzzle velocities for existing guns, published range

[26] For further information concerning Moulton's life and work, we refer the reader to [131] and [319]. The thesis by Craig Stephenson [313], which we will reference, is an account of Moulton's work on lunar theory and adds material to his biography; see also [314].

firing tables based on the new theories, conducted extensive experiments on the effect of center of gravity and rotating bands on projectile performance, involved himself in wind tunnel experiments to observe the forces acting on a projectile, and coordinated efforts to train students of all branches and ranks of the armed forces in the results of the new work.[27] He insisted on strong coordination between theory and experiment, criticizing existing theory as previously mentioned and reporting the results of newly authorized investigations. He also wanted to make these results usable on the field: Alexander was commissioned to design a pocket-sized range and deflection corrector for abnormalities in muzzle velocity, wind, air density, or weight of projectile [255, pp. 69–70]. It is not an exaggeration to say that Moulton had a vision of the entire subject of exterior ballistics, its deficiencies, its possibilities, and its methods; these took place at all three levels we have delineated.

Information regarding Moulton's charge for the section, any notion of how he knew, which problems needed to be attacked first, is lacking. However, one notes that, after having listed the new conditions of battle brought on by the War, he states–

> These problems, and related ones which need not be enumerated, let Major Moulton, in May 1918, to formulate new processes for treating the problem of computing trajectories and making allowances for abnormal conditions. These processes are characterized by their mathematical rigor, their generality, and their simplicity of application. Improvements were made on the basis of practical experience. Other methods were devised by other offers, and many related problems were treated in numerous investigations by the officers and civilians of the Ballistic Branch.

He then lists his own reports, numbers 1 and 1', which between them cover, in ten investigations, his small-arc computation method, differential corrections for variations in the initial conditions and retardation functions, general theory of effects of winds, air-density weighting factors, and the effects of rotation on flight of projectiles. Apparently, his priorities lay with putting in place a way of calculating as wide a variety of trajectories as possible with methods of correcting for as wide a set of conditions as possible.

It is best to quote Moulton himself on his criticisms of his predecessors [255, pp. 9–10]:

> From the mathematical point of view the methods of Siacci and his followers have several imperfections. In the first place, the method of representing the resistance by a series of functions of the form $c_n v^n$ has the defect of giving solutions whose higher derivatives are discontinuous, though doubtless this defect could be remedied by adopting another form for the resistance function $[F(v)]$. In the second place, the method of taking into account the effects of the decrease of density of the atmosphere with increase in altitude has been on no solid logical basis. In introducing the so-called "altitude factor" in the ballistic coefficient the mean value theorem of integral calculus has often been incorrectly applied. And even if no error were committed in this connection, the results obtained by the method would be correct only for the end of the trajectory, and no solid foundation would be laid for

[27] An officer from the Coast Artillery School was assigned to Moulton's office to learn the new methods, and one senses from the account [255, pp. 81–82] the inevitable conflict between the career ballisticians and new mathematical practitioners, though "in the end satisfactory agreements were almost invariably reached, and the discussions were always found valuable."

> taking into account the effects of the wind and other disturbing factors. In the third place, these methods replace a function of the angle the tangent to the trajectory makes with the horizontal by a constant which is ordinarily expressed in terms of initial value, or of the initial and final values, of the angle. The basis of the determination of the function is inadequate and, as in the case of the altitude function, the results would be correct only for the terminal values. In the fourth place, the methods of determining the effects of such disturbing factors as winds, abnormal air densities, and the rotation of the earth were imperfect in the extreme and in some cases even gave results of the wrong sign.

This paragraph not only summarizes some of Moulton's objections to specific aspects of Siacci's theory and practice but introduces often-repeated themes and mathematical issues. There is a raising of the mathematical level of discourse: questions of continuity and differentiability of functions and correct applications of the integral mean value theorem, for example, are posed. Surely, these were not of concern to Ingalls or Hamilton. They were, however, characteristic of the second wave expositors: Dunham Jackson,[28] in his 1919 ordnance textbook, writes "The resistance function which is tabulated for use in applying the method described in the following pages ... has a continuous derivative, in addition to being continuous itself... [207, p. 9]." Thus, the new discussions were taking place at the level of advanced calculus or analysis, as opposed to elementary calculus and differential equations of the first group of reformers. The need for a logical foundation for any ballistic method is stressed; this type of need was repeatedly stated by Moulton on different occasions and in different contexts: "They [the earlier methods] contained defects of reasoning, some quite erroneous conclusions, and the results arrived at by singularly awkward methods. [259, p. 1]." "Reliance will not be placed on intuitions, or insecure methods of reasoning, but a direct treatment of the problem [of the variations of the trajectory] will be taken up. [259, p. 81]" From his 1930 textbook on differential equations: "It would be inexcusable to discuss general theories of differential equations and not to treat them with rigor and completeness that are so characteristic of the present day ... This aim has been inspired, in part ... by the firm conviction that the time is at hand for insisting on equally high standards in the application of differential equations to the physical world [262, vi]." The need to take into account factors which "disturb," or vary, the trajectory is also explicitly brought into prominence,[29] as well as the fact that information about the entire trajectory was now needed. Moulton the astronomer was attempting to deal with the "terrestrial imperfections" of the ballistic science noted by Charbonnier with advanced mathematical tools.[30]

---

[28] Jackson, a recurring figure in our story, was an analyst who obtained his PhD in Göttingen in 1911 after having spent the two previous academic years there. His advisor was Edmund Landau. Jackson worked with Moulton's group, where he is listed as "Professor of Mathematics in Harvard University."

[29] In fact Moulton's first technical report was devoted to his method of dealing with the variability problem, as outlined below.

[30] Moulton's introduction of numerical methods into the solution of the ballistic differential equations needs some attention; it highlights his concerns and style, and was used to compute the differential variations in the trajectory as well as the trajectory itself. It is not central to the concerns of this section, but computation will play a central role in the overall story of the differential variations.

We first describe the form in which Moulton treated the basic equations of motion, because this became the standard in American practice for the years of this study. Locating the muzzle of the gun at the origin of the Cartesian coordinate system, the range direction along the positive $x$-axis, $y$-axis vertical with positive values upward, and positive z-axis to the right of the line of fire (thus there is no longer a "plane of fire"), Moulton set

$$\frac{d^2x}{dt^2} = -Fx'$$
$$\frac{d^2y}{dt^2} = -Fy' - g$$
$$\frac{d^2z}{dt^2} = -Fz' \tag{1.9}$$

where $g$ is the acceleration of gravity, $F = \dfrac{G(v)H(y)}{C}$; the derivatives are with respect to time. In the latter, $v = v(t)$ is the velocity, $G(v)$ a new resistance function established by the French ballisticians at Gâvre in 1883. Moulton forthrightly stated that "It is not certain that $G(v)$ is the same for projectiles of all shapes and sizes; in fact, it is very probable that it depends upon the shape of the projectile ... but the dependence at present is quite unknown [259, p. 84]." Then, $H(y) = e^{-.0001036y}$ is a factor for the atmospheric density, which varies with altitude, and $C$ is the ballistic coefficient.[31] The initial conditions are given by $x(0) = y(0) = z(0) = 0$, $x'(0) = x_0'$, $y'(0) = y_0'$, and $z'(0) = 0$. These equations replace the "Principal Equation" of the Siacci theory.[32] We note in passing the explicit use of functions in contrast with the earlier use of variables and differential notation, a minor indication of mathematical sophistication.

A word needs to be said about Moulton's contribution to the numerical solutions to these equations. Moulton introduced a numerical integration technique based on the Picard iteration procedure to give an approximate solution (1.9); this kind of technique was known at the time as a "small arc" method. The method underwent many modifications by his co-workers shortly after its introduction; see [135] for further remarks. He published a treatment of his method in a 1922 series of volumes issued under the auspices of the Smithsonian Institution [258] and also included a more abstract account as Chap. 2 of his 1926 book [259].

Although this is the mathematical war contribution for which Moulton is most frequently cited, several comments are in order. The idea of using numerical methods of solution goes back to Euler, as Moulton was aware, but even in contemporary times, their application was relatively common in other countries.

[31] Moulton spends no more time on $C$, only occasionally stating that it depends on the shape, size, and weight of the projectile. Absent are the encumbrances used by the earlier ballisticians.

[32] The Siacci equations can in fact be easily derived from Moulton's as shown in [104].

France, for example, had been using numerical methods at Gâvre since 1887, and a recent article [33] describes Karl Pearson's contemporary computational war efforts in England. Moulton himself downplayed this aspect of his work, stating that "The introduction of the method of solving numerically differential equations is so simple and obvious that any one familiar with the general field of differential equations would hardly fail to do substantially what I did" [260, p. 246]. He did claim, though, that "it [numerical integration] seems to have been entirely unknown to pre-war writers on ballistics and to be almost wholly unknown to mathematicians of the present time" [255, p. 18]. Moulton himself had written on the use of numerical methods for astronomical purposes in his book on celestial mechanics [253, p. 425], whose second edition appeared in 1914; here the technique was called "mechanical quadrature." However, of greater importance to Moulton was his proof that the solution of the differential equations exists and that his solution method converges to the sought-after solution to within any desired degree of accuracy. The existence proof involves a two-parameter process, which includes those of Picard and the Cauchy-Lipschitz method. It can be found in the 30 pages of Chap. 4 of [259] and could easily have appeared in a journal of pure mathematics at that time. In a handwritten set of notes taken during Moulton's lectures on "The Modern Theory of Differential Equations" for nnn at the University of Chicago, one finds extensive coverage of these processes; the material was evidently at hand upon the beginning of his war work [254, pp. 65 ff]. Of this work, Moulton also said–

> I laid down for the first time explicit conditions under which the process is valid in a strict mathematical sense. One having any considerable degree of mathematical sophistication would not feel at liberty to ignore the question of the validity of the process on which he bases all his conclusions ... Most of those who have commented on my work either have been blind to the necessity of proving that the process is mathematically sound or have assumed that its validity was established in some indefinite past. [260, p. 247]

Thus, Moulton's concern here as elsewhere was with logical foundations and mathematical rigor, even in the numerical techniques, which provided the entries in the range firing tables. He is in effect distancing himself from the "rough, old, formal" style of ballistics by bringing in the spirit of Weierstrassian rigor as advocated at the University of Chicago, personified, for example, by the analyst Oscar Bolza.[33]

Moulton contributed an article to the JUSA/CAJ describing his method [256]. A brief examination of it not only illustrates his mathematical style but also gives an indication of the difficulty posed by the new level of exposition. He noted that, in typical developments of differential equations, the student assumes that if an equation cannot be solved by separating variables (as the first wave of reformers did, after some approximations), that it cannot be solved at all; he added that this perspective often forces the resistance law to take certain forms that it might not in reality have. In contrast to this, his method was applicable to any differential

[33] See [135] for further information about the history of such numerical methods in the United States prior to and after the World War.

equation, separable or not, it "rests on a solid logical basis" and is very convenient in practice. The results of the use of the method and the results of actual firing were claimed to be "astonishingly harmonious." "Unfortunately for practice, the requirements of modern analysis make it necessary to pay much attention to delicate points of logic in which the practical man has little interest, with the result that the value of a method in applications is obscured by the theory of its correctness." He included a computation illustrating the method, executed by Elizabeth Webb Wilson[34] and Mabel Williard, for the orbit of a planet. Although a good account, readable by a serious student of mathematics, it is interesting that the article was followed quickly by another, authored by First Lt. J. J. Johnson, which contained only the results of a computation laid out step by step. This article did "not presuppose a knowledge of mathematics beyond plane trigonometry," for those who were interested only in the "mechanics" of the method. Moulton may have misjudged the mathematical sophistication his audience. We shall return to the issues of the reception of the "New Ballistics" in the Chap. 1 of this work.

We now turn to Moulton's treatment of the variability problem. He frames the question in terms of the economics of range firing table use and computation at that time:

> It is impossible to use range tables having as arguments all the factors upon which the motions of projectiles depend. Hence the effects of the minor factors are given in supplementary tables [in his newly proposed work] and are applied as corrections to the results for normal conditions ... Trajectories can be computed by the methods given [in an earlier chapter of his ballistics book], whether or not any of the minor factors is present. The differences between the results obtained for normal conditions and those when a minor factor is included are the effects of the minor factor. While this method of determining the effects of factors is theoretically correct, it is very laborious because it makes no use of the fact that the properties of the normal trajectory are known when one differing only slightly from it is being determined. [259, p. 80]

Moulton's method, and Bliss' as well, uses the known trajectory to compute the variations, as we will see below. Thus, the massive amount work involved in computing and using tables suitable for all possible combinations of factors involving the trajectory forced, in Moulton's mind, the decision to treat the differential variations as corrections to normal conditions. This rationale launched the use of the differential variations in American ballistics work. Consequently, when Bliss was called to Aberdeen to work on the subject, its form had already been established. Anticipating slightly, one can see that if the work involved in computing a trajectory subject to all factors could be reduced significantly from that demanded by contemporary hand-computing methods, the overall topic of differential variations might diminish in importance. Variations easily could be incorporated into such a high-speed calculations of the trajectory in question.

Moulton's method is grounded in the theory of linear differential equations and, as presented in Chap. 3 of his book, is straightforward in execution in spite of the 45-page length. (He remarked typically that this material "cannot be treated in such a

[34] See [154] for more on Wilson.

rough and ready manner as has heretofore been employed by writers on ballistics.") His main tool is variation of parameters. Beginning with (1) as the "undisturbed" trajectory, that is, the trajectory given as normal with normal initial conditions, he assumes that the solutions to these equations are known (having been computed by the short arc method), given as

$$x = \phi(t), \ y = \psi(t), \ z = 0 \tag{1.10}$$

He then assumes that any disturbing forces such as wind or abnormal densities of atmosphere acting on the projectile have components represented by $X$, $Y$, and $Z$, and thus sets up the differential equations of the disturbed trajectory as

$$\begin{aligned} \frac{d^2x}{dt^2} &= -Fx' + X \\ \frac{d^2y}{dt^2} &= -Fy' - g + Y \\ \frac{d^2z}{dt^2} &= -Fz' + Z \end{aligned} \tag{1.11}$$

using the same initial conditions as (1.9). The solution of this disturbed trajectory is expressed using the solutions to the original system as

$$x' = \phi' + \xi', \ y' = \psi' + \eta', \ z' = 0 + \zeta' \tag{1.12}$$

where it is assumed that $\xi$, $\eta$ and $\zeta$ are small quantities relative to $\phi$ and $\psi$. The function $F$ is expanded in a power series and all terms of order higher than the first are dropped, since products of $\xi$, $\eta$ and $\zeta$ can be assumed negligible. (On the expansion of $F$, Moulton says that, even if its smoothness cannot be justified experimentally, it can be approximated by polynomials, which have such an expansion.) When (1.12) are substituted into (1.11) and terms cancelled, he obtains a system of differential equations *for* $\xi$, $\eta$ *and* $\zeta$, given by

$$\begin{aligned} \frac{d^2\xi}{dt^2} &= P_1\xi' + P_2\eta' + P_3\eta + X \\ \frac{d^2\eta}{dt^2} &= Q_1\xi' + Q_2\eta' + Q_3\eta + Y \\ \frac{d^2\zeta}{dt^2} &= -F\zeta' + Z \end{aligned} \tag{1.13}$$

where $P_1$, $P_2$, $P_3$, $Q_1$, $Q_2$, and $Q_3$ are functions of $F$, $\psi$, $v$, $\phi$, $G$, and various derivatives of these. We will not list these explicitly. It is important to note that these expressions are assumed known from the original trajectory and have their values at each $t$ already determined from the original trajectory. Moulton's strategy is to solve this system in general for $\xi$, $\eta$ and $\zeta$ by variation of parameters. Then, by specifying $X$, $Y$, and $Z$ for the particular force in question, he derives formulae for $\xi$, $\eta$ and $\zeta$ as functions of $t$ for the entire trajectory corresponding to that force, not just at the terminus. Variants of this technique give the $\xi$, $\eta$ and $\zeta$ for abnormal initial conditions. These three variables are known as the differential variations due to the force considered.

We will only sketch out the remaining steps in the procedure, since it is standard variation of parameters. The third equation is independent of the first two and can be solved on its own. The first two equations are expanded into a system of four equations in the usual way as

$$
\begin{aligned}
\frac{d\xi}{dt} &= \xi' \\
\frac{d\xi'}{dt} &= P_1\xi' + P_2\eta' + P_3\eta + X \\
\frac{d\eta}{dt} &= \eta' \\
\frac{d\eta'}{dt} &= Q_1\xi' + Q_2\eta' + Q_3\eta + Y
\end{aligned}
\tag{1.14}
$$

To effect variation of parameters, four linearly independent solutions to the corresponding homogeneous system are sought, forming a basis, or "fundamental set" in Moulton's terminology, for the general solution. Two of these can be easily found and expressed in terms of $F$, $\phi$, $\psi$ and their derivatives with respect to $t$. The remaining two are those computed trajectories corresponding respectively to a change $\Delta\theta_0$ in initial angle of elevation and $\Delta v_0$ in initial muzzle velocity. Linear independence of the four is verified by computing the Wronskian. These four solutions are then linearly combined with coefficients $C_1$, $C_2$, $C_3$, and $C_4$ to given the general solution to the homogeneous system. The solutions with the $C_i$ regarded as functions are then substituted into (1.14) to get the equations for the $\frac{dC_i}{dt}$, as required by variation of parameters. These derivatives are found using Cramer's rule, then integrated to get the $C_i$. When substituted back into the expressions for $\xi$, $\xi'$, $\psi$, and $\psi'$, they give the required differential variations in terms of $X$ and $Y$ and the functions from the original, undisturbed trajectory. The individual stages required in the computation of the solution were organized in an economical eight-step procedure, which minimized the number of integrations necessary and introduced other shortcuts.

To give an example of the use of the procedure, Moulton states that the normal trajectory assumes that the air is stationary with respect to the earth's surface and that winds can be regarded as an abnormal force. If we assume a wind of constant magnitude and direction given by $(u, s, w)$ at each point of the trajectory, then the differential equations for the disturbed trajectory become

$$\begin{aligned}
\frac{d^2x}{dt^2} &= -F(x' - u, y' - s, z' - w, y)(x' - u) \\
\frac{d^2y}{dt^2} &= -F(x' - u, y' - s, z' - w, y)(y' - s) - g \\
\frac{d^2z}{dt^2} &= -F(x' - u, y' - s, z' - w, y)(z' - w)
\end{aligned} \tag{1.15}$$

Now, the right-hand members are expanded in a power series in $u$, $s$, and $w$ and, since these are small relative to $v$, all terms of the second order and higher are again dropped, giving

$$\begin{aligned}
\frac{d^2x}{dt^2} &= -F(x', y', y)x' + \left[F + x'\frac{\partial F}{\partial x'}\right]u + x'\frac{\partial F}{\partial y'}s \\
\frac{d^2y}{dt^2} &= -F(x', y', y)y' - g + y'\frac{\partial F}{\partial x'}u + \left[F + y'\frac{\partial F}{\partial y'}\right]s \\
\frac{d^2z}{dt^2} &= -F(x', y', y)z' + Fw
\end{aligned} \tag{1.16}$$

In this case, the functions $X$, $Y$, and $Z$ are given by

$$\begin{aligned}
X &= \left[F + \phi'\frac{\partial F}{\partial x'}\right]u + \phi'\frac{\partial F}{\partial y'}s \\
Y &= \psi'\frac{\partial F}{\partial x'}u + \left[F + \psi'\frac{\partial F}{\partial y'}\right]s \\
Z &= Fw
\end{aligned} \tag{1.17}$$

These expressions are then substituted along with the other necessary ones into the integrals to be evaluated in determining the differential variations due to wind.

Of course, this assumes that the winds are constant and known ahead of time. To account for the more realistic situation of winds of varying direction and velocity, which could be known only at the time of battle, it is noted that the integrals that are necessary to express the variations are of the form

$$\int_0^T f_1(t)\,u(t)\,dt. \tag{1.18}$$

for example, where $T$ is the time of flight, $f_1(t)$ is some combination of the known functions from the undisturbed trajectory and $u(t)$ is the (usually varying) wind component in the range direction. Moulton proposed[35] meteorological service would transmit to the battery commander data at various prescribed intervals of time. These data would allow for an approximation of $u(t)$ by a step function having as values those extracted from the message. Thus, (1.18) would be simplified to a sum of integrals, over time intervals, of $f_1(t)$ with the constants, extracted from the message, multiplying them. Since these integrals only involve $f_1(t)$ with the wind components as constants, they could be evaluated beforehand, reducing the work of the battery commander to arithmetic. The integrals without the $u$ present were called "weighting factors"; each disturbing force had its own set. In this manner, an attempt was made to account for the third level of ballistic practice, actual engagement. From this point of view, we can agree with the opinion expressed in [119, p. 32] that "this chapter is probably the most important part of the book practically." These ideas began to be realized upon the inclusion of weighting factors for wind in "some range tables issued by the Technical Staff of the Ordnance Department" [282], according to a JUSA/CAJ article by Joseph F. Ritt, a professor of mathematics at Columbia University who was involved in war work.

Similar considerations applied to forces $X$, $Y$, and $Z$ corresponding individually to abnormal air density, abnormal weights of projectiles (due to "tolerances permitted in manufacture" [259, p. 107]), variations in the velocity of sound,[36] rotation of the earth, and variations in gravity. Special emphasis was given to the fact that these variations held all along the trajectory, not merely at the end.

As with the ballistics of the earlier practitioners, the mathematics used and the surrounding milieu both need to be appreciated. For Moulton, it was important to justify every approximation by reference to an appropriate mathematical result, as we have seen in the case of the series expansion of $F$. While it is true that the level of mathematics used here is higher than that of Hamilton and Ingalls, so is the level of justification. The older ballisticians often defended their approximations by seeing if the agreement between theory and firing warranted them, but Moulton sought mathematical and computational rationales. For example, in the previous discussion of the weighting factor, Moulton expresses concern about replacing $u(t)$ by a constant approximation and references the integral mean value theorem in a general discussion:

> The integrals arising in the determination of [the weighting factors] are strictly speaking of the form [(1.18)], where actually $u$ is variable. It follows from the mean value theorem that if $f(t)$ does not change sign in the interval of integration, then there is a number $u_1$ between the smallest and largest values of $u$ such that
>
> $$\int_0^{t_1} f_1(t)u\,dt = u_1 \int_0^{t_1} f_1(t)\,dt. \tag{1.19}$$

---

[35] It is evident that, here as elsewhere, Moulton's ideas of how the differential variations material would be used in practice, as well as his visions of the overall battle use of the tables, would be considered mere suggestions subject to dismissal by more seasoned military personnel.

[36] "The retardation function $R$ and the function $F$ have rapid changes in value for $v$ near the velocity of sound. The normal tables for $F$ are constructed for a particular value of the velocity of sound ... But the actual velocity of sound, which depends upon the (elasticity temperature) of the atmosphere, is [different]" [259, p. 111].

> This is not in general the mean value of $u$ on the interval [as older ballisticians may have assumed], but if $u$ does not vary too widely and if the zone is narrow enough so that $f_1(t)$ does not vary greatly, then the mean value of $u$ may be taken for $u_1$ without making appreciable error. There would be no difficulty in examining numerically the question of the possible extent of the error in typical cases [259, p. 121].

This expression of concern with mathematical validity of his techniques and computations to back up his assertions is typical of Moulton's work. Five pages of the chapter on differential variations [pp. 102–106] are devoted to considering what would happen if terms of higher order were retained in the expansion of $F$. The section including these pages concludes with the comment "Computations have shown, indeed, that the terms of the first order give the variations with sufficient approximation for initial conditions differing from those that are normal more widely that the variations encountered in gunnery." Details of these computations appear in a technical report entitled *Second and Higher Order Errors in Differential Variations* [255, p. 12]. See Figs. 1.4a and b for portions of a computation sheet, executed by Ruth Voorhees, a computer on Moulton's staff, which supports the conclusions reached in the report. (The upper right hand corner contains Voorhees' signature and the date January 13, 1918.)

Moulton's background in astronomy can be seen in his choice of variation of parameters as a method of solution for the problem of differential variations. He was of course aware of the history (going back to Newton) and application of the technique to problems in mechanics; he made historical comments to this effect in his books on differential equations and on celestial mechanics. In the former, he introduced variation of parameters in the fourth chapter in generality before moving on to linear systems. His first "illustrative problem" has the differential equations of motion of two mutually attracting planets moving in orbit around the sun, where the forces due to the sun are the dominant terms and those between the planets a smaller

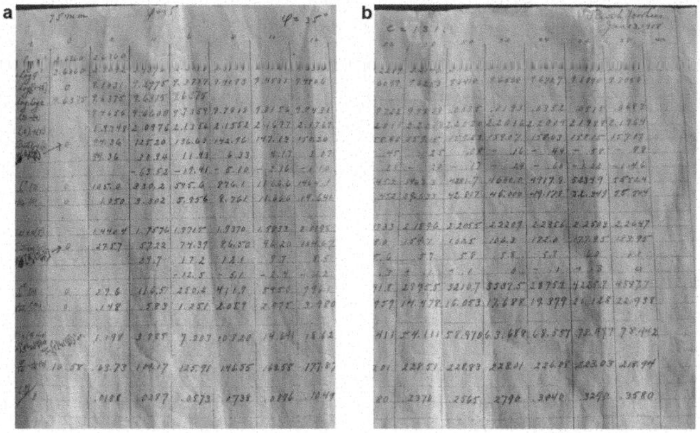

**Fig. 1.4** (**a**) Left portion of computation sheet for *Second and Higher Order Errors in Differential Variations, [NARA Washington]*. (**b**) Right portion of computation sheet for *Second and Higher Order Errors in Differential Variations, [NARA Washington]*

"disturbing" one.[37] The form here is identical to that of (1.11) in his ballistics book. The terminology of the day, as reproduced by Moulton, spoke of the "elements" of the orbits (major and minor axes, angle of inclination) as ballisticians spoke of the elements of a trajectory.

In his book on celestial mechanics, which preceded the ballistics book, Chapter 10 is devoted to perturbation theory, his "illustrative example" considers the vibration of a tuning fork, in which the magnitude of the vibration $x$ satisfies

$$\frac{d^2x}{dt^2} + k^2x = -\mu\left(\frac{dx}{dt}\right)^3 + \nu \cos lt. \tag{1.20}$$

Here the factor containing $\mu$ is said to arise from the resistance to a medium in which the fork vibrates. He considers the left-hand side as the undisturbed vibration and the right-hand side as introducing perturbations, one of which is dependent on velocity, and the other independent. "The first is analogous to the mutual perturbations of the planets, which depend upon their relative position; the second is more of the nature of the forces which produce the tides ..." A discussion of variation of parameters to solve the equations then ensues.[38]

Moulton had several other uses for the differential variations, in the construction of general ballistic tables. "... in view of the great variety of guns and projectiles in modern artillery practice, it is desirable to have general ballistic tables from which all special range tables can be interpolated, or otherwise determined, by a relatively small amount of work," he declared [259, p. 124]. Rather than compute new trajectories for each type of gun together with a set of differential variations for it, why not allow the parameters $C$, $x_0'$, $y_0'$, $\theta_0$ in the equations (1.9) to assume a large set of reasonable values separated by small intervals and compute a comprehensive collection of trajectories once and for all? This being done, any particular gun would yield its values for the parameters, furnishing specifications of trajectory, which would then be selected or approximated from the existing collection. The idea was not new with Moulton.[39] Charbonnier [91, pp. 256–258] describes this idea and its attempted execution in France, which began in 1894. He pointed out, however, that such tables still depend on some resistance law, and existing tables become useless if that law is revised. Moulton proposed to deal with this problem by claiming that when the retardation law is improved, trajectories from the older tables would be

---

[37] This problem is considered in his nn lectures of differential equations.

[38] This chapter includes another passage in praise of rigor. While relating the history of the subject, he pays tribute to the founding work of Euler, Clairaut, Lagrance and Laplace. "Yet there were points at which pure assumptions were made, it having become possible to establish completely the legitimacy of the proceedings, under proper restrictions, only during the latter half of the nineteenth century by the aid of the work in pure Mathematics of Cauchy, Weierstrass, and Poincaré" [253, p. 367].

[39] It is not clear whether Ingalls or Hamilton had this conception. Ingalls published an article "General Ballistic Tables For Mortar Firing" in 1896 [202], but it appears that it is just a summary of how to calculate such a table using a square resistance law and the usual Siacci tables.

corrected by using the technique of differential variations with the law itself allowed to vary [255, p. 78].

In fact, the differential variations were to play a role in the construction of the general tables themselves, a fact noted by later writers. Moulton observed that each trajectory calculated by short arcs from (1.9) simultaneously furnished a whole set of "secondary trajectories." Each such trajectory was found by taking as its initial point a fixed point of the already computed trajectory and following the original trajectory from there. The secondary trajectory corresponds to the flight of a projectile with different $C$, $\theta_0$, and $v_0$ than the original, but these values were easily determined [259, pp. 125 ff.]. Thus, computing a single trajectory could supply an entire family of others, which could then be used for the general tables. This idea cuts down the computing labor. One difficulty in this plan was that the secondary trajectories did not differ from each other by unit multiples of $C$, $v_0$, and $\theta_0$ and so were not necessarily suitable as they stood for table use. Moulton found a remedy for this problem by using his differential variations; it is at least clear that the variations could play a role in such adjustments. We will not go into the specifics of this technique, which are given in [259, pp. 127–129].

The construction of the general ballistic tables proposed by Moulton was begun under his direction by A. A. Bennett (1888–1971),[40] and three volumes were eventually released. The first in appeared 1924, the second in 1931, and the final volume in 1936 [38]. They were helpful "to obtain approximate answers to the questions that always arise during the design of new projectiles," to quote from a history of ballistics in United States [30, p. 21]. The long intervals between the release of these volumes testifies that the work involved in compiling them was massive. The labor involved computing the trajectories by hand; this was done by "computers," whose efforts have been documented by Grier in [154] and [156]. This work was shortened by the acquisition of computing machines, such as the Vannevar Bush's differential analyzer. The story of the emergence from this milieu of the ENIAC, the electronic computing machine, during and after the second World War, is by now well known; in fact, Herman Goldstine's account of the number of calculations necessary to cover all types of ordnance [138, pp. 135–137] reads like a recapitulation of Moulton's similar comments in his reports. When framed in terms of the spectrum of possibilities for range table construction given earlier, one can see the evolution from the gun-specific option of "firing only" to the mathematically created general firing tables, which acted as a sort of universal trajectory bank for which a specific gun provided some parameters. Indeed, the projectiles might have been designed using the tables. The contrast with Ingalls and Hamilton, whose textbook problems were almost invariably tied to a particular gun, is evident.

---

[40] Bennett, a professor of mathematics at the University of Texas at this time, had a background in pure mathematics, having spent a year at Göttingen. His ballistics efforts in both World Wars were many, and he appears in several guises in this account. Most of his career was spent at Brown University.

## 1.6 Henri Poincaré and the Equations of Variation

Moulton's astronomical background has already been emphasized, but there is a single celestial mechanician for whom he expressed the greatest admiration and a single idea due to him, which can be said to form the basis of Moulton's analysis of the differential variations.[41]

In an article in the journal *Popular Astronomy* [252], Moulton wrote an obituary and appreciation of Henri Poincaré, who had died in July of that year (1912).[42] Moulton praised Poincaré and his achievements in the highest terms "... he was one of the greatest geniuses the human race has produced [p. 623] ... there is not a man living who is fully competent even to appraise all this work [p. 624]." He pointed out that the Frenchman's use of mathematical analysis was "positively dazzling" and that "the order of [mathematical] ideas he introduced into celestial mechanics had been hitherto quite foreign to astronomy, and it is not strange that astronomers found it impossible to understand him." Moulton held that "Beyond question his [Poincaré's] celebrated researches on the problem of three bodies, which won the prize offered by King Oscar II, and which were developed more fully in three volumes of his *Les Méthodes Nouvelles de la Mécanique Céleste*, are his most important contributions to the applications of mathematics." The American disciple[43] mentions the effort that he expended reading the volumes in question, likely during his student days, and it is almost certain that he modeled the title of his own 1926 work on ballistics on that of Poincaré's treatise.

In Moulton's mind, Poincaré's work and the issue of mathematical rigor were closely related.[44] An allusion to this is made in the obituary when discussing Poincaré's attack on the problem of the stability of the solar system, a problem considered in contemporary times by Karl Weierstrass, though there were predecessors. "The difficulty in obtaining an answer to the problem arises from the fact that the expressions which have been developed by mathematicians to represent the coordinates of the planets converge either not at all or for only a limited time." [p. 626] Of Poincaré's work on the three-body problem he said: "Since the expressions for the periodic orbits converge for all values of the time ... it is clear that the time is now ripe for building anew the theories of celestial mechanics with full rigor ..." [p. 626].

[41] The author is indebted to Craig Stephenson for bringing to his attention the articles from which the following material comes.

[42] Chapter 7 of Stephenson's thesis is devoted to a much more extensive discussion of Moulton's status as a follower of Poincaré's work.

[43] G. D. Birkhoff was another great admirer of Poincaré.

[44] Relevant to this discussion is the fact that Poincaré's original memoir on the three body problem contained an error "affect[ing] only the discussion of the existence of asymptotic solutions..." [252, p. 625], and Moulton addresses this error, whose importance he tends to minimize, in the obituary. A recent study devoted to this error, and the attendant consequences is [32].

Moulton brought the spirit of proof to his work on ballistics, as has been shown, but was evident in his work on celestial mechanics earlier. In a 1902 paper [250], Moulton begins forthrightly by stating that–

> Mathematical science was not sufficiently developed to enable the founders of celestial mechanics to prove in many cases the validity of the methods which they found it necessary to employ ... In both pure mathematics and astronomy many conclusions were drawn which have since been shown to be erroneous or which still lack proof. Celestial mechanics has been filled with series, comparatively few of which have been proved to be convergent.

As a corrective to this state of affairs, he noted that "The latter half of the nineteenth century witnessed a complete re-examination of the foundations of analysis, the correction of numerous errors, and the introduction of an entirely new spirit of rigor" [p. 119]. Weierstrass' name is not mentioned in connection with this re-evaluation, but it is clear that Moulton knew the major responsible party. "The question is pertinent whether those interested in celestial mechanics should not henceforth adhere more nearly, if not entirely, to methods which are known to be rigorous" [p. 119]. The paper begins with a rigorous discussion of solutions to systems of differential equations in terms of power series and continues this theme with discussions of convergence of the series. The assumed analyticity of the defining functions in systems of differential equations in celestial mechanics was mirrored in the analyticity of the functions in the equations of motion for ballistics; thus, Moulton's preoccupation in the latter with series expressions and their domains of convergence.

Moulton's attitude toward rigor was only one of the outcomes of his encounter with Poincaré's work. He adopted one of the latter's points of view, a view that ultimately was used in a textbook treatment of variability in the ballistic trajectory. The last section of Moulton's paper discussed in the preceding paragraph is entitled "The Method of Small Variations." In it, he considers the following scenario: if a solution for a system of differential equations can be found for particular initial conditions and if the actual initial conditions differ only slightly from the particular ones, then, at least for time, the solutions to the two systems will vary only slightly. He proposes a method of finding the deviation between the two solutions, a method that he refers to as "small deviations." This deviation is to be represented by a convergent power series. The original system is given by

$$\frac{dx_i}{dt} = X_i(x_1, ..., x_n, t) \tag{1.21}$$

where $x_i(0) = a_i$, $i = 1...n$. The solution to this system is $x_i(t) = f_i(t)$ for $0 \leq t \leq T$. The $X_i$ in this system are required to be analytic. He then assumes that the actual initial values of the system are $x_i(0) = a_i + \epsilon\alpha_i$, where $\alpha_i$ are small constants and $\epsilon$ a parameter to be set equal to one in the final result. Let $x_i = f_i(t) + \epsilon\xi_i$ be the solutions under the actual conditions, where the $\xi_i$ are the corrections to be determined.[45] Substituting these last solutions into (1.21) and expanding in powers of $\xi$, he obtains

$$\epsilon\frac{d\xi_i}{dt} = \sum_{j=1}^{n} \frac{\partial X_i}{\partial x_j}\epsilon\xi_j + \text{higher powers in } \epsilon\xi_j. \tag{1.22}$$

He then proceeds to find values for the $\xi_i$ by expanding them in turn in series of powers of $\epsilon$ and finding the coefficients of the powers. The equations that result from dropping the higher power terms in (1.22) and cancelling the $\epsilon$ are what became known as the "equations of variation" for the system (1.21) for the case of differing initial conditions, though Moulton does not single these equations out in his 1902 paper. His interest in pursuing this topic stems from the fact that it is essentially this process that was used by the American astronomer George William Hill in his writings on the lunar theory [181]. Moulton expresses the hope that further work on this problem will produce more satisfactory expressions for exact periodic solutions for orbits in this theory and cites *Les Méthodes Nouvelles de la Mécanique Céleste* [274] in this context.[46]

In Poincaré's treatment of 1892, found in Chap. 3 of *Les Méthodes Nouvelles* [274, pp. 162–165], system (1.21) is laid down without initial conditions, and $x_i = \phi_i(t)$ is given as an arbitrary solution, called the *solution génératrice*, and a second solution, $x_i = \phi_i(t) + \xi_i$ is also given as differing slightly from the first. He then calls the system

$$\frac{d\xi_i}{dt} = \frac{\partial X_i}{\partial x_1}\xi_1 + \frac{\partial X_i}{\partial x_2}\xi_2 + \cdots + \frac{\partial X_i}{\partial x_n}\xi_n \tag{1.23}$$

the "équations aux variations" and proceeds to discuss Hill's work; Moulton merely repeated this discussion. More examples of the use of these equations are given in the chapter. Thus, it appears that Poincaré generalized Hill's special use of the equations of variation from the lunar theory and applied them more widely, and Moulton in turn used them for other problems in celestial mechanics as well. Moulton was to eventually frame the ballistic problems of variation in these terms, as will be described in the next paragraph, and expanded their use to include more general causes for variation of solutions than differing initial conditions.[47]

---

[45] Moulton does have two occurrences of $\epsilon$ here.

[46] In Moulton's 1916–1917 lecture notes cited earlier [254] a discussion of "Hill's Equation" is interrupted by a "Digression on Stability of Periodic Orbits" in which this derivation is repeated and the notion of a stable periodic orbit is defined. As in the case of numerical solutions of differential equations, this material was at hand for Moulton's use.

[47] It would be quite difficult to trace the use of this idea, straightforward as it is, any further back than Poincaré's work. Another contemporary use of these equations was found by Gaston Darboux in his work on the theory of surfaces. He referred to them as "l'équation auxiliare" [100], [102, p. 505].

In his text on differential equations [262], published in 1930, he took the opportunity to use these notions in ballistics. In Section 133 of his book, he reproduces his argument from [250], this time explicitly calling his equivalent for (1.23) the "equations for variations" and borrowing Poincaré's terminology for the "generating solution." There follows a discussion of general solutions to the equation set, and in Section 135, "Illustrations from Ballistics" are given. In this section, the equations for variations from normal initial conditions are derived, the equivalent of (1.14), and two sets of solutions of the fundamental set of four are found, based on principles of his earlier discussion.

In an attempt to cover other possible sources of variation in the ballistic problem, Moulton, in Section 137, proposes the system

$$\frac{dx_i}{dt} = f_i(x_j; \mu; t) \tag{1.24}$$

for $x_i(t_0) = a_i$ and $\mu$ is a constant parameter. He then obtains the equations of variation for this system by allowing simultaneously the initial conditions and the value of $\mu$ to vary. These are

$$\frac{d\xi_i}{dt} = \frac{df_i}{d\mu}\nu + \sum_{j=1}^{n} \frac{\partial f_i}{\partial x_j}\xi_j + \dots \tag{1.25}$$

where $\nu$ is the amount by which $\mu$ is allowed to change. He identifies $\nu$ with a "slight departure from normal atmospheric resistance, due to such things as winds or abnormal air density" [p. 243], perhaps allowing for a different constant in the atmospheric density law $H(y)$. Again, consideration of this type of system had a well-established astronomical precedent. In [250], Moulton had considered systems like (1.24) in his Section 5, stating that "The most satisfactory method [for solution] which has been so far devised is to expand the solutions as power series in the parameters which occur in the differential equations ... If no natural parameters occur, it is sometimes possible to introduce them artificially so as to attain much the same results" [ibid. p. 124]. He states that this method was used by pioneering workers in celestial mechanics and that Cauchy in 1842[48] examined the validity of the method.[49] In the following section, there is an extensive application to the "Computation of the Absolute Perturbations of the Elements" of the orbits of two planets around the sun. (In this section, the original system is dealt with, not the equations of variation.) One can see how the parameters occurring in (1.24) were far from the abstractions they may have been for later pure mathematicians considering these systems.

[48] A reference to the *Collected Works* is given.

[49] In [262, p. 59] Moulton credits Euler with the idea of solutions in power series of a parameter and gives further historical information.

In this manner, Moulton was able in his later work to see the problems of differential variations as a special case of Poincaré's "équations aux variations."[50] These equations became a common topic in modern perturbation theory, where they are variously referred to as "the perturbation equation" or "the variational equation." In Chap. 2 of the present work, we will see how they figured in a line of inquiry in pure mathematics motivated by this War-related topic, a sort of spinoff research effort. This work considered the continuity and differentiability properties of the solutions to systems of differential equations, a topic related to variability questions. (In this line of inquiry the functions occurring in the system, such as the $X_i$ in (1.21), are not required to be analytic.) The group of researchers involved in this work included several we have already met: T.H. Gronwall, J.F. Ritt, and Forest Ray Moulton, as well as Gilbert Ames Bliss.

## 1.7 Remarks on the Work of Forest Ray Moulton on Differential Variations

In concluding our account of Moulton's treatment of the ballistic problems of variation, we might observe that the celestial mechanician, with his classical tools of series solutions and variation of parameters, updated with a healthy dose of twentieth century mathematical rigor, can best be seen as a figure straddling the sister worlds of astronomy and ballistics noted by Charbonnier. Yet with all our emphasis on these factors, it must not be forgotten that his work resulted in practical contributions to the War effort. He entered the Army at a time when, to cite one of his recollections, "...those who were working in ballistics were engaged in prolonged discussions respecting such simple questions as whether a head wind would increase or decrease the maximum ordinate of a projectile" [261, p. 320]. He found these circumstances intolerable and set out to provide definitive answers by invoking the appropriate differential equations and drawing conclusions. In the process, his work was used for the preparation of range tables for 17 guns during his one-year tenure [255, p. 42] as well as a host of other related technical tables. Many of these tables included information on variations obtained from his theory and computed using his numerical methods. His classical mode was to be extended by the work of Gilbert Ames Bliss, a pure mathematician whose work in calculus of variations and the emerging discipline of functions of a line provided not only new methods but a new outlook on these problems. His roles in War-related ballistic matters will continue to be highlighted throughout the remainder of this story, with an encore performance in projectile design, given in the last chapter.

Parenthetically, it may be added that his ballistic labors continued to be appreciated by some military members after the events related in his *History*. In 1920,

[50] In exercises at the end of the chapter containing this material Moulton has the reader deal with the two body problem and pendulum motion using Poincaré's approach.

Moulton published *Periodic Orbits*, a lengthy investigation of the three body problem. It was begun in 1900 and all but the last chapter were finished by 1912. Material for use in the last chapter was completed by 1917, but the beginning of his War work forced the book to be set aside. Picked up after the War, it was finished and subject to review; one such appeared in JUSA/CAJ in 1922 [257]. It was favorable, and its author's military career was recalled: "It [the book] will also be of special interest to students of theoretical ballistics for it will be remembered that Major Moulton successfully applied his knowledge of the orbits of Heavenly bodies to the computation of the trajectory. As a result of his labors and those of many other able mathematicians a practical 'short arc' method was evolved" [257, p. 587]. It was in this manner that Moulton was beginning to be remembered.

Before beginning the story of Bliss' contributions, we finish with some biographical highlights of Moulton after his academic service, which ended in 1926 with his resignation from the University of Chicago. His career was summarized in [131, p. 346] as beginning in astronomy and ending in public service; it also included a wide variety of scientific activities. Some time after 1930, he participated in some research involving estimating the size of the meteorite, which produced Meteor Crater in Arizona. His textbooks from the 1920s and 1930s have already been noted, to these can be added a 1935 book about the trials of Galileo. He was a member of many professional organizations, the American Astronomical Society, American Mathematical Society, and the National Academy of Sciences, to name a few. He took part in many educational projects relating to the instruction of science in general in America, including radio broadcasts. On the commercial side, he was, for a decade following his departure from the University of Chicago, the financial director of the Utilities Power and Light Corporation of Chicago. He was also a trustee of Chicago's "Century of Progress" World's Fair (1933–1934) and wrote extensively on the promise of progress through science. Editing the Bulletin of the American Association for the Advancement of Science during the years 1942–1946 was only one of his many contributions to that organization. It is also to be pointed out that Moulton did author papers on what one would consider topics in pure mathematics; one such, involving solutions to differential equations with periodic coefficients, dates from the year 1911 [131, p. 352]. He died in 1952.

## 1.8 Gilbert Ames Bliss

Gilbert Ames Bliss (1876–1951) was born in the Kenwood section of Chicago in 1876.[51] His father was associated with various electrical firms, indirectly encouraging an interest in his son in the inventions of the day. Bliss entered the University of Chicago in 1893 for graduate work, in the second year of that institution's operation. At first, his course of study was under Forest Ray Moulton in astronomy, work that

[51] References for the life and work of Bliss include [73, 149, 240].

he found "fascinating," and Bliss expressed a debt of gratitude to Moulton for his inspiring teaching and directing him in the writing of his first paper "The Motion of a Heavenly Body in a Resisting Medium." He soon changed his direction of study to mathematics. Among his teachers, he singled out E. H. Moore, O. Bolza, and H. Maschke as the most influential. Oskar Bolza, in particular, "a product of the meticulous German school of analysis led by Weierstrass" [73, p. 595], became his thesis advisor. Bolza had a written record of an important course in the calculus of variations given by Weierstrass in 1879, a record that he allowed Bliss to copy for his own use.[52] This record and Bolza's own lectures pointed Bliss in the direction of this area of analysis as his lifelong interest. His thesis, completed in 1900, was entitled "The geodesic lines on the anchor ring."[53]

Bliss' teaching career before his permanent association of with the University of Chicago took him to various posts. From 1900 to 1902, he taught at the University of Minnesota, a period that included the beginning of an intensive course of self-study in the literature of the calculus of variations; in particular, he read Kneser's newly published book [216]. During nnn, he spend a year in Göttingen, which at that time was led by Klein, Hilbert, and Minkowski.[54] Bliss made the acquaintance of the younger men, including Zermelo, E. Schmidt, Abraham, Féjèr, and Carathéodory; he attended meetings and seminars and presented papers. The year nnn was spent back at Chicago; there followed a year at the University of Missouri. In the spring of 1905 came an invitation to spend a year at Princeton, which he accepted, becoming one the the new "preceptors" in the autumn of that year. In his three years, there he met L. Eisenhart, J. W. Young, R. L. Moore, and, critical for our story, Oswald Veblen. " ... my friendship for them and admiration of their scholarship have remained with me permanently" [73, p. 599]; he wrote in a set of autobiographical notes. In 1908, Maschke of Chicago died, and Bliss was offered a position back at his home institution. Upon accepting this position, Bliss was asked to teach the courses in geometry, which was Maschke's area, and this required much preparation and study time. It may have an effect on his work in analysis: one of his biographers, his student E. J. McShane, conjectured that "He must have visualized clearly; a curve was a picture in his mind, not a system of functions" [240, p. 39]. He was able to return to his own field of analysis upon the departure of Bolza for Germany in 1910. He spend the rest of his career at the University of Chicago, though summer terms were often spent at other institutions.

There have been many assessments of his contributions to and style in the calculus of variations. All include the fact that he was one of the outstanding figures in this field in the first half of the twentieth century. A former student, Herman Goldstine, referred to the "classical" nature of his advisor's work: "There were two schools of thought on the calculus of variations. One of them was a sort of

---

52 We shall have more to say on Weierstrass' contributions to the calculus of variations in Chap. 2.

53 "Anchor ring" is an archaic term for a torus.

54 Göttingen figures in the biographies of several of the American mathematicians in this story; Dunham Jackson, for example.

conservative classical school, of which Bliss I think really represented the last important member ... It was concerned with ... properties of extremal curves on the small. ... Bliss really finished up the classical subject, pretty much" [139]. According to McShane, Bliss

> did not seek maximal generality, but preferred to exhibit the problem's true center of interest with clarity. Thus he did not choose to use Lebesgue integrals in the calculus of variations, presumably feeling that the real interest lay in the behavior of families of smooth curves, and that the extension of the theory, for example to all curves of finite length, could be added afterwards if one wished. [240, p. 39].

Speaking of the subject as a whole, McShane offered the opinion that

> In the early twentieth century the calculus of variations was in danger of losing its appeal because of mounting complexity. How much Bliss contributed to its rescue, as well as to its advancement can be seen by anyone who will compare the compactness and generality of the theory in the *Lectures on the Calculus of Variations* (Bliss' magnum opus, completed in 1946 [72], five years before his death) with the older papers on the same subject. It is a worthy monument. [240, p. 45][55].

His specific uses of material from the calculus of variations in ballistic writings will be discussed in Chap. 2. A summary picture of Bliss' entire work is best found in [149].

The list of his honorary degrees, association memberships both professional and civic, awards, academy positions, honors, expositions and writings on professional duties is long, as can be seen by consulting the references cited. "On the political side, he was a Republican conservative gentleman who lived in the suburbs of Chicago. He was formal but very kind to me," related Goldstine, [139], who considered him like an "uncle," and other accounts concur with the general picture of kindness, enthusiasm, and patience and dry humor.

In the late summer of 1918, Oswald Veblen began writing to Bliss from the Aberdeen Proving Grounds requesting his help in the war effort. Bliss was already involved in teaching prospective naval students at the University, but his faith in Veblen's judgment prevailed. Bliss joined the Proving Ground team that year as a "scientific expert," a standard title given to mathematicians there. He quickly familiarized himself with the state of work, absorbing the inadequate Siacci theory and Moulton's efforts. He found "the theory and application of the so-called 'differential corrections' of trajectories [to be] still very imperfect" [73, p. 601]. This was the area in which he made his major contributions. Some technical reports were written in November of 1918; he was then asked to stay on for a longer period. He did so, but this work was interrupted by the death of his wife in December 1918 from the influenza epidemic. Another period at Aberdeen followed in 1919.

Bliss' contributions to the problems of the differential variations can be said to fall into two distinct categories. The first is a practical method for calculating the variations based on what is called the adjoint of a system of linear differential

---

[55] McShane referred to Bliss' activities in this regard as "a program of saving the calculus of variations from death by elephantiasis" [p. 45]

equations; this technique we will simply call the "Bliss adjoint method." It will be described below and requires only a knowledge of calculus and elementary differential equations. The second is the framing of the problem of differential variations as a topic in the emerging field of "functions of a line," a precursor to what eventually became known as functional analysis. This framing resulted in theorems, which treated the required calculations as the problem of computing the first differential of a functional. In the process, other theorems involving the variation of solutions to systems of differential equations were proved; these theorems considered the effect of changing the functions defining the differential equations as well as changing initial conditions.

In contemporary accounts of the mathematics of the war effort, it is usually the second aspect of Bliss' work, which was cited (from which the applied method followed) with the attached moral that one cannot predict the eventual practical uses of pieces of such abstract mathematics as the functions of a line. And indeed this is a conclusion fairly drawn. But the path from the theorems to useful methods of computation is a convoluted one in this case, more so than can be described in such phrases as used even by Bliss himself: "This [the differential variations] was my particular field of study, and I was astonished to find that the most advanced mathematics which I knew was needed and effective...The calculation of the differentials of such functionals is a principal problem of my special field, the calculus of variations" [73, p. 601]. Those familiar only with the classical problems like the brachistochrone may well wonder why calculating the differential of a functional, which produces a system of differential equations is a part of the calculus of variations. The mapping of this path is to be attempted in Chap. 2, but for now, our focus is on the practical aspects of the adjoint method.

Before closing this summary, something should be said about the almost inevitable comparison of Moulton's and Bliss' general outlook on these problems. Bliss was a pure mathematician and did not hold forth on the benefits of rigorous application of modern mathematical ideas to applied problems as avidly as did Moulton. His remarks in this direction seem limited to the genuine surprise he felt at the unexpected applicability of the functions of a line to the differential variations problem and an occasional foray into longer expositions on pure and applied mathematics such as [66] and [69]. The latter was a 1933 lecture in the General Course of the Division of Physical Sciences at the University and was a discourse on the correlation between mathematical theory and observational data from the experimental sciences. The lecture was dedicated to debunking the notion that "When such a theory has been successful as a correlating agent the conviction is likely to become established that the theory has a unique relationship to nature as interpreted for us by the observations. Furthermore, it is sometimes inferred that nature behaves in precisely the way which the mathematics indicates" [69, p. 472]. Examples are given from non-Euclidean geometry and the more timely relativity theory and quantum mechanics. Although his work in ballistics is not cited, it is an interesting exercise to keep the admonishment in mind as the equations used to describe the variations unfold.

The letter of request for an extension of Bliss' stay at Aberdeen mentions experiments with new types of ammunition, so his contributions were at both the theoretical and firing range levels of discourse. And he was not beyond a blunt financial assessment of the benefits of his efforts:

> ...they [his adjoint methods] reduced by three fourths the labor in computing the most numerous corrections in the range tables, and the cost of these improvements in terms of scientific experts was negligible [during the War] in comparison with savings which would presently have accrued if the war had continued ... An improvement in a range table which might reduce by one the number of shots fired for adjustment [to correct for a variation] from a fourteen inch gun in order to attain a target, would on each occasion save one third or more of the average salary of a scientific expert at 1918 rates. [66, p. 314]

Once again, the cost-saving aspect of new mathematical ballistic theories is stressed.[56]

## 1.9 The Bliss Adjoint Method

The practical aspects of Bliss adjoint method appeared in several forms in the years 1918 and 1919. The first is a package of documents, which forms the technical reports on differential corrections in the record group for Aberdeen Proving Grounds [57]. It contains Bliss's original idea in two installments, each dated November of 1918, an alternative computational scheme for the adjoint method by T. H. Gronwall, dated January 24, 1919, and a technical report by Bliss on adaptation of his idea to anti-aircraft fire, dated February 1919. The latter was never published, though it appears that it was submitted to the Journal. The first two were combined and formed the basis of the paper [61] in JUSA/CAJ with minor alterations but much supplementary material. Two summaries were also printed under the "Professional Notes" [59, 60] section of the same journal, a section set aside for brief descriptions of technical issues of interest. The first of these summaries was so full of errors and misprints that it rendered the article unintelligible, hence the second corrected version. Apparently, the unfamiliarity of the typesetters with the profusion of higher mathematical symbols took its toll; perhaps, also the War-time backdrop provided additional stress. Indeed, the published version [61] still suffered from many typos. In the following account will draw from all of these sources as well as the 1919

[56] In a letter dated October 28, 1919, from the Ordnance Office Technical Staff to the Ordnance School of Application at Aberdeen, copies of papers were forwarded to the latter in order that Bliss' new method be implemented in the School. The letter concludes with observation that "It is thought that a marked reduction in the labor involved in computing differential corrections in range will result from the use of the methods of this paper in the place of those formerly used. On the basis of a comparison made in this office a reduction of approximately thirty-five percent has been estimated" [NARA Ordnance School].

ordnance textbook [207] by Dunham Jackson, a professor of mathematics at the University of Minnesota called for War work, and the paper [161] by Gronwall. The latter contains a more complete treatment of the same material as [61] as well as much new related work but at a higher mathematical level of exposition, which also cites Bliss' papers on ballistics as a topic in the functions of a line.

The second of the Professional Notes [60] is a five-page summary of the method, which incidentally mentions a plan to combine the ten-paper series on the New Ballistics mentioned earlier, of which [61] was a part, in a supplement to the Journal. This plan did not come off, as the articles were issued independently. From the Note, we learn that Bliss was familiar with the activities of range firing, as described in an introductory paragraph, as well as the older and existing ballistic theories. Specifically mentioned here is the use of a firing to determine the value of the ballistic coefficient of a shell. The inadequacy of the Siacci theory is stressed. As we have seen in equation (1.2) of Section 3, in this theory, the problem of variation of the density of the atmosphere with altitude, addressed by Moulton with the introduction of the density function, was dealt with by an "altitude factor" multiplied into the ballistic coefficient; Bliss refers to the latter as "not a particularly happy device." Measurement of current values of the sources affecting the variation are described as taking place by observation of a pilot balloon, in the case of velocity and direction of wind, and airplane observation for temperatures and pressures at different altitudes. This information is not included in [61].

We consider the two papers from the technical report package and [61] simultaneously since they cover roughly the same material. The equations below appear as in [61] itself unless otherwise noted. The passing from the ballistics of the artillerymen to that of the scientific experts, from what we have called the first wave to the second wave of ballistic reformers, was noted poignantly in the editor's introduction to the series on The New Ballistics: "The editors of the Journal confess that the contents of some of these articles appear rather formidable to them ... but at the same time it is realized that there are readers who will distinctly relish these contributions which are decidedly out of the popular fiction class. Seriously, however, it is believed that these articles will break the ground for a more general delving of the ballistic field by Coast Artillery officers ... the Journal suggests that, based on the thorough analytical discussion of this series, there is a demand for a comprehensive presentation of up-to-date ideas on ballistics and gunnery, set forth in simple form and pitched to a mathematical key such that the average artillery officer can carry the theme and keep in tune ... groundwork for American ballistic practice was laid by Artillery officers, notably Ingalls and Hamilton ..." whose work took place "in the face of a discouraging paucity of means of experiment and practical research" [15].

In the introductory section of [61], Bliss explains what he is setting out to do and claims that, with this new method, "...the results explained here enable one to compute a range table by the newer methods with an amount of labor not excessively greater than that required by the Ingalls-Siacci theory" [61, p. 296]. Differential corrections are defined as those changes that cause usually no more than a 3%

change in the range.[57] Bliss begins by assuming that an original, "undisturbed" trajectory had been computed and given using a method of approximation. This is the solution to the system

$$\begin{aligned} x'' &= -Fx' \\ y'' &= -Fy' - g \\ z'' &= 0 \end{aligned} \tag{1.26}$$

where, as before, $g$ is the acceleration of gravity, $F = \dfrac{G(v)H(y)}{C}$; the derivatives are with respect to time. $G(v)$ is said to have been "tabulated from the results of experimental firings and the table has been smoothed with the help of an analytic function which fits it with sufficient accuracy" [60, p. 442]. $H(y)$ is the exponential air density law as used by Moulton. No initial conditions are given, but one can assume $x(0) = y(0) = z(0) = 0$, $x'(0) = x_0'$, $y'(0) = y_0'$, and $z'(0) = 0$. He introduces $K(y) = H(y)/C$. He then assumed all possible types of variations simultaneously:[58] winds parallel to and transverse to the line of fire exist and depend on altitude $y$; these were denoted by $w_x(y)$ and $w_z(y)$, respectively. Variation in the density law was given by $\delta H(y)$, to be added to $H(y)$ to get the new density; $\delta K$ is introduced to account for a variation in the quotient $H(y)/C$ due to a change $\delta H(y)$ in $H$ or $\delta C$ in $C$, or both. On the disturbed trajectory, the new coordinates are represented by[59]

$$\overline{x} = x + \xi,\ \overline{y} = y + \eta,\ \overline{z} = 0 + \zeta. \tag{1.27}$$

The equations on the disturbed trajectory are now

$$\begin{aligned} \overline{x}'' &= -(F + \Delta F)[\overline{x}' - w_x(\overline{y})] \\ \overline{y}'' &= -(F + \Delta F)\overline{y}' - g \\ \overline{z}'' &= -(F + \Delta F)[\overline{z}' - w_z(\overline{y})] \end{aligned} \tag{1.28}$$

where

$$F + \Delta F = G(V)[K(\overline{y}) + \delta K(\overline{y})].$$

and

$$V = \sqrt{[\overline{x}' - w_x(\overline{y})]^2 + \overline{y}'^2 + [\overline{z}' - w_z(\overline{y})]^2} \tag{1.29}$$

---

[57] This percentage is mentioned for the range only, though deflection from the plane of fire is also considered in the paper. A percentage for deflection is not given.

[58] One quantity which Bliss did not treat was the variability of $G(v)$, the resistance law. This was addressed by Gronwall in [161]. Bliss eventually looked at $G(v)$ in [71]. The possible alteration in $G$ is important because it allows for a change in the resistance law due to new experimental firings.

[59] In choosing the overbar notation and these specific Greek letters Bliss is following a tradition in the calculus of variations exemplified by Oskar Bolza, in his classic [83].

is the new velocity expression. Again, no initial conditions are given, but later, Bliss considers a change in initial velocity of $\Delta v_0$ and a change in initial angle of elevation of $\Delta\theta_0$. From these, he proceeds as Moulton did to arrive at a system of differential equations for the variations themselves; subtraction of (1.26) from (1.28) from gives

$$\begin{aligned}\xi'' &= -F[\xi' - w_x(\overline{y})] - [x' + \xi' - w_x(\overline{y})]\Delta F\\ \eta'' &= -F\eta' - (y' + \eta')\Delta F\\ \zeta'' &= -(F + \Delta F)[\zeta' - w_z(\overline{y})]\end{aligned} \tag{1.30}$$

Bliss then linearizes this system. To this end, he writes

$$\begin{aligned}\Delta v &= V - v = x'\frac{\xi' - w_x(\overline{y})}{v} + y'\frac{\eta'}{v} + \dots\\ \frac{\Delta G}{G} &= \frac{G(V) - G(v)}{G(v)} = \frac{G\prime}{G}\Delta y + \dots\\ \frac{\Delta K}{K} &= \frac{K(\overline{y}) + \delta K(\overline{y}) - K(y)}{K(y)} = -h\eta + \frac{\delta K}{K} + \frac{\delta K(\overline{y}) - \delta K}{K} + \dots\end{aligned} \tag{1.31}$$

Again, as with Moulton, if an argument is not displayed, it is assumed that it is evaluated along the original, undisturbed, and already computed, trajectory.

An expression for $\Delta F$ is now given by

$$\frac{\Delta F}{F} = \frac{\Delta G}{G} + \frac{\Delta K}{K} + \frac{\Delta G}{G}\frac{\Delta K}{K} = \frac{G'}{vG}[x'(\xi' - w_x) + y'\eta\prime] - h\eta + \frac{\delta K}{K} + \dots \tag{1.32}$$

After the omission of small terms Bliss arrives at

$$\begin{aligned}\xi'' &= -F(\xi' - w_x) - x'F\{\frac{G'}{vG}[x'(\xi' - w_x) + y'\eta'] - h\eta + \frac{\delta K}{K}\}\\ \eta'' &= -F\eta' - y'F\{\frac{G'}{vG}[x'(\xi' - w_x) + y'\eta'] - h\eta + \frac{\delta K}{K}\}\\ \zeta'' &= -F\zeta' + Fw_z\end{aligned} \tag{1.33}$$

No initial conditions are prescribed here, but later, Bliss points out that since the two trajectories begin at the same point, we have $\xi(0) = \eta(0) = \zeta(0) = 0$.

We have gone into such detail to emphasize the consideration of the order of the terms dropped.[60] Bliss devotes ample attention to the inelegant matter of

[60] There is a technicality here not addressed by Bliss, who was certainly aware of it, but presumably chose to ignore it so as not to bog down a mathematical paper in a military journal with more mathematics. Strictly speaking, the solutions to (1.33) and (1.30) are different; (1.33) as a simplified and linearized version of (1.30), so the solution variables should be different. In [161], this difference is noted, and he uses $\Delta x$ for $\xi$, $\Delta y$ for $\eta$, and so in (1.30), saving the Greek letters for (1.33). The order of magnitude of $\Delta x - \xi$ is then discussed.

justifying dropped terms, and this discussion occupies about two pages of the paper. In reading the treatment, one can see that in addition to showing his strength as a pure mathematician, Bliss operates fully as a firing range member. "It is not easy to discuss the numerical magnitudes of the terms which have been neglected in deducing the equations [(1.33)] except in a very rough way on the basis of experience in the computations" [61, p. 391]. The following is a close paraphrase of Bliss' remarks. In the expansion for $\frac{\Delta K}{K}$, for example, he argues that the value of $h$ is 0.000106 and that a maximum for $\eta$ for a relevant trajectory is 100 meters so the term $h\eta$ is of the order of $\frac{1}{10^2}$, also the value of $\frac{\Delta K}{K}$ for a 1% variation from normal in density. Thus, the first- and second-order terms in $\frac{\Delta G}{G}$ (which had been analyzed in the previous paragraph) and consequently in $\frac{\Delta F}{F}$ have the same relative magnitudes, namely, those of $\frac{1}{10^2}$ and $\frac{1}{10^4}$. Now in (1.30) the terms of $\frac{\Delta F}{F}$ are multiplied by quantities of the magnitude of $Fx'$ and $Fy'$, quantities which do not exceed 100 on the relevant trajectory. Thus the first and second order terms in (1.30) are of the order 1 and $\frac{1}{100}$, and the latter are the ones dropped in (1.33).

It is not just higher-order terms whose omission calls for explanation but others as well. He has replaced $\frac{\delta K(\overline{y})}{K}$, $\frac{w_x(\overline{y})}{v}$, and $\frac{w_z(\overline{y})}{v}$ by their values at $y$, and this is "justifiable because the numerators of the two sets give density and wind components at $\overline{y} = y + \eta$ and $y$ which are less than 100 meters apart, and in practice the corresponding elements of the two sets will differ by exceedingly small quantities" [61, p. 302]. It is fortunate that this replacement exchanges terms nonlinear in the variations with simple nonhomogeneous terms. He also states that variations found from (1.33) agree "with all accuracy requisite" with those obtained by integrating (1.30) directly. (Recall that Moulton had done a similar check on his version of the differential equations for variations.)

The resulting system is replaced by a linear[61] of first order system in the standard manner, so that the final version of the differential equations for the variations is

$$\begin{aligned}
\xi' &= \xi_1 && && && +f_1 \\
\eta' &= && \eta_1 && && +f_2 \\
\zeta' &= && && \zeta_1 && +f_3 \\
\xi_1' &= x'Fh\eta - F(1+x'^2\frac{G'}{vG})\xi_1 && -x'y'F\frac{G'}{vG}\eta_1 && && +f_4 \\
\eta_1' &= y'Fh\eta \quad -x'y'F\frac{G'}{vG}\xi_1 && -F(1+y'^2\frac{G'}{vG})\eta_1 && && +f_5 \\
\zeta_1' &= && && -F\zeta_1 && +f_6
\end{aligned} \tag{1.34}$$

[61] Again, strictly speaking, this is not the linearization of (1.30) due to the replacement of some nonhomogeneous terms mentioned in the preceding paragraph. Also, neither Bliss nor Gronwall refers to any system for the variations as formally the "equations of variation" (of Poincaré, though Bliss uses this terminology in other papers.)

where

$$
\begin{aligned}
f_1 &= f_2 = f_3 = 0\\
f_4 &= F\big(1 + x'^2 \frac{G'}{vG}\big)w_x - x'F\frac{\delta K}{K}\\
f_5 &= \big(x'y'F\frac{G'}{vG}\big)w_x - y'F\frac{\delta K}{K}\\
f_6 &= Fw_z.
\end{aligned}
$$

Note that the quantities that cause the variation in the original trajectory appear only in the nonhomogeneous terms $f_i$. The novelty of the new technique lies in the introduction of the adjoint system to a given linear system of differential equations. Bliss considers a four by four system, for purposes of illustration, given by

$$
\begin{aligned}
\xi' &= a_1\xi + b_1\eta + c_1\xi_1 + d_1\eta_1 + e_1\\
\eta' &= a_2\xi + b_2\eta + c_2\xi_1 + d_2\eta_1 + e_2\\
\xi_1' &= a_3\xi + b_3\eta + c_3\xi_1 + d_3\eta_1 + e_3\\
{\eta_1}' &= a_4\xi + b_4\eta + c_4\xi_1 + d_4\eta_1 + e_4
\end{aligned}
\tag{1.35}
$$

He then defines the adjoint to (1.35), given by

$$
\begin{aligned}
-\lambda' &= a_1\lambda + a_2\mu + a_3\lambda_1 + a_4\mu_1\\
-\mu' &= b_1\lambda + b_2\mu + b_3\lambda_1 + b_4\mu_1\\
-{\lambda_1}' &= c_1\lambda + c_2\mu + c_3\lambda_1 + c_4\mu_1\\
-{\mu_1}' &= d_1\lambda + d_2\mu + d_3\lambda_1 + d_4\mu_1
\end{aligned}
\tag{1.36}
$$

where $\lambda$, $\mu$, $\lambda_1$ and $\mu_1$ are called "auxiliary" variables. The adjoint is always a homogeneous system.

In the case of the linear system (1.34), the adjoint is

$$
\begin{aligned}
\lambda' &= 0\\
\mu' &= -hF(x'\lambda_1 + y'\mu_1)\\
\nu' &= 0\\
\lambda_1' &= -\lambda + F(1 + x'^2\frac{G'}{vG})\lambda_1 + x'y'F\frac{G'}{vG}\mu_1\\
\mu_1' &= -\mu + x'y'F\frac{G'}{vG}\lambda_1 + F(1 + x'^2\frac{G'}{vG})\mu_1\\
\nu_1' &= -\nu + F\nu_1
\end{aligned}
\tag{1.37}
$$

A routine calculus-based derivation shows that (1.35) and (1.36) are related by

$$(\lambda\xi + \mu\eta + \lambda_1\xi_1 + \mu_1\eta_1)' = \lambda e_1 + \mu e_2 + \lambda_1 e_3 + \mu_1 e_4$$

$$(\lambda\xi + \mu\eta + \lambda_1\xi_1 + \mu_1\eta_1)_0^T = \int_0^T (\lambda e_1 + \mu e_2 + \lambda_1 e_3 + \mu_1 e_4)\,dt, \quad (1.38)$$

which is called the "fundamental relation" between the two systems.

Bliss explains this derivation using only terminology involving row, columns, and differentiation, but a modern reader would see it as a result of the following manipulations: if (1.35) were written as

$$\vec{y}' = A\vec{y} + \vec{e}$$

and (1.36) as

$$A^T\vec{z} = -\vec{z}'$$

then

$$\langle \vec{y}', \vec{z}\rangle = \langle A\vec{y} + \vec{e}, \vec{z}\rangle = \langle A\vec{y}, \vec{z}\rangle + \langle \vec{e}, \vec{z}\rangle = \langle \vec{y}, A^T\vec{z}\rangle + \langle \vec{e}, \vec{z}\rangle = \langle \vec{y}, -\vec{z}'\rangle + \langle \vec{e}, \vec{z}\rangle$$

from which it follows that

$$\langle \vec{y}', \vec{z}\rangle + \langle \vec{y}, \vec{z}'\rangle = \langle \vec{e}, \vec{z}\rangle$$

or

$$\frac{d}{dt}\langle \vec{y}, \vec{z}\rangle = \langle \vec{e}, \vec{z}\rangle.$$

In this sense, a known solution $\vec{z}$ to the adjoint system produces an exact differential equation for its inner product with an unknown solution $\vec{y}$ to the original system; it thus acts analogously to an integrating factor for the original system.

Expanded to a six-by-six system, this fundamental relation is given by

$$[\lambda\xi + \mu\eta + \nu\zeta + \lambda_1\xi_1 + \mu_1\eta_1 + \nu_1\zeta_1]_0^T = \int_0^T (\lambda f_1 + \mu f_2 + \nu f_3 + \lambda_1 f4 + \mu_1 f_5 + \nu_1 f_6)\,dt \quad (1.39)$$

The crux of Blis's method is best described by his first example. Suppose we want to know how the range $X$ of the original trajectory is affected by any of the variations already described. It is easy to see (See Fig. 1.4) that $\Delta X$, the change in the range, is approximately given in terms of the original variables by $\xi + \eta\cot(\omega)$

evaluated at $t = T$, where $\omega$ is the angle of fall[62] and $T$ is the time of flight on the original trajectory. We then set the auxiliary variables $(\lambda, \mu, \nu, \lambda_1, \mu_1, \nu_1) = (1, \cot(\omega), 0, 0, 0, 0)$ at $t = T$. A look at the adjoint system and the boundary values chosen allows the immediate conclusions that $\lambda = 1$, $\nu = 0$; it also follows that $\nu_1 = 0$. These equations hold for all domain values of the variables. The *adjoint* system is then integrated numerically backward from $t = T$ to $t = 0$ using these initial values. The resulting values of the auxiliary variables as well as the initial values are substituted into the fundamental relation; with intermediate work this reduces to $\Delta X =$

$$[\xi + \eta \cot(\omega)]^{t=T} = [\lambda_1 \xi' + \mu_1 \eta']^{t=0} + \int_0^T \left\{F(1 + (x'^2 \frac{G'}{vG})\lambda_1 + x'y'F\frac{G'}{vG}\mu_1\right\} w_x dt \tag{1.40}$$

$$- \int_0^T F\left(x'\lambda_1 + y'\mu_1\right)\frac{\delta K}{K} dt$$

Seeing past the complexity of this expression, the important point is that all of the quantities on the right have either already been computed in the undisturbed trajectory or are variables that have been computed in the numerical integration of the adjoint system. All possible sources of variability are contained simultaneously in the right-hand side of (1.40); differences in initial conditions in the first term, range wind in the first integral, and changes in density law or ballistic coefficient in the second integral.

Further straightforward simplifications reduce the first-order range variation expression to

$$\Delta X = [\lambda_1 \xi' + \mu_1 \eta']^{t=0} + \int_0^T (\lambda_1 + 1)\, w_x\, dt - \frac{1}{h}\int_0^T \mu' \frac{\delta K}{K} dt \tag{1.41}$$

A similar calculation for the first-order expression for deflection is done by assuming that the six auxiliary variables have values (0 , 0 , 1 , 0 , 0 , 0, 0) at $t = T$ and following the same procedure. From (1.41), Bliss deduces that the first-order range variation is independent of the cross wind, due to the absence of $w_z$, a fact that is "well known from other sources" [61, p. 306]. The next section deduces formulae for the range variation due to the rotation of the earth; we will omit discussion of this.

We conclude with Bliss' further reduction of (1.41) to basic arithmetic for the use of obtaining numerical results. In the technical report, the computations are arranged in "A form for computing the solution of the adjoint system of differential equations" with the relevant variables in a column on the left in the order in which they are needed and a corresponding column on the right stating how each value is

[62] It is assumed that the angles of fall for the original and disturbed trajectories are nearly the same.

to be obtained.[63] Inclusion of such computation sheets appears to have been routine in the Aberdeen technical reports. In the journal article, the corresponding material [pp. 308–312] can be reduced to four steps for the calculations:

(1) By computing the original, undisturbed trajectory, the computer has available a table of values for $x,\ y,\ x',\ y',\ Fx'$, and $Fy'$ at selected short intervals of time.
(2) Next, we must find the coefficients for $\lambda_1$ and $\mu_1$ in the reduced version of the adjoint system (1.37) consisting of the second, fourth and fifth equations.

$$\begin{aligned}\mu' &= -hF(x'\lambda_1 + y'\mu_1)\\ \lambda_1' &= -\lambda + F(1 + x'^2\frac{G'}{vG})\lambda_1 + x'y'F\frac{G'}{vG}\mu_1\\ \mu_1' &= -\mu + x'y'F\frac{G'}{vG}\lambda_1 + F(1 + x'^2\frac{G'}{vG})\mu_1\end{aligned} \tag{1.42}$$

The computation of these coefficients requires only the values from step 1 at the ends of the intervals of time and tables for values of $\frac{G'}{vG}$ which had been prepared by Moulton's staff.
(3) For the range correction desired one integrates (1.42) with $(\mu,\ \lambda_1,\ \mu_1)^{t=T} = (\cot\omega,\ 0,\ 0)$ beginning at $t = T$ and going backward to $t = 0$. Thus, one obtains a table of values for $\mu$, $\lambda_1$, and $\mu_1$ at the ends of the time intervals selected.
(4) Bliss then gives five explicit formulae for the range corrections $\Delta X$ based on (1.41); these cover a change in initial velocity, a change in angle of elevation, the presence of wind, change in density of air, and change in ballistic coefficient. For example, if one has a following wind of 1 meter per second blowing between $t = t_1$ and $t = t_2$ then $w_x = 1$ in this interval, and if this is the only source of variation, it follows from (1.41) and the Fundamental Theorem of Calculus that

$$\Delta X = \int_{t_1}^{t_2} (\lambda_1' + 1)\, w_x\, dt = [t + \lambda_1(t)]_{t_1}^{t_2}. \tag{1.43}$$

To account for more realistic wind structures, the weighting factor method as used by Moulton is employed. Similarly, for each other individual variation, the suitable portion of (1.41) is simplified. Changes in density only, for example, are accounted for by employing percentage change $\frac{\delta K}{K} = .01$, say, for a 1% change in density over the time interval between $t = t_1$ and $t = t_2$. Again, courtesy of the fundamental theorem of calculus one obtains $\Delta X = \frac{.01}{h}[\mu(t_1) - \mu(t_2)]$. The computation of the

[63] Bliss recommends using either a 20-inch slide rule or an adding machine to do the calculations.

differential variations has now been reduced to arithmetic performed on values of the auxiliary variables.[64]

The range correction for initial conditions is handled by referring to the nonintegral part of (1.41). For example, for a change in angle of elevation $\Delta\theta_0$ from an initial value $\theta_0$, one has

$$\xi'(0) = v_0 \cos(\theta_0 + \Delta\theta_0) - v_0 \cos(\theta_0) = -v_0 \sin(\theta_0)\Delta\theta_0 + \dots$$
$$\eta'(0) = v_0 \sin(\theta_0 + \Delta\theta_0) - v_0 \sin(\theta_0) = v_0 \cos(\theta_0)\Delta\theta_0 + \dots \tag{1.44}$$

neglecting terms of higher order in $\Delta\theta_0$. This yields

$$\Delta X = v_0[-\lambda_1(0)\sin(\theta_0) + \mu_1(0)\cos(\theta_0)]\Delta\theta_0 \tag{1.45}$$

This ends our account of Bliss' published and unpublished work about adjoint method, with the exception of his treatment of the anti-aircraft problem, which will be laid out in the section following the next two.

## 1.10 Questions and Reflections

After one absorbs the complications of the method just outlined, many questions come to mind, both theoretical and practical. From a purely mathematical point of view: How did Bliss think to use the adjoint in this situation? What was it in his mathematical storehouse that suggested this tool? How was the adjoint method a result of the mathematics of the "function of a line" material, an assertion repeatedly made in accounts of the Aberdeen activities? Could the technique be generalized to arbitrary systems of differential equations, as Moulton presented his method as a general case of dealing with equations of variation? Could (1.41) be used as the basis for theorems about $\Delta X$? Note that in this equation the integrals are the canonical form for linear functionals in $w_x$ and $\delta K$; what was the current state of the notion of a linear functional?

From a practical point of view: how does this method compare to Moulton's solution of the differential variations problem? How was it received by mathematicians,

[64] The relation

$$\frac{\delta K}{K} = \frac{\delta H}{H} - \frac{\delta C}{C}$$

shows how $K$ intends to incorporate all changes related to density of air. Remember that the ballistic coefficient $C$ was considered roughly a measure of the ability of the projectile to penetrate the air. Thus, a positive percent change in actual density alone as measured by $H$ was viewed as having the same effect on $\frac{\delta K}{K}$ as the corresponding negative percent change in $\frac{\delta C}{C}$ alone.

ballisticians, and military personnel? Did corrections computed in this manner agree with the results of test firings? How was the data necessary for use of the method, the wind measurements, air density values, and so on collected in battle situations? Was this material taught in military schools?

Answers to these questions will be attempted throughout the following pages, but we will settle here for a few relevant observations. First, at the time of the appearance of the journal article in October of 1919, the method was already in use in computing range tables, according to the second paragraph of the article, and its continued use is documented in ordnance textbooks published in the decades after the War. An immediate selling point of this method is that, other than the presumed already available undisturbed trajectory, only one numerical integration is required, namely, that of the adjoint system, to get all range variations simultaneously. This contrasts with Moulton's scheme, in which separate integrations are required for each variation. This point is made in Dunham Jackson's ordnance textbook [207, p. 27], which outlines each technique. To counterbalance the problem of multiple integrations of a single system, Jackson concedes that Moulton's technique given the values of $\xi$, $\eta$, $\xi'$, and $\eta'$ all along the trajectory, where Bliss, at least in the material cited, can provide values only at endpoints.[65] He scores another advantage for the adjoint system: it can provide a value for the variation of any linear combination of $\xi$, $\eta$, $\xi'$, and $\eta'$ at $t = T$, not just $\Delta X$.

Jackson raised another interesting point: "In behalf of the more direct method [Moulton's] it is to be said that it is somewhat easier to follow, in practice as well as in theory, because of the more obvious physical significance of the quantities involved." Put another way, the auxiliary variables in Bliss' system have no physical meaning, at least in his presentation. It is conceivable that this would put an extra burden on a student of this material, who, already having to contend with difficult mathematical derivations, would also have to deal with variables in the abstract. This could not be said of Moulton's method, nor indeed of any previous ballistic expositions in the United States. Once again, the bar had been raised.[66]

Moulton himself presented Bliss' adjoint method in his 1926 text [pp. 134–141], after having his account reviewed by Bliss. Moulton described it as "characterized by singular elegance and great simplicity in application" [p. 135], noting its advantages and disadvantages as Jackson had. He added a criticism that it is not helpful for determining *all* terminal variations, but did not provide examples of this failure.

[65] In fact Bliss' antiaircraft material remedies this fault, but recall that this material was not published.

[66] This difficulty would be addressed by Roger Sherman Hoar in his ordnance textbook, which will be reviewed in the next chapter.

## 1.11 A Contribution From T.H. Gronwall

Although neither a part of the original technical report package from which the above has been drawn nor given in Bliss' JUSA/CAJ article, a contribution to the adjoint method, which became closely identified with it was made by T.H. Gronwall shortly after the method itself was issued. What follows is a short account of this addition, as well as some information on Gronwall himself.

Thomas Hakon Gronwall was a Swedish-born pure mathematician with an engineering degree from the Charlottenburg Technische Hochschule in Germany, near Berlin, who immigrated to the United States in 1904. After working for various railroad and steel companies, he turned to mathematics again in about 1910 while in Chicago, making the University a "base of operations." It was at the University of Chicago in 1912 that he made his American "mathematical debut," presenting several papers at an American Mathematical Society meeting held there. He had an unsuccessful stint of teaching at Princeton from 1913 to 1915, when he was dismissed for inattention to his teaching duties. The next several years were ones of itinerancy, during which he relocated to New York City and held various engineering positions while making research-related visits to Columbia University. Veblen, who befriended Gronwall at Princeton, kept in touch in his wayward colleague at this time, and Gronwall eventually came to Aberdeen in the Fall of 1918. There he authored several reports [see [134, p. 335] for specifics] and began a study, which eventually was published as [161] in 1920. His time at Aberdeen ended in mid-July 1919 when he was appointed as a "technical expert" with the Office of Ordnance in Washington, D.C.[67] Again, Veblen was instrumental in helping him obtain this position. His immediate supervisor was Dunham Jackson, the author of the ordnance textbook we have frequently cited.[68]

In Washington, Gronwall made a discovery, which he communicated to Veblen in a letter dated August 14, 1919:

> As soon as I arrived here some problems in differential variations were put up to me, and the result was a rather lengthy investigation of various aspects of the theory. A couple of days ago, I made what seems a rather important discovery, namely that the Bliss adjoint system has a first integral, which is $x'\lambda + y'\mu + x''\lambda_1 + y''\mu_1 = \kappa$, in the notation of Bliss' first paper [61]. This reduces the adjoint system from the third order to the second, and I expect that the numerical work in computing the differential corrections will reduce, by the new method the details of which I am now developing, by something like 40 per cent for the range corrections, and 60 per cent for the anti-aircraft trajectories. [Veblen Papers]

[67] Moulton's work in Washington was completed as of April of that year, so Gronwall was not part of Moulton's original group.

[68] By coincidence these two men were linked by their nearly simultaneous discovery a few years earlier of the nonnegativity of the trigonometric sum $\sin(x)+\frac{1}{2}\sin(2x)+\frac{1}{3}\sin(3x)+\cdots+\frac{1}{n}\sin(nx)$ on the interval $0 \leq x \leq \pi$. See [27, Introduction] for this story and references.

Yet again, a mathematical contribution is accompanied by a report of reduction in work it allows. Clarification of mathematical structure goes hand in hand with the economics of computation.

Some words of explanation are in order. By referring to the adjoint system as having order three, Gronwall is referring to system (1.42). His equation is called a "first integral" for the adjoint system, which for our purposes can be defined as a given relation among the variables of the system, a relation that then can be used to reduce the order of the system by one.[69] Gronwall's first integral is a result of an observation to which he devotes an unusual amount of attention in [161, p. 511]. He notes that the original equations for the ballistic trajectory (1.26) do not contain $t$ explicitly and thus are satisfied by $\overline{x} = x(t + \epsilon)$, $\overline{y} = y(t + \epsilon)$ and $\overline{z} = z(t + \epsilon)$ for $\epsilon$ a constant, if $x$, $y$ and $z$ are solutions as well. Subtracting the two solutions and expanding the results in series form and looking only at the first order terms, it is found that $\xi = x'(t)\epsilon$, $\eta = y'(t)\epsilon$, $\xi_1 = x''(t)\epsilon$, $\eta_1 = y''\epsilon$ satisfy equations 1, 2, 4, and 5 of (1.34) with all nonhomogeneous terms set to 0, since both $(x, y, z)$ and $(\overline{x}, \overline{y}, \overline{z})$ are solutions to the same system (1.26). Now since all the nonhomogeneous terms in the differential version of the fundamental relation (1.38) are 0, the first integral follows.[70] The calculations that use the first integral to reduce the order of the adjoint system to two equations will not be given here; Gronwall does this in [161, p. 512].

This discovery was included at the last minute in his supervisor's Ordnance Textbook: Jackson's volume contains a Supplementary Note dated November 29, 1919, which states:

> At the time when the above words [Preface, dated August 30, 1919] with regard to current progress were written, Dr. Gronwall was engaged in working out the practical details of a modification of Professor Bliss's method, by which the labor of computing differential variations is materially reduced. This improvement has been found so effective that a section describing it has been inserted in Chapter II. It is not to be inferred that progress has now come to an end, but, on the contrary, that further advances in one direction or another are to be looked for at any time.

In Section 8 of the book, it is shown directly that $x'$ and $y'$ are solutions to the equations of variation for the trajectory, the first integral is obtained, and its use in reducing order of the adjoint system is demonstrated. This marked the beginning of the inclusion of Gronwall's reduction in ballistic textbooks, which included the Bliss adjoint system.

---

[69] For a more exact definition of a first integral from a contemporary source, one can see [144, pp. 74–76].

[70] Gronwall's first observation is that the nonappearance in (1.26) of $t$ results in the system for the variations of (1.26) having $\xi = x'(t)\epsilon$, $\eta = y'(t)\epsilon$, $\xi_1 = x''(t)\epsilon$, $\eta_1 = y''\epsilon$ as solutions. It is unlikely that this was his original idea but attempts to find its predecessor have not been successful. His second observation is that a particular solution for a linear system results in a linear first integral of the type shown above for its adjoint. This was pointed out by C. G. J. Jacobi in a lengthy 1837 paper, which introduced the adjoint system [209, pp. 403–404]. A contemporary reference is [144, pp. 403–404].

The subject of the differential variations occupied much of Gronwall's time during the rest of his stay in Washington, which ended in 1922 with his move back to New York City. Thus, a list of "Mathematical Papers" from the Washington office [see NARA Washington] includes a list of ten of his papers written in these years with such titles as "Cross Wind Weighting Factors" and "Differential Variations in Range to Any Point on a Trajectory." Another group from which some papers survive lists "Approximate Formulas for the Variations in Range and Deflection due to the Rotation of the Earth." Yet, another topic was the "Range Variations Due to Changes in Gravity," an issue first raised by Moulton. In an attempt to put his own stamp on the entire field, Gronwall wrote a paper of 20 or so pages [159] summarizing the state of the art. In this respect, one could say that one result of Bliss' adjoint method was employment material for a technical expert, perhaps the only pure mathematician of the original Aberdeen and Washington staffs to continue war work in the years after the Armistice.

## 1.12 Bliss' Antiaircraft Work

It was mentioned earlier that one criticism of the Bliss adjoint method as presented so far was that it failed to give variations in horizontal and vertical coordinates at all points of the trajectory of a projectile, as Moulton's method had done, but only a terminal range correction. Since the World War was the first time that information about the entire trajectory was needed due to the presence of airplanes in battle, the phrases "whole trajectory" and "antiaircraft guns" often appeared together in the ballistic literature of the day.

Moulton wrote–

> An absolutely new type of problem was presented by hostile airplanes. In artillery work against airplanes it was not only necessary to fire at all elevations, but the targets were free to move in three dimensions and traveled with great speed, and the projectiles had to be exploded by time fuzes [contemporary spelling]. Moreover, the time for making all the adjustments for firing was limited to a few seconds, and the targets moved so rapidly that one shot gave no basis for making subsequent shots. [255, pp. 11–12]

Concern for these problems resulted in striking graphical displays of the relevant work done by the ballisticians. Perhaps, the best known of these are the illustrated sets of trajectories truncated by curves of constant time. One such, for example, is labeled "THEORETICAL TRAJECTORIES AND TIME CURVES FOR 37mm A. A. GUN, WEIGHT OF SHELL = 1.25 LBS., C [the ballistic coefficient] = 0.491, MUZZLE VELOCITY = 2000 FT/SEC," prepared by the Ordnance Department in 1924. (See Fig. 1.5.) The lettering is hand done in drafting style. It is roughly 53 cm by 56 cm and fashioned on blueprint paper, with the figures appearing in white. Range is given horizontally and altitude vertically in yards. Trajectories for each angle of from 5 to 85 degrees are given. For units of time between 3 and 22 seconds, the corresponding point on each trajectory is marked. For the family of trajectories, the points corresponding to the same time are connected with a

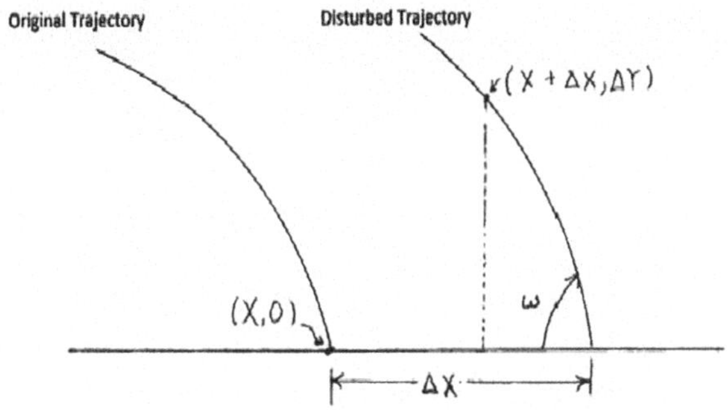


**Fig. 1.5** Diagram for calculation of variation in range [63, p. 99]

smooth curve, resulting in a set of circle-like arcs appearing somewhat orthogonally to the trajectories. Moulton's method for the solution of the differential equations of motion provided the trajectories, and this information together with knowledge of when in the time of flight a projectile might be exploded by fuze settings was incorporated into an inadvertently appealing graphic artifact.[71]

Knowledge of the entire trajectory needed to be accompanied by knowledge of the variability of the entire trajectory. An example of this concern was the appearance of an article [246] published in JUSA/CAJ in 1919 outlining a different derivation of Moulton's trajectory method; the author was a member of Moulton's working group.[72] The article included information about the effect of changing $G$.

Bliss did address the question of variability along the total trajectory in the final portion of his technical report, which appears as a continuation, labeled pp. 16–22 and dated February 1919, of the earlier part. It never appeared in print, though, as mentioned earlier, it was submitted to JUSA/CAJ [63, p. 94, p.104].[73] We will examine Bliss' work on this subject and begin with a caveat that, lest the reader think that the forthcoming equations are far too complicated for use, the report is accompanied by a three-page set of outsized computation sheets that display the use of all the material in the report.

Bliss uses three systems: First, the linearized version for the differential variations $\xi, \eta$:

$$\begin{aligned}
\xi' &= \xi_1 \\
\eta' &= \eta_1 \\
\xi_1' &= x'Fh\eta - F(1 + x'^2\frac{G'}{vG})\xi_1 - x'y'F\frac{G'}{vG}\eta_1 + f_4 \\
\eta_1' &= y'Fh\eta - x'y'F\frac{G'}{vG}\xi_1 - F(1 + y'^2\frac{G'}{vG})\eta_1 + f_5
\end{aligned} \tag{1.46}$$

---

[71] Some of the more complicated nomograms of the day have this same appeal.

[72] For further information concerning the author, W. E. Milne, and his eventual career-long interest in numerical mathematics as a result of his War work, see [135].

[73] An outline of the theoretical steps in the following, unaccompanied by computations, was included in [63, pp. 104–106].

where

$$f_4 = \left(1 + x'^2 \frac{G'}{vG}\right) w_x - x' F \frac{\delta H}{H}$$
$$f_5 = \left(x' y' F \frac{G'}{vG}\right) w_x - y' F \frac{\delta H}{H} \quad (1.47)$$

In this section $\frac{\delta K}{K}$ has been replaced by $\frac{\delta H}{H}$ since the ballistic coefficient is not considered.

Second, there is the system adjoint to this:[74]

$$\lambda' = 0,$$
$$\mu' = -hF(x'\nu + y'\rho)$$
$$\nu' = -\lambda + F(1 + x'^2 \frac{G'}{vG})\nu + x'y'F \frac{G'}{vG}\rho$$
$$\rho' = -\mu + x'y'F \frac{G'}{vG}\nu + F(1 + x'^2 \frac{G'}{vG})\rho \quad (1.48)$$

Finally, there is the adjoint to the adjoint, called the "auxiliary system":

$$\begin{aligned} A' &= C \\ B' &= D \\ C' &= x'FhB - F(1 + x'^2 \frac{G'}{vG})C - x'y'F \frac{G'}{vG} D \\ D' &= y'FhB - x'y'F \frac{G'}{vG} C - F(1 + y'^2 \frac{G'}{vG}) D \end{aligned} \quad (1.49)$$

Due to the nature of the adjoint operation, the auxiliary system is merely the homogeneous version of (1.46). Let[75] $[\lambda_i,\ \mu_i,\ \nu_i,\ \rho_i],\ i = 1 \ldots 4$ be the four solutions to the adjoint system (1.48) satisfying the initial conditions $[\lambda_i,\ \mu_i,\ \nu_i,\ \rho_i]_{t=0} = \vec{e}_i,\ i = 1 \ldots 4$ where $e_1 = [1,\ 0,\ 0,\ 0]$, $e_2 = [0,\ 1,\ 0,\ 0]$, $e_3 = [0,\ 0,\ 1,\ 0]$, and $e_4 = [0,\ 0,\ 0,\ 1]$. Now each of the solutions satisfies the fundamental relation

$$[\lambda\xi + \mu v + \nu\xi' + \rho\eta']_0^t = \int_0^t (\nu' + \lambda)\, w\, dt - \int_0^t \mu' \frac{\delta H}{H'}\, dt \quad (1.50)$$

[74] The variables $\lambda$, $\mu$, $\nu$, $\rho$ used here for the adjoint system became standard in treatments of this material in the years after the World War.

[75] We switch to vector notation in this section for ease of exposition, though Bliss did not employ it in his report.

Noting that $\lambda$ is a constant for any solution, and keeping in mind the initial values of the variables in each solution set, we can write

$$\lambda_1\xi + \mu_1\nu + \nu_1\xi' + \rho_1\eta' = \int_0^t (\nu_1' + 1)\, w\, dt - \int_0^t \mu_1' \frac{\delta H}{H'} dt = m_1$$

$$\lambda_2\xi + \mu_2\nu + \nu_2\xi' + \rho_2\eta' = \int_0^t (\nu_2' w\, dt - \int_0^t \mu_2' \frac{\delta H}{H'} dt = m_2$$

$$\lambda_3\xi + \mu_3\nu + \nu_3\xi' + \rho_3\eta' = \xi'(0) + \int_0^t \nu_3' w\, dt - \int_0^t \mu_3' \frac{\delta H}{H'} dt = m_3$$

$$\lambda_4\xi + \mu_4\nu + \nu_4\xi' + \rho_4\eta' = \eta'(0) + \int_0^t \nu_4' w\, dt - \int_0^t \mu_4' \frac{\delta H}{H'} dt = m_4 \qquad (1.51)$$

where the $m_i$ are the definitions of the middle quantities in each equation. These equations constitute a system of four algebraic equations in the four unknowns $[\xi,\ \eta,\ \xi',\ \eta']$, each a function of $t$. The remainder of the work involves solving this system.

Departing from Bliss' notation, we write the system as

$$\left[\xi\ \eta\ \xi'\ \eta'\right] \begin{bmatrix} \lambda_1 & \lambda_2 & \lambda_3 & \lambda_4 \\ \mu_1 & \mu_2 & \mu_3 & \mu_4 \\ \nu_1 & \nu_2 & \nu_3 & \nu_4 \\ \rho_1 & \rho_2 & \rho_3 & \rho_4 \end{bmatrix} = \left[m_1\ m_2\ m_3\ m_4\right] \qquad (1.52)$$

Thus, solving the system is a matter of finding an inverse (or "reciprocal" in Bliss' terminology) for

$$\begin{bmatrix} \lambda_1 & \lambda_2 & \lambda_3 & \lambda_4 \\ \mu_1 & \mu_2 & \mu_3 & \mu_4 \\ \nu_1 & \nu_2 & \nu_3 & \nu_4 \\ \rho_1 & \rho_2 & \rho_3 & \rho_4 \end{bmatrix} \qquad (1.53)$$

.

To this end, let $[A_i,\ B_i,\ C_i,\ D_i]$, $i = 1 \ldots 4$ be the four solutions to (1.49) satisfying the initial conditions $[A_i,\ B_i,\ C_i,\ D_i] = \vec{e}_i$, $i = 1 \ldots 4$ where $e_1 = [1,\ 0,\ 0,\ 0]$, $e_2 = [0,\ 1,\ 0,\ 0]$, $e_3 = [0,\ 0,\ 1,\ 0]$, and $e_4 = [0,\ 0\ ,0,\ 1]$. It follows from the fundamental relation holding between (1.48) and (1.49) that the desired inverse is

$$\begin{bmatrix} A_1 & B_1 & C_1 & D_1 \\ A_2 & B_2 & C_2 & D_2 \\ A_3 & B_3 & C_3 & D_3 \\ A_4 & B_4 & C_4 & D_4 \end{bmatrix}. \qquad (1.54)$$

Now noting that the only solutions to (1.49) with initial conditions [1, 0, 0, 0] are $A \equiv 1,\ B \equiv 0,\ C \equiv 0,\ D \equiv 0$, on multiplying each side of (1.52) on the right by (1.54) we obtain

$$\begin{bmatrix} \xi\ \eta\ \xi'\ \eta' \end{bmatrix} = \begin{bmatrix} m_1\ m_2\ m_3\ m_4 \end{bmatrix} \begin{bmatrix} 1 & 0 & 0 & 0 \\ A_2 & B_2 & C_2 & D_2 \\ A_3 & B_3 & C_3 & D_3 \\ A_4 & B_4 & C_4 & D_4 \end{bmatrix}. \tag{1.55}$$

At this point, the variation problem has been "solved," in that the values of $\xi$ and $\eta$ for any time on the trajectory have been expressed in terms of the quantities on the right side of (1.55), which ultimately involve the functions in (1.53), (1.54), and the computed quantities of the original trajectory. In this sense, no more need be said, but the three large computation sheets accompanying this portion of the technical report show that the derivation above was more than just an exercise in analysis. Bliss and his associates must have thought that the procedure needed illustration in at least one case and perhaps intended it as a display piece for the entire Bliss adjoint method. In any event, the sheets are stunning. Each blueprint sheet measures roughly 43 cm by 57 cm and is covered with handwritten numbers, variables, and equations in neat rows and columns, again in Prussian blue and white. One wishing for a graphical representation of the essence of the new ballistic methods could hardly do better than admire and study these sheets. We will now proceed to give an account of these pages.[76]

The first sheet bears the title "Differential Corrections for A. A. Guns," lists the weapon as the 3 inch antiaircraft (1917, 1918) with 3 inch fuzed shrapnel. An angle of elevation of 30 degrees and initial velocity of 2400 ft/s = 731.52 m/s is given, along with a ballistic coefficient of C = 1.90 and gravitational constant g = 9.81. In the report, Bliss states "The purpose was to suggest a convenient form for computation, to indicate the degree of accuracy required at each step, and to gain some idea of the time required for the work." The first two-thirds of the sheet, under the title "The Trajectory," is taken up by the computation of the complete trajectory by Moulton's method, producing initially 25 columns of values for a

[76] It is at least worth asking whether such documents as these sheets and the graphical trajectory sheet mentioned earlier played a role in persuading various parties of the effectiveness and power of the new mathematical ballistics work. In a presidential address before the American Mathematical Society in April of 1919 [106], a few months after Bliss' report was completed, Leonard Eugene Dickson surveyed the use of mathematics in the War and the educational efforts of the Allies in relevant mathematical subjects. This address was followed in the afternoon by talks by "five ballistics experts who come to us direct from the two centers of ballistic work in America" [106, p. 297]. Dickson remarked that "Following my request for some information suitable for use in this address, I received from both Washington and Aberdeen huge bundles of blue prints showing hundreds of beautiful trajectories and other curves and many heavy mathematical manuscripts - a sort of long range bombardment which it seemed the part of wisdom to dodge and trust to the direct fire of the newly arrived experts" [106, p. 297].

32-second path followed by an additional 23 columns for differences; values for $t$, $x$, $y$, $x'$, $y'$, $x''$, $y''$, $v$, and $F$ are ultimately found for unequal time intervals from $t = 0$ to $t = 32$ seconds. From these values and tables for $G$ are formed, under the title "Coefficients of the Auxiliary Equations," the required coefficients in 15 columns for the same time intervals. The auxiliary equations themselves are depicted at the top of the second page; see Fig. 1.7. "Match boxes or other markers, moved step by step down the column of the trajectory, were of great service here, as in later parts of the computation. The time required was about ten minutes per column, and the whole computation of the coefficients of the auxiliary equations can be accomplished in 2 1/4 hours."

The second page is headed by "Solutions of the Auxiliary Equations" as shown in Fig. 1.6. Recall that four solutions to this system are required, one for each set of initial conditions $[A_i, B_i, C_i, D_i] = \vec{e}_i$. Since the first solution is trivially $A_1 \equiv 1$, $B_1 \equiv 0$, $C_1 \equiv 0$, $D_1 \equiv 0$, only three solutions are required; the fact that only these three "integrations by mechanical quadratures" were needed for the entire process was a selling point, in Bliss' mind, for the method. The computations for these solutions form the bulk of the second page, giving the remaining 12 variables $A_i$, $B_i$, $C_i$, $D_i$, $i = 2 \ldots 4$ for the same time intervals as the coefficients calculated on the preceding page. The 21 columns of figures took 19 1/2 hours to produce. "This could be much improved upon with a model already at hand, and by an experienced computer." We now have the numerical values for matrix on the

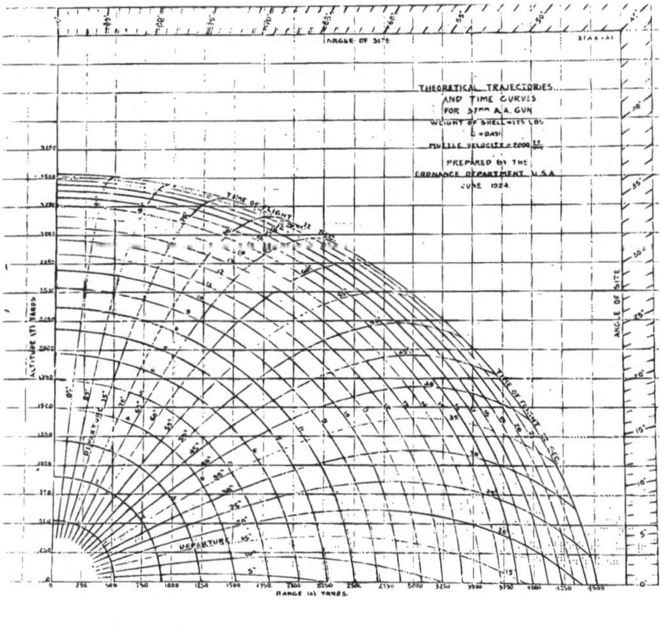

**Fig. 1.6** Trajectories calculated from Moulton's differential equations; [NARA Washington]

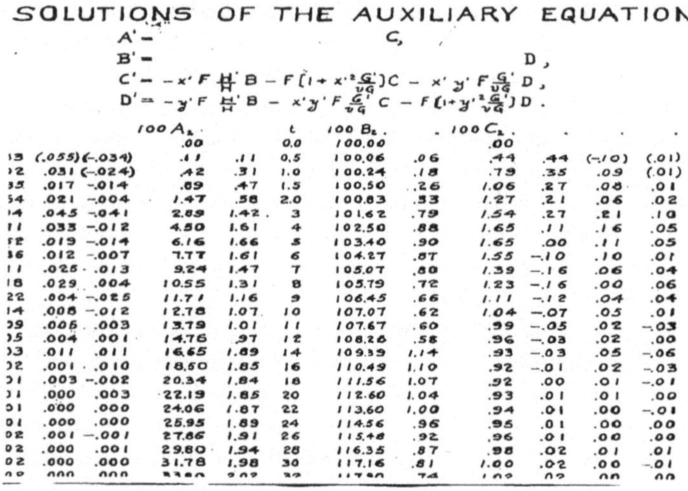

SOLUTIONS OF THE AUXILIARY EQUATION.

A′ = C,

B′ = D,

$C' = -x'F\frac{H'}{H}B - F[1 + x'^2\frac{G'}{vG}]C - x'y'F\frac{G'}{vG}D,$

$D' = -y'F\frac{H'}{H}B - x'y'F\frac{G'}{vG}C - F[1 + y'^2\frac{G'}{vG}]D.$

| | | | $100A_2$ | | $t$ | $100B_2$ | | $100C_2$ | | | |
|---|---|---|---|---|---|---|---|---|---|---|---|
| | | | .00 | | 0.0 | 100.00 | | .00 | | | |
| 13 | (.055) | (−.034) | .11 | .11 | 0.5 | 100.06 | .06 | .44 | .44 | (−.10) | (.01) |
| 12 | .031 | (−.024) | .42 | .31 | 1.0 | 100.24 | .18 | .79 | .35 | .09 | (.01) |
| 35 | .017 | −.014 | .89 | .47 | 1.5 | 100.50 | .26 | 1.06 | .27 | .08 | .01 |
| 54 | .021 | −.004 | 1.47 | .58 | 2.0 | 100.83 | .33 | 1.27 | .21 | .06 | .02 |
| 14 | .045 | −.041 | 2.89 | 1.42 | 3 | 101.62 | .79 | 1.54 | .27 | .21 | .10 |
| 11 | .033 | −.012 | 4.50 | 1.61 | 4 | 102.50 | .88 | 1.65 | .11 | .16 | .05 |
| 52 | .019 | −.014 | 6.16 | 1.66 | 5 | 103.40 | .90 | 1.65 | .00 | .11 | .05 |
| 36 | .012 | −.007 | 7.77 | 1.61 | 6 | 104.27 | .87 | 1.55 | −.10 | .10 | .01 |
| 11 | .025 | .013 | 9.24 | 1.47 | 7 | 105.07 | .80 | 1.39 | −.16 | .06 | .04 |
| 18 | .029 | .004 | 10.55 | 1.31 | 8 | 105.79 | .72 | 1.23 | −.16 | .00 | .06 |
| 22 | .004 | −.025 | 11.71 | 1.16 | 9 | 106.45 | .66 | 1.11 | −.12 | .04 | .04 |
| 14 | .008 | −.012 | 12.78 | 1.07 | 10 | 107.07 | .62 | 1.04 | −.07 | .05 | .01 |
| 09 | .006 | .003 | 13.79 | 1.01 | 11 | 107.67 | .60 | .99 | −.05 | .02 | −.03 |
| 05 | .004 | .001 | 14.76 | .97 | 12 | 108.26 | .58 | .96 | −.03 | .02 | .00 |
| 03 | .011 | .011 | 16.65 | 1.89 | 14 | 109.39 | 1.14 | .93 | −.03 | .05 | −.06 |
| 02 | .001 | .010 | 18.50 | 1.85 | 16 | 110.49 | 1.10 | .92 | −.01 | .02 | −.03 |
| 01 | .003 | −.002 | 20.34 | 1.84 | 18 | 111.56 | 1.07 | .92 | .00 | .01 | −.01 |
| 01 | .000 | .003 | 22.19 | 1.85 | 20 | 112.60 | 1.04 | .93 | .01 | .01 | .00 |
| 01 | .000 | .000 | 24.06 | 1.87 | 22 | 113.60 | 1.00 | .94 | .01 | .00 | −.01 |
| 01 | .000 | .000 | 25.95 | 1.89 | 24 | 114.56 | .96 | .95 | .01 | .00 | .00 |
| 02 | .001 | −.001 | 27.86 | 1.91 | 26 | 115.48 | .92 | .96 | .01 | .00 | .00 |
| 02 | .000 | .001 | 29.80 | 1.94 | 28 | 116.35 | .87 | .98 | .02 | .01 | .01 |
| 02 | .000 | .000 | 31.78 | 1.98 | 30 | 117.16 | .81 | 1.00 | .02 | .00 | −.01 |
| [illegible] | [illegible] | [illegible] | [illegible] | [illegible] | [illegible] | [illegible] | [illegible] | [illegible] | [illegible] | [illegible] | [illegible] |

**Fig. 1.7** Closeup of the middle sheet of the three-sheet computation document accompanying Gilbert Ames Bliss's technical report, [NARA Aberdeen]. The document shows the computation of a trajectory and differential variations for an anti-aircraft firing

right hand side of the solution equation (1.55) over the necessary range of $t$ values (Fig. 1.7).

The matrix of $m_i$'s in the solution require values for the matrix (1.53), from which the values of the $m_i$ are formed. But since this matrix and (1.54) are inverses, one need not integrate in order to find them, one can use the cofactors of the already computed (1.54) and the determinant of (1.53), which is denoted $\Delta$. One only needs six of the functions in (1.53) for the corrections, namely $\mu_2$, $\mu_3$, $\mu_4$ and $\nu_2$, $\nu_3$, $\nu_4$; these are all found by formulae such as

$$\mu_2 = \Delta \begin{vmatrix} C_3 & D_3 \\ C_4 & D_4 \end{vmatrix} \tag{1.56}$$

where $\Delta$ is found from the known determinant $\frac{1}{\Delta}$ of (1.54) "by using a Barlow's table of reciprocals." Computations of the determinants for the six functions are crowded onto the right side of sheet 2; they took 1 1/2 hours to execute. The expressions for the six functions are given in the top corner of page 3.

The last page is devoted to "The Differential Corrections" themselves, which consisted for the usual time interval, of values for $\xi$ and $\eta$ corresponding to, individually, wind and change in density over a subinterval, and change in angle of elevation and initial velocity. We will consider the density correction. For a 1% variation in density from normal, $\frac{\delta H}{H} = .01$, and using the exponential law for density, $\frac{\delta H}{H'} = \frac{\delta H}{H}\frac{H}{H'} = \frac{.01}{-.000106} = -94.3396$. Now if this change in density holds between $t = t_0$ and $t = t$, the values of the $m_i$ simplify to

$$m_i(t) = \frac{\delta H}{H'}[\mu_i(t) - \mu_i(t_0)] \tag{1.57}$$

for $i = 1 \ldots 4$. Together with the already calculated values of the $A_i$ and $B_i$, the matrix product in (1.55) gives the following, dutifully appearing at the top of page 3:

$$\xi(t) = \frac{\delta H}{H'}[A_2(t_0)-A_2(t)]\mu_2(t_0)+A_3(t_0)-A_3(t)]\mu_3(t_0)+[A_4(t_0)-A_4(t)]\mu_4(t_0)]$$

$$\eta(t) = \frac{\delta H}{H'}[1 - B_2(t)\mu_2(t_0) - B_3(t)\mu_3(t_0) - B_4(t)\mu_4(t_0)] \tag{1.58}$$

for $t = 0$ this simplifies to

$$\xi(t) = 94.3396A_2(t)$$
$$\eta(t) = 94.3396[1 - B_2(t)] \tag{1.59}$$

Similar expressions for the other factors occupy the top portion of the page. Next follow columns of values for all variables used in the expressions for $\xi(t)$ and $\eta(t)$, including the differences $A_i(t_0) - A_i(t)$ for two-second intervals from $t = 0$ to $t = 32$. The last quarter of the final sheet gives the wind corrections followed by the density corrections for the same time interval. The cascading form allows reading, at each two second value, the $\xi$ and $\eta$ for the disturbing factor holding from that value to any time further along the interval. These are the desired data.

Several comments concerning methods of shortening the computations are made, and the document concludes by stating the total time for the three sheets combined was about 30 hours. "This would probably be less than that necessary for a computer on his first attempt, but more than would be required on his second. Parts of the work could be accelerated by more than one half if two persons worked together." The exposition shows the tedium of hand-done calculations done at this time but also testifies to the practical use to which Bliss thought his method could be put. Need it also be added that it made the need for faster methods of calculation apparent?

We conclude this account by asking the reader to ponder the similarity between Bliss' manipulations and the familiar derivation, which follows, stated and solved in the more standard form of left multiplication of a column of variables by a square matrix:

Let $A$ be an $n$ by $n$ matrix of known functions of $t$, $\vec{f}$ a column of length $n$ of known functions, and $\vec{y}$ a similar column of unknown functions, all defined on an interval $[0, t_1]$. Then, we propose to solve the linear system of differential equations

$$\vec{y}' = A\vec{y} + \vec{f} \tag{1.60}$$

This can be thought of as the linearized system of differential equations for the variations. Let $M$ be the $n$ by $n$ matrix whose row $i$ is the solution $\vec{z}_{i,j}$, $j = 1...n$ to the adjoint system

$$\vec{z} = -A^T\vec{z} \tag{1.61}$$

with initial values $\vec{e}_i$ the standard basis vectors. Thus M is analogous to the transpose of (1.53). We then have

$$M\vec{y}' = MA\vec{y} + M\vec{f} \tag{1.62}$$

or

$$M\vec{y}' = -M'\vec{y} + M\vec{f} \tag{1.63}$$

from which it follows that

$$M\vec{y}' + M'\vec{y} = M\vec{f} \tag{1.64}$$

or

$$\frac{d}{dt}M\vec{y} = M\vec{f} \tag{1.65}$$

.

Now integrating both sides from 0 to $t_1$ gives

$$[M\vec{y}]_0^{t_1} = \int_0^{t_1} M(t)\,\vec{f}(t)\,dt \tag{1.66}$$

or

$$M\vec{y}(t_1) = \vec{y}(0) + \int_0^{t_1} M(t)\,\vec{f}(t)\,dt \tag{1.67}$$

since $M(0) = I$, the identity matrix. Finally, multiplying both sides by $M^{-1}$ gives

$$\vec{y}(t_1) = M^{-1}(t_1)\vec{y}(0) + M^{-1}(t_1)\int_0^{t_1} M(t)\,\vec{f}(t)\,dt. \tag{1.68}$$

This last expression is analogous to a more compactly written version of Bliss' solution for the variations: the integral expression corresponds to the integral portion of the $m_i$ in (1.51) multiplied by (1.54) and the $M^{-1}\vec{y}(0)$ to the initial conditions also multiplied by (1.54). In this manner it is seen that the matrix $M$ of solutions to the adjoint system acts as an integrating factor, making the original equation exact.

## 1.13 Conclusion

This concludes our account of Bliss' contributions to the practical problem of the differential variations. We have seen the subject evolve from purely empirical observation of deviation due to cross winds to an attempt, in the Siacci theory, at accounting for these factors by loading the ballistic coefficient with various constants. Moulton was the first to attempt a rigorous account of the "secondary effects," an account influenced by his astronomical experience and familiarity with Poincaré's notion of the equations of variation. His first technical paper was devoted to this topic, and his method continued to be used alongside Bliss'. To him is due the credit for changing the whole tone of the subject from the assignment of apparently *ad hoc* constants to the question of solution of systems of differential equations by numerical methods. A greater shift in mathematical tone could hardly be imagined, though that alone was no guarantee of the correspondence of theory with practice, a goal deeply sought by Moulton. Bliss' innovation was the use of the adjoint system, which will be shown to have arisen from his research in calculus of variations and possibly influenced by contemporary use of the adjoint in solving partial differential equations. His methods were demonstrably more efficient than Moulton's though some preference was shown for one or the other method in certain circumstances. He too was interested in the accuracy of his methods, and the computations done using his method were included in range firing tables as were Moulton's.[77] The material presented here makes no explicit use of the ideas of functions of a line, though one can see the notion of a differential of a functional lurking behind some of the equations, such as (1.51). It will be shown in Chap. 2 that his work had, simultaneously, permanent ramifications in pure mathematics in the variability theory of differential equations and the new discipline of the functional analysis. In Chap. 3, it will be related that the adjoint method continued to be of use in unexpected settings long after its service in the ballistic context ceased. In this sense, the story of Bliss' work is far from complete.

## Archival Materials (Abbreviations)

[Veblen Papers] Oswald Veblen Papers, Archives of the Library of Congress, Library of Congress, Washington, D.C.
[NARA Aberdeen] Record Group 156-866, National Archives and Records Administration, College Park, Maryland

[77] Curiously, Moulton makes no mention, in published work considered by the present author, of the anti-aircraft method for obtaining the differential variations for an entire trajectory. His *Methods* of 1926 has only the terminal variation material, and his exchange of comments with range table-constructor Roger Sherman Hoar, described in the next chapter, shows no sign of familiarity with the idea. As noted earlier, Bliss' only published account is a brief sketch in his paper [63], in the *Transactions of the American Mathematical Society*, which was perhaps missed by Moulton.

[NARA Washington] Record Group 156.7.18, National Archives and Records Administration, College Park, Maryland

[NARA Ordnance School] Record Group 156 - 1512A, Box 2 , National Archives and Records Administration, Philadelphia, Pennsylvania

[NARA ENIAC] Record Group 156-1512A, Box 9, National Archives and Records Administration, Philadelphia, Pennsylvania

[Columbia University] University Archives and Columbiana Library, Columbia University in the City of New York

[Princeton University] University Archives Project, Seeley G. Mudd Manuscript Library, Princeton University, Princeton, New Jersey

[MIT] MIT Institute Archives and Special Collections, Massachusetts Institute of Technology, Cambridge, Massachusetts

[University of Chicago] Special Collections Research Center, University of Chicago Library, Chicago, Illinois

[Concord] Hoar Family Papers, Concord Free Public Library, Concord, Massachusetts

[Harvard] Office of Harvard War Records, Harvard University, Cambridge, Massachusetts

# Chapter 2
# The Reception, Development, and Comparisons of the Work of the Second Wave of Ballistic Reformers

## 2.1 Introduction

After the account given in the preceding section of new material developed by the second wave of ballistic reform, the material dubbed "The New Ballistics," one naturally wonders about its reception. In this chapter, we look at this question from the point of view of several audiences. We see what influence it had on the course of research for pure mathematicians. We tell the story of how the material was incorporated into military instruction in such institutions as the Coast Artillery School and the Ordnance School of Application. The contributions of practicing mathematicians to the literature of instruction are noted. Critical responses by certain members of the military are discussed, as the inevitable change from old to new methods occurred. We shall see how courses in the new material were initiated in established universities. It will be demonstrated how the example of the successful use of advanced mathematical methods in the ballistic context was cited in a larger literature devoted to the promotion of mathematics itself. A final section summarizes comparable work by the French, English, and German ballisticians.

Let us note for completeness that among the many mathematical lines continued or begun in America around the period of the War, four stand out as having been developed, incorporated, and discussed by various audiences in print. In no particular order of importance, the first is the numerical solution of differential equations, as instituted by Moulton. In [465], one finds an abbreviated history of the precursors and followers of this topic as it gradually became an object of rigorous study in the United States in the decades after the War. The second was the differential variations, also initiated by Moulton and further developed by Bliss and Gronwall. As previously mentioned, this material found a certain measure of acclaim for its connection to functions of a line, a connection we have yet to explore. Third was the study of the rotating projectile and resulting drift factors. This can be thought of as a variation of a different sort and will be the subject of a section of Chap. 3. Finally, the topic of probable error has already been noted in the Preface.

 
A. Gluchoff, *From Fréchet Differentials to Firing Tables*,
https://doi.org/10.1007/978-3-031-68267-4_2

## 2.2 American Mathematical Work Connected with the Differential Variations

It must be admitted that the response to the new work on differential variations on the part of the mathematical community was limited. To G. H. Hardy's familiar assessment of ballistics,[1] we might add the words of the prominent applied mathematician Clifford Truesdell in a review of an American ballistics textbook [581]: "To most readers, the subject is dull, and its practical side is not likely to make it any more attractive in times of necessity" [657]. These quotes speak against the appeal of ballistics as an area of pure mathematical research. Bliss' two 1920 follow-up papers [386] and [387] constitute the bulk of the attention paid to the differential variations material at this time and will be considered in the next chapter. They contain the natural generalizations of the adjoint method to arbitrary systems of differential equations whose defining functions are allowed to vary, and they introduce new types of convergence of sequences of functions naturally appropriate to the situation.

The remaining work done by pure mathematicians specifically related to the differential variations falls into three categories. The first two featured the work of a familiar figure-T. H. Gronwall. Gronwall's first efforts involved issuing two papers [490, 491] in which he considered what he termed "the qualitative properties of the ballistic trajectory," that is, theorems that could be proved about the solutions to Moulton's differential equations of motion under various theoretical assumptions on $G, H, C$, and the initial conditions. In this study, the adjoint system was invoked. His second contribution was the expansion of Bliss' adjoint method to finding variations in the trajectory other than range changes. These included changes in time of flight and maximum ordinate.[2] The third category consisted of a series of papers on the equations of variation in a general context, work done by Bliss, Joseph Ritt, Moulton, and Gronwall. We begin with Gronwall's "qualitative properties" studies.

---

[1] "It is true that there are branches of applied mathematics, such as ballistics and aerodynamics, which have been developed deliberately for war and demand a quite elaborate technique ... They are indeed repulsively ugly and intolerably dull; even Littlewood could not make ballistics respectable, and if he could not, who can?" [507, p. 140].

[2] Gronwall's role in developing these variations is a bit murky. Bliss' work both published and unpublished at this time considered only range variations. It was not until his 1944 summary ballistics volume *Mathematics For Exterior Ballistics* [395] that other variations were treated by him. In that book, Gronwall is not specifically credited with initiating this work. In an official account of the history of the Aberdeen Proving Grounds [352, p. 5], however, it is stated flatly that "Dr. Gronwall, who also was a member of the Range Firing Section, elaborated this method [the adjoint method] to determine the effects of such variations on time of flight and the maximum ordinate...." In the technical report [T. S. 101] on differential variations dated August 1919 (mentioned in the previous chapter), Gronwall notes in the introduction that "The method of treatment is that developed by Bliss for the special case of range corrections and which is here extended to the corrections of time of flight and maximum ordinate." Both the first and second of Gronwall's post-armistice efforts considered here appeared in [491]. Under these circumstances, it seems reasonable to attribute the new variations to him.

It is not necessary to dwell on the contents of [490] other than to summarize a few typical results and tools. Moulton's equations are written out, and, for example, hypotheses involving the existence and boundedness of the partials of $F$ with respect to $x,\ y,\ x',\ y'$ are given. Conclusions following from this are such as "the horizontal distance $x$ increases steadily from zero as $t$ increases from zero" and "the trajectory is concave downward" and "lies below the trajectory in vacuum corresponding to the same initial velocity components." "At two points of equal altitude, the slope of the trajectory is numerically greater on the falling than on the rising branch." The trajectory is treated as a purely mathematical object defined by differential equations. Several well-known theorems are used: Tchebychef's integral inequalities for monotone functions, Schwartz's inequality, and Cauchy's Mean Value Theorem all appear. It is essentially a workout at the first level of ballistic discourse, and its results are compared by the author with those of classical and practicing theoretical ballisticians.

In [491], the paper in which his new first integral is presented, and work of this nature is continued. His reduction of the fourth-order adjoint system to a second-order system by means of the two first integrals results in the pair

$$\mu' = -hE(x'\nu + y'\rho)$$

$$(x'\nu + y'\rho)' = (1 + \frac{vG'}{G})E(x'\nu + y'\rho) - 2(x'\lambda + y'\mu) \tag{2.1}$$

for range variations,where $E$ is the new symbol for $F$, $h = h(y) = -\frac{H'(y)}{H(y)}$. (Here the density function $H(y)$ need not be the exponential form used in all preceding work.) Gronwall has gone to some trouble to write the second equation as it appears in order to introduce the expression $x'\nu + y'\rho$ as a variable of interest, which appears frequently in Bliss' work [see (1.42), for example]. It is assumed as in the applied situation that a trajectory under "normal" conditions has been computed with resulting variables $x,\ y\ ,x',\ y'$ found and that the four adjoint variables have been found from the adjoint system by integrating backward from the terminal conditions $(\lambda,\ \mu,\ \nu,\ \rho)^{t=T} = (1,\ \cot(\omega),\ 0,\ 0)$. Thus the terminal conditions for (2.1) are $\mu = \cot(\omega)$, $x'\nu + y'\rho = 0$ when $t = t_\omega$, a new symbol for the time of flight. From here, we will summarize his arguments, which reach some conclusions about the range of the projectile.

Gronwall considers a special case for Moulton's equations; he assumes that $G' > 0$ and $h(y) \geq 0$. The first assumption speaks to the rapidity of increase of the retardation $\frac{vG'}{G}$ as a function of velocity and the second that the density of the atmosphere never increases with altitude. On p. 517, it is stated that "The fundamental qualitative property of the range variations is expressed by the inequality

$$x'\nu + y'\rho > 0 \tag{2.2}$$

for $0 \leq t < t_\omega$." Absent any interpretation for $\rho$ and $\omega$, it is difficult to attach any meaning to this quantity or consequences of its positivity. Easily one has $x' \geq 0$

throughout and $y' \leq 0$ on the descending branch and $\nu = \rho = 0$ at the terminus, but what can one conclude from this? Gronwall proceeds to prove the fundamental property. To this end, he notes that the second equation of (2.1) is linear first order in $x'\nu + y'\rho$ and consequently satisfies

$$e^{\psi}(x'\nu + y'\rho) = 2\int_t^{t_\omega} e^{\psi}\,(x' + y')\mu\,dt \tag{2.3}$$

where $\psi$ is the integrating factor

$$\psi = \int_t^{t_\omega} (1 + \frac{vG'}{G})\,E\,dt \tag{2.4}$$

He then argues that the portion of the integrand in parentheses decreases to 0 on $t_\omega - \epsilon \leq t \leq t_\omega$. From this the desired positivity of $x'\nu + y'\rho$ follows at least on that interval. The positivity throughout is then established by an intricate argument involving the behavior of the adjoint variables on the entire time interval. In the process, Gronwall proves that $\mu$ decreases to its terminal value $\cot(\omega)$ on $0 \leq t \leq t_\omega$.[3]

An easy consequence of the positivity of $x'\nu + y'\rho$ follows from the last integral of (3.165) in the case of a change of ballistic coefficient: under these circumstances, we have

$$\delta X = \int_0^{t_\omega} E\,(x'\nu + y'\rho)\,\frac{\delta C}{C}\,dt. \tag{2.5}$$

Thus, an increase in the ballistic coefficient increases the range. This coincides with the notion of the ballistic coefficient as a measure of the ability of a shell to penetrate the atmosphere. Another such consequence follows from the expression for the range variation due to a change in initial velocity

$$\delta X = (x'\nu + y'\rho)_0\frac{\delta v_0}{v_0} \tag{2.6}$$

namely, the range increases when the initial velocity increases.

For more on this connection with initial velocity, Gronwall integrates the second equation in (2.1) from $t$ to $t_\omega$ and after some manipulation obtains

$$x'\nu + y'\rho = 2(x_\omega - x) - 2y\mu - \int_t^{t_\omega} (1 + \frac{vG'}{G} - 2hy)\,E\,(x'\nu + y'\rho)\,dt \tag{2.7}$$

[3] Interestingly, on the three-page computation sheet accompanying Bliss anti-aircraft report, the calculated values for the adjoint variables for the specific sets of initial conditions show monotone behavior: two are positive and monotone increasing, and one is negative and monotone decreasing. These do not correspond to the case at hand since Bliss' $\mu$ values correspond to standard basis vector initial conditions and Gronwall's to the standard range correction terminal values.

where $x_\omega$ is the terminal value of $x$. Now by definition $E > 0$, and the last quantity in the integral is also positive; thus, if we assume that

$$2hy < 1 + \frac{vG'}{G} \tag{2.8}$$

over the entire trajectory, the integral term is negative. Since $x(0) = y(0) = 0$, it follows from the preceding equation that $(x'\nu + y'\rho)_0 < 2x_\omega$. Substituting this into (2.6) yields

$$\frac{\delta X}{X} < 2\frac{\delta v_0}{v_0} \tag{2.9}$$

From this, it follows that, for a "low" trajectory, one in which (2.8) is satisfied everywhere, the range increases less rapidly than the square of the initial velocity. Gronwall of course notes the well-known range formula in a vacuum, namely, $x_\omega = \frac{v_0^2 \sin(2\alpha)}{g}$ for comparison.

We now turn to the extension of Bliss' adjoint method, the second of Gronwall's investigations, and second of the three categories previously mentioned. Gronwall's two-equation adjoint system (2.1) is a special case of his general one given by

$$\begin{aligned} \mu' &= -hE(x'\nu + y'\rho) \\ (x'\nu + y'\rho)' &= (1 + \frac{vG'}{G})E(x'\nu + y'\rho) - 2(x'\lambda + y'\mu) - \kappa \end{aligned} \tag{2.10}$$

where $\kappa$ is the constant appearing in his first integral[4]

$$\kappa + x'\lambda + y'\mu + x''\nu + y''\rho = 0 \tag{2.11}$$

He also obtains the general linear combination of $\xi, \eta, \xi', \eta'$ from the fundamental relation as

$$(\xi\lambda + \eta\mu + \xi'\nu + \eta')_{t=t_0} = (x'\nu + y'\rho)_0 \frac{\delta v_0}{v_0} + (x'\rho - y'\nu)_0 \delta\alpha + \tag{2.12}$$

$$\int_0^{t_0} (\lambda + \nu')\, w_x\, dt + \int_0^{t_0} (\mu + \rho')\, w_y\, dt - \int_0^{t_0} E\,(x'\nu + y'\rho)\,(\frac{\delta G}{G} + \frac{\delta H}{H} - \frac{\delta C}{C})\, dt.$$

This form is similar to that used by Bliss but contains a term for a vertical wind $w_y$, allows for a change in the drag law $G$ alone, and holds for any initial conditions for the adjoint variables as well as any value of $t_0$, $0 < t_0 < t_\omega$.

[4] Gronwall wrote $x'\lambda + y'\mu + x''\nu + y''\rho = \kappa$ in his letter announcing this result, see Sect. 2.11 of the previous chapter.

He then develops expressions in the adjoint variables for the change in time of flight and change in maximum ordinate. In the first case, for example, he defines the new trajectory from the old by $\overline{x} = x + \Delta x,\ \overline{y} = y + \Delta y$ and the new time of flight as $t_\omega + \Delta t_\omega$, where one has

$$y(t_\omega + \Delta t_\omega) + \Delta y(t_\omega + \Delta t_\omega) = 0. \tag{2.13}$$

Then, by the mean value theorem, one has $y(t_\omega + \Delta t_\omega) = y(t_\omega) + y'(t_1)\Delta t_\omega$ where $t_1$ lies between $t_\omega$ and $t_\omega + \Delta t_\omega$. Then since $y(t_\omega) = 0$ it follows that

$$\Delta t_\omega = -\frac{\Delta y(t_\omega + \Delta t_\omega)}{y'(t_1)}. \tag{2.14}$$

Remember that Gronwall's presentation distinguished between, for example, the actual change in y-coordinate $\Delta y$ and the corresponding variable $\eta$, which was defined as the solution to a set of linearized differential equations for the actual $\Delta$ quantities. He analyzed terms like $\Delta y - \eta$ and, using results from [386], established that (2.14) could be replaced by the "first-order term in $\Delta t$, or the *differential variation* $\delta t_\omega$ *in time of flight*" given by[5]

$$\delta t_\omega = -\frac{\eta(t_\omega)}{y'(t_\omega)} \tag{2.15}$$

By a similar analysis, Gronwall finds that the expression for the change in maximum ordinate is

$$\delta y_s = \eta(t_s) \tag{2.16}$$

where $t_s$ is the time at which the summit is reached on the original trajectory and $y_s$ is the maximum ordinate on the original trajectory.

Now all the preliminaries are in place. Taking as an example the change in maximum ordinate, Gronwall notes that (2.16) is in fact the linear combination $\xi * 0 + \eta * 1 + \xi' * 0 + \eta' * 0$ for $t = t_s$. Thus, to obtain the required change, we must integrate (2.10) with the initial conditions $\lambda = 0, \mu = 1, \nu = 0, \rho = 0$ at $t = t_s$ backward to $t = 0$, as previously done for range corrections. Recalling that $\lambda$ is always a constant, we find $\lambda = 0$ throughout, and since $y'(t_s) = 0$ the first integral (2.11) gives $\kappa = 0$. Thus (2.10) becomes

$$\begin{aligned} \mu' &= -hE(x'\nu + y'\rho) \\ (x'\nu + y'\rho)' &= (1 + \frac{vG'}{G})E(x'\nu + y'\rho) - 2(y'\mu) \end{aligned} \tag{2.17}$$

---

[5] The $\delta X$ appearing on the left of (2.5), (2.6), and (2.9) should be interpreted in the same manner.

with initial conditions $\mu = 1$, $x'\nu + y'\rho = 0$ for $t = t_s$. With this integration performed, the needed values are substituted into (2.12), and the required variations in maximum ordinate due to separate causes are obtained as in the case of range variation.

Qualitative properties of the differential variations for maximum ordinate are then developed as in the case of range variations. In an interesting footnote, Gronwall gives an explanation of why his techniques fail in the case of analysis of variations in the time of flight, acknowledging that this information is more important from a practical point of view. He comments "Moreover, numerical computation shows that the behavior of the time of flight variations to be so much more complicated and irregular than in the cases of range and maximum ordinate, that it appears difficult even to formulate conjectures in respect to their qualitative behavior" [491, p. 521]. His theoretical results were apparently checked against computational work.

It is difficult to say how much of this work on variations was put to use. Bliss' *Mathematics for Exterior Ballistics* provides an account of them and others such as variation in terminal angle of impact. But the present author has not encountered any computational work based on them. They remain an impressive demonstration of the "elasticity" (to use Bliss' word) of the original adjoint method.

We conclude this section with an acknowledgment that the division between what we have called the applied work on differential variations during the War and the reception to this work by mathematicians afterward is somewhat artificial. Gronwall, for example, had worked continually on these problems from his War-time arrival at Aberdeen through his departure from Washington in 1922. But it seems natural to consider Bliss' work as seminal and Gronwall's first integral, for example, as part and parcel of it, especially since it appeared as such in ordnance textbooks shortly after the war and beyond. This in spite of the chronology placing the work nearly at the same time as the qualitative properties and extensions investigations. The latter have somewhat the character of considerations after the fact. The same is not true of the final line of research, discussed next.

## 2.3 Joseph Ritt, Weighting Curves, and Smoothness of Solutions

Joseph Fels Ritt was born in New York City in 1893.[6] He attended City College for two years and then took an appointment at the Naval Observatory in Washington to help his financial situation as well as that of his parents. While in the city, he attended classes at George Washington University, where he finished his undergraduate work. Developing an interest in mathematics, he then began classes in his home town at Columbia University, where he earned a Doctorate in 1917 in

[6] See [566] and [647] for more on Ritt.

the subject of differential equations of infinite order. After his thesis, there followed "a brief period of orientation" [566, p. 310] in his research and publication before he produced important work on permuting rational functions in 1922 and 1923.[7] To this period of orientation belongs a paper that treats the iteration of rational functions [621]; it was presented to the American Mathematical Society on December 27, 1917, and February 23, 1918. At some point prior to the latter date, he met T. H. Gronwall, who in his pre-Aberdeen period was a frequent visitor to Columbia; one result was a suggestion on Gronwall's part to generalize an aspect of the iteration paper (see [621, p. 354]).

Ritt became a member of the Aberdeen Proving Ground mathematics group; an undated letter from a weary Veblen to E. H. Moore at the University of Chicago states that "My work here is so absorbing and pressing that I don't find time or energy to think of anything else. Within the last fortnight I have organized a Ballistic Computers division at the P. G. and called in Ritt and F. L. Carmichael and others as computers." [Oswald Veblen Papers] Ritt then switched to Moulton's Washington group, in which he served from May 24, 1918, to April 2, 1919; his entry in [594, p. 4] lists him as "Instructor in Mathematics in Columbia University, Master Computer." During his time at these two ballistic centers, he must have had contact with Gronwall again: in his contribution to the original Bliss technical report "On the Computation of Differential Corrections," Gronwall mentions simplifications of certain of Bliss' expressions "which follow from a remark made by Dr. Ritt" [381, p. 1]; this document is dated January 24, 1919, before Ritt's departure.

In addition to his computing duties, Ritt wrote a technical report and a JUSA/CAJ article on weighting factor curves. These curves are an intermediate step between the calculation of range variations due to, for example, a constant tail wind over a time interval, and the variable winds actually encountered in nature. They were envisioned as being used in battle, when current additional atmospheric data was also being transmitted. Either a gun battery commander or the meteorological service would combine the information to lay a gun for fire under conditions of variable winds.[8] In the JUSA/CAJ article [620] several of these curves appear.

Ritt's explanation of the construction of the curves will suffice (see Fig. 2.1):

> Having been given the muzzle velocity of the projectile, and the quadrant elevation for which the calculation is to be performed, the trajectory of the projectile for the normal atmospheric conditions is computed by any of the recently developed short-arc methods. The next step is the determination, for every altitude $y$ less than the maximum ordinate of the undisturbed trajectory, of the increase in range $\Delta R_y$, caused by the action of a rear wind of one meter per second while the altitude of the projectile is less than $y$, both on the ascending branch of the trajectory and on the descending branch. The air is assumed to be quiet above altitude $y$ The determination of $\Delta R_y$ is effected by the method of Moulton and Bliss. Denote by $\Delta R_0$ the increment in range produced by a rear wind of one meter per second acting throughout the flight of the projectile and by $y_0$ the maximum ordinate of the trajectory. Put $p = \frac{\Delta R_y}{\Delta R_0}$, $k = \frac{y}{y_0}$. A graph is constructed with $p$ for abscissa and $k$ for

[7] See [566] for further information on this work.

[8] For more information on how these curves were to be used, see [652], pp. 567 ff.

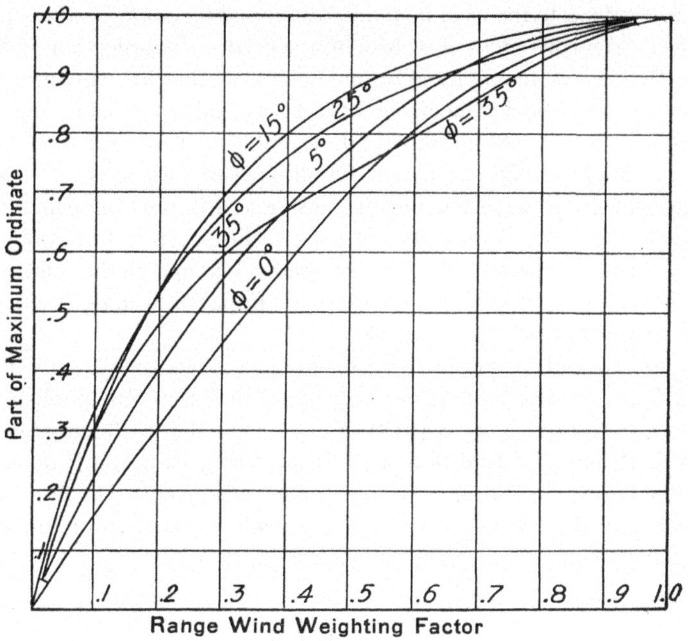


**Fig. 2.1** Ritt's weighting factor curves, [618]

> ordinate . . . The curve obtained is the weighting factor curve for range wind for the given trajectory. For the zone of air between *a* percent and *b* percent of the maximum ordinate the weighting factor is obtained as the of the abscissae corresponding to *a* and *b*. [620, pp. 405–406]

One sees where Bliss' material needs to be used in this practical ballistic instrument. Ritt's contribution to the subject was to notice that, as the angle of elevation of the gun approached zero, the weighting factor curves approached a limiting curve. This curve has a simple form, algebraic in $p$ and $k$, in each of the cases of range variation caused by range wind, cross wind, and density of air. The formulae were determined by Ritt's lengthy analysis in his technical report. The advantage of using the limiting curves is that they are good approximations for the actual curves for low elevations, or what was called "flat fire," and were of a simpler form. The approximation was especially good in cases of guns with large caliber and high velocities; the 16-inch Panama Rifle and the 12-inch Navy Gun were mentioned as particular examples. Again, a mathematical contribution served as a simplifying agent. In the technical report [618], published February 25, 1919, one sees, on page 4, at the beginning of the derivation for the curves a version of the linearized system for the differential variations given by Moulton.

This work of Ritt's qualifies as a mathematical extension of the differential variations material and also serves to document his familiarity with this aspect of Moulton and Bliss' work, as do his comments on the original technical report. It also illustrates Ritt's distaste for what he terms "empirical formulas" and his preference for the newer, "rigorous method of Moulton and Bliss," a reflection of Moulton's own views. He cites as an empirical formula one that takes the weighting factor for any zone to be proportional to the height of that zone and, after dismissing it as being undependable, introduces Moulton's solution to the problem, based on "a rigorous system for calculating weighting factors for all types of variations ... His method is based on the numerical integration of two systems of differential equations" [618, p. 404]. It is again evident in the titles of some of his seven technical reports listed in Moulton's history [594, p. 13]: "Note on the Second and Higher Order Terms in Differential Variations," "On Certain Asymptotic Expansions in the Theory of Wind and Air Density Variations," for example.

This familiarity with higher the level mathematical work apparently also served him in another capacity. During his stay in Washington, Ritt published, in the *Annals of Mathematics*, a note [619] entitled "On the Differentiability of the Solution of a Differential Equation with Respect to a Parameter," dated April 1, 1919 (the day before his termination from Moulton's group). The note is his only purely mathematical publication for 1919 and made no mention of ballistics. In it, he considered the simple first-order differential equation

$$\frac{dy}{dx} = f(x, y, a) \tag{2.18}$$

where $a$ is a real parameter in an interval $(\alpha_1, \alpha_2)$. He wished to investigate the existence and nature of the differentiability of the solution of (2.18) with respect to $a$. In the well-known memoirs from the 1890s [608, 609], Picard had proved the existence of solutions to differential equations under certain conditions on their defining functions. His proof by the method of successive approximations became standard fare in discourse on differential equations for decades after its publication.[9] It then became natural for pure mathematicians to inquire after the properties that the solutions had and the conditions on the defining functions, which would guarantee these properties. Moulton and his group had always considered the functions like $f$ above to be analytic, given their connection to ballistic problems, which featured smoothly changing quantities. But mathematicians had no such restrictions.

There was in fact an extensive literature on this problem and its companion problem of the of solutions to differential equations as functions of their initial conditions, already in existence. In [372] from 1905, Bliss had offered theorems dealing with the latter subject, declaring that "As the literature dealing with these ...questions is somewhat scattered and the proofs complicated, it is proposed in

[9] Perhaps not surprisingly, Cauchy had earlier proposed a general method for differential equation solution, in his case the sequence of approximating functions consist of a number of segments of straight lines. The method, given in 1844, was later treated in depth by Lipschitz and has come to be known as the Cauchy-Lipschitz process. Details of the method and references can be found in Moulton's book on differential equations [600, Ch. 13], where comparisons to the Picard approach are also made.

the following pages to collect and simplify as far as possible the results of different writers" [372, p. 49]. In his 1917 paper [380], he improved on his own methods and generalized them.[10]

The connection with the problems of variability of the trajectory is clear: the differential variations treat the changes in the trajectory, a solution to a set of differential equations, as the defining functions of the equations change. Some of these functions may be viewed as depending on a parameter, as Moulton showed, for example, by treating the exponential density law in this fashion, the parameter being the constant multiplying factor of $t$ in the exponent. The corresponding theoretical problem might then become: what is the nature of the dependence of the solution $y$ to a system of equations on a parameter in the defining functions? Would continuity of such functions with respect to a parameter transfer to continuity of the solution with respect to the parameter? What about differentiability? Moulton himself addressed this problem in a paper "issued in photostat form by the Technical Staff of the Ordnance Department," according to Ritt [619, footnote p. 289]. Very unfortunately, the author has been unable to obtain this paper.[11]

The problem that interested Ritt had been addressed already, but Ritt found fault with each of the proofs he examined: every proof made use of the Picard approximations to the solution. He sought a proof that avoided the use of these iterates. To this end, he set the equation (2.18) and assumed that $f(x, y, a)$, $f_y(x, y, a)$, $f_a(x, y, a)$ continuous in the region $x_0 \leq x \leq x_0 + h$, $y_0 - k \leq y \leq y_0 + k$, $\alpha_1 \leq \alpha \leq \alpha_2$. He also assumes that for every $a$ in the interval $(\alpha_1, \alpha_2)$ there is a solution to (2.18) defined on the interval $(x_0, x_0+h)$ assuming only values which lie in $(y_0-k, y_0+k)$ satisfying the necessary initial condition.

With this out of the way Ritt produces the equation

$$\frac{d}{dx}\left(\frac{y_1 - y}{a_1 - a}\right) = \frac{f(x, y_1, a_1) - f(x, y, a_1)}{y_1 - y}\,\frac{y_1 - y}{a_1 - a} + \frac{f(x, y, a_1) - f(x, y, a)}{a_1 - a} \tag{2.19}$$

where $y_1$ and $y$ are solutions to (2.18) corresponding to parameter values $a_1$ and $a$ respectively, both satisfying the required initial condition. This equation holds for $y_1 \neq y$ and $a_1 \neq a$; in order to extend it to these values Ritt defines $U = \frac{f(x,y_1,a_1)-f(x,y,a_1)}{y_1-y}$ for $y_1 \neq y$, $U = f_y(x, y, a_1)$ for $y_1 = y$; $V = \frac{f(x,y,a_1)-f(x,y,a)}{a_1-a}$ for $a_1 \neq a$ and $U = f_a(x, y, a)$ for $a_1 = a$. One can then say

[10] This paper was a stepping stone to [386], one of two theoretical Bliss contributions which drew from his ballistics work.

[11] Moulton considered the problem of differentiability of the solutions with respect to a parameter in his 1916–1917 lectures and included in the lecture of January 12, 1917, a proof of this property under weaker conditions on $f$ than analyticity [593]. This proof was reproduced in [600, pp. 194–197]. Considering that the date of the lecture precedes the beginning of Moulton's service, it is likely that it is this argument that is included in the technical report. It is also possible that Bliss' second paper on this subject, presented on December 29, 1917, had its origin in discussions with Moulton and his associates about the problem.

$$\frac{d}{dx}\left(\frac{y_1 - y}{a_1 - a}\right) = U\frac{y_1 - y}{a_1 - a} + V \tag{2.20}$$

for all values. Arguing that $U$ and $V$ as defined are continuous in all variables, Ritt has a first order linear differential equation for $\frac{y_1-y}{a_1-a}$ with initial condition $\frac{y_1-y}{a_1-a} = 0$ for $x = x_0$. The standard formula for the solution yields

$$\frac{y_1 - y}{a_1 - a} = e^{\int_{x_0}^{x} U\,dx} \int_{x_0}^{x} e^{\int_{x_0}^{x} -U\,dx} V\,dx \tag{2.21}$$

A delicate limiting argument as $a_1$ approaches $a$ now gives

$$y_\alpha = e^{\int_{x_0}^{x} f_y\,dx} \int_{x_0}^{x} e^{\int_{x_0}^{x} -f_y\,dx} f_a\,dx \tag{2.22}$$

where $y_\alpha$ is the partial derivative of $y$ with respect to $\alpha$. Ritt has not only proved that the partial exists, but found an explicit formula for it. He finishes by noting that, by simple differentiation $y_\alpha$ satisfies its own linear differential equation:

$$\frac{d}{dx}y_\alpha = f_y y_\alpha + f_a. \tag{2.23}$$

Two questions remain. First, can this proof be generalized to systems of differential equations? Second, is there any significance to the differential equation satisfied by $y_\alpha$? Ritt was unable to give a positive answer to the first. But his associate T. H. Gronwall was able to fill this gap, display the curious final linear differential equation in a familiar form, surely known to Ritt, and provide an unexpected bonus.

## 2.4 The Equations of Variation and Gronwall's Inequality

Before discussing Gronwall's paper, which generalized Ritt's argument, some background material should be provided. Suppose

$$\frac{dy_1}{dx} = f_1(x, y_1, \ldots, y_n, a), \ldots, \frac{dy_n}{dx} = f_n(x, y_1, \ldots, y_n, a) \tag{2.24}$$

is a system of $n$ differential equations. Without imposing any conditions on the $f_i$, formally differentiate both sides of each equation with respect to $a$, and reverse the order of differentiation on the left-hand side. Then one obtains

$$\frac{d}{dx}\frac{\partial y_i}{\partial a} = \sum_{j=1}^{n} \frac{\partial f_i}{\partial y_j}(x, y_1, \ldots, y_n, a)\frac{\partial y_j}{\partial a} + \frac{\partial f_i}{\partial a}(x, y_1, \ldots, y_n, a) \tag{2.25}$$

Now setting $p_i = \frac{\partial y_i}{\partial a}$, it is seen that the previous equation has the form

$$\frac{d}{dx} p_i = \sum_{j=1}^{n} \frac{\partial f_i}{\partial y_j}(x, y_1, \ldots, y_n, a) p_j + \frac{\partial f_i}{\partial a}(x, y_1 \ldots y_n, a) \tag{2.26}$$

This is a form previously encountered; it is a slight modification of Moulton's version of the equations of variation for the system (2.24), see (1.25) in the Introduction. What has been demonstrated is that, provided all operations are justified, $\frac{\partial y_i}{\partial a}$ is a solution to the equations of variation. In the case of Ritt's single differential equation (2.18), these equations reduce to (2.23). From now on the phrase "equations of variation" will refer to (2.23), and it is clear that any generalization of Ritt's result must include the equation (2.23).

The argument just given occurs, for example, in Hadamard's *Leçons sur le calcul des variations* of 1910 [496],[12] and Bliss among presumably many others was aware of it [450, p. 376] by the time he came to Aberdeen. Hadamard had uses for these equations in his study of the calculus of variations, which will be discussed later. Suffice it here to say that he rigorously justified the operations and in the process, of course, needed to prove the existence of the $\frac{\partial y_i}{\partial a}$, which he in fact did without recourse to the successive approximations of Picard. Another occurrence of this argument can be found in [475, p. 13] from 1923, and it persisted in the next decades, for example, in the classic [366, pp. 123–124] and in texts on stability theory such as [629, p. 138]. Bliss did not refer to the equations of variation in [372] and [380], however.

Next, an outline of Gronwall's argument.[13] He sets the system as (2.24) with initial conditions $y_i = y_i^0$ for $x = x_0$, and assumes $\frac{\partial f_i}{\partial y_j}$ and $\frac{\partial f_i}{\partial a}$ are all continuous for $x_0 \leq x \leq +h$, $y_i^0 - k \leq y_i \leq y_i^0 + k$, $a_1 \leq a \leq a_2$. He also assumes, like Ritt, the existence of a unique system of solutions with the given initial conditions with domain and range similar to what Ritt had declared. His conclusion is that all $\frac{\partial y_i}{\partial a}$ exist and satisfy the equations of variation for $x = x_0$ (this last condition had not been discussed by Ritt.). Gronwall works with the integral version of the differential equations rather than the equations themselves, as had Ritt. He sets

$$y_i = y_i^0 + \int_{x_0}^{x} f_i(x, y_1, \ldots, y_n, a)\, dx \tag{2.27}$$

and with a similar expression for a solution $y'$ with parameter $a'$ and satisfying the same initial condition he gets

[12] "C'est ce que M. Darboux appelle le système *auxiliaire* et M. Poincaré les *équations variations*" [496, p. 25], to mention the two previously cited men who addressed these variations.

[13] Gronwall's paper followed directly after Ritt's in the *Annals of Mathematics*. There are several minor misprints.

$$y_i' - y_i = \int_{x_0}^{x} [f_i(x, y_1', \ldots, y_n', a') - f_i(x, y_1, \ldots, y_n, a)]\,dx \tag{2.28}$$

from which it follows from the multivariable integral mean value theorem that

$$\frac{y_i' - y_i}{a' - a} = \int_{x_0}^{x} [\sum_{j=1}^{n} \bar{f}_{ji} \frac{y_j' - y_j}{a' - a} + \bar{f}_{0i}]\,dx \tag{2.29}$$

where $f_{ji} = \frac{\partial f_i}{\partial y_j}$ and $f_{0i} = \frac{\partial f_i}{\partial a}$ and the bar indicates that the partials are evaluated at values between $y_i$ and $y_i'$ and a parameter value between $a$ and $a'$. One can see in this equation the kernel of the equations of variation provided the appropriate limits can be taken. Gronwall's idea is to use an existence theorem to obtain the unique solution set $Y_i$ for the desired equations of variation

$$\frac{d}{dx} Y_i = \sum_{j=1}^{n} \frac{\partial f_i}{\partial y_j}(x, y_1, \ldots y_n, a) Y_j + \frac{\partial f_i}{\partial a}(x, y_1, \ldots y_n, a) \tag{2.30}$$

with initial conditions $Y_i = 0$ for $x = x_0$, and show that $\frac{\partial y_i}{\partial a} = Y_i$. Combining (2.29) and (2.30) gives

$$\frac{y_i' - y_i}{a' - a} - Y_i = \int_{x_0}^{x} [\sum_{j=1}^{n} \bar{f}_{ji} (\frac{y_j' - y_j}{a' - a} - Y_j) + \sum_{j=1}^{n} (\bar{f}_{ji} - f_{ji}) Y_j + \bar{f}_{0i} - f_{0i}]\,dx \tag{2.31}$$

to measure the distance between a difference quotient for $\frac{\partial y}{\partial a}$ and the known solution to the equations of variation. Note that the left hand side is contained in the summation part of integral of the right hand side. Since these are the quantities to be made small, Gronwall defines, for a fixed $x$, $a$, $a'$, $z(x, a, a')$ = the largest of the quantities $\left|\frac{y_i' - y_i}{a' - a} - Y_i\right|$. Next, he wants to use the triangle inequality to bound the integral expression on the right of (2.31), and to this end, he defines $M$ as an absolute constant such that $n\left|f_{ji}\right| < M$ for all $i$, $j$. He appeals to the continuity of various expressions to claim that $\left|\sum_{j=1}^{n} (\overline{f_{ji}} - f_{ji}) Y_j + \overline{f_{0i}} - f_{0i}\right|$ can be made uniformly small over $i$, that is, given $\epsilon > 0$, one can get $\left|a' - a\right|$ so small that $\left|\sum_{j=1}^{n} (\overline{f_{ji}} - f_{ji}) Y_j + \overline{f_{0i}} - f_{0i}]\right| \leq \epsilon$. Using then the triangle inequality gives and the definition of $z(x, a, a')$ one has

$$0 \leq z(x, a, a') \leq \int_{x_0}^{x} (M z(x, a, a') + \epsilon)\,dx \tag{2.32}$$

for $\left|a' - a\right|$ sufficiently small, and $M \geq 0$ and $\epsilon \geq 0$.

This is a curious situation, in that $z(x, a, a')$, the quantity to be made small for a given $x$, has been shown to be less than the integral of a linear function of itself, with vanishing intercept as $\left|a' - a\right|$ approaches zero. Gronwall broke the circularity by having proved the following, stated as it appears in the paper (appellation is the present author's not Gronwall's):

Lemma: When, for $x_0 \le x \le x_0 + h$, the continuous function $z = z(x)$ satisfies the inequalities

$$0 \le z \le \int_{x_0}^{x} (Mz + A)\, dx \tag{2.33}$$

where the constants $M$ and $A$ are positive or zero, then

$$0 \le z \le Ahe^{Mh}, \tag{2.34}$$

for $x_0 \le x \le x_0 + h$.

The proof will not be given here. Applying this lemma to (2.32) with $\epsilon = A$ gives

$$\left| \frac{y_i' - y_i}{a' - a} - Y_i \right| \le \epsilon h e^{Mh} \tag{2.35}$$

uniformly for $a, a'$ in $(a_1, a_2)$. This finishes the proof, for simultaneously, the partials have been proved to exist and satisfy the equations of variation. The remainder of the paper shows similarly that $\frac{\partial y_i}{\partial y_j^0}$ and $\frac{\partial y_i}{\partial x_0}$ all exist, with each set satisfying equations of variation. The desired generalization of Ritt's proof has been achieved without resort to the iterates.

The lemma isolated above has come to have a history and development all of its own, totally out of proportion to the modest place it occupies in the proof. Variously known as Gronwall's Lemma or Gronwall's Inequality, it has spawned a huge literature, which continues to grow to this day. It has been generalized in many ways [606] and has become part of standard presentations in stability theory. The following example of the use of the inequality [668] is essentially what Gronwall did in his paper [487, p. 294 top] to establish the uniform continuity of his solutions with respect to the parameter $a$.

Let $y_1(t)$ and $y_2(t)$ be solutions to $\frac{dy}{dt} = f(y, t)$ with $y_1(t_0) = y_1,\ y_2(t_0) = y_2$. Then using the integral form for the differential equation yields

$$y_1(t) - y_2(t) = y_1 - y_2 + \int_{t_0}^{t} [f(y_1(s), s) - f(y_2(s), s)]\, ds. \tag{2.36}$$

Then, by the triangle inequality

$$|y_1(t) - y_2(t)| \le |y_1 - y_2| + \int_{t_0}^{t} |f(y_1(s), s) - f(y_2(s), s)|\, ds. \tag{2.37}$$

A standard hypothesis on $f(y, t)$ is Lipschitz continuity in the dependent variable $y$; this is a condition that, for example, guarantees the existence of a unique solution in Picard's iteration proof. Asserting this condition gives

$$|y_1(t) - y_2(t)| \leq |y_1 - y_2| + L \int_{t_0}^{t} |y_1(s) - y_2(s)| \, ds. \tag{2.38}$$

where $L$ is the Lipschitz constant. Now using the slight modification of Gronwall's lemma proved in [668], one obtains

$$|y_1(t) - y_2(t)| \leq |y_1 - y_2| \, e^{L|t-t_0|} \tag{2.39}$$

This inequality gives simple bounds simultaneously on the proximity of two solutions in terms of their initial values and on time $t$.

It seems justifiable, given the chronology and the milieu in which the papers were written, to conclude that Gronwall's inequality is a result in pure mathematics coming from war-related work, a product of the reception of the differential variations by members of the mathematical community.

After his service in Moulton's group, Joseph Ritt returned to Columbia University in 1921 where he remained as a research mathematician for the rest of his career. He served as head of the mathematics department during the years 1942–1945. He died in 1951.

## 2.5 The Differential Variations as a Contribution to the War Effort: The Mathematicians' View

Given that the pure mathematical topics with origins in War-related matters elicited limited interest as research material, one can still ask for the responses to this work when viewed as a contribution to the War effort and later military practice. It is interesting to ask about the reception it received among both the mathematicians as a group and the military as the users of the new developments. In the next few sections, we turn our attention to these responses, beginning with that of the mathematical community.

The first and most obvious perspective from which to view the mathematicians' reception is that of simple community pride. In particular, if we accept the thesis that the coming of age of the research community in the United States only had occurred roughly two decades prior to the American entry to the War,[14] the "community" of pure mathematicians as such was relatively new at this time. The War was its first opportunity to offer its services to the military as an organized whole, and Bliss' achievement was occasionally singled out. "It was a period of nascent energy in

[14] This is a major theme in [607].

American mathematics" [665, p. 257], according to Norbert Wiener, a participant in the Aberdeen group. His ebullient assessment, often quoted, never fully analyzed, is perhaps the high point of praise:

> One thing is clear, however: we accomplished the task expected of us ... we Americans were behind neither our enemies nor our allies. ... In fact, in the matter of interpolation and the computation of the corrections of the primary ballistics tables, Professor Bliss of Chicago made brilliant use of the new theory of functionals. Thus, the public became aware for the first time that we mathematicians had a function to perform in the world. We still did not count in its eyes as magicians comparable with the chemists and engineers ... our newly won prestige contributed appreciably to our salaries and our ease of finding employment... [665, p. 256].

Usually, the expressions of pride were more modest. Veblen's post-war tour of European countries[15] to report on current procedures for range firing table construction concluded that "The most satisfactory general method [of dealing with differential corrections] seems to be the one devised by Professor Bliss at the Aberdeen Proving Ground. The English have arrived after a long process of evolution at what amounts to the same method (Doodson's method), but their mathematical treatment of it is far from being as clear as that given by Bliss" [660, P. 4].[16] A similar attitude of quiet satisfaction was taken by reviewers of ballistics texts in which Bliss' method was included. Thus, W. E. Milne's review of Bliss' 1944 *Mathematics For Exterior Ballistics*, Bliss' summary volume of ballistics material dating from the first World War, with some updates, stated "It is interesting to observe how the subject of adjoint differential equations, once an esoteric theory of 'pure mathematicians,' has now become a commonplace tool of practical computers [586]." Remembering that Milne was in Moulton's group, and his comment, though made at the time of the second World War, was as appropriate as when the adjoint was first introduced. As a follow-up, we may note that by 1954 Clifford Truesdell, in his review of [581], apparently felt it unnecessary to elaborate on the nature of Bliss' work, so well-established had it become, that he merely noted "The next section [of the book] contains Bliss' proof of the existence of functional differential effects, arising from perturbations of the empirical coefficient functions. ... Thereafter the methods of Bliss and Moulton for calculating these effects are developed" [657, p. 174]. Lying as it does among accounts of the book's treatment of dimensional analysis,[17] the motion of a rotationally symmetric rigid body in a resisting medium, and aerodynamic coefficients, the mention of Bliss' material stands out for its "pure" nature and long-lasting value.[18]

---

[15] The countries visited are not mentioned in the report of this visit [660], but material from French, English, and Italian sources are listed.

[16] For more on Doodson, see [355].

[17] That is, the determination of the units of measurement attached to the variables occurring in ballistic theory.

[18] On the other hand, Moulton's and Bliss' methods for the differential variations were not always appreciated by those with limited mathematical training. The same critic who praised Moulton's treatment of the spinning projectile in *New Methods* felt that "The subject of differential variations

Some physicists contributed to the praise of the mathematicians' work. In an address to the American Association for the Advancement of Science given in 1920, Gordon F. Hull, a professor of physics at Dartmouth College who served in Moulton's group, attempted a summary of the science behind the "warfare with guns, projectiles, and bombs" [531, p. 222], a subject he felt had not been adequately described publicly. He claimed therein that "For two hundred years or more the subject of exterior ballistics has been regarded as belonging to pure mathematics" [531, p. 222] and attempted a reclamation. But he spoke in the highest terms of Moulton's mathematical achievements,[19] especially the short-arc method, and added his praise of the differential variations work: "Professor Bliss gave an inclusive method of computing variations in range, altitude, and time due to changes in air density, winds, muzzle velocity. Dr. Gronwall greatly simplified and extended the work by Bliss, and made other important contributions" [531, p. 225]. He also had reservations about the simplifying assumptions necessary to construct these theories, but concluded that "... leaving out these factors the mathematical basis for finding the trajectory of a projectile is secure" [531, p. 225]. These were words in which the mathematicians could take some pride.

Any consideration of the reception of the new ballistics material among mathematicians in the post-First World War era must take into account the heavily pure research-oriented nature of that group. Simply put, among many research, mathematicians applications were of little interest.[20] In the words of Garrett Birkhoff, writing in 1977:

> During the 1920s and 1930s, few American mathematicians paid much attention to contemporary engineering developments. Although they used the new concepts of relativity and quantum mechanics as stimulus for mathematical speculations, the solution of special problems bored them. This led to a paradox: while American industry was growing by leaps and bounds, our academic mathematicians were taking less and less interest in it. [367, p. 62].[21]

We may include ballistics as presenting "special problems" capable of boring the pure mathematicians, as Truesdell remarked. This indifference took several forms and had many consequences. One form was a casual assurance that some

---

as presented in Chapter IV will be interesting to the mathematician, but is not elementary ... The adjoint system of differential equations is introduced as if it were such a familiar portion of mathematics that this could be done without explanation" [626, p. 231].

[19] Again one encounters the suspicion of the empirical as apparently held by Moulton and Ritt "For his courage in setting aside the long-established, revered but rather empirical method [of range table construction] in use in the War Department, and in introducing a logical, simple method of computing trajectories ... he [Moulton] deserves great commendation" [531, p. 225]. See Ritt's remarks on "empirical methods" in the material following the illustration of his weighting factor curves earlier in this chapter. Moulton's distaste for empirical methods may be inferred from his repeated criticisms, voiced in [594] of the pre-World War methods has having "no logical basis."

[20] See [644, 645] for analyses of this situation.

[21] In the same piece one finds mention of Bliss as a leader of American mathematics in the period 1921–1941 and mentions his contributions of the adjoint method at Aberdeen [p. 55].

use will eventually be found for abstract research: "The recently developed subject of integral equations has sometimes been confronted with the inquiry: Why develop this theory? Will not differential equations serve the purpose? But the mathematician goes calmly ahead with the development of those things which interest him just as he did formerly; and in the new case he anticipates with confidence the same triumphant justification in the event which has uniformly crowned his labors in the past." [413, p. 468] according to an address delivered in January 1917 before a chapter of Sigma Xi.

Consequences of this attitude included some dissatisfaction by practicing engineers with the perceived irrelevance of the mathematicians efforts. In a pre-War "Appeal to Producing Mathematicians" [605], the author, a civil engineer, criticized the research community:[22]

> A perusal of modern mathematical treatises and dissertations for degrees makes the lay mathematician ... despair of ever keeping up with the modern trend of mathematical thought ... Hardly any treatise has attempted to discuss or analyze the serious problems of the applied science professions. The profession of civil engineering is teeming with problems awaiting the solutions of a St. Venant, a Laplace, a Newton. [605, pp. 127–128]

They also included efforts on behalf of industrial mathematicians to win over pure researchers to industry; this was especially true in the heavily mathematical venue of electrical engineering; see [410], for example, in which the author, a mathematician in the employ of the Bell Telephone System, makes a pitch for the applied work done at this time in telephony.

One response to the mismatch of industrial need and academic indifference was the insistence by the latter, already noted, that their work did indeed eventually yield practical ends. Veblen adopted this attitude in a fundraising effort for the American Mathematical Society in the early 1920s for which a pamphlet entitled "Our Debt to Mathematics" was published. Loren Butler Feffer [448] has written about this work, and quotes from it support the attitude just described: mathematics develops "by the devoted labors of highly trained scientists who pursue their studies with indefatigable patience, unnoticed by the world, undeterred by the supercilious pity of those who are unable to appreciate their work." [448, p. 484], according to the author of the appeal. Veblen was able to back up this scenario with his experience at Aberdeen. In a newspaper interview given in connection with a 1924 meeting of the American Association for the Advancement of Science, Veblen described his War work and concluded with the statement that

> It was very noticeable that the men who were of the most use were the men of the active research type. The reason that the mathematicians could deal with the problems I have been talking about, and many others, was that we have a large group of men who are continually making new discoveries in the higher mathematics. And I must say that, to me, these discoveries are much more exciting and interesting than any of the war problems that I was asked to talk to you about [661, p. 338].

[22] This quote will be considered again in the last chapter.

Among the men he mentioned by name were those familiar to this account: "This achievement [introducing new methods for range table construction] was, of course, a great deal more complicated than it sounds, particularly the problem of allowing for the varying effects of the wind and weather. In working these matters out further important contributions were made not only by Prof. Moulton, but also by another Chicago mathematician, Prof. Bliss, and also by Dr. Gronwall, of New York." [661, p. 338].[23]

It is against this background that one encounters defenses of the labors of pure mathematicians in terms of the long-term possibility of the use of these labors. Dunham Jackson, again a member of Moulton's group and author of the Ordnance textbook cited in the previous chapter, made this case in a 1928 Mathematical Association of America address:

> Collectively and in the long run we are finding out things that the world wants to know, however ephemeral most of our individual contributions may be. To mention specific instances, the methods in exterior ballistics developed by Moulton and Bliss during the war depend on the Picard process of successive approximations and the introduction of an adjoint system of differential equations, familiar products of modern mathematical research. [544, p. 410].

But the most important of these defenses came from Bliss himself, in the same article, from 1927, in which he pointed out the economic virtues of his method in the differential variations [390, pp. 313–314]. His remarks were made in the context of a pure mathematician addressing the relations of mathematics to industry. This is perhaps the quote most encountered in short descriptions of his work, equal in frequency to Wiener's: "I should like to relate to you an experience of my own which shows that sometimes a really intimate acquaintance with a problem may be necessary before its nature can be well-appreciated." There follows a description of the problems of range table construction and the problems of the variations in particular.

> I do not wish to enter here into mathematical details farther than is necessary to indicate that the range as a function of the winds is a function of a new type. It is not function of a single variable, such as one commonly meets in application of mathematics, but it is a function of another function, or, as mathematicians sometimes call it, a function of a line. The computation of the first differential of this function is the computation of the wind correction to the range. Now the theory of functions of lines is on the farther boundary of the explored mathematical domain to-day, and I for one should never have expected it to have an application in ballistics. Its methods were effective, however ... It seems to me that this bit of applied mathematics is interesting in itself and that it illustrates principles upon which I have insisted here before. No one can foresee what portions of the science of mathematics of to-day may have applications in the future. [390, pp. 313–314].

[23] This account appeared in an issue of *Army Ordnance*, a military journal of the day which will be discussed later.

The insistence that his methods were based on the concepts from functions of a line needs clarification and will be addressed in the next chapter,[24] but the substance of the remarks fall in line with those we have already cited.

Occasionally, a reference was made to Moulton's parallel work in astronomy, as a means of comparing the scientific level of his ballistics work with that of a well-respected discipline. Thus, in a short article by A. A. Bennett in a military journal devoted to the study of ordnance, one finds these concluding remarks:

> The finest example of motion subject to complicated laws but solved in practical form by the use of "higher" mathematics is that of the heavenly bodies within our solar system. Here again we have as a first approximation, in the case of the moon, motion in a conic section about the earth as one focus. Except for illustrative purposes, the disturbing attraction of the sun cannot be ignored, and although resistance in the ordinary sense does not exist, the mathematical problem is essentially analogous to that of the projectile in the air. One is scarcely surprised therefore to learn that the methods now in use in ballistics were suggested by recently current methods in astronomical computation. They were first pushed through in this country by Prof. F R. Moulton, Professor of Astronomy in the University of Chicago, while he was serving with the Ordnance Department during the recent war
>
> One of the larger undertakings at present under way by the technical Staff is the construction of tables for exterior ballistics using these astronomical methods. ... It is to be expected ... that something like astronomical accuracy may be secured by their use provided that the corresponding improvement in matériel[25] is sustained. Such accuracy as that with which eclipses may be predicted is perhaps beyond all attainments of artillery If a more or less permanent peace becomes probable, and target practice with major caliber artillery continues to increase in cost, the trajectory of a large projectile may become as rare but as dependable a phenomenon as a reappearance of Halley's Comet. [360, p. 40]

A final perspective from which the reception of the new mathematical ideas can be viewed is the ascendency of the specialist. It was noted in the opening section of this work that in the early years of the twentieth century concern was expressed over the number and complexity of new military weapons, a diversity that required constant education on the part of officers. It has also been demonstrated that that complexity was reflected in the increasing sophistication of ballistic methods. Such concern naturally gives rise to calls for higher education generally and the notion that only an expert is capable of mastering such knowledge. This opinion was expressed in usual forceful language by Moulton at the conclusion of [594]:

> The feature of ordnance work most striking to the civilian scientist is the almost complete lack of exact knowledge of its fundamentals. ... Consider cannons and projectiles and the hundreds and even thousands of millions of dollars spent for them! It is doubtful whether in the history of the world any other expenditures comparable to these have ever been made on so slender a basis either in well-conducted experiments or sound theory. ... Considered in a broad way, the development of fundamental scientific principles is the most important and the cheapest war preparation than can be made. ... The most obvious lesson from the applications of science in the war is that specialists are required. ... [In the past] a man could be an expert in everything. But those days are past, not only in war but in other things.

[24] The adjoint system, for example, is not a foundational concept in functions of a line.

[25] Note the use of the French word in the original.

> ... What probability is there, therefore, that West Point or Annapolis will make a universal expert out of a man who would not think of being one in civil life? [594, pp. 86–87].

Proposals for higher education are then given. Moulton initiated a course in ballistics at the University of Chicago; the first such class met in the fall of 1919. In a complementary move: "Permanent civilian experts should be employed by the War and Navy Departments" [594, p. 88]. Moulton was not alone in these views; virtually, the same language was used on the first page of the Presidential Address of 1918 [434, p. 289] cited earlier for its description of ballistic graphs. A survey of all the ways in which mathematics was used in the War included the differential variations [434, p. 297], noting that this work required "the elaboration of the theory of differential equations which takes into account not only the resistance of the air but also its temperature and its decrease in density at higher altitudes, as well as corrections for the wind." This is a job for the specialist in differential equations!

A single dissenting view should be aired. This came from Joseph Eugene Rowe, a student of the geometer Frank Morley and one-time president of Clarkson College who had done consulting work on ballistic problems. Several of Rowe's individual criticisms have been already mentioned, and they show an interesting mix of mathematical knowledge and concern for the needs of the people to whom the innovations were addressed. Though crediting Moulton with putting ballistics on a sound mathematical basis and listing his book among several outstanding mathematical "products" of the war, he called into question the admiration that the new theories were receiving. His objections were basically those recognized by Moulton: that the methods involved approximations, as must any ballistic method, that the resistance function itself was subject to revision, and that the ballistic coefficient only approximates the behavior of the projectile. He found the method of numerical integration too difficult for a typical army officer "of any nation." "Their training must be broad and practical in the extreme, and they want mathematics presented in as simple and usable form as possible." Perhaps Rowe's most generous overall comment was that "they [Moulton and his associates] must have credit for giving an impetus to mathematical research in the Army and Navy that has had excellent results" [626].

One final remark from Bliss seems appropriate. In his 1927 address, he faced head-on the sometimes rocky relationship between engineering and mathematics.

> I find that practicing engineers, and members of engineering faculties, frequently show great reluctance in admitting that mathematics plays an essential role in engineering problems. ... I was much interested recently to find in print an exposition of the engineer's mistrust of mathematics, in the last chapter of a well-known book on the strength of materials ... The keynote of the chapter is the adjuration to use common sense in avoiding mathematics where mathematics is inapplicable ... This is wisdom to which I should heartily subscribe. But in deference to mathematics I should like to add one further principle. It is that common sense should also be scrupulously avoided in places where common sense does not apply. No amount of common sense unaided can predict the motions of the heavenly bodies, or construct a range table... [390, p. 316].

Bliss' experiences with the latter indicated the absurdity of attempting to use "common sense unaided" in the finding of a trajectory of a projectile, so complex had the task become.

Yet, one wonders about the role left for judgment in gunnery. Hadn't the earlier books on the subject often talked about the "art and science of gunnery?"

## 2.6 Expositions of the New Ballistics for Military Use

It has already been seen how many of Moulton's personnel had written expositions of their work for the JUSA/CAJ in an attempt to reach the military audience. This is not the entire list of such publications. So numerous were the articles and books written by the Ballistics Branch that one can believe that they had been given a mandate to communicate their new methods and that the editors felt it prudent to allow these new views to gain exposure. In this section, we round out the literature of the Branch by summarizing the remaining pieces. All material but one ordnance textbook were written by members of Moulton's office.

A total of five books on ballistics and ballistics related topics was produced in the United States in the years after the War. Four of the five were written by research mathematicians. Moulton's own 1926 text was a result of the classes he taught at the University of Chicago after he left Washington and bears the stamp of university level teaching. Officers were invited to attend these classes. We have used it as a reference for Moulton's approach to the differential variations and discussed its chapter on rotating projectiles. Reviews of this book in the JUSA/CAJ and *Army Ordnance* caused an fierce debate, which needs its own space to discuss; this will be done in a later section. It need not be emphasized that this is the most mathematically advanced of the texts.

In October of 1919 Dunham Jackson issued *The Method of Numerical Integration in Exterior Ballistics, Ordnance Textbook*, [543]. It is divided into two sections, the first dealing with numerical integration and the second with the differential variations. The latter is so well written that it could have replaced Bliss' own JUSA/CAJ article. The calm, deliberate tone throughout makes it probably the most accessible of the mathematical sources for these two topics. It was used as a text in the Ordnance School of Application, an institution on par in importance with the Coast Artillery School. The history of that school will be discussed below.

In 1921, A. A. Bennett released *Introduction to Ballistics* [359]. It is notable that it was "an extensive revision of the publication by the Field Artillery School entitled 'Exterior Ballistics Subject Number Fifty-five'" [359, p. 5]. The mathematicians were encroaching on territory formerly held by the military.[26] This 255 page "pamphlet" was also meant for use as a textbook, and the Preface states unequivocally that "Practical facility in ballistic computations is an accomplishment

[26] On page 101 the idea of a function is defined.

whose value for the student artilleryman is hard to overemphasize." Bennett wished to present the subject with "an intelligent grasp of basic principles [and] a minimum of mathematical machinery. Emphasis is laid on the physical rather than the mathematical side of the subject matter" [359, p. 4]. The book contains the most complete coverage of the entire subject of ballistics among the five works noted here. On pages 169–179, one encounters a nonmathematical discussion of the variation problem. The adjoint system and differential equations are mentioned only in passing, on p. 213, while approximate formulae for variations are given. From a reading of the book, one concludes that of all the authors considered, Bennett had the best grasp of the practical aspect of ballistics and gunnery.[27] The dual nature of the mathematical tendencies and practical explanations are exemplified by the following: a numerical integration of the differential equation $x' = -x^2 - 9$ with $x(0) = 2$ is presented on pages 154–155, but, in an unrelated context on p. 81, one encounters the statement "It is very important for the student to think of the partial ballistic coefficient, not as an abstract mathematical symbol appearing in ballistic formulae, but as a 'live' physical quantity bearing somewhat the same relation to a projectile as horse-power does to an engine." Many similar examples could be given.

Lester Ford contributed *Elementary Mathematics for Field Artillery* in 1919 [453]. The word "Elementary" in the title is key, since the memorandum covers only a portion of today's pre-calculus material. Beginning with basic arithmetic, there are sections on decimal fractions, algebra, geometry, logarithms, graphs, and nomograms. In Moulton's *History*, he states "In the rapid expansion of the army the demand for artillery officers became greater than it was possible to supply. One of the difficulties was to get men who could learn the necessary mathematics and then to provide them with competent instruction. When the great Artillery School was being formed at Camp Taylor, Dr. L. R. Ford was released from this office to organize there the instruction in mathematics" [594, p. 81]. Ford's book bore the heading of Camp Zachary Taylor and was evidently used for remedial or introductory instruction. It is likely that one of the other ballistics volumes was used in addition to Ford's book.

The last of the volumes is 1921s *A Course in Exterior Ballistics* [522] by Roger Sherman Hoar, the only writer not associated with the Ballistics Branch. His colorful history will be described later in this section, as will this idiosyncratic piece. It was based on a series of lectures given by the author at the Ordnance School of

[27] On the other hand, the importance of the basic scientific and mathematical principles of ballistics is emphasized not only as a necessity for intelligent use of the weapons but for their development, to which the artilleryman may be asked to contribute: "While it is true that the artilleryman is accustomed to use the range tables as they come to his hand, without concerning himself unduly as to how they were constructed, it is nevertheless obvious that, without some knowledge of the scientific principles which underlie the practice of artillery, he can neither attain a thorough understanding of gunnery and the possibilities of his weapon, nor hope to do his share toward improving his materiel and advancing the technique of his profession" [359, pp. 3–4]. Thus, a role beyond the use of the gun itself, namely, development, is envisioned for artillerymen.

Application during the years 1919–1920 and features the most complete treatment of actual range table construction available among the five books. This process used the differential variations in a novel way to construct the tables themselves, in addition to appearing as entries in the tables. As in the case of Bennett, physical understanding was given a prominent role, even in the introduction of the adjoint system. Bliss' system was introduced in an interesting new way, and the topic of differential variations in general was discussed in depth using as much intuition as possible. It resembled Jackson's book in that the two basic topics were numerical integration and the differential variations, along with other useful information.

These men, of varying mathematical background, thus responded to the New Ballistics by providing a range of textbooks geared for different audiences; in particular, the differential variations appeared in some guise in all but one. With the exception of Moulton's book, there were no reviews of them in the American mathematical journals. For some of the educational institutions in which these books may have been used, see [347, pp. 249–255].

We shall finish this section by considering material which appeared in *Army Ordnance*, a journal that began publication in 1920 and played the same role for the American Ordnance Association that JUSA/CAJ did for the Coast Artillery School. A. A. Bennett authored a short paper [360] outlining in general terms what the problem of exterior ballistics was, giving a history up through the Siacci method and concluding with Moulton's work. H. S. Beckman, a B. A. in mathematics from the University of Pennsylvania, offered a detailed exposition of "The Elements of Firing Table Construction" [356], an account that bears a close similarity to the method used by Hoar. As part of the close reading of Hoar's book given below, this method will be explained.

A curious final offering to the military from the Ballistics Branch was the battery commander's Range and Deflection Corrector, designed by the Princeton topologist James W. Alexander. As described in Moulton's *History* [594, pp. 69–70], it was to be a pocket-sized device "to give by mechanical means the corrections in range and deflection setting . . . for any abnormalities in (a) muzzle velocity (b) air density, (c) wind components, and (d) weight of projectile." The Frankford Arsenal, a large, active arsenal in Philadelphia, responded by constructing the Corrector, now called the Range, Deflection and Wind Correction Computer, in 1924, as reported in *Army Ordnance* [337]; see Fig. 2.2. The particular model was designed for use with the 75 mm gun Model of 1897 (French), was eight inches in diameter, and was carried in a canvas pouch. As constructed, the computer was more ambitious than conceived, having scales for muzzle velocity, air density, weight of projectile, powder temperature, air temperature, elasticity, drift and cross wind, and rear and head wind. Other factors such as barometric pressure were accounted for. It resembled a circular slide rule with several layers of discs, which were secured and could be disassembled by thumb nuts. A certain degree of flexibility was provided by a design, which allowed for discs suited for a particular gun and ammunition. It was the embodiment of the differential variations material in metal and celluloid.

**Fig. 2.2** Range Computer, [337]

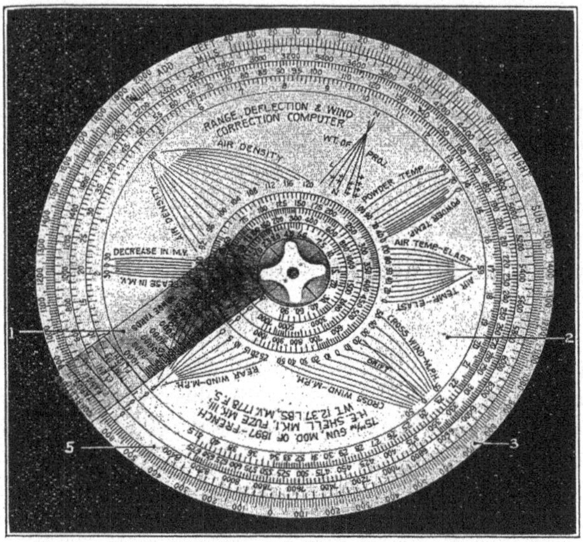


Fig. 6. Range, Deflection, and Wind Correction Computer for 75-mm. Gun, Model 1897

## 2.7 The Differential Variations as a Contribution to the War Effort and Military Practice: Military Views and Uses

As we begin the change of perspective from the mathematicians' to the military's views of the New Ballistics, it is evident that a good starting place is the pages of the military journals to which the mathematicians' efforts were submitted. The journals were an exchange medium: papers of a mathematical nature were occasionally submitted by research practitioners as well as mathematically inclined military and read by interested officers. And indeed we find there some useful opinions expressed by the audience to which the pieces were addressed.[28]

First, one must acknowledge that the revision of the basis on which the range firing tables were constructed would be a topic of limited interest. In comparison with the analysis of a battle in the War, a report on the capabilities of new weapon, or the description of a new manufacturing process, the mathematics behind calculating trajectories may seem irrelevant. An amusing expression of this relative indifference is found in the issue of *Army Ordnance*, which contains a review of *New Methods*: "...ballistic theories are not usually gloated over as one sips his morning coffee nor does the *ignoble volus* rush for such an ingenious treatise as it does for a baseball extra" [340]. But this observation is followed by the respectful tone that usually accompanied discussions of the work of Moulton and others.

[28] From time to time the covers featured artistic renderings of the machinery of military production, as shown in Fig. 2.3. The artist was allowed exclusive civilian access to a manufacturing facility to make sketches of significant scenes.

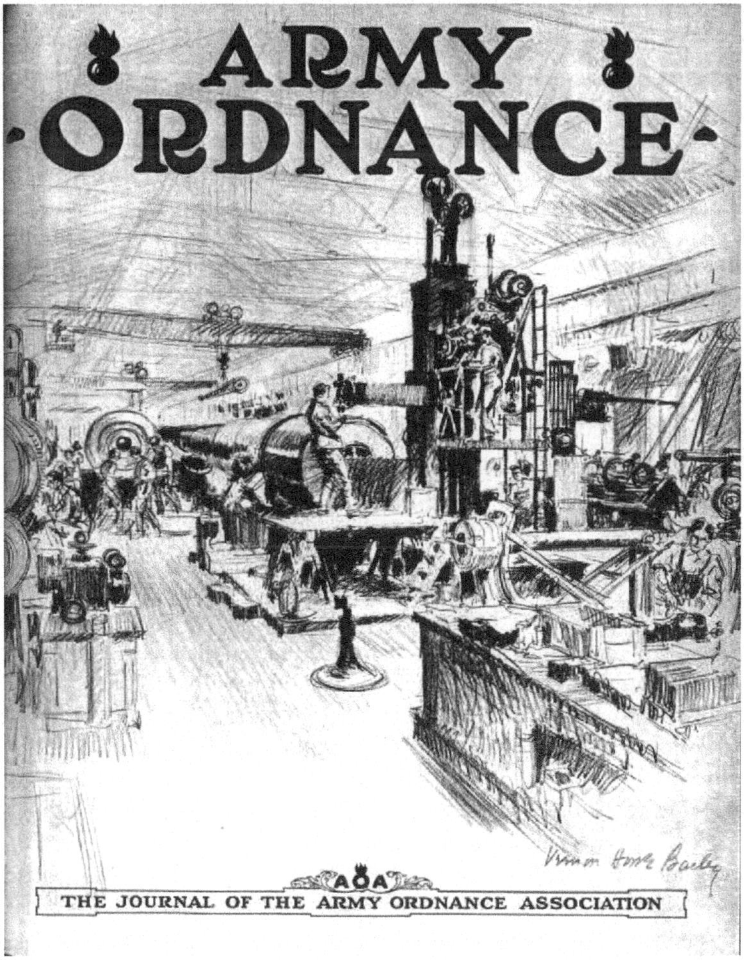


**Fig. 2.3** Cover illustration by Vernon Howe Bailey

One would also expect a degree of ambivalence in the reception of this new material, especially given the level of sophistication of a book like *New Methods*, and comments of this nature can be found. Thus, while the review of the 1926 book in JUSA/CAJ does justice to its merits, it is admitted that "It will appeal particularly to the ballistician; but, in general, the method of presentation is beyond the comprehension of the casual reader in ordinary artillery circles" [342]. And it has already been noted that the editor in charge at the time of the New Ballistics set of articles expressed concern at the "formidable" nature of the material. But a place for the ballistic work was found: "It is neither possible nor necessary that every Coast Artillery officer should be an expert ballistician. Nevertheless, every Coast Artillery officer should be a master of gunnery, and to interpret all the conditions that will confront him in various gunnery problems, he must be familiar with ballistic principles, although he need not know involved ballistic processes" [335].

The offhand negative remark does appear. In the midst of an article in the series devoted to probable error and adjustment of fire in JUSA/CAJ, the author, after having argued the complexities of the topics, submits the unrelated opinion that "... in 'trajectory computation' it is not unfair to say that our army has never seen finer ballisticians than our own fellow artillerymen, Colonel Ingalls and General Hamilton" [414]. A reviewer of the first post-War ordnance and artillery text to include the new ballistics material praised the book, expressing a grudging respect for numerical integration and the progress, which it engendered: "To the average person, mere mention of this method of integration is likely to result in the feeling that it is something entirely too tedious for consideration. Truly, it does require concentration and accuracy for its application, but the method has opened up new fields of knowledge concerning the behavior of a projectile. The recent advances in exterior ballistics are due in part to this method of calculation, and its importance must not be minimized" [343].

But the overall attitude was one of acceptance. There is even evidence that, in some military quarters, the views of the Technical Staff, the name for the permanent office, which succeeded Moulton's Washington group, were being adopted. In particular, the often criticized empirical attitude was mentioned: the Editorial of the New Ballistics introductory series admitted that, while Coast Artillery officers [Ingalls and Hamilton] had done much to "establish ballistics and gunnery as an exact science," "There is still much to be done to divorce ballistics from the realm of the empirical. Major Moulton and his colleagues have opened several new doors" [519].

Stepping away from the journals, it has already been mentioned that tables produced by Moulton's methods for specific guns were issued in the years following the World War. The 1921 75 mm gun tables previously discussed incorporated the corrections computed by differential variations among their entries as well as specifying the protocol for fire preparation, including the meteorological message. The 60-page *Provisional Range Tables* for the British 75 mm Gun, Model 1917 described by Moulton in [594], has sections for French shrapnel, American shrapnel, short and long fuzes, each separated by different colors for easy use; again, these divisions were due to the differential variations for type of ammunition. (See Fig. 2.4.)

In 1929 came *Textbook of Ordnance and Gunnery* [577], previously noted, which offered post-War updates on interior and exterior ballistics for the cadets at West Point. Self-described as relying heavily on already published material, it was addressed to future officers of the line. Written in the pre-War comprehensive style, it has 21 chapters with two devoted to exterior ballistics and range firing tables. The example of Moulton's short-arc method first given in Dunham Jackson's book is repeated on pp. 422–427. Four of the five ballistics books written by the second wave reformers (Ford's elementary book was excluded) are cited. Although no mathematics related to the development of short-arc methods or differential variations is given, the contents of range firing tables are discussed. The presence

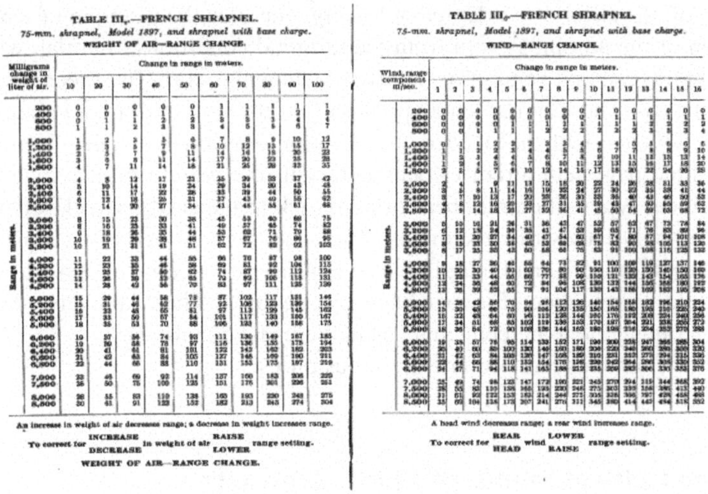

TABLE III$_v$.—FRENCH SHRAPNEL.
*75-mm. shrapnel, Model 1897, and shrapnel with base charge.*
WEIGHT OF AIR—RANGE CHANGE.

An increase in weight of air decreases range; a decrease in weight increases range.

To correct for INCREASE / DECREASE in weight of air RAISE / LOWER range setting.

WEIGHT OF AIR—RANGE CHANGE.

TABLE III$_w$.—FRENCH SHRAPNEL.
*75-mm. shrapnel, Model 1897, and shrapnel with base charge.*
WIND—RANGE CHANGE.

A head wind decreases range; a rear wind increases range.

To correct for REAR / HEAD wind LOWER / RAISE range setting.

**Fig. 2.4** Pages from *Provisional Range Tables* [614] one of a set of range firing tables assembled by Forest Ray Moulton's group

of the differential variations is felt in a single extended exercise, which requires the quadrant elevation and deflection settings for a 75-mm gun given data on position relative to target, matériel (projectile weight, related matters), and weather data given in the actual coded form in which it would be received from the meteorological message. In two and one-half pages, the final protocol containing all possible factors affecting the range and deflection is transformed into an algorithm which shows how to correct for each. The individual corrections are algebraically combined. A worksheet is provided to assist in the task, which involves differential variation charts labeled $A$ though $K$.

The new methods, including differential variations, were adopted as a standard method for computing range firing tables. A 1934 volume entitled *Computation of Firing Tables For the U.S. Army* [517], by H. P. Hitchcock, is a comprehensive compilation of all aspects of this activity as they were practiced at that time.[29] The book lists many of the technical reports upon which this account has drawn, including Bliss', and many other reports. It references the same four ballistics texts as McFarland's book does. It has Moulton's equations of motion on page 40 and explains the numerical integration technique. Page 224 begins an explanation of the differential variations. The mathematics is given in a style similar to Bliss' JUSA/CAJ article and continues till page 239. The highlight of these pages is a 16 inch by 20 inch fold-out computation schedule for the differential variations for the 12 inch Model 1895 gun. It accompanies a trajectory computed by numerical integration for the gun (muzzle velocity 2500 ft/sec, ballistic coefficient $C = 12.788$

[29] The copy of the volume obtained for this study bears the marking of the Frankford Arsenal.

and angle of elevation $\phi = 5$ degrees). Finally, one sees the example of a backward integration of the adjoint system from the terminal conditions to initial, which was required in the original version of the Bliss technical report. The sheet is read from left to right with the terminal data appearing on the left, and the numbers worked systematically to yield the needed initial conditions on the right. In spirit, it is akin the three-page anti-aircraft document, with the white on blue style replaced by mechanical typewritten numbers on plain white paper, the Greek letters $\mu$ and $\rho$ being the only handwritten symbols. The chart is more organized than the anti-aircraft sheets; the left-hand margin lists quantities to be computed in a fixed order, which would be replicated for each such computation. The results give the needed changes in range due to percent changes in the usual quantities.

The New Ballistics had become the status quo.

## 2.8 The Critical Stance of Philip Schwartz

Although we have cited isolated instances of criticisms by the military and others of the second wave reformers' work, the most sustained critical assessment of this work came from a new member of the armed forces, Philip Schwartz. In unpublished memos and publications in JUSA/CAJ and *Army Ordnance*, he proved to be an astute observer and analyst of ballistic theory and practice, one whose work never easily fell into easy acceptance or wholesale rejection of the new ideas.

Philip Schwartz (1895–1989) was born in Jersey City, New Jersey, in 1895. He attended Columbia University, graduating with a degree in mathematics "with honors" in 1917. Among the advanced courses taken by Schwartz were differential equations and an advanced calculus course, which used the Goursat-Hedrick *Mathematical Analysis* [474] as a text, as well as the usual physics and chemistry requirements. The "honors" aspect of the degree is explained in the *Columbia College Announcement* of 1914–1915 [Columbia University] as requiring "less prescribed work, but more independent reading and more thorough study in chosen subjects than are required of general candidates for such a degree." Schwartz thus took two honors conferences with selected topics in calculus, analytic geometry, and differential geometry, which involved much supplementary reading. Thus, was established early a desire to acquire learning through independent study, a characteristic that marked Schwartz's career throughout the years we consider. We will list his other educational achievements and ballistics experience up to 1923 before entering into his often controversial involvement with ballistic topics.

According to Grier [484], Schwartz, upon entering the Army, instructed mathematics at Fort Hancock in Sandy Hook, New Jersey, where Veblen was conducting his range table construction using the Siacci methods. Schwartz wasrecruited to help in this job also, which took place in the winter months of 1917–1918.[30] Veblen's diary for this period notes on March 12 "To sandy hook[31] coast for tables. . . worked with Alger, Schwartz, and Cobb." [Oswald Veblen Papers] This work continued at Aberdeen, where Schwartz and Cobb among others did the work of range firing. From May 24 to July 1 of 1919, Schwartz worked in Moulton's office in Washington as a computer of trajectories, presumably using the short-arc methods of Moulton. He also attempted to further his education by attending Columbia again for one semester in 1920, though his course of study is not known; he attended the Graduate School of Arts and Sciences [Columbia University]. For reasons unknown, this study was discontinued, but he attempted further work at Princeton University, where he is listed in the *Princeton University Catalogue* as a graduate student in electrical engineering in the years 1920–1921 and 1921–1922 [Princeton University]. It is not completely clear whether he graduated from Princeton with a degree or merely left[32] His last formal educational achievement was the completion of the course of study at the Ordnance School of Application at MIT,[33] while he was stationed at the Watertown Arsenal, in 1923. Here, the course work included an Ordnance Engineering sequence in which gun design, recoil and counter-recoil mechanisms, and stresses in parts of the guns were studied; this material Schwartz passed with honors [MIT]. Also included was a course in differential equations.

From the foregoing, it is clear that Schwartz had a high degree of exposure both to classroom education and practical experience in the first and second wave methods of exterior ballistics by 1923. The determined acquisition of this education was coupled with a distinctive independence of mind, which found early expression in two documents: a 1920 survey written while at Aberdeen Proving Grounds and a 1924 letter to Veblen describing his opinion on the contemporary state of education in ballistics. We will discuss these two items before moving on to his post-1924 efforts to rouse interest in a subject, which had by then become somewhat dormant in the years since the War, in his opinion.

"A Critical Survey of the Advances in Practical Ballistics in the Ordnance Department, U.S. Army Since 1917" [630] was dated September 1920, issued from the Aberdeen Proving Grounds. It was apparently an unsolicited study and was never published. Schwartz had read the reports of Moulton, Bliss, and Gronwall among others in preparing the survey, and he also shows his awareness of the first wave material. In one paragraph, he describes mathematical research in the Hamilton-Ingalls era: "As for the advancement of ballistic theory very little was done. The air resistance laws determined by foreign countries were adopted. On account of the lack of appropriations for extended firing it was found convenient to let foreigners

[30] He maintained a close association with Veblen through correspondence which lasted into the post-World War II years.

[31] Uncapilalized in the original. Sandy Hook is a peninsula extending northward from the Atlantic coast of New Jersey.

[32] Letter to author from Seeley-Mudd Manuscripts Collection, Princeton.

[33] This institution is discussed below.

do the firing and have the officers of the United States Army spend their time in mathematical research. The mathematical research was of a recondite [Veblen's word] nature ... making more or less accurate extensions of the Siacci Theory." In this memo, Schwartz also reveals himself to be a critic of the second wave ideas and one who is familiar with at least the opinions of artillerymen who were in battle, if not battle itself. There are many interesting comments: "On account of the assumptions involved in constructing the air resistance law and in the representation of the projectile by a particle little can be said concerning the actual practical accuracy of the other computed characteristics of the trajectory such as maximum ordinate, time of flight, striking velocity, etc., except that within the limits of the assumptions these characteristics can be computed as accurately as desired. The gain in the known accuracy of these characteristics is not of extreme importance." Bliss's adjoint system is described as an "ingenious system of computing differential corrections to Moulton's trajectories ... The mathematical theory involved in the derivation is of a very high order." The superiority of the short-arc method over the Siacci method is questioned in several paragraphs. An extended quote is appropriate:

> Ballistic tables based on short arc computations are now being computed, and when they are published will undoubtedly do away with the use of Ingalls's tables. But until the time when the new tables are available for general use it is believed that Ingalls's tables may be used to solve most of our problems even for fire up to 45 degrees elevation ... The benefits of the results obtained from computations by short arc methods may be easily overestimated by those who have not looked into the matter critically. Mathematicians may say that the Siacci method is based on poor and complicated methods and that army officers have been kept away from the study of ballistics on this account [a reference to Veblen's report], but a casual glance at the published papers on the short arc method and the method of computing differential corrections, will make one think that the Siacci mathematics is much simpler. A complete understanding of the mathematics of the new methods including the differential corrections, involves an understanding of a great deal of higher mathematics. The opinion has been expressed that the pre-war methods were to make firings at all elevations and thus really not need any good mathematical theory - merely a table for interpolation purposes [Mouton's criticism]; but this procedure is still being followed and will be followed even after the new tables are published, for ranges are not know[n] any better now from a priori reasoning without firing, nor does firing at one elevation determine such a short arc value of the ballistic coefficient that it can be used accurately for all elevations. [630, p. 6][34]

But the survey is much broader in its scope and brings up practical considerations that display the mathematics of exterior ballistics as one of a multitude of considerations in range firing and battle. Schwartz notes, for example, that the theory of the rigidity of the trajectory, discarded by the second wave reformers, "is still sufficiently exact for many purposes since battle-fire is still at fairly low elevations and the target is not usually on a level very much different from that of the gun ..." The time spent on the corrections for the rotation of the earth "is of no particular importance in more than 95 per cent of battle firing." The issue of how a specific gun can be expected to behave when compared with table predictions is

[34] This criticism will be seen relevant in the discussion of range table construction given by Roger Sherman Hoar below.

raised: "Also, range tables are usually based on firings in one gun or in a limited number of guns and for this reason do not apply exactly to any gun. Each gun has a calibration error and this error may run as high as 2 per cent at the maximum range ... Recent firings at Aberdeen have shown considerable variation in the range from gun to gun of the same model and caliber when fired at the same time under exactly similar conditions." On the same page, however, Schwartz states that "It appears that a knowledge of the differential corrections is of primary importance in ballistic tables and that one of the measures of the utility of any ballistic table is the accuracy with which the differential corrections are known." Distinctions between the needs of the coast artilleryman and the field artilleryman are noted: he claims that the latter actually needs the differential corrections data even more than the range-elevation data. Also, "Field artillerymen in the United States have not yet brought themselves to the point where they believe, that under battle conditions, it is advisable to make ballistic corrections before firing; they prefer to fire first without making any ballistic corrections and then to correct by observation where possible. Coast artillerymen believe that correcting before firing is a good thing, but they object to adding any burden on the battery commander's shoulders...." This is the type of analysis one desires and seems to find nowhere else: exposure to all methods of range table construction and knowledge of the practices of different branches of the military.

This memo foreshadows many of Schwartz's writing activities in the 1920s, the impetus for which was explained in a letter to Oswald Veblen dated March 12, 1924 [Veblen Papers]. The letter is on Coast Artillery Board stationery, Schwartz's post at that time being the ballistics member of the Board. He had asked the editor of the JUSA/CAJ whether the journal had any interest in publishing some notes based on Veblen's European ballistics-related post-World War tour, a suggestion that Veblen himself had made to Schwartz at Princeton in 1921. From there, the letter becomes an expression of dissatisfaction with the present state of ballistics as a subject of instruction in the Army (in contrast to the Navy). At Fort Monroe, Virginia (the site of the Coast Artillery Board), "the Board was the center of ballistics activities before the war, but this condition has completely disappeared since the Ordnance Department took over the ballistic problem." His perception was that "ballistics has lost considerable prestige in Army Schools since the war. I doubt that any service school has a course in ballistics at present. The Ordnance School had a course for about two years but dropped it. The Coast Artillery School had quite an extensive ballistics course before the war but now has none." He continues–

> Although Bennett's book 'Introduction to Ballistics' was written primarily for the use of students in ballistics at the Field Artillery School, I believe that they too have dropped the subject. I think this condition needs remedying and I intend to start something shortly, if I can think of some way of stirring things up ...Of course the Artillery Schools teach gunnery, but this is done in as practical way as possible, and with a minimum of theoretical explanation. Ballistics is still a part of the course in Ordnance and Gunnery at West Point but I doubt that very much is absorbed by the cadets. I had a letter from the Ordnance office a short time ago in which I was informed that the Ordnance Department may no longer send an officer to Chicago University as a student in Moulton's ballistics course. From all this you can see that something has happened to the attitude of the Army toward ballistics

> ... The tendency seems to be to clutter up the Army Schools with so much material that is of secondary importance that some more important things are [b]eing gradually eliminated by the 'practical' people ... By publishing such a series of professional notes in the Coast Artillery Journal as I have already mentioned, it is possible that interest in the subject may be awakened.

The letter is remarkable in several ways. Schwartz, in the phrase used to describe Moulton's efforts, seems to have taken it upon himself to stir up interest in a subject once the center of intense activity but now relatively ignored. One can also detect an ambivalence about the role of mathematics in ballistics-related matters. Based on his 1920 report, one would be tempted to think of Schwartz as dismissive of contributions of mathematical ballisticians, but in his correspondence with Veblen, he always admits the importance of the mathematical approach to problems and is here critical of the cluttering of courses with overly 'practical' material.[35] It is this ambivalence that makes Schwartz such an intriguing character in this story, as his struggle to reconcile all aspects of ballistics is revealing of the impact the of New Ballistics on Army practice. The promised professional notes were delivered: Schwartz wrote on many subjects. In "The English Long Range Gun" [635], he compiled notes from Veblen's trip to Europe, which contained a description of this weapon. In "Studies of the Motion of a Projectile" [634], he described new work by the Ordnance Department on the precession of a projectile considered as a rigid body, not a particle. In "Weighting Factors" [636], he provided a computation showing how these factors were used to determine the effects studied by Moulton and Bliss. "Accuracy Firing" [632] attempted to examine firing data to determine which factors of the traditional range-table variables (angle of elevation, ballistic coefficient, or initial velocity) have the predominant influence on the dispersion of shots fired from each of five types of guns at different ranges.

Of particular interest for the variability problem are "Experimental Investigation of the Effect of Wind Upon the Motion of a Projectile" [633] and "Remarks Concerning the Use of Firing-Table Values" [637]. In the former, Schwartz reported on the results of tests conducted over a two and a half year period by "The Ordnance Department, with the moral support of the Coast Artillery Board" to test the accuracy of the wind correction factors as deduced by Bliss and Moulton. In this series of experiments, the most important question about the work of Moulton and Bliss was finally taken up: how accurate are the theories? The article is noteworthy for is scholarly beginning: a page is devoted to a short history of the problem of measuring the effect of wind on range. The test is described: two new French 75-mm guns, model 1897, were fired side by side and also at opposing positions across Chesapeake Bay (on which the Aberdeen Proving Grounds is located) and data collected. The results: "(1) the method which is now in use for making range and deflection corrections due to wind, has a theoretically correct foundation. (2) Errors

[35] In a letter to Veblen from the 1940s, when Schwartz's son was considering studying nuclear physics at Princeton, Schwartz related that his son needed a strong understanding of the basic mathematics behind the subject.

in the application of the wind correction at the battery are due for the most part to the unavoidable approximations required for the successful use of artillery, and are not caused by faulty ballistic theory, except as affected by error in the fundamental retardation law." A list of formulae "which this test has shown to be satisfactory" is given, among which are Bliss' and Moulton's expressions.

In the latter note, the accuracy of the values for quantities appearing in range tables is discussed, resulting in a "table in which the relative accuracy of each quantity has been tabulated"; a report card, as it were, for the accuracy of the tables. Again, a topic that would seem fundamental to military practice but somehow fell between the cracks is addressed. Schwartz states that the table is based on his own experience and does not represent the opinion of the Ordnance Department. The quantities receiving the highest "grade" of A are the range and time of flight. The effects of change in elevation, muzzle velocity, and air density on range all receive B's. The worst grade of F is given to the probable error in range or deflection, the measure of the "inevitable" dispersion of shots, dispersion not attributable to the usual sources. A special note is made of this: "...the probable error is very inaccurate, and in using this quantity, allowance should be made for a large variation from the firing table value." Schwartz thus set himself the task of evaluating the accuracy of the theoretical work of Moulton and Bliss, providing a necessary check, albeit limited in execution, to the numbers pouring out of the computations.

Schwartz continued in his task of evaluating the accuracy of the range firing tables until the end of the 1920s. It is good to recall some of the remarks, quoted in the Preface, from "The Fork" [640, pp. 608–609]: "the U. S. Army firing tables may be considered to be as accurate as any firing tables published by either the Navy or any foreign countries. The inaccuracy [in various quantities] may all be attributed to lack of information which is being sought at present through research at the proving ground and elsewhere ... It must be realized that the firing tables constitute a tool which is prepared by the Ordnance Department and furnished to the using branches to help in firing ... To use this tool properly it is necessary that the user have some notion of its limits of accuracy." Surely, this is a reasoned, considered judgment, one made after long study in the wake of the absorption of the techniques of the New Ballistics by the Army. None of the originators of this material, including Veblen, Moulton, and Bliss, concerned themselves with this task, and they spent little time in considering how the results of their efforts should be used in practice. In this way, they differed greatly from Ingalls and Hamilton, whose work, however inelegant, was at least the result of men with artillery experience.

Schwartz also attempted to use the differential variations in an argument involving how to test the accuracy of a weapon, with special attention to determination of when a gun was sufficiently worn out to be considered useless. In a 1924 paper [632], he defined accuracy of a gun:

> One gun is more accurate than another, if when fired: a, from the same mount; b, by the same crew; c, at same time or under same weather conditions; d, using same type and lot of ammunition; e, using the same angles of elevation, one has a smaller dispersion in altitude, range, and deflection, at all battle ranges, than the other. [632, p. 521]

He reproduced portions of range firing tables for several weapons and included for each the range, angle of departure, and changes in range for each of 1 degree change in angle, 10 ft/s change in initial velocity $V$, and 1% change in the ballistic coefficient $C$. He argued that "An eroded gun causes a reduction in muzzle velocity and in ballistic coefficient, and also affects the uniformity of these two quantities" [632, p. 522]. Noting that greater changes in range occurred for each weapon for a change in $V$ and $C$ at higher angles of departure, then argued that testing for accuracy ought to occur a longer ranges than those currently used in practice. The table results were used for recommendations for other tests as well.

Though not part of the differential variations material as such, Schwartz participated in the prolonged JUSA/CAJ debate on probable error, spotting and fire preparation by contributing two articles, [638] and [641], which showcased his mathematical background and were the only entries, which contained attempts at calculating and analyzing quantities like standard deviation and variance. Given the state of understanding of these concepts and their applications in the military at this time, Schwartz's papers showed a sincere effort at their employment. Criticisms of these pieces were addressed more at their lack of knowledge of military practice than at their mathematical limitations. Upon consideration of this topic, one is led to natural speculations: in a proving ground firing how can one distinguish variation in range caused by measurable effects such as wind and that due to the "inevitable" dispersion measured by probable error? These realities were not addressed by Bliss or Gronwall, for example.

Philip Schwartz achieved the rank of colonel in the United States Army, becoming an ordnance officer for an airborne unit in the Second World War. During this time, he escorted Veblen on trips around England. In a letter to Veblen dated August 5, 1946, Schwartz relates that he was present at the Bikini island group on July 1, when a nuclear weapon was dropped as part of the Army's Operation Crossroads. Weapons had indeed undergone a radical transformation since he began working with Veblen on range firing tables at Sandy Hook in 1918, far more so than in the 30-year period referred to by Lissak in our introduction.

## 2.9 The Advocacy of Roger Sherman Hoar

At the opposite end of the spectrum from Schwartz's cool analyses of the New Ballistics stands the participant who most likely coined the phrase and was its most enthusiastic supporter, other than Moulton himself. Roger Sherman Hoar (1887–1963) is also the most colorful of the characters in this story, a lawyer by training with strong mathematical interests who through a series of turns ended up authoring a book on ballistics and instructing the subject. His nonstandard path was quite different from the others we have met. (See [529] for an account of this path.)

Roger Sherman Hoar was born in Waltham, Massachusetts, on April 8, 1887. He came from a distinguished legal and political family: an ancestor, Sherman

Adams, helped compose and signed the Declaration of Independence; a grandfather was Attorney General under President Ulysses S. Grant; Hoar's father was a congressman. Hoar graduated A.B. from Harvard University in 1908 and LL.B. from Harvard in 1911; however, during his last year in Law School, he served in the Massachusetts State Senate, passing the exams for the law degree in spite of this commitment. We will add out of sequence that, motivated by his mathematical experiences with the army, he completed an MA in mathematics at Harvard in 1921. His accomplishments as of 1910 earned him a *New York Times* profile entitled "Young Leonardo," which listed even more achievements [Concord]. He worked as a law clerk for Louis D. Brandeis, who later became associate justice of the United States Supreme Court. After forming his own law firm in 1913, Hoar also served as Assistant Attorney General of the State of Massachusetts and became legal advisor to the state Constitutional Convention of 1917–1918. By this time in his career, he had published two books, one on constitutional conventions, and several articles on legal and philosophical topics. His service with the army began in March of 1918, when he enlisted as a private. From there, he achieved the rank of captain by 1920. In April of 1918, he was sent to Fort Monroe, where he served in different capacities until June of 1919.

Hoar's letters home to his family from 1918 reveal a restless and eccentric mind: one from September of that year consists of nothing but the quote "Ah, but my computations, people say, Reduced the year to better reckoning...," attributed to "Omar," a reference to an English translation of the Rubaiyat of Omar Khayyam. Perhaps, he was involved in some computational activity at this point. But with the Armistice in November came a concern with his future. In that month, he wrote to an uncle, revealing that, in spite of his successful law career, he had at heart really desired a life in engineering and inquired about possible positions. In doing so, he listed his qualifications: "...present position at the head of the most technical Department (post graduate) for the Coast Artillery School, together with having taught plane, geodetic, and railroad surveying at Harvard, having specialized in engineering law...and having taught Advanced Orienteering here for five months..." He found teaching congenial, appreciated that it allowed for "mathematical research on the side," and he found the prospect of life in the Army at that point satisfying financially, in lieu of other offers. In another letter to his mother, he criticized one aspect of the practice of law: "Elva [his wife] likes the army life and agrees with my choice. She alone knows how the law used to get on my nerves. I have an exact mathematical mind, and therefore prefer a profession where two and two make four, instead of one in which two and two may make anything from zero to six and a half, depending on whether some Judge has had a good dinner, or doesn't like the looks of my face" [Concord]. On June 1919, he was detailed to the Technical Staff at Aberdeen Proving Grounds, from which his involvement with the second wave of ballistic reform began.

Hoar wrote about and was involved with several technical aspects of military matters as well as a few pure mathematical endeavors, some of which came after his contributions to ballistics. In May of 1921, he wrote an article for JUSA/CAJ on

"The New Artillery Grid Maps" [520] for coastal defense, based on the "polyconic method of projection" rather than the rectangular system of co-ordinates. (At the time he was stationed in Boston and given the task of dealing with coastal defenses.) In "The Pickering Polaris Attachment" [518] he describes a device, co-invented with Edward Pickering of the Harvard Astronomical Observatory, which was able to detect true north at any hour of the night. The device was an accessory to a surveyor's transit, and required only one setting of the transit to make the determination. At the time of the article, May 1919, the device was being tested at the Coast Artillery School for orientation of heavy artillery. He wrote a note to the *American Mathematical Monthly* in 1922 [523] on proofs by mathematical induction, exhibiting familiarity with E.V. Huntington's book "The Continuum," probably obtained as a result of the course taken with Huntington at Harvard. In 1921, he proposed an elementary problem [521] for the *Monthly*, which was solved by Philip Franklin, whose work on ballistics was later used by Hoar in a manner to be described shortly. These notes and articles show an talent and appreciation for engineering and mathematical topics, albeit on what one would have to consider an elementary level.

Hoar's involvement with ballistics began in the summer of 1919 at Aberdeen, where he was selected as one of three Coast Artillery officers to learn the New Ballistics. In September of 1919, he became a spokesman of sorts for this material when he published an article [519] in JUSA/CAJ with that title, the first in the ten-paper series mentioned previously. The article relates in a nontechnical way ("stripped of deep mathematical exposition," stated the editor) the goals of the new efforts and their advantages as well as a brief history of earlier work. There is little here that we have not already dealt with, and the heavy reliance on Moulton's summary [594] is obvious. A pitch is made for the new general ballistic tables under construction by Bennett as answering all possible trajectory questions. Hoar's familiarity with the work of Ingalls can be inferred from his use of a certain unusual word. He described analytical expressions used to approximate the piecewise differentiable resistance function as "operose," as had Ingalls in his ballistics book. Hoar's championing of the new methods is evident as he lists reasons for the inadequacy of Ingall's tables. (One of these is the rejection of the "rigidity of the trajectory," the rule of thumb used by the older ballisticians.) On the other hand, speaking from firsthand experience, he declares that the methods used by Ingalls and Hamilton will not be completely discarded, since within the domain of direct fire they are sufficiently accurate. He states that, though the computation of trajectories using the new methods "requires a resort to very high mathematics, yet the use of these formulas requires little more than arithmetic. The teaching of practical Ballistics will therefore be much simpler under the new system than under the old." The author of the article was about to embark on such a teaching assignment, as he was given the task of instructing the first course in exterior ballistics based on the new methods for the Ordnance School of Application, during the winter of 1919–1920. A word needs to be said about this institution, another of the advanced officer training schools of the era.

## 2.10 The Ordnance School of Application

In 1902, the Office of the Chief of Ordnance established the Ordnance School of Application at the Sandy Hook Proving Ground, Sandy Hook, New Jersey, in order to provide a course of instruction in design and construction of ordnance for army officers.[36] By the end of the decade, courses in electricity, chemistry, differential equations, and machine and power appliance repair were added.[37] A more practical Ordnance School of Technology was created in 1905 at the Watertown Arsenal in Massachusetts, and both schools continued in operation until the United States entered the World War in 1917, upon which instruction was allowed to lapse. In 1919 the School of Application was revived at Aberdeen Proving Ground, where Hoar instructed the first course in Exterior Ballistics; the School of Technology was also revived at Watertown. A sometimes combative correspondence between representatives of the Ordnance School of Application and the Technical Office in Washington (Moulton's group) indicates a desire on the part of the former to amass a first-class mathematical library, consisting in part of the collection of the recently deceased Maxime Bôcher, professor of Mathematics at Harvard [NARA Ordnance School]. Bôcher's collection as well as the other books assembled included such authors as Hermite, Hadamard, Poincaré, a book on the electric current by Victor Karapetoff, and many other contemporary mathematical and physics classics.[38] Reprints of the technical reports from Moulton's Washington office were requested by the Ordnance School. Wishing to extend the schools to students in civilian life and to take advantage of existing civilian facilities, the School of Application and the School of Technology merged in 1921 into the Ordnance School, based in Watertown, and arrangements were made with the Masschusetts Institute of Technology (MIT) to help with instruction. A great deal of attention was paid to the upcoming motorization of the field artillery, and opportunities for enlisted men to assist in maintenance duties were instituted. The development of new weapons, a factor in the reconstitution of ballistic theory, also resulted in the large need for maintenance personnel. The mathematics courses given at MIT were described as "calculus, including differentiation, differential properties of curves and differential equations of the first order" in a column in JUSA/CAJ Professional Notes [338].

[36] See [648] for further information about this institution.

[37] The differential equations coursebook of 1907 [435] contains mostly routine exercises in the standard types of first- and second-order equations and equally routine verbal problems, many on electrical and mechanical topics. No effort is made to introduce the ballistic trajectory as an example.

[38] The ubiquitous T.H. Gronwall was put in charge of organizing this library.

## 2.11 The Contributions of Philip Franklin

Before discussing Hoar's ballistics text created for the Ordnance School of Application, a technical report by Philip Franklin, heavily relied upon by Hoar, will be described [455]. The report is dated August 1, 1919, and was written at Aberdeen. In several pages, it outlines a new method for the adjoint system, which makes novel use of some concepts in the calculus of variations.

Philip Franklin, the last of the three young men of New York City featured in this chapter, was born in that metropolis.[39] He graduated from the City College of New York with a degree in mathematics in 1918, having worked as a computer at Aberdeen for the previous year. The paper used by Hoar was dated about two months prior to his enrollment in the Master's of Mathematics program at Princeton, from which he received a PhD in 1921. (In a letter dated June 7, 1921, Veblen describes Franklin as "the best man we have had in several years." [MIT, Norbert Wiener Papers, letter to H.W. Tyler])

The impetus for the report by the recent college graduate is not known, but the result was a new derivation for the differential equations of the Bliss auxiliary variables. This derivation avoided the idea of the adjoint system, replacing it with physical definitions of the variables (lacking in Bliss' original treatment) and employing differentiation and some unstated principles in the calculus of variations. The physical definitions were seized upon by Hoar, who incorporated them in his book. This, to Hoar's mind, removed the obscurities in Bliss' derivation while adhering to a perspective that Hoar repeatedly stressed: "Throughout this book the attempt is made to explain as much as possible from the viewpoint of physics rather than from the viewpoint of abstract mathematics" [522, p. 9], as he stated in the introduction. Even more emphatically in a 1928 paper, he declared, "It is axiomatic that a mathematical explanation of any physical fact should be, if correct, susceptible of a physical interpretation; and conversely that a physical explanation of any physical fact should, if correct, be susceptible of expression in rigorous mathematical terms. In spite of the obviousness of the foregoing, every ballistics book or article which treats of the rotation of the earth [the subject of the article] keeps the physical explanation and the mathematical explanation in two water-tight compartments…" [524, p. 377]. The expression for the associated differential correction had been derived by Moulton and others at the time of the New Ballistics, but had no physical interpretation associated with its individual terms. Though hardly a new dichotomy and one surely appreciated by Moulton and Bliss, Hoar used it in his own way as rallying point for his book.

We will first quote Franklin's calculations, supply some explanation, and then show how Hoar attempted to fill in the remaining gaps. The goal is the derivation Bliss' system of differential equations for the auxiliary variables without resorting to the adjoint system. Franklin begins by setting Moulton's equations of motion.

[39] See [457] for information concerning Franklin's life.

Then: If at any time $t$ the values of $x,\ y,\ x'$ and $y'$ are changed by small variation, the trajectory being defined by [Moulton's equations] and the new values of these quantities from this point on, the disturbed trajectory will have a slightly different range from the original range, and

$$\delta R = \lambda\delta x + \mu\delta y + \lambda_1\delta x' + \mu_1\delta y' \tag{2.40}$$

where $\delta u$ denotes the variation in $u$, terms of order higher than the first are neglected, and the quantities $\lambda,\ \mu,\ \lambda_1,\ \mu_1$ may be *defined* [emphasis added] as the changes in range for unit change in $x,\ y,\ x',\ y'$ respectively at the time $t$, and are evidently functions of the time whose values at $T$ (the ending time of the trajectory) are 1, $\cot(\omega)$, 0, 0. Since the range of the disturbed trajectory does not depend on the particular time at which the variations are made, provided their proper values are taken, $(\delta R)' = 0$, and upon differentiating we obtain;

$$0 = \lambda'\delta x + \lambda(\delta x)' + \mu'\delta y + \mu(\delta y)' + \lambda_1'\delta x' + \lambda_1(\delta x')' + \mu_1'\delta y' + \mu_1(\delta y')' \tag{2.41}$$

or since

$$\begin{aligned} &(\delta x)' = \delta x', \qquad (\delta y)' = \delta y' \\ &(\delta x')' = \delta x'' = \frac{\partial x''}{\partial x}\delta x + \frac{\partial x''}{\partial y}\delta y + \frac{\partial x''}{\partial x'}\delta x' + \frac{\partial x''}{\partial y'}\delta y' \\ &(\delta y')' = \delta y'' = \frac{\partial y''}{\partial x}\delta x + \frac{\partial y''}{\partial y}\delta y + \frac{\partial y''}{\partial x'}\delta x' + \frac{\partial y''}{\partial y'}\delta y' \end{aligned} \tag{2.42}$$

[it follows that]

$$\begin{aligned} &(\lambda' + \lambda_1\frac{\partial x''}{\partial x} + \mu_1\frac{\partial y''}{\partial x})\delta x + (\mu' + \lambda_1\frac{\partial x''}{\partial y} + \mu_1\frac{\partial y''}{\partial y})\delta y + \\ &(\lambda + \lambda_1' + \lambda_1\frac{\partial x''}{\partial x'} + \mu_1\frac{\partial y''}{\partial x'})\delta x' + (\mu + \mu_1' + \lambda_1\frac{\partial x''}{\partial y'} + \mu_1\frac{\partial y''}{\partial y'})\delta y' = 0 \end{aligned} \tag{2.43}$$

[Equation 2.41 follows from 2.40 by the product rule, 2.43 from 2.41 by substitution of 2.42 into 2.41. Equations 2.40 and 2.42 themselves are based in the calculus of variations, to be justified shortly.]

Since $\delta x,\ \delta y,\ \delta x',\ \delta y'$ are independent, their coefficients must vanish separately, and making use of the values

$$\frac{\partial x''}{\partial x} = 0, \quad \frac{\partial x''}{\partial y} = -x'\frac{\partial F}{\partial y}, \quad \frac{\partial x''}{\partial x'} = -F - x'\frac{\partial F}{\partial x'}, \quad \frac{\partial x''}{\partial y'} = -x'\frac{\partial F}{\partial y'} \tag{2.44}$$

$$\frac{\partial y''}{\partial x} = 0, \quad \frac{\partial y''}{\partial y} = -y'\frac{\partial F}{\partial y}, \quad \frac{\partial y''}{\partial x'} = -y'\frac{\partial F}{\partial x'}, \quad \frac{\partial y''}{\partial y'} = -F - y'\frac{\partial F}{\partial y'}$$

obtained from [Moulton's equations] and the fact that $x''$ is independent of $x$, we obtain the equations:

$$\begin{aligned} \lambda' &= 0 \\ \mu' &= \frac{\partial F}{\partial y}(x'\lambda_1 + y'\mu_1) \\ \lambda_1' &= -\lambda + \lambda_1 F + \frac{\partial F}{\partial x'}(x'\lambda_1 + y'\mu_1) \\ \mu_1' &= -\mu + \mu_1 F + \frac{\partial F}{\partial y'}(x'\lambda_1 + y'\mu_1)\text{“} \end{aligned} \tag{2.45}$$

This ends the direct quote from the memo.
After calculations which replace the partials of $F$ with respect to $y$, $x'$ and $y'$ with partials of $G$ and $H$, he arrives at

$$\begin{aligned} \lambda' &= 0 \\ \mu' &= \frac{F}{H}\frac{\partial H}{\partial y}(x'\lambda_1 + y'\mu_1) \\ \lambda_1' &= -\lambda + F\lambda_1 + \frac{x'}{V}\frac{F}{G}\frac{\partial G}{\partial V}(x'\lambda_1 + y'\mu_1) \\ \mu_1' &= -\mu + F\mu_1 + \frac{y'}{V}\frac{F}{G}\frac{\partial G}{\partial V}(x'\lambda_1 + y'\mu_1) \end{aligned} \tag{2.46}$$

The equations (2.45) are Franklin's version of Bliss's system (3.162) for the auxiliary variables.

Before continuing, we can note that Franklin has reversed Bliss's order: he writes out the relation for $\delta R$ *first* as equation (2.40) with interpretations for the coefficients, *then* derives the system of differential equations (2.45) satisfied by the coefficients, avoiding use of the adjoint concept.

The final step which we sketch is Franklin's derivation of the $\delta R$ in terms of the auxiliary variables. Here, rather than have them appear as a result of the fundamental relation of the adjoint system as Bliss does, Franklin relies on the definitions exclusively. For example, it follows immediately from the definitions of the variables that, for a change in range due to a change in initial velocity only,

$$\begin{aligned} \delta R &= \lambda_1(0)\delta x' + \mu_1(0)\delta y' \\ &= \delta V(\lambda_1(0)\cos\phi + \mu_1(0)\sin\phi) + V\delta\phi(-\lambda_1(0)\sin\phi + \mu_1(0)\cos\phi) \\ x_0' &= V\cos\phi, \quad y_0' = V\sin\phi \end{aligned} \tag{2.47}$$

$$\delta x_0' = \delta V \cos\phi - V \sin\phi \, \delta\phi,$$

$$\delta y_0' = \delta V \sin\phi + V \cos\phi \, \delta\phi,$$

which gives a final expression for $\delta R$ that agrees with that obtained by Bliss for this case. Other expressions obtained Bliss are justified similarly, and the remainder of the document outlines the usual instructions for computation.

Hoar's treatment of the differential variations can be viewed as an attempt to provide a commentary to this text. But before describing his work, we will relate Franklin's derivations to contemporary literature on the calculus of variations, since Hoar makes no mention of them and Franklin's use of them seems rather novel. Of the available books from the era, the treatment of Kneser [555] (written in German) is the easiest to reference; the first several pages closely parallel Franklin's and Hoar's style. (Remember that Bliss read Kneser's book as part of his project to familiarize himself with the literature on the calculus of variations.) The traditional problem of the subject is to find the curve $(x(t), y(t))$ which maximizes or minimizes integrals of the form

$$J = \int_{t_0}^{t_1} F(x, y, x', y')\, dt \tag{2.48}$$

where it is assumed that all curves considered have the same initial and final point. The integral could represent, for example, the surface area generated by rotating the curve $(x(t), y(t))$ between the given fixed limits around an axis: which curve $(x(t), y(t))$ gives a minimum area of the object of revolution? Answering this problem requires perturbing $(x(t), y(t))$ to $(x + \delta x, y + \delta y)$, where $\delta x$ and $\delta y$ are functions of $t$ which are small relative to $x$ and $y$, to see how the integral changes. Now $x + \delta x,\ y + \delta y,\ x' + \delta x',\ y' + \delta y'$ are substituted for $x,\ y,\ x'$, and $y'$ in (2.48), and the difference between the two integrals is taken, in a manner similar to the formation of the numerator of a difference quotient. The goal is then to replace this difference of integrals by an expression corresponding to a derivative, which is then set equal to zero, as in standard calculus optimization. Here, however, as in the functions of a line it helped to inspire, the variables on which the derivative-type expressions depend are the functions $x(t)$ and $y(t)$, not scalars. Since this difference involves the change $F(x+\delta x, y+\delta y, x'+\delta x', y'+\delta y') - F(x, y, x', y')$, an expression for this change in $F$ is required. That provided by the Taylor series expansion is used, with terms of the higher order neglected. Thus, Kneser writes [555, p. 5]

$$\delta F = F_x \delta x + F_{x'} \delta x' + F_y \delta y + F_{y'} \delta y'. \tag{2.49}$$

which is substituted into the difference integral, the result set to zero, and the analysis for the optimization proceeds from there.

Franklin was not concerned with integrals, but considered $x,\ y,\ x'$, and $y'$ as the usual trajectory quantities, where $(x(t), y(t))$ is a parametrization of the trajectory curve. He then framed his exposition in the language of calculation of variations. Equation (2.40) is his version of (2.49), where the range $R$ is the function dependent on $x$, $y$, $x'$, and $y'$. Thus, the change in range due to disturbances in the latter four quantities is given by (2.40), modeled after (2.49), and the partial derivatives in (2.49) give Franklin his interpretation of the auxiliary variables as rates of change.

Part of the preliminary work of the calculus of variations is proving rules for manipulation of the quantities, like $\delta x$, which mirror and extend those of the traditional calculus. Thus, equation (2.41) is the result of a product rule, which needs to be proved: Kneser gives a proof of a corresponding quotient rule [555, p. 4]. Hoar included a similar proof of the quotient rule in his book [522, p. 21] Another fact needed is the commutativity of $\delta$ and $d$; that is, that $\delta x' = d/dt(\delta x)$, which Franklin uses in (2.42). This identity is dismissed in three lines by Kneser [p. 4], which seems typical of texts of the era; see the last section of this chapter for the proof. The equations in (2.44) also follow from the rules of differentiation, using the original differential equations of motion given by Moulton. Hoar incorporated (2.44) as exercises in his book [p. 27].

The sources for Franklin's work are not known, but calculus of variations was often taught at that time as an additional topic in the calculus, for example, in Todhunter's text [654] and Bolza's works [406, 407], though having no discernible influence, were available. Given the short time between the date of Franklin's memo (August, 1919) and the commencement of Hoar's course (October, 1919), a likely scenario is that Hoar and Franklin discussed the material extensively during this time, with Franklin originating the derivation, based on whatever knowledge he had obtained, and Hoar developing his exposition based on their conversations. Hoar may even have suggested to Franklin that he try to provide an alternative derivation, in light of the perceived difficulties with Bliss' scheme. We conclude by mentioning that the lack of exact definitions of the quantities used by Franklin as well as the absence of proofs of rules for their manipulation will be addressed in the last section of this chapter.

## 2.12 Hoar's *A Course In Exterior Ballistics*

Hoar's main contribution to the ballistics of the era was the 1921 textbook "A Course In Exterior Ballistics" [522], which he wrote as a result of his course for the Ordnance School of Application. That such a course existed at all was no doubt due to the notion that the New Ballistics represented a progressive topic, which needed to be taught as much as the latest developments in other, more practical fields. The quality of the mathematical library assembled by the school certainly speaks to a vision of scientific sophistication of the highest order. But Hoar's book was anything but a mathematical treatise, or a text in the style of Moulton or [395]. It was a highly individual book that attempted to relate to the officers the ideas of the New Ballistics without resort to any mathematics beyond elementary calculus and some improvised concepts derived from the calculus of variations. Like the books

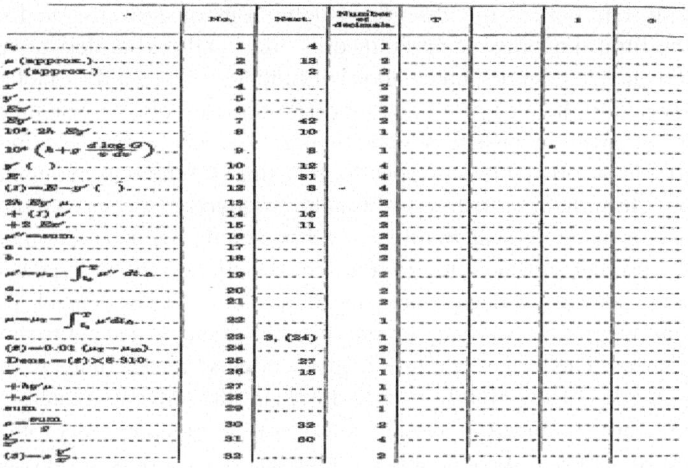

| | No. | Next. | Number of decimals. | T | | 1 | 0 |
|---|---|---|---|---|---|---|---|
| $t_0$ | 1 | 4 | 1 | | | | |
| $\mu$ (approx.) | 2 | 13 | 2 | | | | |
| $\mu'$ (approx.) | 3 | 2 | 2 | | | | |
| $x'$ | 4 | | 2 | | | | |
| $y'$ | 5 | | 2 | | | | |
| $\Sigma x'$ | 6 | | 2 | | | | |
| $\Sigma y'$ | 7 | 42 | 2 | | | | |
| $10^4$, $2h\ \Sigma y'$ | 8 | 10 | 1 | | | | |
| $10^4\left(h+g\frac{d\log G}{v\,dv}\right)$ | 9 | 8 | 1 | | | | |
| $y'(\ )$ | 10 | 12 | 4 | | | | |
| $E$ | 11 | 31 | 4 | | | | |
| $(J)=E-y'(\ )$ | 12 | 8 | 4 | | | | |
| $2h\ \Sigma y'\ \mu$ | 13 | | 2 | | | | |
| $+(J)\ \mu'$ | 14 | 16 | 2 | | | | |
| $+2\ \Sigma x'$ | 15 | 11 | 2 | | | | |
| $\mu''=$sum | 16 | | 2 | | | | |
| $a$ | 17 | | 2 | | | | |
| $b$ | 18 | | 2 | | | | |
| $\mu'=\mu_T-\int_{t_0}^{T}\mu''\,dt.a$ | 19 | | 2 | | | | |
| $a$ | 20 | | 2 | | | | |
| $b$ | 21 | | 2 | | | | |
| $\mu=\mu_T-\int_{t_0}^{T}\mu'\,dt.a$ | 22 | | 1 | | | | |
| $a$ | 23 | 3, (24) | 1 | | | | |
| $(z)=0.01(\mu_T-\mu_{t_0})$ | 24 | | 2 | | | | |
| Dens. $=(z)\times 8.310$ | 25 | 27 | 1 | | | | |
| $x'$ | 26 | 15 | 1 | | | | |
| $+hy'\mu$ | 27 | | 1 | | | | |
| $+\mu'$ | 28 | | 1 | | | | |
| sum | 29 | | 1 | | | | |
| $\rho=\frac{\text{sum}}{v}$ | 30 | 32 | 2 | | | | |
| $\frac{y'}{v}$ | 31 | 60 | 4 | | | | |
| $(z)=\rho\frac{y'}{v}$ | 32 | | 2 | | | | |

**Fig. 2.5** Computation sheet for differential variations included in Roger Sherman Hoar's *A Course in Exterior Ballistics*, [522]

of the older school of ballisticians, it was tied specifically to the goal of constructing range firing tables but using the techniques of numerical integration. Computation sheets ("schedules") for this purpose were included as part of the text, with detailed instructions on their use. (See Fig. 2.5.) If he did not create them himself, he certainly knew them step by step. (An entire appendix of the book is devoted to the care, which must be taken in sign usage when integrating the system for the auxiliary variables backward from $T$ to 0, as required by Bliss' technique.) They are a considerable accomplishment, showing hands-on familiarity with numerical integration, the equations of the ballistic trajectory and differential variations.[40]

The introduction of the book begins with a description of the work of the "ballistic computer," suggesting the focus of the work. Hoar at the time of teaching this course was "in charge of the Ballistic Section of the Proof Department at Aberdeen." [Concord, Harvard Alumni Survey] Though this description allows a wide latitude of interpretation, parts of the book make it clear that he had experience in range firing and table construction. A five-page section is devoted to the intricacies of this task, including data collection and use. Hence, we move from the first-level theoretical ballistic work of Bliss to the second level, where firing is conducted, tables are compiled, and theories tested. We will split our discussion of the book into two parts, the first a paragraph consisting of a few general observations, the second confined to Hoar's unique view of differential variations, which is the distinguishing feature of the book.

[40] As evidence of the increasing use of such schedules in civilian life, Form 5041, the differential variations worksheet, follows the same bar and dotted line construction as Form 1040, the first Return Of Annual Net Income Of Individuals, of 1913.

In general, one notices the absence of definitions of derivative and integral in terms of the limit, integrals being defined as areas and partial derivatives as one-variable derivatives (assumed understood) of functions of several variables.[41] Much space is devoted to the idea of a function of several variables in general, preparing the reader for specific instances in ballistics. The chain rule for functions of several variables is proved in an Appendix by using differentials and undetermined coefficients. Thus, the need for a proof is acknowledged, but not given prominence, and the result is justified by an intuitive, algebraic method. Each section of the book has assigned problems, which run from routine calculations to material used in later portions of the book, as well as probing questions on the section at hand. The process of numerical integration is broken up into two parts: successive approximation and finite differences. The former attempts to give students the idea of approximation by working examples of iteration in root finding, and the latter introduces Simpson's rule. The two are combined in an elementary way to give the basis for computing trajectories, to which a distinct lengthy chapter is then devoted. Pedagogically, this has certain clear advantages, since numerical integration is a special case of approximation. A chapter on the history of ballistics is included, unlike Moulton's and Bliss' work. It also appears that Hoar advocated and won acceptance for the curved coordinate system in which the rectangular coordinate $x$ is replaced by the distance measured on the Earth's surface. The equations of motion in both the rectilinear and circular coordinate systems are given and compared. At times the prose borders amusingly on the legalistic: "Integration by finite differences is based upon the principle that inasmuch as a derivative is a rate of change, the value of an integral of any smooth function can be computed step by step, if successive values of its derivative are known at sufficiently close intervals" [522, p. 26]. In general, one can say that Hoar attempts to get his points across in an intermediate area between informal discussion and formal mathematical exposition. He was familiar with the latter, as his somewhat ostentatious reference to William Osgood's *Differential and Integral Calculus* in the first chapter as a sourcebook for the basics of calculus shows, but the book is geared toward the job at hand: anything to make the task of range table construction accessible is preferred.

In Hoar's book, the material on differential variations takes up Chapter 3 and parts of Chapters 9 and 10. It is clear that he wanted to develop a sense for this material early, since it precedes even that of the differential equations of motion themselves. He assumes no background in the subject and does everything from basics. The descriptions are geometric where possible. In Chap. 2, two types of disturbances of a trajectory are considered and illustrated: that occurring at a given instant and that occurring over a short period of time. The treatment here stretches over three pages. The four quantities $x,\ y,\ x'$ and $y'$ are considered as variables, which act independently, and $\delta u$ is defined informally as the instantaneous change in $u$, where $u$ is any of the four quantities listed, or as the "total disturbance up to any instant" [522, p. 19] in the case of a continuous disturbance. Great care is

[41] This informal aspect was to raise the ire of Moulton in his exchanges with Hoar, related below.

taken to distinguish $\delta x$ from $dx$, the differential. All formal rules of manipulation of $\delta u$ quantities are stated, some left as exercises, and a unique proof of the relation $d/dt(\delta u) = \delta(u')$, based solely on geometry, is given. Though it is at times difficult to follow, it stands alone among the proofs available from other sources as an attempt to relate the physical meanings of the changes considered. Equation (2.49) is presented and given a rather weak supporting physical argument, with Taylor series unmentioned. In Chapter 9, the substance of Franklin's memo is presented with an attempted justification for each step. Franklin's definitions, in Hoar's hands, do unravel some of the mystery behind the auxiliary variables. For example, $\lambda$, as defined by Bliss, satisfies the differential equation $\lambda' = 0$ in the adjoint system, making it a constant whose value turns out to be 1. Since in Franklin's scheme $\lambda$ is the change in range due to a unit change in $x$, "It is obvious that since $x$ does not enter into the differential equations of motion of a projectile, a change in $x$ at any point can have no effect other than to shift the value of $x$ at all subsequent points by the same amount. Therefore $\lambda$ equals unity" [522, p. 53]. All of the effects that can change the range are ultimately thrown together in one equation for $\delta R$, which includes the special cases computed by Franklin and is comparable to those of Bliss.

The reception of *A Course in Exterior Ballistics* is hard to judge. The volume was not published by a commercial publishing house or university press, as Moulton's and Bliss' work had been and remained as War Document 1051. This naturally led to limited exposure; the *American Mathematical Monthly* merely listed it as having appeared, together with Bennett's ballistics textbook, with no comments, in 1921. Hoar admitted in one of his letters of exchange of 1928 with Moulton that his book was most likely out-of-date itself (in addition to Moulton's book). According to [449, p. 30], Harvard University was apparently considering having Hoar teach a course in exterior ballistics in 1921, but this did not happen, most likely a result of post-war lack of interest in the subject. The book's narrow focus that of range table construction (an evolving activity) and the mathematics needed to effect it, probably restricted its appeal. The interpretations of the $\delta$ operator advocated by Hoar did not find acceptance in later works. The book remains, for all the effort it took to produce it, a curiosity of exposition of a certain time and place, much like the individual styles one finds in calculus textbooks in America of the late nineteenth and early twentieth centuries, as discussed, for example, by George Rosenstein [624].

There is something in the work of Hoar that also brings to mind David Roberts' account [622] of the American geometer Albert Henry Wheeler, a figure from this era. Roberts shows how Wheeler's work in producing and promoting models of polyhedra can be viewed from the three perspectives of "high-brow, middle-brow, and low-brow." The high-brow response to Wheeler's work, that of the mathematical geometers, was virtually nonexistent, the middle-brow response; that of fellow high-school and college instructors was enthusiastic, the low-brow response; and that of the students, was fascination and admiration. Wheeler had a conception of "research"-the investigation and production of his models-which was out of touch with the genuine research in geometry done at that time. Roberts finds Wheeler's place in the contemporary mathematical society by referring to these three strata of individuals. Making allowances for the type of material and the intended audience, Hoar may be viewed in the same manner as Wheeler, as an amateur mathematical ballistician of sorts. His notion of research-his advocacy of circular coordinate systems, adaptation of Franklin's derivation of differential variations for practical use, and the Polaris-Pickering attachment-was surely not those of a research mathematician, even in the limited sphere of exterior ballistics. He relied on others-Moulton, Bliss, and Franklin-to supply the mathematics, which he then interpreted and presented. Moulton's dismissal of Hoar's work, described in the next section, is a more violent form of the neutrality of the research community to Wheeler's efforts. Hoar's book most likely found its admirers, though the work of range firing table construction was surely not as widely practiced or as appealing as polyhedral modeling. At the low brow level, obituaries of Hoar never fail to mention his World War service and now and then, if obscurely, attribute certain military contributions to him, such as a "ballistic system." Unlike Wheeler, however, Hoar moved on from his subject matter after a few years.

## 2.13 Roger Sherman Hoar versus Forest Ray Moulton

It is an irony of this story that the most heated exchange of views in print occurred not between representatives of the mathematical and military communities, but between the two most enthusiastic supporters of the New Ballistics, Roger Sherman Hoar and Forest Ray Moulton. The controversy played out in the pages of both *Army Ordnance* and the *American Mathematical Monthly* during the years 1927 and 1928. It was occasioned by the appearance of *New Methods in Exterior Ballistics* in 1926 and the reviews that it received. The arguments in the letters show in clear terms the differences and expectations with which the two second wave reformers had to deal. The exchange consisted of two nearly simultaneous reviews of *New Methods* by Hoar, one in each of the above journals, in 1927, and the replies to each, roughly one year later. We will treat each review and its reply as a single unit.

The controversy began with a laudatory review [493], with accompanying photo of Moulton, of *New Methods* by James L. Guion, an officer who had been sent by the War Department to take Moulton's ballistics course at the University of Chicago. The review appeared in *Army Ordnance* in the September-October 1926 issue. All aspects of the book are analyzed, and its virtues emphasized. In the March 1927 issue of the same journal came two replies to the review. The first is by R. H. Kent,[42] basically agrees with Guion's positive review but offers criticism of the book's treatment of rotating projectiles. The other, by Hoar, begins the debate proper.

Hoar's criticisms of *New Methods* are framed in terms of its alleged obsolescence. He begins by proclaiming his admiration for Moulton and his achievements during the 1918–1919 period. He noted that Moulton showed how the equations of

[42] Robert Harrington Kent (1886–1961), who was a physicist whose career was spent at Aberdeen. He entered the military in and was active in answering ballistic questions before Moulton's office was opened; he eventually joined Moulton's group.

motion could be integrated by a short-arc method, surpassing the achievements of his predecessors, whose contributions consisted of "devising approximate algebraic expressions whereby to wrench these equations into solvable form" [526, p. 395]. Moulton thus "laid the cornerstone for an entirely new science of ballistics." There follows the dubious compliment that the book is an "invaluable addition to the history of mathematical physics" and a comparison of Moulton's work with that of Morse and Bell, whose inventions progressed beyond their excellent groundwork. Guion's review had prompted Hoar to point out the similar progress since Moulton's corresponding groundbreaking work.

A list of new developments is given: [p. 396 for all that follows]

(1) Moulton's numerical integration method had undergone many improvements since its introduction.
(2) In 1919, what Hoar terms the "Bliss-Franklin-Gronwall differential corrections" were "substituted" for Moulton's; he claims that Moulton, while treating Bliss' adjoint system, placed too much stress on its limitations. Such limitations were in fact nonexistent, in Hoar's opinion.
(3) Hoar claims that in 1920 the rectangular coordinate system was dispensed with and replace by a curvilinear one; he takes Moulton to task for failing to mention this change.
(4) Moulton did not mention Ritt's limiting weighting factor curves.
(5) Moulton's chapter on spinning shells contains equations, which "bear very little resemblance to the complete theory which Captain Kent is working out at Aberdeen."
6) Moulton's symbols were discarded in 1919 and replaced by some agreed upon by the Technical Staff, the Ordnance School, and the Ballistic section of the Proving Ground. Perhaps, sensing the frivolity of this last complaint, Hoar insisted that "They become serious when involved in the teaching of officers whose practical work and further research must be done under totally different symbols and conventions."

Concluding once again with the tactless assertion that *New Methods* is historically valuable, Hoar expresses hope that a course such as Guion took from Moulton would in the future be taught in modern conventions and symbols, "in order to render it of practicable application to what is going on at Aberdeen." Moulton's reply [599] to this version of Hoar's critique was contained in the March-April 1928 issue of the journal, a year later. He first replies to Kent's criticisms with calm politeness; the men had met during their time in the military, and Moulton expressed "high regard for Mr. Kent's judgment respecting physical things" [599, p. 320]. No such respect was accorded Hoar. About the latter's concern for new notations, he offers a paragraph of dismissive comments which concludes with the statement "I cannot get warmed up on so trivial a matter" [p. 320 for the following remarks]. He precedes his reply to Hoar's advocacy of the Bliss method over Moulton's by stating that "In my first paper, written shortly after I entered the service, I seized the problem of computing differential variations and pulled it out of the fogs in which it was involved and placed it on a solid foundation." He then recounts Bliss' introduction of the adjoint system "for use in limited purposes" and states that the account of this method in *New Methods* was reviewed and approved by Bliss, including Moulton's assessment of its limitations. Moulton admits that it would have been possible to substitute a curvilinear coordinate system for a rectangular one but that would have come at the cost of "sacrific[ing] all the powerful machinery I built up, based on fundamental sets of solutions"; said machinery being put in the service of calculating "families of trajectories and their variations" as opposed to the isolated trajectory.[43] The notion of fundamental sets of solutions, mentioned one time earlier in the review, seems to be the focal point both in this reply and the next for Moulton's appraisal of Hoar's mathematical abilities, for he concludes with the blast: "... such things as proving the validity of the methods employed, the use of fundamental sets of solutions of linear differential equations, and the dynamics of rotating projectiles, which cost me more than everything else combined, are apparently completely beyond his [Hoar's] horizon."[44]

Hoar replied woundedly to this broadside in the July-August 1928 edition of *Army Ordnance* [527]. He admitted the "relative unimportance" of the notation issue, but explained that he meant it as an indication that Moulton was unaware of post-War developments such as the "work of Kent, Ritt, Gronwall, and Franklin." He insists that Bliss' adjoint method has far more applicability than is admitted by Moulton and enlists Franklin's derivation of the adjoint system to prove this point. (No remarks on the relevance of Franklin's derivation in this matter is given.) He concludes by repeating his admiration for Moulton and his accomplishments, including the irritating remark as to the value of *New Methods* as an invaluable historical addition to the literature, and an expression of regret that Moulton did not keep up with current practices.

The other exchange began with Hoar's second review of *New Methods* in the June-July 1927 issue of the *American Mathematical Monthly* [525]. Coming so shortly after his *Army Ordnance* review, one would not expect much difference in content, especially since Moulton's reply to the first review had not yet appeared. And this expectation is borne out. The initial remarks concerning Moulton's achievements, the comparison to Morse and Bell, and Moulton's lack of current knowledge, all appear.[45] Ditto the list of specific objections from the first review. He does claim that Franklin's derivation of the Bliss system is "much more interpretable

---

[43] These families were probably a reference to his work on the general ballistics tables.

[44] This statement also implicitly answers Hoar's objection to Moulton's spinning shell theory; Hoar is pronounced incapable of rendering judgment on the matter.

[45] Hoar's flowery expressions of admiration for these achievements drew the suspicion of J. E. Rowe, who cited these remarks in *his* reply to this review of Hoar. "There has been a failure on the part of some properly to appraise the work done in this country during the War by our mathematical men who were interested in ballistics. The contributions that they made to the subject were advertised in most complimentary terms at meetings of the national mathematical organizations and elsewhere." ... It came to be usual to say certain things about these researches in ballistics. ... " [626, pp. 229–230] There follows a quote from Hoar's present review. "... it must be said that it [the quote] embodies the attitude that has become all too prevalent regarding this work and which has not been corrected" [626, p. 230]. His criticisms follow.

physically than Bliss' adjoint system" that Gronwall discovered a new first integral "which revolutionized the computations" and references a 1922 demonstration that Bliss' ideas can be adapted to anti-aircraft work. But this review remains essentially a copy of the first.

The second reply, also about a year later (May 1928), is equally if not more scathing. Moulton begins with a disclaimer about the method of numerical integration, denying that his astronomical experience contributed anything to his special knowledge of this topic, claiming that anyone familiar with differential equations could have introduced the same type of technique into ballistics.[46] As previously noted, he laid emphasis on his proof of the convergence of his method, and in an apparent warm-up, asserted "Unfortunately for those interested in the applications the proof of the validity of the method involves many of the niceties and much of the close reasoning characteristic of modern analysis, and it would be wholly beyond the appreciation of a beginner" [598, p. 247]. He again relates the history of his and Bliss' involvement with the differential variations and his work on rotating projectiles.

But the crux of the reply is contained in the next two pages. Moulton asserts that the difference in purpose between his and Hoar's volumes is so great that "it would be difficult even for an impartial person to compare then justly" [598, p. 248]. Though not stated explicitly, the claim, justifiable to a degree, is that Hoar's was written for a course in ballistics at the Aberdeen Proving Ground and Moulton's for a university level audience interested in a treatment of mathematical ballistics and ready for the analysis (in the mathematical sense) that such a course would entail. Moulton spares no pains to make his point. He quotes the introduction to Hoar's course: "It is assumed that the student is thoroughly grounded in algebra and plane trigonometry, and knows enough calculus to appreciate the meaning of a derivative, a differential, and a definite integral." To which Moulton replies: "It seems to me that the sentence quoted excellently defines the scientific level of the book." Another example of the low level of mathematical exposition is given when Moulton reproduces Hoar's discussion of the definite integral, an example "of the reasoning employed in Mr. Hoar's book" which is found to be "illuminating":

> Let us now imagine a plotted curve with $u$ for its ordinates and $t$ for its abscissas. Then $\int_{t_n}^{t_m} u\,dt$ is the area between the curve, the $t$ - axis, and the ordinates at $t = t_n$ and $t = t_m$. Such an area exists for every continuous curve, and hence every continuous curve has an integral, even though that integral is not expressible by means of the ordinary elementary functions. [598, p. 248]

This must of course be judged against the rigorous treatment of the Riemann integral and the nature of integrable functions, which was in current circulation, of which Moulton was surely aware. In this comparison, Hoar is bound to come up short.

Moulton repeats his belief in the crucial role played by the general theory of differential equations:

---

[46] His reply has been quoted verbatim in the previous chapter.

> The theory of differential variations depends upon linear differential equations. The very heart of any adequate theory of them obviously is attached to that of a fundamental set of solutions of the differential equations. In my treatment of the problem I went straightaway to a fundamental set of solutions and made essential use of them in most of the subsequent discussions of the chapter. Mr. Hoar uses no such general and powerful machinery - he does not have a syllable on it. [598, p. 249]

The story of Moulton's inclusion of Bliss' adjoint system in his book is repeated. The reply ends with the same expressed indifference to the matter of symbols used. No mention is made either in this note or its predecessor of Franklin's derivation of Bliss' system; it apparently elicited no great interest.

Seeing beyond the specifics of Hoar's criticisms of *New Methods*, it is clear that the fundamental issue at stake here is the difference in level of exposition and mathematical sophistication between the two authors. This difference highlights the first occurrence in the American literature of a university treatment of ballistics; this topic had previously been reserved for military audiences only. The latter audience did not require the subtleties of analysis enjoyed by Moulton; they would have proved an insurmountable obstacle. Franklin's technique, regardless of its success or failure, was specifically designed to address the problem of the difficulty ordnance students would have with this idea, a difficulty that Rowe specifically mentioned in his remarks. Hoar's entire view was predicated on the motivation by physical considerations at the expense of the purely mathematical. Concerns with notation, coordinate systems, and new aspects of numerical integration are relevant to someone who had to get his hands dirty in the actual construction of range firing tables. Despite Moulton's assertion that his book was inspired by the need to make these tables, one gets no sense of the real job from his pages. In this conflict, the differential variations play a key role: Moulton makes repeated references to the need to understand linear differential equation theory, the ground on which his theory is built, makes no apologies for its difficulty. He acknowledges Bliss' work but notes what appears to him as limitations of it. Hoar takes Bliss' work as fundamental and tries to make it as accessible as possible by physical means. How successful this latter effort was is impossible to judge.

As a finale to this quarrel, we mention that the correspondence analyzed above must have renewed Hoar's interest in matters ballistic, after some time away from the military. He began to express interest in the differential variations again in the late 1920s. His specific interests were the correction to be made in firing to account for the rotation of the earth and the curvilinear coordinate system. Bliss had settled in his technical report of 1918 for the corrections as devised by W. H. Roever of Washington University, St. Louis, Missouri. In Moulton's 1926 book, these equations are derived, based on a technical report he wrote during the War, and were substantially the same as Roever's. Moulton prefaced his exposition with the remark that "The treatments of the problem hitherto given have generally been defective and have often led to erroneous results" [597, p. 12].

In 1927, Hoar organized a weekly "ballistics research group" at Marquette University, located near Hoar's employer, The Bucyrus Company, in South Milwaukee, Wisconsin. The group included several professors of mathematics and engineering at Marquette as well as some employees of Bucyrus and was run under the auspices of the College of Applied Sciences and Engineering, with the cooperation of the War department. The object of study was what Hoar now termed the "Moulton-Bliss-Gronwall-Franklin-Hoar System," according to a notice in Army Ordnance [339]. From this group first came a pair of papers ([528, 625]), one by Hoar and one by William Roth, a doctoral student at the University of Wisconsin, Madison, describing the equations of motion of a projectile in the curvilinear coordinate system. The original mathematics was done by Roth, with Hoar writing a comparison of the curvilinear and rectangular equations. This pattern was similar to his work with Phillip Franklin. Differential corrections in general were discussed.

Hoar then wrote another article [524] "The Rotation of the Earth," based on work done by the group, in which he attempted to justify Moulton's formulae for these corrections by a set of physical explanations, which would account for each term. The circular coordinate system was used. Hoar stated, "...no set of physical explanations which I have yet seen printed is susceptible of reduction to the universally accepted Moulton formulas... [p. 377]." These efforts paralleled his attempts to provide physical meaning for Bliss' auxiliary variables. Acting on a suggestion of R. H. Kent and Hoar, the group were able to interpret the terms as due to "conservation of moments of momentum" and the effect of the direction of fire on the centrifugal acceleration of the projectile. "These results are identical to those obtained by pure mathematics by the Moulton method. [p. 386]" Criticisms of other explanations, including one given in *Gunnery For Heavy Artillery*, were given.

Philip Schwartz came across the article, and wrote an essay in response, [639] "Much About Little." The Editor's comment: "...the author of this article attempts to explain in non-mathematical and non-technical language the influence of the rotation of the earth upon a projectile in flight." Schwartz: "The subject usually appears to be a mysterious one when first mentioned; after a limited amount of study the complications involved become noticeable, and after an extended study the interested person is usually willing to admit that the mathematical method of determination is correct [p. 291]." Hoar's paper is mentioned but classified as "Greek letter mathematics" for most readers of the JUSA/CAJ. The need to understand the ideas is established by the fact that three and a half pages of *Gunnery For Heavy Artillery* are devoted to it. "Every Coast Artillery officer will therefore, at some time in his career, be interested in the subject, either from the point of view of the student, the instructor, or both [p. 392]." Schwartz then offers two explanations for the effects of the rotation, accompanied by simple geometric diagrams and trigonometry. Significantly, Schwartz actually used this adjustment: in his 1924 letter to Veblen, he listed among his problems the responsibility of deciding whether "a rotation of the earth correction based on 35 degrees N. latitude is satisfactory for all Coast Artillery batteries [Veblen Papers]." The battle for a physical explanation for mathematical formulae had continued. Hoar resigned from the army in 1921 and took an MA at Harvard University in mathematics. His course load for these years consisted of two semesters of Advanced Calculus from G. D. Birkhoff (including calculus of variations), probability, two semesters of applied mathematics, and a course from E. V. Huntington of a foundations/philosophy nature [Concord]. He then took the position of Assistant to the Secretary of the Bucyrus Company of Milwaukee, Wisconsin, "the world's leading manufacturer of excavating machinery," according to Harvard alumni survey he filled out. Eventually, he became the commercial attorney for this company, which, incidentally, often advertised in *Army Ordnance*.[47] He eventually became the head of the Legal and Patent Department of the Bucyrus-Erie Company, retired from that company in 1954, and entered the private practice of patent law.[48]

Let us now turn to the work for which both Moulton's and Hoar's books were written, the range firing table.

## 2.14 How To Construct a Range Firing Table

Given all that has been said about range firing tables, the question of exactly how they are constructed has not been addressed yet. We have discussed the two theoretical extremes of firing only and theory only, cited numerous articles and books such as Hitchcock's *Computation of Firing Tables*, mentioned the difficulties caused by determining a value for the ballistic coefficient, and explored Philip Schwartz's comments on the range of options for the task. But what about the specifics of the process as practiced at this time?

Hoar devoted five pages of his *Course* [pp. 96–100] to an outline of one of the methods of table construction as it was done under his supervision. While acknowledging that this activity is complex enough to occupy the contents of a compendium like Hitchcock's, it is important to have some idea of its execution in one venue, to see the mixture of theory and reality that is in the nature of ballistic practice. The fact that it was described by a participant in conjunction with a text, uses the New Ballistics, and is relatively brief, makes it an appropriate choice. It also has the advantage of displaying a final practical use for the differential variations.

A few preliminary points need to be made. First, ever since it was possible to account for the differences, which secondary factors made in the trajectory, a certain set of conditions needed to be accepted as standard for firing. The list of such "normal" conditions is long, but a few relevant ones can be stated as they appear, for example, in Hoar's *Course*, pp. 34–35:

(1) the earth is motionless
(2) gun and target are at the same altitude
(3) no wind
(4) exponential density law with 750 mm mercury, 78% saturation, 59°C
(5) uniform gravity 6) standard drag law $G(v)$.

[47] J. J. Arnaud, the master computer involved with the general ballistic tables project, also found employment with this firm.

[48] He also became a writer of science fiction stories under the pseudonym Ralph Milne Farley, beginning with the "Radio Man" series in 1924.

There are other more technical ones, but this should suffice to give the general idea. The goal then is to find the range for a projectile, which correspond to these conditions, an initial velocity $V$, angle of departure $\phi$, and ballistic coefficient $C$, using data at time of firing, which will differ from standard. This procedure is called "reducing the data." Every ballistic theory had methods to accomplish the task. It should be evident that this is an opportunity for the differential variations to play a role, since the theory addresses this scenario.

The raw firing data was collected on a "firing record," an official sheet containing the results and conditions of firing. Hitchcock's book has a foldout sample of such a sheet. Much information is contained therein; one has, for example, the date and time of firing, number of guns, number of rounds fired, initial velocities as chronographed, angles of departure, ranges, all information about secondary factors, model of gun, and signatures of participants in the firing. Any irregularities in the firings were recorded, and the functioning of all components such as gun carriage mentioned. Detonation of shells and details of fuze settings were also noted. Reams of such records were kept at Aberdeen and other proving grounds and arsenals.

From this information and use of the new theories, Hoar describes the process of table construction. The following summary of his account, itself a summary, will attempt to impose some structure on a narrative, which is somewhat rambling. It agrees in many points with the much longer account given in [356]. The architect of the method is not credited in any available publications.

(1) First, obtain the data of firing. About 10–20 rounds are fired at several elevations, say 5, 15, 25, 35, and 45 degrees. Ranges are determined, meteorological conditions are recorded as observed from airplanes and pilot balloons, all nonstandard conditions are noted, and information about the projectile, initial velocities, and angle of elevations are given. More than one gun of the same type may be used.

(2) Next, obtain, from the raw data, values for initial velocity V, angle of departure $\phi$ (this is technically different from angle of elevation), range $X$, and the secondary factors. These values, or altered versions of them, will be used in computations. The values are obtained by averaging, for groups of shots at the same angle of elevation. So one averages, for shots at a given angle of elevation, all initial velocities, weight of projectiles, observed ranges, and similar quantities. The meteorological data is also averaged (details of methods used here are not given) to provide values for computation. The angle of departure $\phi$ for each group is obtained by mechanical considerations of the relative positions of the axis of the bore and the quadrant measuring device. An alternative method of finding initial velocities is given in the event that they were not recorded at the time of firing. At this point, one has for each elevation group, a value for $\phi$, $V$, $X$, and the secondary factors such as wind.

(3) The next step is to estimate, for each elevation group, the ballistic coefficient. First, for each group, the range value $X$ must be "roughly corrected," that is, reduced, for the nonstandard conditions, which actually held during the firing. Hoar offers six possible sets of charts to aid in doing this, but the process is

not described. Let us refer to the reduced value of $X$ as $X_s$, for standard. One now has values of $\phi$, $V$, and $X_s$, which correspond roughly to values on a standard trajectory; what is missing is the $C$ for this fictitious trajectory. One gets this value by entering the three values $\phi$, $V$, and $X_s$ in any one of three specified charts from the Siacci theory and finding the corresponding $C$. This value of $C$ is not the one needed; however, it, too, must be corrected. Again, two alternative methods are used to find the corrected $C$, which we may call $C_s$: one by charts and the other by formulae. One finally has the four values $\phi$, $V$, $X_s$, and $C_s$, which correspond to a fictitious trajectory obtained under normal firing conditions. The values of $C_s$ as a function of $\phi$ for the elevations used are then plotted, and a smoothed out curve of this function produced (method not described) for further use $X_s$ is used only to find $C$, which yields $C_s$, it serves no further purpose.

(4) Now using the values $\phi$, $V$ and $C_s$ found in (3) for each elevation, compute a trajectory under standard conditions using Moulton's short-arc method. Call the range as computed here $X_C$. Then, using this trajectory and the recorded secondary effects data, compute the differential variations in the range for the computed trajectory. Apply these range corrections to the *observed* ranges $X$ from step 1 as they appear on the firing sheet. The result is the first approximation to the range under standard firing conditions corresponding to $\phi$, $V$, and $C_s$ and to be entered in the table. This approximation is compared to $X_C$, the value of $X$ which came from the computed trajectory, and accepted as final if the difference between the two is less than a certain tolerance, given here as five per cent. If the difference is unacceptable, "correct" the values of the $C_s$ (method not given) and recompute the trajectories and the differential variations and iterate the comparison procedure. Though not stated, the procedure must at some point terminate, and the final value of $X$ entered.

(5) The values for ranges obtained in this way, for a given $\phi$, $V$, corrected $C$, and final $X$, are then recorded along with the differential variations used in the process. "These variations [have been] computed for a double purpose; first, for correcting the results of the firings to standard conditions and second for insertion in the range table for use in the field" as stated in [356]. In other words, the differential variations have been used to reduce the firing data to normal conditions, though much else transpired in the process.

Commentary on this construction method could be lengthy. We will only cite a few points. The "outdated" Siacci theory is still quite in use, here to obtain preliminary values for $C$. The averaging procedures used to obtain values for the variables are not subject to any criticism of a statistical nature; the dispersion of shots is not addressed. The short-arc method plays an unusual role; together with the associated differential variations, it provides the corrections to the observed ranges. In fact, the interweaving of computed trajectories and actual firings is quite intriguing. One may ask if the process as a whole is guaranteed to converge, or even terminate within the 5% difference desired. Above all, no logical justification is given for the algorithm; the meanings of the intermediate values obtained for the $X$ and $C$ are not described.

Nevertheless, this was one contemporary version of range table construction. Moulton's algorithm for computing the differential variations as presented in *New Methods* may indeed be the most useful part of that book, but it is at quite a different level than one finds here. Perhaps, the unanswered questions and the looseness of the procedure speak to much that which was still judgment-dependent in ballistics, calling into question the "exact" solutions to ballistic problems sometimes promised by the second wave reformers.[49] But the new critical attitudes of the reformers at least made them aware of the limitations of their tools.

One last remark can be made regarding the beleaguered ballistic coefficient. One finds in the writings of the second wave reformers a dichotomy in attitude toward $C$. The theoreticians, including Moulton, merely introduce it in the defining equations of motion of the projectile with remarks such as Bliss' in [385]: "...$C$ is the ballistic coefficient which enables the formula $vG(v)/C$ to be made to represent the retardation of projectiles other than a standard one under similar conditions..." [p. 298]. Again Gronwall: "...the ballistic coefficient $C$ depends on the weight and shape of the projectile" [491, p. 506]. After a lengthy discussion of the form that the drag law must take, Moulton in *New Methods* proceeds in his chapter on differential variations to set the law of motion in its usual form with the comment "...and $C$ is the 'ballistic coefficient', depending on the shape, size and weight of the projectile." In none of these writings is a dependence on the angle of fire mentioned, and one leaves with the impression of its constancy on all flights, though Gronwall considers its variation along with the other quantities. As one considers the works that were more closely addressed to students of table construction, this attitude changes. Thus, Jackson remarks: "The ballistic coefficient is constant throughout the trajectory, and in general has a fixed value for any one projectile, except so far as minor modifications may be necessary in practice to secure the closest possible agreement with actual firings at different angles of elevations and different velocities" [543, pp. 11–12]. Hoar states "$C$ is conceived of as an empirical function of the characteristics of the projectile, and is constant over any given trajectory" [522, p. 43]. By omission, it would appear that it can vary with elevation, and the algorithm described above shows that that is indeed the case. Thus, Moulton's criticism, mentioned earlier at the beginning of "Introduction", regarding the flexibility of this "constant" could still be made. However, in a thorough attempt to reconcile all possible uses and definitions of $C$, a 23-page technical paper on this matter by Bennett was appended to his ballistics book. Thus, all parties could be satisfied.

[49] The same criticism could be leveled at Moulton's short-arc method itself. Though thought by some to be the final word on solutions to nonlinear differential equations and proved convergent by Moulton, its application leaves room for judgment calls. Anyone who thinks the procedure is a razor-sharp method for cutting unambiguously to the solution of the equations of motion should read the account in [596], which describes choices in how to begin the process.

## 2.15 The Differential Variations: A Summary of All Uses

A great deal of attention has been paid in these paragraphs to the styles of expositions associated with the differential variations. But this attention is warranted not only because it highlights the changes, which ballistics in the United States underwent during the years of the War, but also the need to get the new ideas across to differing audiences. In the process one encounters mathematical and military journal articles, ordnance textbooks written at all levels, the first university level treatment of ballistics, as well as Hoar's highly individual book. These sources give differing interpretations of the variations: as disturbing factors in a real trajectory, variations in the defining functions or initial conditions of a system of differential equations, or solutions to a set of their own linear differential equations. Looking at the literature that adopts any of these interpretations gives a cross-section of the widely differing attempts to communicate to students, be they army officers, college students, mathematicians, or physicists.

It will be fitting to end this discussion, which has been devoted to the reception of the differential variations by the mathematical and military communities, with a list of the uses to which they were put by these parties. They have proved to be quite a multifaceted topic and versatile tool. They began as a theoretical addition to Forest Ray Moulton's account of the trajectory of a particle-like projectile, attempting to take into account secondary effects on range in a systematic way. Moulton was able to run with this idea by using the equations of variation more generally in his *Differential Equations* text, where their use in solving other physical problems was suggested. Moulton also advocated their use as a technical tool in constructing the massive general ballistic tables. They gave rise to table entries in the range firing tables, where they were part of the protocols for preparation of fire needed for a more comprehensive account of secondary effects. Philip Schwartz used them to make recommendations for testing conditions associated with determination of accuracy of a weapon. The differential variations were suggested as a means of dealing with the constant problem of the accuracy of the drag law $G(v)$: a change in this law could be accommodated by using the variations as a means of altering the range table values to correspond to a new $G(v)$. In the realm of theoretical ballistics, Gronwall was able to prove theorems about the ballistic trajectory itself using the variations. Hoar and his associates used them as a means of reducing the data of firing to standard conditions in the construction of range firing tables. They became the focus of attention in the quarrel between Moulton and Hoar, whose different emphases on the topic brought out the spectrum of educational options in ballistics in the 1920s. It remains to show how they influenced Bliss' investigations of the variability of the solutions to a general system of differential equations as the functions which define the system vary, how they fit into his studies of the function of a line, and the abstract ideas to which they gave rise. It is to these final topics that we turn in the next chapter.

## 2.16 Franklin's Derivation and the Calculus of Variations

Mention has already been made of the use of some notions of the calculus of variations in Franklin's derivation of the adjoint system. But the reader familiar with this subject will note the absence of the traditional format: the problem of optimizing an integral of a set of functions and their derivatives. The lack of rationale for the some of the steps is troubling as is the missing definition of the symbol $\delta$. Hoar does not dwell on the calculus of variations as such in his book: "The operator $\delta$ is the operator employed in that branch of mathematics known as the calculus of variations ... No further understanding of the principles and methods of this branch other than here given is necessary to the purposes of this book" [522, p. 20]. Franklin's style, however, mirrors that of the first publication of Lagrange on the calculus of variations, *Essai D'une Nouvelle Méthode Pour Déterminer Les Maxima et Les Minima des Formules Integrales Indefinies* [556]. This 1762 publication was available in German translation in an 1894 volume [649] of important papers on the subject. It is difficult to say if either of these sources would have been available to Franklin.[50] Lagrange is credited with introducing the $\delta$ quantity in this paper, whose title promises a new method of dealing with the classical problems considered by the Bernoulli brothers and developed further by Euler. The following is a capsule presentation of the initial section of Lagrange's work in English translation by the present author. The reader may be struck by the similarities with Franklin's calculations.

The first paragraph introduces a brief history of the relevant problems. His goal, not surprisingly for Lagrange, is to take the methods of the Bernoullis and Euler and replace any geometric considerations by a treatment in purely analytic terms. Lagrange states "Now here is a method that demands only a very simple application of the principles of the differential and integral calculus, but first of all I must warn you that, since this method demands that the same quantities vary in two different manners, I have, in order not to confound these variations, introduced into my calculations a new characteristic $\delta$. Thus $\delta Z$ will express a difference of $Z$ that will not be the same as $dZ$, but which will however be formed by the same rules; so that having an equation $dZ = mdx$ one will equally have $\delta Z = m\delta x$, and similarly for others." He then considers his first problem: letting $Z$ represent an arbitrary determined function of $x$, $y$, and $z$ and their differentials (French = "differences," German = "Differentiale") $dx, dy, dz, d^2x, d^2y, d^2z, \ldots$, he wishes to find the relations among the variables (including the differentials), which maximizes or minimizes the quantity $\int Z$. His solution is to follow the known method of maxima/minima in calculus: Apply his new "characteristic" $\delta$ (French = differentielle, German = Differential) to $\int Z$ and set the resulting expression equal to

[50] In Franklin's application for graduate work at Princeton University he reports that he can read "French and, (with difficulty) German." He may have had prior familiarity with the calculus of variations from reading material included in his college textbook on integral calculus [666], which included several chapters on this material.

zero. "Therefore indicating these variations by $\delta$, one will then have, for the equation of maximum or minimum,

$$\delta \int Z = 0 \tag{2.50}$$

or, what is equivalent,

$$\int \delta Z = 0.'' \tag{2.51}$$

No limits of integration appear on the integral, nor is anything said about them. This is all that is forthcoming with regard to the meaning of $\delta$, and there is no justification given for the interchange of the order of operations $\delta$ and $\int$. He then presents for $\delta Z$ an infinite linear combination of the variations:

$$\begin{aligned}\delta Z =& n\delta x + p\delta dx + q\delta d^2 x + r\delta d^3 x + \dots \\ &+ N\delta y + P\delta dy + Q\delta d^2 y + R\delta d^3 y + \dots \\ &+ \nu\delta z + \omega\delta dz + \chi\delta d^2 z + \rho\delta d^3 z + \dots\end{aligned} \tag{2.52}$$

where the coefficients will be allowed values, which involve the variables and their differentials, though this is not mentioned. Upon integration, $\int Z$ becomes

$$\begin{aligned}&\int n\delta x + \int p\delta dx + \int q\delta d^2 x + \int r\delta d^3 x + \dots \\ &+ \int N\delta y + \int P\delta dy + \int Q\delta d^2 y + \int R\delta d^3 y + \dots \\ &+ \int \nu\delta z + \int \omega\delta dz + \int \chi\delta d^2 z + \int \rho\delta d^3 z + \dots = 0;\end{aligned} \tag{2.53}$$

"But one easily understands that"

$$\delta dx = d\delta x,\ \delta d^2 x = d^2\delta x \tag{2.54}$$

Again, no justification is given for the reversal of the order of operations. There follows a group of integrations by parts:

$$\int pd\delta x = p\delta x - \int dp\delta x \tag{2.55}$$

$$\int qd^2\delta x = qd\delta x - dq\delta x + \int d^2 q\delta x \tag{2.56}$$

$$\int rd^3\delta x = drd\delta x + d^r\delta x - \int d^3 r\delta x \tag{2.57}$$

and similarly for the other variables. The equation for the optimization now becomes

$$\int (n - dp + d^2q - d^3r + \ldots)\delta x + \int (N - dP + d^2Q - d^3R + \ldots)\delta y + \tag{2.58}$$
$$\int (\nu - d\omega + d^2q\chi - d^3\rho + \ldots)\delta z$$
$$+ \text{(terms involving the boundary expressions)} = 0$$

From this, Lagrange concludes that the sum of the three integrands is zero:

$$(n - dp + d^2q - d^3r + \ldots)\delta x + (N - dP + d^2Q - d^3R + \ldots)\delta y + \tag{2.59}$$
$$(\nu - d\omega + d^2q\chi - d^3\rho + \ldots)\delta z = 0$$

He concludes also that the sum representing the boundary terms is zero and adds some explanation regarding how the limits of integration are to be understood in said equality.

It is at this point that a conclusion similar to that drawn by Franklin is stated, with some justification: Now, in order to free the equations obtained from the indeterminate 'differences' $\delta x$, $\delta y$, $\delta z$, $d\delta x$, $d\delta y \ldots$ one examines whether, by the nature of the problem, there is some given relation among them, then, having reduced them to the smallest possible number, one equates the coefficients of each to zero. If they are absolutely independent of each other, then [one has] immediately the following three:

$$n - dp + d^2q - d^3r + \ldots = 0$$
$$N - dP + d^2Q - d^3R + \ldots = 0$$
$$\nu - d\omega + d^2q\chi - d^3\rho + \ldots = 0.\text{"} \tag{2.60}$$

There then follows an application of this procedure to the brachistochrone problem in $xyz$ space, and consideration is given to the boundary terms, which are null also in the optimization process. The rest of this paper as well as another by Lagrange is discussed in [470], who both praises and criticizes its contents. He pronounces part of the proof discussed here as "curious" [p. 115].

Another possible source of ideas for Franklin is Kneser's volume [555] cited earlier. If one is interested in just basic manipulations possible with $\delta$, Kneser deals with these on pp. 4–5 of his book. Significantly, he begins with the case in which $x$ and $y$ are both functions of $t$, forming a curve $B$, introduces $\delta x$ and $\delta y$ as continuously differentiable functions; the curve $(x + \delta x, y + \delta y)$ is called $B^0$. He then describes a limiting process involving $\frac{d(x+\delta x)}{dt}$ and $\frac{d(y+\delta y)}{dt}$ that results in the conclusion that $\Delta x' = \delta x' = \frac{d\delta x}{dt}$, $\Delta y' = \delta y' = \frac{d\delta y}{dt}$. Thus the reversal of the order of these operations is proved. An argument for reversal of $\delta$ with integration is given at the bottom of page 5.

On page 5, the equation

$$\delta F = F_x \delta x + F_{x'} \delta x' + F_y \delta y + F_{y'} \delta y' \tag{2.61}$$

is derived, where $F$ is a general function of $x,\ y,\ x',\ y'$; this of course was the beginning point of Franklin's argument.

Thus, some of the ingredients in the memo appear early in [555]. Although running to more than 300 pages and having as its goal the presentation of the most recent research problems, in its first pages, it contains identities useful to the young accidental ballistician. Other contemporary texts have this material scattered throughout their pages, most downplaying the importance of the analogy of differentiation and variation. While the matter of Franklin's sources is not deeply relevant, it is worth examining the available literature on the calculus of variations to see different treatments and in particular the handling of $\delta$, whose meaning underwent significant changes in the years from 1900 to 1940. In the next chapter, we choose Oscar Bolza's 1904 *Lectures on the Calculus of Variations* [406] as a launching pad for a discussion of Bliss' work, a book in which rigorous definitions and theorems are the norm.

After his war work was completed, Philip Franklin took master's and PhD degrees from Princeton University and, beginning in 1924, spent the rest of his career at MIT. He figures in one other offshoot of the work on variations, to be described in the next chapter.

## 2.17 Some Naval Responses

Among the parties from whom one would expect a response to the ballistic innovations lies the US Navy, a sister service. Indeed, the Navy had its own tradition of ballistic study and practice and its adoption of the Siacci methods paralleled the Army's course of study with the volume [553], whose appearance in 1885 was nearly co-incident with Ingall's work. Successor naval works followed, leading to series of books by Philip R. Alger, a graduate of the Naval Academy at Annapolis and a representative of the "New Navy," a movement, which stressed new designs of steel ships and guns. Alger, who became a professor of Mathematics at the Academy, introduced his own version of the Siacci theory in a manner similar to Hamilton, in 1917 [344], though he authored several ballistics texts prior to this date without this material.[51] The emphasis here is on practicality, given the urgency of the ongoing war, and many references are made to the limited amount of time the midshipmen have to absorb new knowledge of any kind.

[51] A debt to Ingalls is acknowledged on the title page, where one finds "Revised and Extended to Include the Formulae and Methods of Colonel James M. Ingalls, U. S. Army."

Before turning to the introduction of the new methods, one might note that the Navy had military problems of a different nature to confront involving, for example, fire control protocols for battleships. The complexities of firing at a moving target from a vessel in motion demanded much technological innovation: new instruments, gyroscopes, rangekeepers, and onboard communication systems were introduced; see [530], Chap. 2 for a discussion of this material. Though culminating over time in concepts of feedback and control, whose mathematical expression eventually found form in Wiener's cybernetics, it appears that no new mathematical ideas were introduced at this time from these problems. On a technologic-mathematical note, mechanical integrators were used to calculate certain relevant expressions.

What follows is a sampling of material related to the reception of the New Ballistics by the US Navy, culled from articles in the *Proceedings of the U. S. Naval Institute* and a standard post-War exterior ballistics text. No claim to completeness is made, for a glance at the articles from the *Proceedings* shows that the Navy had its own versions of contemporary issues paralleling those already explored in the Army: officer training, the rise of the specialist, proliferation of new vessel construction techniques, and innovations of all kinds dating from before the turn of the twentieth century were all addressed. Some of these bore on ballistic concerns, but will not be mentioned.

On page 82 of Moulton's *History* [594], a two-page section devoted to "Relations with the Navy" begins. Moulton lists several naval officers who gave the reception of the new methods a "gratifying response." (Among these officers was one Lieutenant Commander Kirk, who expository work is treated below. Kirk was said to have taken up the new approaches "with enthusiasm.") Beginning with material on long-range trajectories, the blueprinted reports from Moulton's office were sent to Navy Ordnance, and Naval personnel were sent to Moulton's office to see the production of results firsthand. The process was reciprocated as the Navy supplied the Army with results of its own experiments and designs for new instruments, including several papers by Kirk.

A series of post-War papers in the *Proceedings of the U.S. Naval Institute* tells an interesting story of the reception of the new ideas. The first, [442], by Captain E. F. Eggert of the naval Construction Corps,[52] was an effort independent of the stream of events involving Moulton, Kirk, and the War effort in general. According to a follow-up article [443], Eggert was studying armor protection of vessels and needed an answer to a specific problem: in a comparison of the performance of a 14-inch gun and a 16-inch gun, "What I wanted to know was whether the 16 inch gun was really superior to the 14 inch gun at a given long range, what the approximate difference in range, for a given elevation was, and what were the final velocities and angles of fall..." [443, p. 1761]. Working in an unreported but apparently isolated location, he obtained a copy of the appropriate 16-inch gun tables constructed by the Alger method used in [344] and attempted an extrapolation

[52] This agency was responsible for supervising the construction of new ships and inspection of existing ones.

from the data, but this process yielded no useful information. He was familiar with "mechanical integration" (his term for numerical integration) "from study in other directions" and over an extended period of time was able to apply it to the trajectory problem. He finally reached satisfactory conclusions, which answered his questions and simultaneously showed the inadequacies of Alger's methods, especially for long-range fire. "I determined to publish the results, more to indicate the possibilities of mechanical integration, and to call attention to the great errors in the old method, than to claim finality of precision" [443, p. 1762].

One sees here some interesting details: that the mechanical integration technique was readily available to interested parties, as has been stressed in the present work, not an esoteric topic in mathematics. Another noncombat use for the range tables finds expression, namely, their use in problems of armor protection, a problem in terminal ballistics. (One might assume that under these circumstances calculation is preferable to experiment for deriving the desired conclusions.) And a fine independent validation of the shortcomings of the Siacci method and introduction of a remedy was given, this time by a member if the military. As we give a short summary of Eggert's paper, a final observation will be some examples of the "rough, old, formal" ballistics, the use of which caused so much trouble in long-range trajectories.

Eggert begins by devoting a section to demonstrating the inadequacy of Alger's methods at long ranges, tracing the difficulties to two variables which are treated, as an approximation, as constants throughout the trajectory. One of these constants was introduced to account for the density of the air along the trajectory; this was taken by Alger to be the value corresponding to the mean height of the trajectory. Alger assumed that on low angles of departure the trajectory was a parabola of vertical axis with the mean height equal to two thirds of the maximum ordinate. By replacing these constants by more accurate values and recomputing, Eggert concludes that the older methods give, for the 16-inch gun, ranges that are too long by from 50 to 2600 yards. "These corrected values [of the constants] cannot however, be predicted, so that Alger's method cannot be used for longer ranges" [442, p. 75]. Pointing out that no such assumptions are needed for mechanical integration, he then sets out the assumptions for his method and the method itself. (One of his assumptions is an older resistance law than that used by the army, due to Mayevski.) His plan does not use Moulton's equations, but proceeds by a straightforward consideration of the velocity components and their changes over short intervals of time. A computation sheet is given for the method, and an example worked out; the results of several other computations are given, as are results for varying interval lengths. The final page lists his practical conclusions, a nearly final one reads "At ranges above 30,000 yards, any one of these shells [with varying weight and coefficient of form specified] will easily penetrate the armored decks on any ship afloat, and at such ranges the decks offer a target roughly ten times as large as the side armor above water" [442, p. 84]. Throughout this paper, neither Moulton nor Bliss are mentioned. One is reminded of Alston Hamilton's pioneering attempt among the Aberdeen documents, which ultimately found no follow-up.

One result of Lieutenant Commander Kirk's "enthusiasm" with respect to the new ballistic methods of Moulton and Bliss was his 1919 article [551], which consisted of an introduction of these methods in the context of a point by point comparison with Eggert's scheme, unfortunately often to the detriment of the latter. Kirk points out that Eggert's work is "essentially similar" to the army Ordnance methods and gives kudos to Eggert: "And it is gratifying to find that from within the regular navy should come a scheme whose general principles are identical" [551, p. 1375]. But the compliment is followed by a qualification: "Major Moulton's method, however, has several refinements of detail so that it will be of interest to reproduce it. Likewise, Professor G. A. Bliss, at the Aberdeen Proving Ground, has developed a system of differential corrections to the undisturbed trajectory. . . From these. . . the calculated trajectories can be corrected accurately for all variations from standard conditions-muzzle velocity, angle of departure, wind,. . . " [551, p. 1375] (Presumably, this was mentioned to contrast with Eggert's consideration of the density of air only.) Kirk then begins his exposition of the work of Moulton and Bliss.

The only significant difference presented in the assumptions is the use of the more recent Gâvre resistance function. In execution, the main improvement pointed out is the use of second differences rather than Eggert's use of first differences only; Kirk claims that, at least at the beginning, significant errors could creep in without the former. The trajectory computed by Eggert is then computed using Moulton's method, and the results declared comparable, though Kirk adds "The two short arc computations would scarcely show much variation with so large a ballistic coefficient and at so short a range. It is, however, in the field of providing suitable means to correct a disturbed trajectory that Major Moulton's method has its great advantage" [551, p. 1384]. The next section begins a presentation of Bliss' method, used to account for six possible variations. Bliss' integral formulae are presented without derivation, and about ten pages are devoted to the practical implementation of the differential corrections.[53] In the final section is given an application to the same 16 inch gun considered by Eggert and himself, and a reference is made to how one may use this new method to amend the older Siacci theory.[54]

After such an introduction, one would expect a reply from Eggert, and one did come in the form of [443], from which the information on the origins of his interest in mechanical integration was taken.[55] He wishes to say a few words in criticism of Moulton and Bliss' technique, "even though it was developed by a

[53] Among these considerations is found the assertion that "it is necessary to know what the wind is doing in the upper atmosphere. This information was obtained daily on the Western Front and its source and distribution became a highly organized service. A special name was given to this data, and it was called 'Sondage"' [551, p. 1390].

[54] Final statement: "It is not deemed probable that the method has any direct application to shipboard conditions" [551, p. 1395].

[55] The irritation felt by Eggert must have been exacerbated by his note of the presumably inadvertent inclusion, following his article, of a "special statement [by the Board of Control], generally to the effect that during the war it had been obliged to publish practically any material

professional astronomer, of long experience, and with a large staff, unlimited time, and great experimental facilities available" [443, p. 1762]. A list of objections follows: Eggert recomputes the recurring 16-inch gun trajectory, this time using the Gâvre function, and finds a difference of only 21 yards out of nearly 20,000 between the range calculated his way and by Moulton's method, Eggert not including second differences. He also points out that Kirk had not used a varying value of the gravitational constant $g$, and that using second differences does not significantly alter the trajectories of small arms, which entail smaller ballistic coefficients. He declares Moulton method of short arcs to be more cumbersome, taking nearly twice the time as his own. Additionally comes the point that "The apparent precision is hardly worth while, since we all know that, after making all possible corrections, we still have a mean dispersion in range, on actual firing, of from 40 to 80 yards, and we know also that actual measurements of ranges, when made by different methods, vary by 10 to 20 yards" [442, p. 1763]. As for the Bliss formulae, it is considered an "awkward complication," and a simpler method is proposed for variations of any sort, "This correction can be determined in a minute, with a slide rule" [443, p. 1763]. Bliss' treatment of the wind is considered inadequate, in particular since no account is taken of drift of the projectile. As a final general remark about precision, Eggert states, after giving some concrete examples: "In discussing precision of calculation, the precision of measurement must be a guide, and there is no use in carrying precision farther in calculation than in measurement" [443, P. 1764]. This comment was made in the context of errors committed in numerical integration as opposed to measured variation, but serves as a general principle not stressed by Moulton and his associates.

Leaving aside the awkwardness of Kirk's introduction of the new methods, Eggert's reply is a valuable and sobering rejoinder to the occasional bombast with which these methods were presented. Much of the virtue of the new material was its emphasis on precision, to use Eggert's final term; certainly, the presence of abstruse mathematical equations could belie the difficulty in applying the formulae in the presence of such disturbing factors as measurement and dispersion. The importance attached to the magnitude of difference made in using a newer resistance law is not dealt with by the "New Ballistics," except perhaps in theoretical work of Gronwall; here, Eggert provides one answer: let precision of measurement be the guide. Other sources of "variation" mentioned by Eggert include such mundane but genuine phenomena as errors in tables and calculation itself, how do these compare with real variations in the trajectory? The fact that these objections were raised by a military man with no advanced training in mathematics but much knowledge of naval vessels is telling. He is not so quick to accept the Moulton/Bliss treatment of the wind corrections, for example: "We are still far from knowing the effect of the wind, and we still have much to learn by experiment. . . .until we can make projectiles that are and remain dynamically balanced on their axes, and that rotate in exactly the same

---

received, however worthless. I should regret it very much if my article could not rise above that mediocre plane. . . " [443, p. 1761].

way, we cannot hope to solve satisfactorily the question of either wind or drift, which are closely related" [443, p. 1763]. This exchange may be viewed as a complement to the Moulton/Hoar debate, with the onus now falling on the Kirk to justify the elaborate mathematical machinery brought to bear on trajectory computation.

In 1920, Kirk published [552] in the *Proceedings*, a straightforward introduction to the new methods "originally prepared for the midshipmen at the Naval Academy" [552, p. 1778]; a debt to Dunham Jackson's textbook and the introduction to Bennett's ballistics tables was acknowledged. Moulton is credited with the introduction of "short arcs," and the last two pages are devoted to Bliss' "ingenious system of differential equations" for handling variations. No technical details are provided, evidently unnecessary due to Jackson's work. "Thus it will be seen that this new method of computation is more exact; free from mean values of constants; capable of applying more accurate corrections for variations from standard conditions; and able to compensate for more variables" [552, p. 1784]. Eggert's work was not mentioned.

Two final articles from the *Proceedings* describe an attempt at novel application of mechanical integration and introduce an interesting new character. They are both authored by Reverend Alan S. Hawkesworth, described as a "Mathematician" working at the naval Office of Ordnance, and a Fellow of the Royal Society of the Arts. Information on Hawkesworth is difficult to obtain; the most abundant source is [432]. He is described as a "mathematician and clergyman," producer of "100 new theorems in geometrical conics," a "member of the Fourth International Mathematical Conference in Rome," a "perpetual fellow of the Circolo Matematico," and an "ordained deacon of the Protestant Episcopal Church." He served in various religious capacities in posts worldwide and wrote several theological works but was employed by Naval Ordnance during the period 1917–1923. The circumstances surrounding this last position are unknown, as is much detail of his mathematical education. A cursory search of his publications yields several articles on conics in the *American Mathematical Monthly*. His contributions bring to mind the efforts of Roger Sherman Hoar, the mathematically educated intermediary between the academic world and the classroom of military officers, though Hawkesworth's efforts were not addressed to students.

The first paper, from 1921, is devoted to "Underwater Trajectories" and opens with the following question: "As a concrete case, were a 6-inch shell-weight 105 pounds, coefficient of form .61, muzzle velocity 3000 ft/s-fired horizontally into a water tank 6 feet above its floor, how far will this shell travel under water before it strikes the floor of the tank?" [509, p. 207]. Additional information about the trajectory is also desired.[56] His method of solution uses formulae similar to those of Siacci, with a modified underwater ballistic coefficient, and employs an "empirical

[56] In a footnote following the title of the article, one finds a comment by the Bureau of Ordnance to the effect that the article has been checked by the Bureau and has valid assumptions and procedures, but "The conditions chosen are such as to make the conclusions of interest mainly from a theoretical point of view ... the effect of entry into the water at varying angles of impact found in practice is not considered" [509, p. 207].

formula," derived at the Indian Head Naval Proving Ground, for distance traveled under water. The answer being obtained, he goes on to state that "...a similar correction[in ballistic coefficient] will make available, for this purpose, the exterior ballistics method of Moulton and Garnier [a French ballistian]" [509, p. 212]. He adds that variations in certain factors can be accounted for by adjusting the ballistic coefficient by the old school methods and concludes by providing a simple method of adapting his work to the entry problem footnoted by the editors.

The article was severely criticized in a Discussion comment by J. J. Arnaud, the Master Computer for the Army Ordnance Department [348], who cited "manifest errors in theory in that article [which] compelled the discarding of Dr. Hawkesworth's methods" [348, p. 1417]. The rationales for rejection are less important than the revelations that Arnaud at this time was engaged in determining the time required for "drop bombs," striking the water at a given velocity, to achieve a given depth; this problem drew Arnaud to Hawkesworth's work. The former proposed an altered version of Moulton's basic equations of motion suitable for underwater travel and stated other details of his techniques. "The subject is now being investigated by the Technical Staff of the Ordnance Department, U. S. A." [348, p. 1418]. Hawkesworth gave a rejoinder to this criticism, assuring that the correct equations were now in the possession of the Navy Ordnance.

In considering this new debate, the details of validity of methods and assumptions, while of course important, matter less in the present account that the evident fact that in the post-War world extensions of numerical integration to other settings took place, in spite of flaws in attempts to do so. The fact that Hawkesworth's paper had fallacies is undebatable, but the presence of a mathematical "amateur" attempting to make a contribution to ongoing research is evidence of a vanishing phenomenon, as discussed previously in the case of Hoar and Wheeler.

An additional attempt at a contribution, also dating from 1921, was Hawkesworth's "New Ballistic Corrections" [510]. In this paper, motivated by the longer trajectories achieved by the Paris canon, five corrections to the ballistic trajectory are treated. Some of these had been considered previously under the category of rotation of the earth and curvature of the earth, though the author erroneously asserts that they had not. Others, such as a varying gravity coefficient due to location of the projectile, were not prominently featured in earlier work. The methods used involve elementary applications of trigonometry and cartography as well as material on conic sections, largely an exercise in geometry in which physical phenomena are neglected. Other versions of Moulton's equations of motion of a projectile with a modified gravitational constant for each correction are used. This effort may be viewed as similar in nature to Philip Schwartz's essay on a similar topic in "Much About Little" [639], another piece with limited mathematical scope, which tries to demystify the formidable mathematical equations derived by Roever and Moulton, though these are not mentioned by Hawkesworth. The work of Hoar's study group from the late 1920s on this topic may also be mentioned in this respect.

We conclude with a brief account of the basic post-War exterior ballistics text used by the Naval Academy, *Exterior Ballistics* by Ernest E. Herrmann[514].[57] Herrmann had an impressive resume prior to his first attempt at a text in 1926; he was a combat veteran in the War and served in the Yangtze campaign. He became an instructor at the Naval Academy in 1929, a position he held on and off in the following years. The text went through two revisions in 1930 and its final version in 1935; the latter will be discussed here.

In the preface, the author sets the tone for the book by acknowledging that the volume was created in awareness of the limited amount of time allotted to the subject in the Academy curriculum (recall Schwartz's criticism of the status of ballistics in the army schools) and of the mathematical preparation of the students. The material is restricted to the newer methods, the Siacci style being considered "obsolete"; the former however is treated in an appropriately elementary manner. A novel reason for learning the methods is given: "Also, the possibility of having to use a ship's guns in a manner not provided for by available range tables should not be overlooked. Such instances have occurred in the past. Complete mastery of a ship's armament, under any circumstances that may arise, demands the ability to prepare firing data from the ground up, if necessary" [514, p. 2].

Moulton's method is given for undisturbed trajectories, with references to Jackson's and Moulton's books given for more mathematical details and further examples. There are many references to Cranz and Becker's volumes [426] as well, though none to Charbonnier's. Limitations on the accuracy of the examples due to drift and spinning are frankly acknowledged. The bulk of the book is taken up by material on variations. Each such has a column entry devoted to it in the standard form for a range table. For some columns, the "direct method" of simply computing a trajectory with a different value for the variable in question, say, initial velocity, is used, in spite of the effort entailed in doing so. In a footnote at the beginning of the variation material, the notion of differential variations produced by partial differentiation of the fundamental equations of motion is verbalized, but no derivations are given; the reader is referred to Hoar's book and the other already mentioned sources. A combination of formulae derived in this manner as well as alternate elementary considerations fill out the remaining columns. One previously untreated naval variation occurs: effect of motion of the gun itself and motion of target; changes due to other practical effects are discussed, for example, the effect of firing from a cold versus a warm gun. Old standbys like the rotation of the earth display the use of a reduction of the results of the mathematicians to approximate expressions, and the source is a note attributed to Gronwall.[58] All in all, one may characterize the book as containing the essential formulae and simplified techniques for range table construction with no mathematical elaboration; such derivations are given by cited sources. A glance at the Bibliography reveals citations of the

[57] Moulton's *New Methods* was listed under "Books" in the *Proceedings* of 1927; no review was given.

[58] The technical report cited is missing from the appropriate collection.

above mentioned works, as well as Eggert's paper, Hitchcock's massively detailed computation manual, but none of Hawkesworth's work or Bliss' material.

In conclusion, one may say that the Navy adopted the New Ballistics and created its own expositions of it in the *Proceedings*, but the considerable amount of existing material both in the early post-war years and later generated by the army made further work of that nature unnecessary. The pioneering work of Eggert shows how numerical integration could be embraced by a naval officer in the absence of any high-level expositions and highlights ongoing struggle between theory and practice by using practical examples. The introduction of the New Ballistics took place with simple declarations of its advantages over the Siacci and straightforward adaptations of its results. Absent was any claim to its higher status due to a greater scientific, mathematical, or otherwise rigorous standards, claims which characterized the work of Moulton and his students. With few exceptions, one does not encounter original elaborations of ballistic methods in the usual printed sources during this time.[59]

## 2.18 A Comparison With Continental and British Ballistic Efforts

It is perhaps natural to ask at this point how the theory and practice of American ballistics, so far as it has been discussed in these first two chapters, compare with those of the other participants in the World War. An extensive contrast is out of the question, but can any interesting points be raised?

Instances on borrowing and influence among the parties are easy enough to find. Extensive range firings to collect ballistic data have already been mentioned with regard to the first wave of ballistic reforms; the United States used data from the Krupp firings at Meppen.[60] Textbooks on ballistics by authors from other countries were freely used by Ingalls and Hamilton; the Siacci method prevailed in all armies prior to the World War. Around 1916, as new methods of trajectory computation were being developed in England and France, pamphlets devoted to these methods began finding their way to America in an effort to help the ally catch up.[61] [360, p. 38]. In [358, p. 4], one finds a description of how the air resistance function $F(v)$, to be used in the large post-War general ballistic tables project, was that developed by the French at Gâvre, based on range firings conducted there in 1888 and followed by the development of explicit smoothing approximation formulae. A post-War paper of great importance to the design of projectiles [454], which will be treated in Chap. 3, was published by English ballisticians working at the proving grounds of H.M.S. Excellent. The volume [426], whose co-author Carl Cranz was

59 One exception is [664], not mentioned in Herrmann's bibliography.

60 A list of important firings in various countries can be found in [358, p. 16].

61 [613] is an example of one of these, owned by the Military History Institute, Carlisle, Pennsylvania.

Professor at the Technical Military Academy at Berlin-Charlottenburg, stood as a classic work in ballistics in both pre- and post-War editions; it was cited frequently in articles in JUSA/CAJ, and its presentation of probable error was taken as the basis for that given in 1924 *Gunnery For Heavy Artillery*, [494]. In [550, p. 204], one finds that there existed a policy among the allies in the World War to make their ammunition interchangeable. This list could easily be lengthened.

A beginning for comparison can be made by summarizing relevant contents of Veblens' report on his post-War European trip. Its purpose, evident from its title, was to survey the existing state of *Ballistic Theories and the Preparation of Range Tables*. [660] Veblen found support among the French and British for the abandonment of Siacci methods, though the realities of the War made newer methods difficult to implement. New tables had already been prepared by the French, and it was suggested that the Americans adopt them until able to produce their own. "Senator" Volterra[62] (the mathematician Vito Volterra) also approved of the French tables, though the Italian services continued to extend Siacci's theories to higher angle fire. Volterra reported that Jacques Hadamard was attempting to study the resistance law from an experimental view, though no publications resulted as yet.[63] Veblen offered the opinion that Moulton's numerical method, which "is obtained by adapting methods already known in astronomy" seemed the "most expeditious" [660, p. 2] of the methods adopted by the allies. In the realm of differential variations, it has previously been noted that, according to the report, "the most satisfactory general method" was Bliss', and the importance of another source of variation was revealed in the following story, which has since remained to some degree in the popular consciousness:

> There is a report current among the British to the effect that the defeat of the British in the battle of the Falkland Islands was due to the fact that the effect on drift of the rotation of the earth is different in the southern hemisphere from what it is in the northern, and that allowance had not been made for this difference. The theory of this effect has been studied by Major Moulton in Washington and by Dr. Roever at the Proving Ground.

"In view of the fact that in the future, more, rather than fewer, corrections [for variations from standard conditions of fire] will undoubtedly be required, attention [was] invited..." [660, p. 6] to a discussion by Lieutenant [R. H.] Kent, an American, on the variety of opinions among the allies as to how to effectuate these corrections. The consensus leaned toward having weather stations relay the data directly to the battery, where calculations would be performed. A French officer displayed a new table to facilitate such calculations and worked out an example in detail.[64]

---

[62] Volterra was named a senator of the Kingdom of Italy in 1905.

[63] Hadamard's work on the Cauchy problem in linear partial differential equations and its possible relation to Bliss' research will be discussed in the next chapter. Volterra himself also played a role in Bliss' thinking, which will be explained.

[64] Recall the discussion of this issue in the Preface, as well as later American textbook attempts to accomplish this feat by establishing step-by-step procedures.

The form that American firing tables should take was discussed at length, with samples of British, French, and Italian variants submitted with the report, the majority consisting of new French tables computed by short-arc methods. The account concluded with a few paragraphs devoted to the air density and air resistance laws. It was noted that the present law of air resistance, including the favored French law mentioned above, was based on firings with projectiles of obsolete types. Attention was paid to the mathematical form of the law; in the process, one discerns the evolution from simple monomial dependence on the velocity to monomials of varying degree for different ranges of velocity arranged in a continuous curve, to a smoothing of the last by a well-fitting single expression, as done by the French. The desire for smooth, single expression was driven by the awkwardness of inevitable "corners" in any piece-wise law for computation of trajectories by short-arcs. A recent, possibly superior, English law was discussed. Similar computation-driven issues arose with the air density law, which had certain mathematical advantages in simplifying the work involved. "... it will have to be decided whether to use an exponential law of air density as has been done by the French and is favored by the British officers who have been consulted, or whether the exponential law should be replaced by one which is closer to the facts of nature" [660, p. 11].

It is thus fair to say that, by the end of the War, these parties stood in roughly equal positions, with one or the other possibly lagging behind in some area. In Norbert Wiener's words: "we [Americans] accomplished the task expected of us... we Americans were behind neither our enemies nor our allies..." [665, p. 256]. But this is not to say that they always had such a status, as should be evident from "Introduction". It is instructive to ask how the ballistic practices compared in years prior to the War, since some features of the American evolution are best revealed by comparison with those of another country. Again, consideration of all countries and practices is impossible. Here, we choose France as a specific country for comparison. This choice is made for several reasons; to begin with, accounts of some French contributions to the American ballistic effort through the years have already been given, such as the retardation law developed at Gâvre. More materially, certain guns, like the 75 mm Gun, Model 1917, of British construction, fired both French and American ammunition, thus tables covering both cases had to be computed. [594, p. 39] Proper Charbonnier, whose words were used to describe the relation of exterior ballistics to astronomy, acted, in a sense, as spokesman for French mathematical War efforts. He was also a prolific writer on histories, theories, and practices of ballistics in his country covering all eras. These writings, admittedly tending toward the theoretical, were familiar to some members of the American military, as will be shown below. He wrote a classic multi-volume treatise on exterior ballistics, the first edition appearing in 1907, with a revision spanning the 1920s. [419, 420] This treatise, not without its American critics (including Moulton), was a touchstone for discussions of the subject. His histories of the Gâvre invite comparison to American institutions and their role in ballistic developments. Other prominent French figures appear in his accounts. Though no individual should be the source of all relevant information, he offers a guideline one can follow.

In a communication to the 1924 International Congress of Mathematicians in Toronto [418], Charbonnier gave his take on the role of French ballisticians in the recent War. Addressing mathematicians, his comments were largely directed to the theoretical aspects, but some interesting historical issues emerge. He began by claiming that, prior to the War, many artillerymen had little appreciation for ballistics due to the fact that their chief weapon, the French 75 mm, was a precision instrument designed for direct fire at short distances, a scenario that excluded the need for ballistic science. The War brought long-range fire with its attendant problems. His institution, the Gâvre, founded in 1830 as proving ground for naval artillery, had always had to deal with this type of issue and could without much difficulty adapt their methods to firing on land.[65] Noting that the press had ridiculed the theoreticians for their alleged ignorance of, for example, the law of diminishing air density with altitude, a fact that accounted for the success of the German "Berthas," these circumstances had in fact been anticipated by those mathematicians. Citing the work of the (unnamed) ballisticians of the eighteenth century, and of de Saint Robert[66] from around 1860, he pointed out that this problem had already been analyzed. Furthermore, the short-arc method adopted in 1887, together with firings from 1892 at Gâvre (where initial velocities of 1200 meters per second were achieved), and use of a variable density law had allowed the calculation of trajectories for which the projectile passed easily over "Mont Blanc and the Pas-de-Calais." "The fire of the Berthas, from the point of view of exterior ballistics, does not represent anything unexpected, as long as we admitted the possibility of realizing initial velocities on the order of 1600 meters per second" [418, pp. 315–316].[67] Later pages were devoted to contributions of French ballisticians at Gâvre to the variability problem, making references to familiar names, Poisson among them.

His opening paragraphs thus present a long-established institution devoted to both theoretical and practical ballistic matters, adapting well-established mathematical techniques, available science, and routine firings to at least anticipate the major new offensive weapon of the enemy. Exactly how the adaptation of naval techniques to land firings played out was not discussed, but the contrast with the emergency-based response of American mathematics community circa 1917 is evident.

The Gâvre and its historical role in French exterior ballistics was the subject of a 1906 paper by Charbonnier from the *Revue Maritime*, which appeared in translation in two parts in issues of JUSA/CAJ in 1907 and 1908. [416, 417] These articles flesh out the interplay of theory and application alluded to in the 1924 address, an interplay not achieved in American ballistics to any serious extent until the War. It was exemplified by the collaboration of two men: Felix Hélie

[65] This was the reverse of events in America, where, as explained by Moulton in his *History*,his office initially had no contact with the Navy, and an effort had to be made to establish it, especially in light of the similarity of their mutual problem, as noted by Charbonnier. [594, pp. 82–83]

[66] Paul de Saint Robert (1815–1888), was an Italian scientist who made contributions to several disciplines.

[67] Translation by the present author.

(1834–1885) and Pierre Hugoniot (1851–1889), the first "a superstitious adherent of experiment... [who] possessed a reasoning mistrust for theory" [416, p. 239] and the second "an incomparable mathematician, a remarkable physicist, a broad theorist" [416, p. 240] whose work was just beginning to be studied by Hadamard. They co-authored *Experimental Ballistics*, a book whose second edition appeared in 1883,and which became a landmark study used for many years by French ballisticians. Formulae used by these authors in the study of trajectories were also developed, based on "experiments" conducted in the 1830s and 1840s. The fact that firings were referred to as experiments underlines the scientific attitude of the enterprise at the Gâvre, where each firing was subject to explicit rules.[68]

Charbonnier characterizes this era, in spite of its collaborative aspect, as being dominated by the experimental attitude at the expense of theory, citing lack of relationships between the variables used, the encroaching new artillery developments, which invalidated some of the old formulae, their increasing disagreement with the performance of even existing pieces, and inadequate methods of computing described quantities. A crisis of sorts occurred around 1884, forcing a rethinking of the approaches used and resulting in an uneasy compromise described in detail [416, pp. 245–250]. "However the science of artillery was daily making fresh advances, dragging after it a despised ballistics of tortured formulas, which was no longer fit for any but the most ordinary needs." One way out of this crisis was presented in the volume of Héle and Hugoniot of 1883, which contained an adaptation of Euler's short-arc method, used by him for the quadratic power law of resistance, to any law of retardation dependent on velocity. Gâvre adopted this method in 1887, and it became the basis of computation for all range tables from then on, as well as providing a boost to the theoretical studies: "In thus deliberately consenting to the tedium of years of long, wearisome, and inelegant computation of trajectories by arcs, the Commission thoroughly realized the practical consequences of its emancipation from empirical methods, of its return to the natural sources of ballistic science, namely, the possibility of computing *a priori* any trajectory of any gun; reduction to the lowest limit of the amount of practice needed to establish a range table ... determination of accurate value of non-observed elements of the trajectory" [416, p. 252].

It is important to emphasize that Charbonnier attributed to the short-arc method not only a tool for calculating any trajectory, thus making the construction of range firing tables easier and reducing the number of firings to those needed, for example, for the ballistic coefficient. He also saw it as a boon to theory. "To all the Gâvre officers the method of trajectory by arcs appeared provisional and incomplete, a process of computation and not a theory; individual attempts to substitute for it something which should be more satisfying to their intellect, which should be an

[68] 1. Never fire an unnecessary round (a repetition, or one whose results could be anticipated by theory). 2. Never fire a necessary round without having first endeavored to foretell by theory all the experimental results to be expected. 3. Perform all necessary experiments under the most severe conditions, that is, in such a way that a disagreement between theory and experiment may have the fullest opportunity to manifest itself. [416, p. 243]

explanation of ballistic phenomena, consequently became very frequent..." [416, p. 25]. And: "It is truly a vexation of the spirit to realize that numbers seen unrolling indefinitely under tabular form have definite mathematical relations with one another and not to know what these relations are" [416, p. 258]. Charbonnier's account concludes with attempts to use the computed trajectories to guide the production and analysis of relevant differential equations.

There are several contrasts between this history and that of the American development of the science. Steady government support from the early decades of the twentieth century and cooperation with scientists and mathematicians allowed for in-depth study of ballistics in these years, a depth again only achieved in America beginning at Aberdeen. The introduction of numerical methods appeared as a use of a self-produced tool, after years in possession, to resolve a developmental impasse in a science, with twin benefits to theory and practice. The method had roughly a quarter of a century to be absorbed before its use in the World War. Again, by contrast, Moulton's activities took place under extreme pressure, and the luxury of study of computed trajectories for theoretical purposes occurred only well after the conclusion of the War. Perhaps, most obvious, ballistics was considered, at least by Charbonnier, as a science first,[69] not as a military activity subject to scientific study; many of his references attest to this. He discusses "natural philosophy," "experiment," "rational mechanics," "empiricism," quotes Francis Bacon on philosophy, and cites Riemann, Newton and Euler often. "...exterior ballistics offers a very clear example of the way in which a problem in natural philosophy can be reduced to a simple question of rational mechanics..." [417, p. 52] begins one section.[70] Attention to these issues does not occur in American ballistic discourse at this time; Moulton comes closest to discussing them. It does, however, provide a framework for his division of exterior ballistics in to primary and secondary problems, the trajectory and those factors giving small alterations to it, again in analogy with astronomy. The details of the above stated reduction follow immediately after the given quote; the road leads to the principal equation of motion (2.183.1) in "Introduction".[71]

Regarding the solution of these problems and the mathematics used therein, Charbonnier spared the military readers many details. After stating the equation

---

[69] A reference to Hadamard's work [495], a work that occurs in the next chapter, is an example of this attitude.

[70] The reduction follows the lines: "The only problem which exterior ballistics treats is the *study of the motion in the air of the oblong projectiles of modern artillery.* [emphasis in the original] This leads to the principal and secondary problems, the principal being *To investigate the motion of a HEAVY MATERIAL point in a MEDIUM OF REST of CONSTANT DENSITY which opposes to it a TANGENTIAL RESISTANCE, itself a function of the VELOCITY of translation. The earth is supposed to be FLAT and MOTIONLESS; Gravity is CONSTANT in MAGNITUDE and in DIRECTION.*" [417, pp. 52–53]

[71] This kind of subdivision extends even to interior ballistics: "In certain sciences, in interior ballistics for example, the number of secondary problems will be so great, and their study so complicated and so little advanced that no account can be taken of them except as a whole" [417, p. 53].

he does comment that its solution forms a curve called the "hodograph," a function of the variables $v$ and $\theta$, finding the hodograph becomes "the true key of ballistic researches" [72] [417, p. 56]. Quoting a 1904 paper of Liouville regarding the type of the differential equation, he admits that the general solution is unknown as of today, "and even if it were, it would doubtless not be, in the actual state of analysis, of any use to ballisticians" [417, p. 56]. Two comments may be made here: first, any classical analysis of the principal equation of the kind mentioned here by Charbonnier was alien to American ballistics; even Gronwall never dealt with it, and examination of the hodograph, so important in Charbonnier's writings, was similarly avoided. Second, we note that Charbonnier, though largely theoretical in his orientation here, had limits as to what he considered relevant to ballistics: regarding the monomial forms of the resistance laws, for example, he remarks "But all researches of this sort tend to stray farther and farther from real ballstics and to retain only the character of exercises in pure analysis" [417, p. 62].

We finish the account of this JUSA/CAJ paper by noting that the secondary effects, most relevant to the differential variation issue, are sequentially summarized. Curvature of the earth: correction easily made; rotation of the earth: classical problem solved first by Poisson, Saint-Rober's solution preferable; variation of the density of air with altitude: solution by series available; atmospheric wind: solutions available by General Didion[73] (geometric) or Siacci (analytical); drift of oblong projectiles: some analysis of questionable use available, theory not well developed yet. As of 1907, then, there were at hand at the Gâvre mathematical solutions of the problems, which occupied Moulton and Bliss, and one that did not: drift.

A more complete account of the differential variations could be found in a "book" (section) of the second volume of a treatise [415] authored by Charbonnier appearing the same year as this JUSA/CAJ paper.[74] Here, the above secondary effects are attacked with full mathematical force. Choosing the density of the atmosphere as an example (recall the criticism a decade or so later of the Gâvre ballisticians for their neglect of this factor), Charbonnier treats the case of constant density first, then, in a move differing from Moulton, considers the addition of a variable density function itself as a secondary effect; after much discussion of the form of such a law, he settles on $(1-hy)^m$, for $y$ the altitude and $h$ a small constant, $m$ a positive constant. An analysis of the resulting principal equation with this factor added ensues, and a collection of "theorems" follows. One such reads: "On the descending branch of the trajectory, there exists, after the minimum of speed, a point where the speed of the projectile passes through a maximum."[75][415, p.

[72] In an interesting foreshadowing, Charbonnier in [418, pp. 319–320] called attention to an "very curious" machine, designed by a captain, which calculated general trajectories for any given air density law by mechanically integrating the principal equation.

[73] Isidore Didion (1798–1878), French ballistician.

[74] This treatise was cited at the end of 1919s article [435], in a "Minimum List of Books On The Mathematical Theory of Ballistics" compiled by Gronwall, Bliss, and one "Professor Haskins."

[75] Translation by the present author.

28] (Much importance was attached to this result in the Toronto address, as the phenomenon described here appeared in the calculation by short-arcs of Bertha-like trajectories.) Effects of this factor on ascending and descending branches of the trajectory for different types of fire are analyzed in depth; many diagrams of the resulting hodograph appear. Other factors are treated similarly, drift in particular is analyzed extensively.

Despite the heavy-going mathematics, the book was anonymously reviewed in JUSA/CAJ in the same issue of the first installment of Charbonnier's article [334]. The review is most respectful and ends with the assertion that "The present volume is of great practical value both to the artillerist and to the officer on the proving ground" [334, p. 311]. It will be shown that this opinion is quite at odds with that of the 1927 revised version expressed by A. A. Bennett in the Bulletin of the American Mathematical Society. The present edition was not reviewed by that serial.

The much mentioned Jacques Hadamard himself in 1920 delivered a report [497] to the Académie Des Sciences on research done since 1917 on topics in mathematical ballistics, papers both published and classified. The papers were broken down by subject, with one section devoted to[76] "Properties of trajectories. Perturbations" [334, pp. 441–442], a section notable both for its nods to French theoretical traditions and modern realities of conditions of fire. The opening: "We return to the mathematical point of view with the study of the geometric properties of the trajectories and especially, in conformity with the modern mathematical point of view of Poincaré, that of their mutual relations" [334, p. 441]. And the third paragraph: "The circumstances of war brought attention to the *small perturbations of trajectories*. This subject, classic from the mathematical point of view, showed itself of great importance from the practical point of view, since it rules the corrections of all kinds in support of fire" [334, p. 441]. Comments are made on solutions supplied by four different authors, and the section concludes with "These studies were moreover destined to be used at the front, consequently their results had to be made material in a manner permitting numerical calculation with maximum speed. The number of abaques [nomograms][77] and devices conceived towards this end is very great" [334, p. 442]. One such device, due to an engineer at the Gâvre, was displayed; one recalls J. W. Alexander's design for the battery commander's device.

Before concluding this section with a return to Charbonnier's treatise (revised), attention will be paid to nomograms, the "abaque" mentioned by Hadamard above, a graphical means of calculation with deep French roots. One of the chief expositors and promoters of this method, Maurice d'Ocagne (1862–1938), made a contribution

---

[76] All translations by the present author.

[77] The word "abaque" here will be used synonymously "nomogram"; d'Ocagne's first edition (1899) of his *Traitè de Nomographie* has the subtitle *Théorie des Abaques*. "In all the treatise, d'Ocagne uses the word 'abaque,' following the French tradition inaugurated by Lalanne and continued by Massau, Lallemand and Soreau. In the 1921 edition, the word 'abaque' has disappeared in the subtitle, and is replaced almost everywhere in the text by 'nomogramme.'" [Communication from Dominique Tournes to present author, 12/3/2020.].

to the War effort by organizing and directing the Nomographic Section of the Army, in charge of providing the armed forces with charts for rapid calculation of all sorts. Briefly put, the nomogram for, say, a given relation among three variables were of two kinds. An alignment nomogram, the idea for which d'Ocagne himself took credit, consisted, in its simplest form, of three parallel linear scales, each carrying the values of a single variable in the relation. A fourth line, when crossing all three, marked a value from each scale; the three values thus produced formed a solution to the given relation. The second kind, called an intersection nomogram, was older, and allowed calculation of the relation by assigning to each of two variables an axis on the standard Cartesian plane; the third variable was read off on a level curve corresponding to the value of the relation when the first two variable assignments were substituted into the relation. Many were the variations of these basic types; d'Ocagne's post-War volume [603], a revision of his 1899 edition, cataloged the theory and practice of nomography.[78] The construction of the nomograms themselves involved mathematics of various levels of difficulty.

Nomograms were finding a foothold in American calculation at this time; one account [437, p. 23] claims that during the War American and French ballisticians met and the latter introduced nomographic techniques to the former.[79] After writing, in 1912, a massive mathematical study on theoretical aspects of nomography, Gronwall created a nomogram for computing range and deflection variation; it was his last technical report at Aberdeen, dated July 1919. The creation of nomograms formed part of Moulton's efforts, as noted previously. An article in the *Coast Artillery Journal* devoted to nomography appeared in 1923 [548], testifying to the increasing use in the military of this method.[80] In this manner, the War served as an impetus to the reception of a new style of computation, a miniature version of the introduction of electronic computing in second World War.

In [603, pp. 352 ff.], d'Ocagne presents several nomograms relevant to preparation of fire; one, for example, graphs a rational function relating density of air to pressure of the atmosphere and to temperature. A striking compound nomogram appears on page 360 (see Fig. 2.6). The first three parallel horizontal lines display the relation just mentioned, with one scale each for air pressure, temperature, and density (on the middle scale); a value of the last having been found, a line extends down cutting each of two intersection nomograms, whose scales contain empirically determined values of range, variation in range, and longitudinal component of wind velocity. Corrections are read from the appropriate scales. D'Ocagne notes that use of these devices reduced from about 15 minutes to less than 3 minutes the time necessary to compute the initial elements of fire. The charts displayed were collected

[78] D'Ocagne was the director of the library under whose auspices Charbonnier's 1907 treatise on ballistics was published.

[79] In [437]; however, it is also argued that the introduction and development of nomography was already well underway in America at the time of the War. It offered a graphical alternative to numerical techniques; efficiency was a selling point for the "charts."

[80] The article discusses nomograms used in 1916 range firings; the deflection in the trajectory was to be tracked using them.

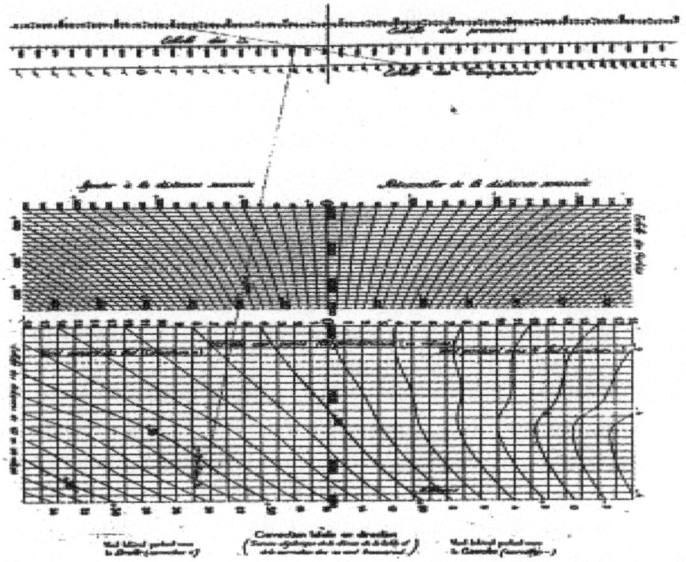

**Fig. 2.6** Nomogram for fire control. From *Traité de Nomographie* by d'Ocagne, [603]

from copies obtained from among those scattered across the front during the last two years of the war. Thus, one has a new method to attempt to put theory into practice, subject to the usual qualifications about the realities of engagement.

In the post-war years of 1921 and 1927, Charbonnier published revised versions of his treatise [419, 420]. Six volumes in all were envisioned, but only these two appeared. In particular, the variation material previously discussed was not revisited. However, in [419, p. 459–470] from 1921, one finds material treating deviations in initial velocity, angle of fire, and ballistic coefficient. The material here was of two types. The first problem: let $\tau$ be the "angle of inclination" which the trajectory makes with the horizontal, $\tau$ is the independent variable, not time $t$. The variable $u = vcos(\tau)$ is introduced, where $v$ is velocity, also dependent on $\tau$. Then, let an arc of a trajectory with initial data $(\alpha, u_0, c)$ be given, where $\alpha$ is the initial angle of inclination and $c$ the ballistic coefficient. If $\tau$ is a fixed later angle value, suppose one gives an "alteration" of $(\delta\alpha, \delta u_0, \delta c)$ and of gravity $\delta g$, what are the perturbations $(\delta u, \delta x, \delta y, \delta t)$ undergone by the initial variables, where $x$ and $y$ are the usual Cartesian coordinates?

Charbonnier outlines a solution to the first problem given by two military personnel, each in his own manner, which involves giving a perturbation $\delta u$ to $u$ in the principal equation, and chasing the result down to a complicated integral expression for $\delta u$ in terms of $\delta c$ and $\delta g$ which is in turn substituted in to the variation expressions for $t$, $x$ and $y$; $\delta g$ is involved in the last three equations. Variations in range, time of flight, and angle of fall are then given.

The second problem involves the production of three theorems, which relate various partial derivatives of $t$, $V_0$ ( = initial velocity), $x$, $y$, $c$ with respect to each other. The resulting sets of equations are called "formulas of connection" and were used to check values of these quantities calculated by other approximations or by small arcs.

One curious feature of these problems: neither appeared anywhere in the 1907 edition of these volumes. The two solvers of the first problem also occurred among the four contributors listed by Hadamard; the author of the second collection is uncredited. It would appear that the ballistic demands of the War gave rise to interest in new solutions to these problems.

The appearance of the two volumes of Charbonnier, the second at roughly the same time as Moulton's *Modern Methods*, added to the flurry of reviews of ballistics volumes in the Bulletin of the American Mathematical Society at the end of the 1920s. Gronwall reviewed the 1921 volume in 1927 [492]. Ever the theoretician, he finds the treatment of the trajectory of a heavy particle through a resisting medium, which varies with velocity to be of particular interest to the mathematician, and speculates on its use in an advanced course in differential equations. He finds the theorems correct as stated. He also praises the exposition of a 1920 paper [438], which classifies all possible integrable forms of the resistance function, a paper that runs roughly 100 pages in length.[81]

In contrast, A. A. Bennett had a quite critical view of the second volume; the review also ran in 1927 [361]. After summarizing contents and the circumstances surrounding Charbonnier's efforts, Bennett remarks that "This should ensure respectful consideration from the few technical scientists seriously interested in ballistics" [361, p. 517]. He also states frankly: "[Works on ballistics] can hope to be of significance chiefly to artillery officers, who, however, seldom have the preparation, interest, or opportunity to study them" [361, p. 517]. But his most trenchant points should be quoted at length:

> Ballistics is inherently a branch of engineering. Not only are there a distractingly large number of physical factors present in the problem, but the number of independent variables that are actually taken into account in firing is so large, and these variable factors are to a great extent so uncontrollable, that simple accurate formulas are universally regarded as out of the question. Rules based on the normal probability curve are used in actual field conditions... In the sense of "rational mechanics," there is no rational ballistics.[82] For ballistics, as for every other branch of engineering, there is a group of present-day problems involved on the one side in seeking to test present-day assumptions, and if possible to establish these on a rational basis, and on the other side in devising means for application of these accepted principles with an accuracy consistent with the data and yet with the greatest possible simplification concordant therewith. ... [The present volume] divorces ballistics at this stage from its only logical bases, historical and physical. We meet, consequently, for at least the first seven hundred pages of this volume, a collection of mutually contradictory theories. These are given no historical justification and one cannot make comparisons among them since the author himself fails to compare their relative accuracy and withholds as yet all physical data from the reader in the effort to be strictly rational. [361, pp. 517–518]

[81] Grownall's nearly simultaneous (1920) work in [490], in which theorems follow from hypotheses on the ballistic quantities, is closest in nature to Charbonnier's. It gives generalizations of some results of de Saint-Robert and an alternative proof of a theorem of Charbonnier [490, p. 45].

[82] This comment followed one acknowledging the rich French tradition in rational mechanics.

Bennett's review ends by using Charbonnier's own words against him, "...a propos of the artificial solutions [given in the 1920 paper mentioned by Gronwall] of Drach."

> These solutions [of Drach] ...of some interest as they are, do not fulfill the simultaneous conditions that are demanded of the ballistician: 1) to be applicable to an acceptable form of the [resisitance] function $F(v)$, that is, sufficiently resembling the experimental air resistance function 2) to lead, for the other elements of movement, to a formal solution susceptible of being used for numerical applications.[83] [361, p. 518]

Thus, one is confronted by a scramble of opinion and influence. In a review of Charbonnier's central work in a military journal, it is praised as being of great value to artillery officers at the proving ground, yet ballistics is criticized as being low on the list of interesting subjects to those same officers and the book as being removed from the realities of engineering by a mathematician in a mathematical journal. It was, however, listed among the essential books on ballistics by Aberdeen mathematicians Bliss and Gronwall and particularly appreciated by the latter. Perhaps, a lesson to be drawn is that the first decade following the conclusion of the War in the United States saw an enrichment of the collection of ballistic workers to include theoreticians and mathematicians with a genuine appreciation for the difficulties of the subject, as well as military people whose interest in theory was raised.

This concludes the comparison. A similar account could easily have been given with German practice, including the highly regarded book by Cranz, or with British developments. A more complete story would contain details of government involvement in military planning and the role of various agencies within that government, but that is beyond the scope of this book.

## Archival Materials (Abbreviations)

[Veblen Papers] Oswald Veblen Papers, Archives of the Library of Congress, Library of Congress, Washington, D.C.

[NARA Aberdeen] Record Group 156–866, National Archives and Records Administration, College Park, Maryland

[NARA Washington] Record Group 156.7.18, National Archives and Records Administration, College Park, Maryland

[NARA Ordnance School] Record Group 156–1512A, Box 2 , National Archives and Records Administration, Philadelphia, Pennsylvania

[NARA ENIAC] Record Group 156–1512A, Box 9, National Archives and Records Administration, Philadelphia, Pennsylvania

[Columbia University] University Archives and Columbiana Library, Columbia University in the City of New York

[83] Translation by the present author.

[Princeton University] University Archives Project, Seeley G. Mudd Manuscript Library, Princeton University, Princeton, New Jersey
[MIT] MIT Institute Archives and Special Collections, Massachusetts Institute of Technology, Cambridge, Massachusetts
[University of Chicago] Special Collections Research Center, University of Chicago Library, Chicago, Illinois
[Concord] Hoar Family Papers, Concord Free Public Library, Concord, Massachusetts
[Harvard] Office of Harvard War Records, Harvard University, Cambridge, Massachusetts

# Chapter 3
# Bliss' Contributions in Light of the Pure Mathematics of the Era

## 3.1 Introduction

We now turn to an account of the pure mathematics within which Bliss' ideas were developed and to which they contributed. In doing so very little of Moulton's work is mentioned, but it is not to be inferred from this that the latter's efforts on differential variations were of lesser consequence. The scope of Moulton's investigations was vast: the studies of numerical integration, the accuracy of differential approximations, weighting factor tables, and muzzle velocity studies, not to mention the co-ordination of these activities, were all his initiatives. Most important for this account is his introduction of the differential variation idea itself. But it is fair to say that the mathematics involved in Bliss' work, taken as a whole, is part a of larger mathematical story, having connections with ongoing pure researches of the day and contributing to the development of abstract implicit function theory and analysis of functionals. It was also eventually used to address different military problems related to ballistic missiles, work which continued until the early 1960s, as will be discussed in the next chapter. Thus our emphasis on Bliss' mathematics is not to be construed as conferring any "ultimate honor" on him as a figure with one foot in each of the pure and applied realms but rather as investigating the rich mathematical context of his work.[1]

Gilbert Ames Bliss was first and foremost an analyst, a pure mathematician of his era. Had his mathematical contributions to ballistics of the World War been limited to the publications in military journals, technical notes, and resulting changes in firing table construction, this story would be fairly complete as it stands. But Bliss wrote two papers [718, 719] appearing in the *Transactions of the American*

[1] On the other hand, Moulton's work on rotating projectiles featured in the last chapter of [914], and cited by him as the most personally mathematically demanding of his efforts, constitutes an in-depth study in applied differential equations which finds no counterpart in Bliss' work.

 
A. Gluchoff, *From Fréchet Differentials to Firing Tables*,
https://doi.org/10.1007/978-3-031-68267-4_3

*Mathematical Society* in 1920, which were inspired by his ballistics work on variations. These papers, which were cited in the mathematical literature as late as the mid-1960s [807], connect his war-time efforts with another world, the world of research in pure mathematics in the first half of the twentieth century. It is to this world that we now turn.

In this chapter, an effort is made to place Bliss' innovations in the context of the developments in analysis which preceded and to a certain extent informed his work and to describe the effect that his papers had on the progress of these developments. For the period in which Bliss and his immediate predecessors and successors worked, which we will take as 1900–1944, was a time of intense innovation in mathematical analysis. It is possible to relate his pure contributions to such new areas as the flowering of the calculus of variations, integral equations, implicit function theory, existence theory for ordinary and partial differential equations, potential theory, and that line of inquiry which begins with Volterra's functions of a line, continues with Hadamard's functional calculus, and ends with Banach's functional analysis.

In doing so two groups of American analysts emerge as characters in the story. The first is a loosely connected set of analysts headed by Maxime Bôcher, a student of Felix Klein based at Harvard University, who worked in several areas related to Bliss' research. The other group, more important for our story, is the collection of mathematicians, based at the University of Chicago, which chronologically begins with Bliss' advisor Oskar Bolza and includes Bliss himself and his students and colleagues.[2] It was this dynasty which was responsible for the cultivation of an American brand of functional analysis centered on notions of differentiation of functionals, an abstract notion applicable to solutions of systems of differential equations (Cf. Wiener's praise of Bliss' work as a "brilliant use of functionals.")

A cautionary note should be sounded. There was much work done by European mathematicians, such as Hilbert and Banach, on topics which relate to the development of the broad fields of calculus of variations and analysis of functionals during this era. Hilbert's work on integral equations and his approach to calculus of variations can be cited as particular instances. These and other efforts are not considered here, not because they had no general influence but rather because they were not part of the particular stream of ideas that were specifically mentioned as relevant for the participants in this study. Thus Banach's *Opérations Linéares* [687] was the culmination of his work until 1932 and a touchstone for further work in functional analysis, but Banach's influence on the work of the American mathematicians discussed here can be explicitly documented only by references in their papers to his work, accompanied by a comment such as "see also." Similarly, Hilbert's contributions to the calculus of variations as described, say, in Chapter 7

[2] Among these was Herman Goldstine, a young researcher who participated in the development of the ENIAC computer during the Second World War, as well as authoring books on the history of the calculus of variations [796] and numerical analysis. His work will be cited several times in what follows.

of [796], were of great general significance, but the difficulty of tying them to the stream described here seems outweighed by the benefits of a larger picture. Those desiring such a larger picture are referred to the general expositions on the history of calculus of variation functional analysis given below. For this account, the direct contributions of Vito Volterra, Jacques Hadamard, and Maurice Fréchet, among others, are chosen for their evident influence.

Among the hallmarks of modern mathematics in any division are rigor, generalization, and abstraction. The gradual appearance of each of these in the last decades of the nineteenth century, especially abstraction, will become obvious in what follows. Moulton's apostolic fervor on the subject of rigor has already been noted: the mathematicians in this chapter would take rigor as a matter of course. These men worked in the shadow of Karl Weierstrass, whose role as the initiator of a second wave of rigor (the first having been due to Cauchy) has been widely noted. Generalization-especially the loosening of hypotheses on theorems as much as possible to get a result-is evident as well. An example of this is the movement away from the assumption of analyticity of functions occurring in differential equations to weaker hypotheses such as continuous differentiability. Analyticity would have been considered appropriate for the equations which described certain physical problems such as the orbit of planets, and even Weierstrass assumed that the functions in his treatment of the calculus of variations were analytic. But one sees in his successors a chipping away at this type of hypothesis. Moulton is an interesting figure here: his differential equations text begin with the assumption of analyticity on the defining functions of these equations and present proofs, versions of Cauchy's original arguments, based on power series expansions. But later on, these are replaced by generalized hypotheses as exemplified in the preceding chapter by his theorem on the dependence of solutions on a parameter.

But above all the phenomenon of abstraction-the seeking of a commonality in what are perceived as individual instances of a general theory-will be illustrated by the recasting of the discourse on the areas of mathematics mentioned in the first paragraph in functional analytic terms. Bliss' two papers could be viewed from a variety of perspectives-as a chapter in perturbation theory, a contribution to mathematical ballistics, for example. But the view adopted here stresses the influence of functional analytic ideas, some of them not yet expressed as such, in these works. In this regard, it is part of the abstraction process influencing other areas of analysis, indeed other areas of mathematics in general, as well.

We begin with a short account of Bolza's two books on the calculus of variations, which sets the stage for Bliss' work on existence and smoothness[3] theorems for systems of differential equations and his concerns with implicit function theory. These questions provide Bliss' entré into the problems of variations in the ballistic trajectory. There follows a discussion of his two *Transactions* papers. The remainder

[3] In what follows, "smoothness" refers to continuity or degrees of differentiability of the solution functions with respect to the constants of initial conditions or parameters of the defining differential equations.

of the chapter is devoted to embedding this work in the topics of analysis noted above, showing how they both served as sources of his ideas and in turn how his ideas were used in their continued growth. A section is devoted to the idea of the adjoint in the setting of partial differential equations. The chapter concludes with an account of Bliss' 1944 book on exterior ballistics [727], in which the treatment of differential variations is given almost entirely in functional analytic terms.

## 3.2 Bolza and the Calculus of Variations in America

We have noted several times that Bliss' area of research was the calculus of variations and that he effectively established a school of study in this subject at the University of Chicago which lasted through the first half of the twentieth century. His advisor and fellow progenitor of this school was Oskar Bolza, who came to the university at its opening in 1892.[4] Though not originally a researcher in the calculus of variations, Bolza became one after presenting a series of papers on the subject at an AMS Colloquium[5] in Ithaca, New York, in August of 1901. These lectures were expanded and published in 1904 as *Lectures On The Calculus of Variations* [738], and Bolza became a leading researcher in the subject for the rest of his career. His 1904 work was in turn greatly enlarged and published in German as *Vorlesung Über Variationsrechnung* [739]. This became a standard reference for the subject at the time. Along with Kneser's book, Hadamard's 1910 work *Leçons sur le calcul des variations* [821] rounded out the three most often cited references of the era, and one encounters Bolza's and Hadamard's books linked several times, for example, "One might well introduce himself to the calculus of variations by reading the brief introduction to the theory in Goursat's *Cours d'Analyse* and afterward the fascinating books of Bolza and Hadamard" [720, p. 180]. In the Goursat *Cours* in question one also finds a salute to the three works followed by the comment "I have been well-served by the two last books [Bolza and Hadamard] in writing this chapter. [801, p. 3]"

We will assume that the reader is familiar with the early history of this subject; [871] provides an excellent summary [pp. 382 ff. and pp. 745 ff.]. Also [796] contains accounts of classic papers from Newton through Bolza.[6] Owing its existence to early problems of optimization such as those involved with the brachistochrone, the

[4] For a discussion of Bolza and his time at Chicago see [924, pp. 393–396].

[5] The American Mathematical Society Colloquia series, begun in 1896, consisted of "courses of lectures delivered by specialists on selected chapters of their fields of work. Their purpose is to enable the members of the Society to keep in touch with the most recent advances of mathematical science and to stimulate a wide interest in its development" [905, p. iii]. Several of the mathematicians in our story, including Bliss himself, contributed to these Colloquia.

[6] The problem of Newton, often considered the first problem in the calculus of variations, involved determining the shape of an object moving through a resisting medium: how to design the object so that it encounters the least resistance. The relevance to ballistics is evident.

set of ideas developed as mathematicians sought necessary and sufficient conditions for the maximum or minimum of integral expressions of the sort $\int_a^b f(x, y, y')\,dx$, where $y$ is a function of $x$, or $\int_a^b f(x(t), y(t), x'(t), y'(t))\,dt$; the quantity that varies is the *function* $y$ or $x(t)$ and $y(t)$. But throughout this process, there was a certain looseness in the rigor of the work. This is surely understandable in the infancy of the subject, as the gaps in Lagrange's paper presented in Chap. 1, for example, show.[7] But the lack of rigor had been acknowledged also in the nineteenth century and finds expression in several places in Bliss' time. Bliss himself wrote in his 1925 Carus Monograph [720, p. 177]

> The memoirs and treatises on the calculus of variations up to the latter part of the nineteenth century frequently leave one in doubt as to the validity of the methods and the precise character of the results which they contain. Errors are not infrequent, even among the ablest writers, and the vagueness in the statement and discussion is common. The feeling of uncertainty concerning these writings is not a modern one only, based upon the exacting requirements of the logic of the present-day analysis; it was shared repeatedly by earlier writers themselves, as the literature plainly shows. Weierstrass (1815–1897) had a very great influence in the development of precise thinking in the calculus of variations, as in other important domains of mathematics. He formulated is problems with great care...

Bolza was in an excellent position to introduce the Weierstrassian rigor, having attended the latter's lectures on the calculus of variations in 1879 (recall he loaned a copy of these notes to Bliss). In Bolza's *Lectures*, he notes similarly: "Until rather recently a certain vagueness has prevailed with respect to the fundamental concepts of the Calculus of Variations [738, p. 9]." Finally, a much later remark [1989] by Bliss' student E.J. McShane[8] on method "In the eighteenth century ... If it could be shown that any curve that minimized some functional had to satisfy a certain condition, and a curve could be found that did satisfy that condition, it was accepted without comment that that curve did furnish the minimum. Nor has such a feeling quite disappeared. [897, p. 917]" Our strategy is to summarize the first few sections of Bolza's *Lectures*, noting the places where existence and smoothness theorems appear. Due to the elementary cases considered, the theorems will be the simplest instances of the results addressed by Bliss in his future work and key to his involvement in ballistic topics. For reference, a list of the "final versions" of these theorems, that is, the most intricate and evolved versions used by Bliss himself in his 1946 volume [728], is given as an Appendix to this chapter.[9]

At the outset Bolza announces: "The emphasis lies entirely on the theoretical side. I have endeavored to give clear definitions of the fundamental concepts,

[7] Goldstine remarks on the Lagrange paper: "The whole procedure shows, however, how immature was the real understanding of some aspects of analysis in the eighteenth century and yet how bold and how inspired were its best practitioners" [796, pp. 119].

[8] McShane was the author of a biographical memoir of Bliss [896].

[9] Bliss remarks in his introduction to these theorems that [p. 269] "As a result of their use in the calculus of variations the theorems themselves have undergone extensions and improvements which have not always found their way into the textbooks on the theory of functions of a real variable."

formulations of the problems and rigorous demonstrations. [738, p. vi]" Also: "For a rigorous treatment of the Calculus of Variations the principal theorems of the modern theory of functions of a real variable are indispensable, these I had therefore to presuppose... [ibid., pp. vi-vii]" His introduction ends with an expression of thanks to Bliss "for valuable suggestions and criticisms. [ibid. p. viii]" The first ten pages or so are devoted to notation, definitions, and the basic ideas of the subject. In what follows, we shall not repeat every hypothesis given in order to avoid tedium, but the reader may rest assured that there are no unstated conditions, and reasons for the steps are given, often from several points of view. Also, references to the literature given by Bolza to support the use of the real analysis results will be omitted here.

In the chapter on the first variation, a set of curves is given, defined on an interval $[x_0, x_1]$, a function $F(x, y, y')$ is given and the definite integral $J_C = \int_{x_0}^{x_1} F(x, y, y')\,dx$ is presented. The problem is the determination of the collection of all admissible curves which maximize or minimize the integral, with the usual definitions of absolute and relative minimum given.[10] Extra conditions satisfied by the curve are announced: that it must pass through two given points $A(x_0, y_0)$ and $B(x_0, y_0)$. To a given curve, $y = y(x)$ is added a function $\omega(x)$, where $\omega(x_0) = \omega(x_1) = 0$, called a variation, and the total variation of the integral $J$ is defined as

$$\Delta J = \int_{x_0}^{x_1} [F(x, y + \omega, y' + \omega') - F(x, y, y')]\,dx \tag{3.1}$$

From here on, without loss of generality, minimization only is considered. An application of Taylor's formula changes $\Delta J$ to

$$\Delta J = \int_{x_0}^{x_1} (F_y\omega + F_y'\omega')\,dx + (1/2)\int_{x_0}^{x_1} (\bar{F}_{yy}\omega^2 + 2\bar{F}_{yy'}\omega\omega' + \bar{F}_{y'y'}\omega'^2)\,dx \tag{3.2}$$

where the arguments of $F_y$ and $F_y'$ are $x$, $y$ and $y'$ and those of the functions in the second integral are $x$, $y + \theta\omega$ and $y' + \theta\omega'$, for $\theta$ a constant between 0 and 1. Now Bolza follows a later work of Lagrange, cited on Bolza's p. 15, in which special variations $\omega = \epsilon\eta$ are introduced, where $\eta$ is a function vanishing at the endpoints of the interval. This puts $\Delta J$ in to the form

$$\Delta J = \epsilon[\int_{x_0}^{x_1} (F_y\eta + F_{y'}\eta')\,dx + (\epsilon)] \tag{3.3}$$

[10] This is the simplest possible case of the problem, as later presentations deal with $F(x, y_1, y_2, \ldots, y_1', \ldots)$.

where $(\epsilon)$ is a quantity which goes to 0 and $\epsilon$ goes to 0. From this it follows that

$$\int_{x_0}^{x_1} (F_y\eta + F'_y\eta')\,dx = 0 \tag{3.4}$$

for all such variations $\eta$. An integration by parts on the integral (cf. such a step in Lagrange's first paper) gives

$$\int_{x_0}^{x_1} \eta(F_y - \frac{d}{dx}F_{y'})\,dx = 0 \tag{3.5}$$

for all such variations. From this follows the first fundamental necessary condition for an extremal (so named by Kneser): Every function $y$ which minimizes the integral

$$\int_{x_0}^{x_1} F(x, y, y')\,dx \tag{3.6}$$

must satisfy the differential equation

$$F_y - \frac{d}{dx}F_{y'} = 0 \tag{3.7}$$

This differential equation was due to Euler in 1744 and in his honor is called Euler's equation.

Under "Miscellaneous Remarks" on Euler's equation, Bolza points out that the equality is in fact a second-order differential equation, as can be seen by writing out

$$F_y - F_{y'x} - y'F_{y'y} - y''F_{y'y'} = 0 \tag{3.8}$$

This implies that the general solution contains two arbitrary constants: $y = f(x, \alpha, \beta)$.

But his next remark is the key point from our perspective: how does one guarantee that *in general* there is a solution to Euler's equation passing through a given point $(a, b)$ with a given finite slope? This can be done provided that $F_{y'y'}(a, b, b') \neq 0$, where $b'$ = tangent of the angle of the slope of the solution. This is justified by quoting [ibid. footnote p. 28] the existence and uniqueness theorem for systems of differential equations first given by Cauchy as applied to the system

$$\begin{aligned} \frac{dy}{dx} &= y' \\ \frac{dy'}{dx} &= \frac{(F_y - F_{y'x} - y'F_{y'y})}{F_{y'y'}} \end{aligned} \tag{3.9}$$

Bolza concludes that if $F_{y'y'}(a, b, p) \neq 0$ for every finite value of $p$, then "one extremal can be drawn from $(a, b)$ in every direction, except the direction of the $y$-axis. A problem for which $F_{y'y'}(x, y, p) \neq 0$ for every point $(x, y) \ldots$ and for every point $p$ is called, according to Hilbert, a regular problem [ibid. p. 29]." This is the first appearance of an existence theorem of the sort we have repeatedly mentioned, put in to address the obvious (to the modern reader) but unanswered question of how one knows in general that there are solutions to Euler's equation.

Two more examples of this sort of citation, more important to the development of the theory, will be presented. Further necessary conditions for a minimum are desirable; to this end, Bolza returns to expression (3.29) and applies Taylor's theorem again, stopping at the terms of the third order. If the curve $y = f_0(x) = f(x, \alpha_0, \beta_0)$ in the integral expression is an extremal (called $E_0$ by Bolza), and we set

$$F_{yy}(x, f_0(x), f_0'(x)) = P \tag{3.10}$$

$$F_{yy'}(x, f_0(x), f_0'(x)) = Q \tag{3.11}$$

$$F_{y'y'}(x, f_0(x), f_0'(x)) = R \tag{3.12}$$

then we have

$$\Delta J = (1/2) \int_{x_0}^{x_1} (P\omega^2 + 2Q\omega\omega'^2 + R\omega'^2)\, dx + \int_{x_0}^{x_1} (\omega, \omega')_3\, dx \tag{3.13}$$

where $(\omega, \omega')_3$ is a homogeneous function of $\omega$ and $\omega'$, $\omega$ again being an arbitrary variation. Specializing to $\omega = \epsilon\eta$ gives

$$\Delta J = \epsilon^2[(1/2) \int_{x_0}^{x_1} (P\eta^2 + 2Q\eta\eta'^2 + R\eta'^2)\, dx + (\epsilon)] \tag{3.14}$$

Arguing as before gives that for a minimum, it is necessary that the integral in (3.14) is $\geq 0$ for all variations.[11] Legendre [877] proceeded to explore such conditions by transforming integral in the equation just given to

$$\epsilon^2(1/2) \int_{x_0}^{x_1} [(P + \omega')\eta^2 + 2(Q + \omega)\eta\eta' + R\eta'^2]\, dx \tag{3.15}$$

where $\omega$ is an arbitrary function of $x$.

[11] McShane: "The guiding principle during the eighteenth century and more than half of the nineteenth seemed to be that if a minimizing function is sought for some functional, then by inventing more and more necessary conditions for a minimum, we can feel steadily more confident that a function that passes all the tests is in fact the minimizing function sought" [897, p. 918].

Now Legendre defines $\omega$ to be the solution of the differential equation formed by setting the discriminant of the quadratic form under the integral sign equal to 0, that is,

$$(Q+\omega)^2 - R(P+\omega') = 0 \tag{3.16}$$

This $\omega$ being found reduces the integral to the form

$$\epsilon^2 \int_{x_0}^{x_1} R(\eta' + \frac{Q+\omega}{R}\eta)^2\,dx \tag{3.17}$$

from which he infers that $R$ does not change sign in the interval and that the integral always has the same sign as $R$. He is headed toward arguing that $R(x) > 0$ on the interval.

But this conclusion is flawed, since it assumes that $R \neq 0$ in the interval and that the differential equation for $\omega$ has a solution.[12] Bolza rescues a part of Legendre's argument and proves that a necessary condition for a minimum is that $R \geq 0$ on the interval. This he does by assuming that $R(c) \leq 0$ for some $c$ in $[x_0, x_1]$, using an existence theorem to obtain an $\omega$ such that

$$\frac{d\omega}{dx} = -P + \frac{(Q+\omega)^2}{R} \tag{3.18}$$

valid on a subinterval $(c-\delta, c+\delta)$ on which also $R < 0$, and using it to construct a variation $\eta$ for which Legendre's transformed integral is negative. This contradicts the assumption that the extremal minimizes $J$. Thus the existence theorem here plays a role in correcting an error in reasoning by providing a function needed for a correct proof.

One final example of the use of existence theorems concerns a further exploration of Legendre's equation, due to Jacobi [866]. In this investigation, the object of which need not concern us, Jacobi uses a substitution

$$\omega = -Q - R\frac{u'}{u} \tag{3.19}$$

into (3.16) creating a new differential equation in $u$ (referred to as Jacobi's differential equation):

$$(P - Q')u - \frac{d}{dx}(Ru') = 0 \tag{3.20}$$

as well as an operator

$$\Psi(u) = (P - Q')u - \frac{d}{dx}(Ru') \tag{3.21}$$

[12] Lagrange in 1797 had already pointed this out, according to Bolza.

a linear second-order equation. He makes a discovery: a substitution of the general extremal $y = f(x, \alpha, \beta)$ into Euler's equation followed by a formal differentiation with respect to $\alpha$ shows that $f_\alpha(x, \alpha, \beta)$ is a solution to Jacobi's equation, similarly for differentiation with respect to $\beta$. However, one is faced with the existence of these partials. Bolza states a local result: if $E_0 = f(x, \alpha_0, \beta_0)$ is an extremal, then there is a neighborhood of $(\alpha_0, \beta_0)$ for which the corresponding extremals $f(x, \alpha, \beta)$ have the required partials: in a footnote [p. 54]: "...[the existence of the partials] are a consequence of our previous assumptions concerning the function $F$ [$C^3$ in all variables], the extremal $E_0$ [$C^1$ in $x$], and the function $R$ [$R(x) > 0$ for all $x$]. This follows from the theorems concerning the dependence of the general solution of a system of differential equations upon the constants of integration ...[ibid, p. 54]." There follows a list of references for the theorems, the first and principal of which is also cited by Bliss in his 1905 paper [704].[13] Thus Jacobi's results are on firm ground, and an example of smoothness of solutions with respect to a parameter is concluded.[14]

This concludes our examples of the appearance of existence and smoothness theorems; there are several other occurrences. But one question remains: what happened to the $\delta$, the operator introduced by Lagrange? None of the results given above depends on manipulations with or properties of $\delta$, essential to Lagrange's presentation. The answer is that Bolza does introduce and use it, but it appears only as a notational device. On p. 16 he states: "We now make use of the following permanent notation introduced by Lagrange (1760)." In a general setting, he lets $\phi(x, y, y', y'', \ldots)$ be a function of $x$, $y$, and some derivatives of $y$[15] with smoothness conditions on $\phi$. He then replaces $y$ by $y + \epsilon\eta$ in all arguments and expands the result as a finite power series in $\epsilon$, namely,

$$\bar{\phi} = \phi + \frac{\epsilon}{1}\phi_1 + \frac{\epsilon^2}{2!}\phi_2 + \ldots + \frac{\epsilon^n}{n!}\phi_n + \epsilon^n(\epsilon) \tag{3.22}$$

[13] There is some confusion about the meaning of "constants of integration." The title of Bliss' (1905) paper refers to solutions as functions of their "initial values," that is, the fixed co-ordinates through which a solution passes. But the section in which the existence of the solution partials with respect to the initial values is proved is called *Partial derivatives with respect to the Constants of Integration*. The same existence results in his 1918 paper are called smoothness with respect to "constants of integration." Bolza's first introduction of $f(x, \alpha, \beta)$ in the book [p. 28] refers to $\alpha$ and $\beta$ as "arbitrary constants," though in his example on the next page, they are clearly constants that are the result of two successive integrations. The present author infers from Bolza's use of "constants of integration" without definition in his footnote that Bolza is asserting the existence of the derivatives with respect to $\alpha$ and $\beta$; however, they are to be interpreted, follows from that of the derivatives with respect to initial conditions.

[14] The footnote also addresses the existence of higher-order partial derivatives of $f$ and the reversal of the order of differentiation necessary to prove Jacobi's assertions.

[15] The number of arguments of $\phi$ is presumably finite, though Bolza does not mention this.

and

$$\phi_1 = \phi_y \eta + \phi_{y'} \eta' + \phi_{y''} \eta'' \ldots \tag{3.23}$$

$$\phi_2 = \phi_{yy} \eta^2 + \phi_{y'y'} \eta'^2 + \ldots + 2\phi_{yy'} \eta \eta' + \ldots \tag{3.24}$$

He then denotes $\epsilon\phi_1, \epsilon^2\phi_2, \ldots$ as the first, second, and so on variations of $\phi$.[16] And similarly he defines $\delta^k J$ to be the term of order $k$, multiplied by $k!$, in the expression in powers of $\epsilon$ of

$$\bar{J} = \int_{x_0}^{x_1} F(x, y + \epsilon\eta, y' + \epsilon\eta')\, dx \tag{3.25}$$

Thus

$$\delta J = \epsilon \int_{x_0}^{x_1} (F_y \eta + F_{y'} \eta')\, dx \tag{3.26}$$

$$\delta^2 J = \epsilon^2 [1/2 \int_{x_0}^{x_1} (F_{yy} \eta^2 + 2F_{yy'} \eta\eta' + F_{y'y'} \eta'^2)\, dx] \tag{3.27}$$

for the particular variation $\epsilon\eta$, and the above exposition could evidently have been written in terms of these $\delta$ quantities. But, for example, the equation

$$\delta J = \int_{x_0}^{x_1} \delta F(x, y, y')\, dx \tag{3.28}$$

is merely a consequence of the definition of each side and involves no further argument. The "operational" nature of $\delta$, which was important enough to Lagrange to require some comment, is gone here.[17]

## 3.3 Bolza's *Lectures on the Calculus of Variations*—Implicit Function Theorems

The existence and smoothness theorems for systems of differential equations have an obvious relevance to the ballistic questions considered by Moulton and Bliss, but the role of the implicit function theorems is not so clear. There are several reasons for

[16] Thus, though not stated as such by Bolza, $\phi_1$ is a directional derivative of $\phi$ at $(x, y, y', \ldots)$ in the direction of $(0, \eta, \eta', \ldots)$, using the variable $\epsilon$ in the definition.

[17] But we may note that (3.16) can be interpreted as saying that the $\delta$ operator applied to the functional $\bar{J} = \bar{J}[y(x), \epsilon\eta(x)]$ of $\eta$ for a fixed function $y$ and increment $\epsilon$ yields the linear functional of $\eta$ on the right, more suggestively written as $\epsilon \int_{x_0}^{x_1} (F_y[\eta] + F_{y'}[\eta'])\, dx$. The stage is thus set for the $\delta$ to be considered as an operator, akin to differentiation, in the functional analytic sense.

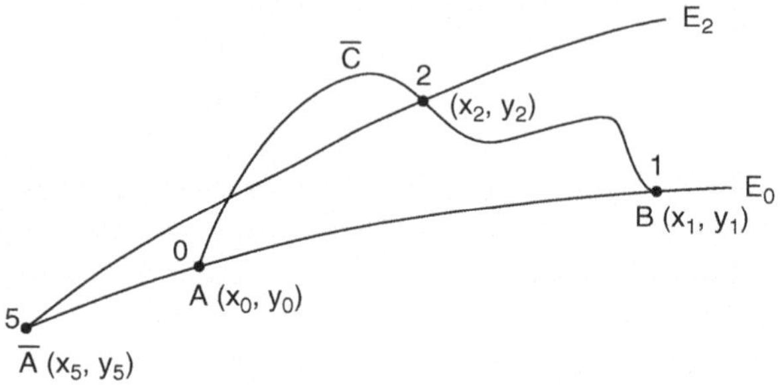


**Fig. 3.1** Rendering of figure on p. 86 [738]

interest in them. Bliss used an abstract version of such a theorem in one of his 1920 papers in a most ingenious way, to be described below. In a Colloquium talk given in 1909 [709], Bliss considered similarities between the two types of theorems. Finally, an abstract version of the implicit function theorem, still in use today, was a major achievement of the Chicago school of functional analysts; it followed from some of the work presented in this chapter. We will consider the principal use of this type of theorem by Bolza in this section.

In the preceding section, necessary conditions that an extremal $E_0$ for a problem be an actual minimum were discussed. Sufficient conditions were known, but the major result of this sort came again from Weierstrass and involved the idea of a field of extremals. The notion is defined [ibid., pp. 78–79] as follows (with less rigor than Bolza): Suppose $\phi(x, \gamma)$ is a one-parameter family of extremals in which $E_0$ corresponds to $\gamma = \gamma_0$. Let $S_k$ be the strip in the $(x, y)$ plane swept out by the extremals as $x$ and $\gamma$ run through their domain values. Then $S_k$ is called a field of extremals about $E_0$ if through every point $(x, y)$ of $S_k$ there passes but one extremal of the set. Furthermore, the inverse function from $(x, y)$ of the strip to the defining $\gamma$ parameter of the extremal is to be continuously differentiable.[18]

The proof of the sufficient condition given by Weierstrass and explicated by Bolza is quite involved; we attempt a summary here, highlighting the role of the implicit function theorem. Given an extremal $E_0$ passing through two given points $A(x_0, y_0) = 0$ and $B(x_1, y_1) = 1$, Bolza assumes that it is possible to construct a field of extremals containing $E_0$, based on $[x_0, y_0]$, consisting of elements passing through a point $\bar{A}(x_5, y_5) = 5$ on a continuation of $E_0$. Now suppose $\bar{C}$ is any other curve connection $A$ and $B$. (See Fig. 3.1. It was customary in the literature of the calculus of variations to abbreviate points in the plane with numbers.)

The goal is then to demonstrate that $\Delta J > 0$, where $\Delta J$ is the difference of the values of the integral

$$J_C = \int_{x_0}^{x_1} F(x, y, y')\,dx \tag{3.29}$$

taken over $E_0$ and $\bar{C}$.

---

[18] Bolza ascribes this notion in a limited sense to Weierstrass, with others generalizing it; see footnote on p. 79.

Let $2 = (x_2, y_2)$ be any point on $\bar{C}$, and $E_2$ the unique extremal through 5 and 2.[19] Two functions are introduced by Bolza:

(1) $S(x_2) = J_{52} + \bar{J}_{21}$ where the first expression is $J_C$ taken from 5 to 2 along $E_2$ and the second is $J_C$ taken from 2 to 1 along $\bar{C}$. The co-ordinate $x_2$ is considered a variable, $x_0 \leq x_2 \leq x_1$.
(2) $E(x, y; p, \tilde{p}) = F(x, y, \tilde{p}) - F(x, y, p) - (\tilde{p} - p)F_{y'}(x, y, p)$, sometimes called the "excess function," due to Weierstrass. It is the difference, for fixed $x$ and $y$, between $F(x, y, \tilde{p})$ and the first Taylor polynomial of $F$ centered at $p$ evaluated at $p = \tilde{p}$.

The basic equation from which the sufficient condition follows is then

$$\Delta J = -[S(x_1) - S(x_2)] = \int_{x_0}^{x_1} E(x_2, y_2; p_2, \bar{p}_2)\, dx_2 \tag{3.30}$$

where $p_2$ and $\bar{p}_2$ are the slopes of the curves $E_2$ and $\bar{C}$ at 2, respectively; the bulk of the argument lies in proving this equation. The fact that some of the paths of integration are along extremals simplifies some of the expressions. A simple application of Taylor's theorem to $F(x, y, \tilde{p}) - F(x, y, p)$ relates the excess function to $F_{y'y'}(x, y, p)$:

$$E(x, y; p, \tilde{p}) = (\tilde{p} - p)F_{y'y'}(x, y, p^*) \tag{3.31}$$

where $p^* = p + \theta(\tilde{p} - p)$ for some $\theta$, $0 < \theta < 1$.

From this a sufficient condition is provable: Suppose there exists a field $S_k$ around $E_0$ on $[x_0, x_1]$ such that at every point $(x, y)$ of $S_k$ and every finite value $\tilde{p}$ one has $F_{y'y'}(x, y, \tilde{p}) \geq 0$. Then $\Delta J \geq 0$; equality implies that $\bar{C}$ coincides with $E_0$. The uniqueness of the field element passing through a fixed point, the property which distinguishes a field from a simple parametrized family of extremals, is used for the proof of the final statement. The introduction of the quantity $F_{y'y'}$ connects this result with earlier necessary conditions discussed in the preceding section.

Thus the focus shifts to conditions which guarantee the existence of a field of extremals containing $E_0$, since further statements regarding sufficient conditions for a minimum are basically those required to produce a field. The field of extremals used by Bolza is actually a family each of whose elements passes through the point $\bar{A}$, the point on a continuation of $E_0$. It must be proved possible to construct such a field. Suppose for the purposes of this paragraph, we relabel $\bar{A} = (x_0, y_0)$. Then assume for the given optimization problem that the entire set of extremals is given by $y = f(x, \alpha, \beta)$, the two-fold infinite collection of solutions to Euler's equation. Let $\alpha = \alpha_0$ and $\beta = \beta_0$ be particular values of $\alpha$ and $\beta$ for which $y_0 = f(x_0, \alpha_0, \beta_0)$, and let $E_0$ be the extremal $y = f(x, \alpha_0, \beta_0)$ passing through $\bar{A}$. The idea is to

[19] Points 3 and 4 were also given, lying, respectively, to the left and right of 2 on $\bar{C}$, for purpose of proving the basic equation which follows.

create a one-parameter family of extremals using the slope of a solution through $\bar{A}$ as the parameter. To this end, Bolza introduces a new variable $\gamma$ be defining $\gamma = f_x(x_0, \alpha, \beta)$ and sets $\gamma_0 = f_x(x_0, \alpha_0, \beta_0)$. There are then three variables $\alpha$, $\beta$, and $\gamma$ connected by the two equations

$$f(x_0, \alpha, \beta) - y_0 = 0 \tag{3.32}$$

$$f_x(x_0, \alpha, \beta) - \gamma = 0 \tag{3.33}$$

These equations are satisfied at $\alpha = \alpha_0$, $\beta = \beta_0$, and $\gamma = \gamma_0$. Bolza shows that the Jacobian of the two functions is nonzero at $(\alpha_0, \beta_0, \gamma_0)$ which allows by the implicit function theorem a solution $\alpha = \alpha(\gamma), \beta = \beta(\gamma)$, continuous in a neighborhood of $\gamma = \gamma_0$ satisfying $\alpha_0 = \alpha(\gamma_0)$ and $\beta_0 = \beta(\gamma_0)$. The one-parameter family of extremals through $\bar{A}$ is then defined by $\phi(x, \gamma) = f(x, \alpha(\gamma), \beta(\gamma))$; one has $y_0 = \phi(x_0, \gamma)$ for all $\gamma$ sufficiently close to $\gamma_0$ and $E_0 : y = \phi(x, \gamma_0)$. To verify that this family forms a field, Bolza notes that this follows if one can show that $\phi_\gamma(x, \gamma_0) \neq 0$ throughout the interval of discussion; this condition guarantees that the extremals trace out a strip as $\gamma$ varies in one direction. The final statement of Weierstrass' sufficiency condition [ibid. p. 96] involves, in addition to the hypothesis $F_{y'y'}(x, y, \tilde{p}) > 0$ throughout a neighborhood of $E_0$, a technical condition which implies the nonvanishing of $\phi_\gamma$.

Thus the implicit function theorem plays a role in the parametrization of a family of extremals surrounding $E_0$ which becomes a field under the appropriate conditions. One example of this is described in the next section.

## 3.4 Bliss' Publications on Existence, Smoothness, and Implicit Function Theorems Up to 1919

Bliss of course was familiar with the contents of the material just presented, and in a string of publications from the early years of the century up to the War, he published related work. In this section, we review some of this work and show how it provided the backdrop to his key 1920 papers. We shall proceed roughly chronologically.

In a 1904 paper [703], Bliss generalized a theorem of Picard regarding a solution of the second-order differential equation

$$\eta'' = f(\xi, \eta, \eta') \tag{3.34}$$

which joins two points $p$ and $q$. In reviewing the results for this equation, he points out that the theorems of Picard guarantee the existence and uniqueness of a smooth solution $\eta = \phi(\xi, \xi_0, \eta, \eta_0)$ passing through a given point $(\xi_0, \eta_0)$ and provides the usual references for the existence of derivatives of these solutions with respect to the constants. Then if a second point $(\xi_1, \eta_1)$ is given, he observes that if $\eta_0'$ can be determined so that

$$\eta_1 = \phi(\xi_1, \xi_0, \eta_0, \eta_0'), \tag{3.35}$$

then the given solution will pass through both points. He then imposes conditions which allow the equations just given to be solved for $\eta_0'$ in terms of the remaining variables using the implicit function theorem, "and solutions of [4.1] will be determined which pass through any two points in the neighborhoods of the two originally considered" [703, p. 29]. In the remaining sections, he generalizes this result to the equation

$$\frac{x'y'' - x''y'}{(x'^2 + y'^2)^{3/2}} = 1/\rho = G(x, y, x', y') \tag{3.36}$$

in order to minimize an integral of the form

$$I = \int_{x_0}^{x_1} F(x, y, x', y')\, dx \tag{3.37}$$

In providing this generalization, the use of the implicit function theorem is again made, and a proof of the differentiability of the solutions with respect to constants is again given. This paper, which filled a gap in a lecture of Weierstrass, shows then at the least an absorption of the techniques reviewed by Bliss and their application to a new problem.

This work was followed by a key paper of 1905 [704] in which Bliss addressed proofs of the existence and smoothness of solutions to systems of differential equation. The introduction declares that such proofs for a single equation

$$\frac{dx}{dt} = f(t, x) \tag{3.38}$$

where $f$ is "expansible" in a power series, are given in textbooks on differential equations, and methods which allow weaker conditions on $f$ are known. As we have seen, "In many applications however the mere existence theorem is of little value unless accompanied by a proof of the continuity and differentiability of the solution considered as a function of the initial constants." Then, as previously noted, "As the literature dealing with these latter questions is somewhat scattered and the proofs complicated, it is proposed ... to collect and simplify as far as possible the results of the different authors" [704, p. 49].

The existence proof for the single equation closely follows Picard's in his *Traite d'analyse*, and uniqueness is given an unusually lengthy treatment. In the third section is Bliss' first expression of the idea of "imbedding,"[20] though the

---

[20] Also referred to as "embedding"; Bolza: "einbettung."

term is not used here. "…it can be shown that the solution through the point $(\tau, \xi)$ can be surrounded by solutions corresponding to neighboring initial values $(\tau + \Delta\tau, \xi + \Delta\xi)$ and that these go over continuously into the former when $\Delta\tau$ and $\Delta\xi$ approach zero." [ibid. p. 55][21] Use is not made of this notion in this paper, but it is mentioned in later work and reminds one of the more stringent notion of a field of extremals, which requires a parametrization of the family. Existence of the first- and higher-order partial derivatives of the solutions with respect to the constants is proved, the latter under the hypothesis of higher-order differentiability of $f$.

The results and proofs of this first part of the paper are then generalized immediately in the second by the use of vector and matrix ideas. Vectors are referred to as "complex numbers," and credit for their introduction into these problems is given to Peano, who used them in several papers in the preceding decades. After listing the necessary properties of vectors and matrices, including new definitions of modulus, the proofs are made to resemble those of the first section, and so the results follow elegantly with a minimum of effort. Differentiability of the solutions with respect to arbitrary parameters of the equation, as presented in [704], is not addressed, however.

This paper is cited repeatedly in Bolza's *Vorlesung* of 1909, and it is possible that it was in a sense commissioned for him for just such a purpose. It serves as Bliss' first original contribution to this subject. The use of it was made immediately in a 1906 joint publication with Bliss' close friend Max Mason [730], where it was called on to provide a family of solutions for a differential equation involved in the minimizing of an integral with discontinuous integrand. The two authors collaborated on a paper a few years later [732] in which fields of extremals in three space were constructed using a generalization to higher dimensions of the implicit function theorem due to Bolza but proved independently by the authors. It was perhaps on account of this work that Bliss presented his own proof of a general implicit function theorem in 1912 [708].[22]

That Bliss' interest in these two types of theorems transcended their use as a tool is shown in a 1909 series of lectures given at an American Mathematical Society Colloquium at Princeton University and published in 1913 [709]. The theorems are introduced together in the opening paragraphs, and some historical background is given. In an acknowledgment of the influence of E. H. Moore's General Analysis, it is pointed out that the proof of the implicit function theorem offered in the lectures[23] can be generalized to allow the independent variable to lie in a "general range," a notion of Moore's. The bulk of the lectures is devoted to trying to free the hypotheses and conclusions of this type of theorem from the "local" nature; to this end Bliss introduces some new ideas which are not relevant to this discussion. As for differential equations, his results of 1905 are reproduced, and their "immediate

[21] The phrase "go over" echoes Bolza's "ubergehen."

[22] These theorems are not identical; hypotheses and conclusions vary. But an account of the differences would lead us too far afield.

[23] This is the same proof as in [708].

consequence" [709, p. 87]. Bolza's imbedding theorem, [739, p. 178], is noted. But of particular interest is the remark that Bolza's theorem is "an analogue for differential equations of the fundamental theorem for implicit functions proved in section 1" [709, p. 87]. A similarity between the theorems, perhaps obvious, is thus pointed out: roughly speaking, each deals with a situation in which the existence of a single solution for an equation will, under certain circumstances, lead to that solution being surrounded by a family of solutions.

This is more intriguing given the presentation, earlier in the lecture, of a proof of the implicit function theorem by Goursat, [799], which uses successive approximations, the basis of Picard's existence proof. Bliss presented this because "it affords a direct method of finding the values of implicit functions, and justifies computations sometimes used in the applications of the theory" [709, p. 16]. Earlier in the paper, Bliss showed the practical use of the differential equations approximation technique by giving an application to solving Kepler's equation: "This procedure is closely allied to a method of approximation due to Goursat ... suggested apparently by Picard's treatment of the existence theorem for differential equations." [ibid. p. 2] (It would appear then that Bliss knew something of the numerical approximation techniques at this time.) But he also notes Goursat's reference to Picard, though not quoting the line "One knows the beautiful results obtained by M. É. Picard in the study of differential equations and partial differential equations, thanks to the method of successive approximation. This method applies equally with great ease to the theory of implicit functions" [799, p. 184].

We conclude this section by citing the previously mentioned 1918 paper, in which Bliss presents proofs of the 1905 results "by a method which seems more natural and simpler than those which have hitherto been published" [712, p. 5]. The notion of imbedding is made more precise here [ibid. p. 16], and the analogy to the implicit function theorem is repeated. The uniqueness proof is brief. Vectors are now called "multipartite numbers." A reference to Moore's General Analysis is again given when a definition of the modulus of a matrix is presented;[24] connections with definitions of such a modulus by Hilbert and Peano are mentioned. Smoothness of solutions with respect to arbitrary parameters is cleverly dispatched at the end of the paper in a single paragraph. But overall the proof methods are the same as those of 1905 in their use of Picard's technique and its generalizations to vectors and matrices, with one interesting exception.

Bliss starts with the simple differential equation

$$\frac{dx}{d\tau} = f(\tau, x) \tag{3.39}$$

and a point $(\tau_0, x_0)$ in a suitable region of the plane, and by the usual sequence of Picard iterates obtains the solutions with the appropriate properties. When it comes

[24] For a vector $y$, mod $y$ is the usual Euclidean length; for a square matrix $A$, mod $A$ = maximum of $\sqrt{< Ay, Ay >}$ over vectors $y$ with mod $y = 1$.

to the derivatives of the solution with respect to, say, $x_0$, he differentiates both sides of the successive approximation equation y

$$v_{m+1}(\tau, \tau_0, x_0) = x_0 + \int_{\tau_0}^{\tau_1} f(\tau, v_m)\, d\tau \tag{3.40}$$

with respect to $x_0$ to get

$$\frac{\partial v_m}{\partial x_0} = 1 + \int_{\tau_0}^{\tau_1} f_x(\tau, v_{m-1}) \frac{\partial v_{m-1}}{\partial x_0}\, d\tau \tag{3.41}$$

By successive substitution, he gets a finite series for $\frac{\partial v_m}{\partial x_0}$ which limits out to and infinite series convergent to $\frac{\partial v}{\partial x_0}$:

$$\frac{\partial v}{\partial x_0} = 1 + \int_{\tau_0}^{\tau} f_x(\tau, v)\, d\tau + \int_{\tau_0}^{\tau} f_x(\tau, v)[\int_{\tau_0}^{\tau} f_x(\tau, v)\, d\tau]\, d\tau + \ldots \tag{3.42}$$

notation as in the paper [p. 20].

More interesting than the formula itself is the method of obtaining it. Though one would not need to be familiar with the subject of integral equations in order to give this argument, it bears a striking resemblance to a solution method in that currently active area of research, an area summarized in a 1909 volume with which Bliss was familiar. Since this subject will play a role in future developments, we shall take an opportunity to say a word about it. The volume in question [734] was a slim introduction by Maxime Bôcher. The field had much the same backstory as the calculus of variations: originating in the consideration of certain simply stated physical problems, it passed through a rigorization process until reaching a certain mathematical maturity by 1900. Bliss reviewed Bôcher's book [707], which played a role similar to his own 1905 paper: "…the sources of information concerning the theory have remained widely scattered and none too easily accessible to any but the specialist in the subject." [p. 207]

One of the types of integral equations dealt with bôcher is of the following form: with $f(x)$ and $K(x, y)$ being known functions and $u(x)$ the unknown, one considers

$$u(x) = f(x) + \int_{a}^{x} K(x, \xi)u(\xi)\, d\xi \tag{3.43}$$

Liouville in a paper of 1837 [880] devised a method to solve a particular case of the second type of equation, a method which came to be known as successive substitution.[25] This technique is merely the substitution of the right-hand side into the expression for $u$ in the integrand for $u(\xi)$. Repeating this process yields

[25] Note the difference between successive substitution and the process of Picard iteration.

ultimately an infinite series, similar to that of Bliss', which is proved to be the solution to the equation under certain hypotheses on $K$ and $f$.

In conclusion, we note that versions of Bliss' proofs of the implicit function theorem, existence and smoothness theorem for differential equations, and embedding theorems were bundled as a typewritten set of notes [723], dated "Chicago, 1928," to be used in his classes in the calculus of variations. The notes appeared almost verbatim as an appendix in his final volume on the subject, [728].

We finish this section with a brief account of the published version of the dissertation, mentioned [p. 2] in [709], of one of Bliss' students, Charles Fischer [776].[26] This paper points to a beginning of the influence on Bliss' work of the functional-analytic chain of ideas begun by Volterra's functions of a line. The thesis concerns the Volterra derivative (to be defined in a later section of the present chapter) of a curve $y$ which is connected with a system of differential equations and a perturbation $\eta$ of it. At a certain point, an implicit function theorem of a more abstract nature is proved:

Let $H$ given by $\eta = \eta(x)$ be a curve in the $x\eta$ plane, $\alpha = (\alpha_1, \ldots, \alpha_m)$ real variables, and $\phi_i(H; \alpha_1, \ldots, \alpha_m) = \phi_i(H, \alpha)$, $i = 1, \ldots, m$ a set of real-valued functions. Suppose that $\phi_i(H, \alpha) = 0$, $i = 1, .., m$ has $(H_0; 0, \ldots, 0)$, where $H_0 = \eta(x) = 0$, as a solution. Suppose in addition that at $(H_0; 0, \ldots, 0)$ the Jacobian $|\frac{\partial \phi_i}{\partial \alpha_j}| \neq 0$. Then under appropriate local continuity and differentiability conditions on the $\phi_i$, there will be a neighborhood of $H_0$ such that for each element $H$ of it there corresponds a solution $(H, \alpha)$ of $\phi_i(H, \alpha) = 0$, $i = 1, \ldots, m$ where $\alpha$ is restricted to a neighborhood of $\bar{0} = (0, \ldots, 0)$. The resemblance to the usual implicit function theorem is evident, but the distinct difference is that the first arguments of the $\phi_i$, "the infinity of independent variables" [p. 373] are given by $\eta(x)$.[27] This necessitates a new definition of "neighborhood" of $H_0$, given as an $\epsilon$—ball in the sup norm. The proof of the theorem follows closely Bliss' proof of the real-valued implicit function theorem given in the Colloquium lectures.

## 3.5 Bliss 1920—DECAF

Bliss' two ballistics-inspired pure mathematics papers appeared consecutively in Volume 21 of the *Transactions of the American Mathematical Society*. The first, *Differential Equations Containing Arbitrary Functions*, will be denoted as DECAF in what follows. It makes no mathematical references to the ballistics questions from

[26] Fischer was the first of several students of Bliss who explored the more functional-analytic generalizations of implicit function theory and the calculus of variations. More of these students will be featured in later sections, and Fischer's work will be seen in what follows to contain more topics of interest than the theorem dealt with here.

[27] Regarding a function defined on an interval as an infinite collection of variables rather than a single point in a space is characteristic of Volterra's view in his "functions of a line."

which it originated and could be read as a seamless follow-up to its predecessors of 1918, a further development of the topic of smoothness of solutions to systems of differential equations. The level of abstraction, however, has now been raised to the point of considering the *functions* involved in defining the equations as variable, a step beyond anything previously attempted and an obvious result of the absorption of some literature on the function of a line. Those functions are supplementing the role previously played by scalar parameters or initial condition constants.

One of the goals of the paper is to prove, under certain hypotheses on the system of differential equations, the existence of differentials of those solutions with respect not only to the constants of integration but also to the functions used in defining the equations. This is how the gist of the paper is described by those who later cited it. The differentials in question were of a new type, "... differentials in a sense analogous to that of Fréchet or Volterra in the theory of functions of a line" [718, p. 80]. The existence of these differentials serves two purposes: The first is to describe in purely mathematical terms this new form of smoothness of solutions. The second was to prove mathematically the existence of the differential variations used in ballistics, the variations to which Moulton, Bliss, Franklin, and Hoar devoted so much practical attention. This last point, perhaps of questionable value to some, was not viewed as such by some of his mathematical successors. We shall return to this point after considering in some detail the contents of the paper.

The first paragraph of the paper contains a reference to ballistics: "A recent study of the problem of computing differential corrections for a trajectory has led the writer to the present investigation of the properties of solutions of differential equations considered as functions of other arbitrary functions which enter into the differential equations themselves" [718, p. 79]. These equations, the equations of the motion of a projectile in application, are given by

$$\frac{dx^{(i)}}{d\tau} = f^{(i)}(\tau, x^{(1)}, \ldots, x^{(p)}),\ i = 1, \ldots, p \tag{3.44}$$

and a known particular solution $x^{(i)} = u^{(i)}(\tau)$, $\tau_1 \leq \tau \leq \tau_2$, $i = 1, \ldots, p$ corresponding to a particular set of functions $f^i$ is given. Considering the functions $f^{(i)}$ as variable[28], the solutions are functions of these functions as well as the initial conditions $(\tau_0, x_0)$, where $x_0 = (x_0^{(1)}, \ldots, x_0^{(p)})$. Bliss uses the notation $x^{(i)}[\tau, \tau_0, x_0, f]$ to show the dependence on all four quantities. The use of matrices and "multi-partites" continues from 1918; "one-partites" are given by Greek letters, "p-partites" by small Roman type, and matrices by capital Roman type. Thus the previous system is again abbreviated by

$$\frac{dx}{d\tau} = f(\tau, x) \tag{3.45}$$

[28] "... the presence of arbitrary functions as in the case of ballistics." [p. 80]

with initial solutions $x = u(\tau)$ corresponding to the special set $f'$ of functions $f$, the prime does not signify differentiation. The graph of this solution is denoted by $C$. The members of $u$ have continuous derivatives on $[\tau_1, \tau_2]$. There is also a neighborhood $C_\epsilon$ of $C$ consisting of all points $(\tau, x)$ satisfying the conditions $\tau_1 \leq \tau \leq \tau_2$, $\text{mod}[x - u(\tau)] < \epsilon$, in which all functions $f$ considered are continuous and satisfy the Lipschitz condition $\text{mod}[f(\tau, x') - f(\tau, x)] < \kappa \,\text{mod}(x' - x)$.

Bliss then defines the usual Picard sequence of iterates with the set $f$ included, calling them $v_m[\tau, \tau_0, x_0, f]$:

$$\begin{aligned} v_0[\tau, \tau_0, x_0, f] &= u(\tau) + x_0 - u(\tau_0) \\ v_{m+1}[\tau, \tau_0, x_0, f] &= x_0 + \int_{\tau_0}^{\tau} f(\tau, v_m)\, d\tau \end{aligned} \tag{3.46}$$

(Note that although $u(\tau_0)$ is not necessarily equal to $x_0$, the equality holds for each $v_m$ and their limit $v$.)

Proceeding as in [712], he derives a domain $R$ in which the iterates $(\tau, v_m)$ are well-defined and lie in $C_\epsilon$ The domain is given by

$$R : \tau_1 \leq \tau \leq \tau_2, \;\; (\tau_0, x_0) \in C_{\delta/2}, \;\; \text{mod}(f - f') < \frac{\delta}{2\lambda} \tag{3.47}$$

where $\lambda$ is the length of the interval $[\tau_1, \tau_2]$ and $\delta = \epsilon e^{(-\kappa\lambda)}$. (Note that $\kappa$, $\lambda$, and $\epsilon$ are all prescribed ahead of time.) The account of the values of the constants is less important than the fact that the "region" $R$ is an immediate generalization of that in [712] merely by adding the condition on $f$. Then the proofs that the sequence of "functionals" $v_m[\tau, \tau_0, x_0, f]$ converges on $R$ to a unique solution $v[\tau, \tau_0, x_0, f]$ "are now precisely those of [712] cited above, and they need not be reproduced here. [718, p. 82]" This is the first use of the word "functional" by Bliss in the paper.

The first attempt at defining the smoothness of solutions $v[\tau, \tau_0, x_0, f]$ is given by referring to *continuity of type I*: let $\phi$ be any of the $v_m$ and $\Delta\phi$ be given by

$$\Delta\phi = \phi[\tau + \Delta\tau, \tau_0 + \Delta\tau_0, x_0 + \Delta x_0, f + \Delta f] - \phi[\tau, \tau_0, x_0, f]. \tag{3.48}$$

Then $\phi$ has continuity of type I if for every positive constant $\epsilon'$ there is a second constant $\delta'$ such that $\text{mod}(\Delta\phi) < \epsilon'$ whenever the sets $[\tau + \Delta\tau, \tau_0 + \Delta\tau_0, x_0 + \Delta x_0, f + \Delta f]$ and $[\tau, \tau_0, x_0, f]$ are in $R$ and $|\Delta\tau|$, $|\Delta\tau_0|$, $\text{mod}(\Delta x_0)$, $\text{mod}(\Delta f) < \delta'$, the last inequality assumed to hold uniformly for all points $(\tau, x)$ in $C_\epsilon$. Again, the conditions are natural extensions of those in [718]. It is then proved that the solutions $v[\tau, \tau_0, x_0, f]$ have continuity of type I in $R$.

In order to address the smoothness of the solutions with respect to the initial conditions, Bliss introduces *continuity of type II*:

Let $\phi[\tau, \tau_0, x_0, f]$ be a functional. Then $\phi$ has continuity of type II if for every positive constant $\epsilon'$, there is a second constant $\delta'$ depending only on $f$ and $\epsilon'$ such that $\text{mod}(\Delta\phi) < \epsilon'$ whenever the sets $[\tau + \Delta\tau, \tau_0 + \Delta\tau_0, x_0 + \Delta x_0, f + \Delta f]$ and $[\tau, \tau_0, x_0, f]$ are in $R$ and and $|\Delta\tau|$, $|\Delta\tau_0|$, $\text{mod}(\Delta x_0)$, $\text{mod}(\Delta f)$, $\text{mod}(\Delta f)_x <$

$\delta'$ hold. The last two inequalities hold for all arguments $(\tau, x)$ of $\Delta f$ and $(\Delta f)_x$ in $C_\epsilon$.[29]

For the remaining part of the theorem, Bliss restricts the functions $f$ to those having continuous first derivatives in $C_\epsilon$ with a common upper bound. Then there will exist a constant $\kappa$ such that these functions satisfy the Lipschitz condition above with $\kappa$. The existence of the matrix of partial derivatives denoted by $\frac{\partial v}{\partial x_0}$ and the multi-partite $\frac{\partial v}{\partial \tau_0}$ follows again as in [712], with the matrix and the multipartite of partials being given by the same infinite series. It is then proved that $v$ and $\frac{\partial v}{\partial \tau}$ have partial derivatives with respect to $\tau$, $\tau_0$ and $x_0$ in $R$ The first section of the paper is then concluded with some remarks about the extension of these results to the case where the $f$ involve parameters.

It is apparent that these results not only generalize those of Bliss' earlier papers but also involve a shift in perspective from merely proving the smoothness of solutions under certain circumstances to accounting for the ways in which changes in the input data $[\tau, \tau_0, x_0, f]$ effect changes in the output functionals $v[\tau, \tau_0, x_0, f]$ and their derivatives, the changes in each case being measured by newly introduced "metrics" of a topological nature: the continuities of types I and II.

Type II continuity is sufficient to generalize earlier results on initial conditions, but it is not the best that can be done. The second section of DECAF begins by announcing "Under the assumptions made in the last paragraph of the theorem of the preceding section[30] that the functionals $v[\tau, \tau_0, x_0, f]$ have not only continuous first derivatives with respect to the arguments $\tau, \tau_0, x_0$, but also differentials with respect to the arguments $f$ in a sense analogous to that of the theory of functions of lines" [718, p. 87].

The first step is to prove the existence of what Bliss calls a *difference function for v with respect to f.* The definition of this object is as follows: for each pair of "elements" $[\tau, \tau_0, x_0, f]$ and $[\tau, \tau_0, x_0, \bar{f}]$ of the region $R$, there exists a matrix $L[\tau', \tau, \tau_0, x_0, f, \bar{f}]$ such that the differences $\Delta v = v[\tau, \tau_0, x_0, \bar{f}] - v[\tau, \tau_0, x_0, f]$ are expressible in the form

$$\Delta v|\tau = \int_{\tau_0}^{\tau} L[\tau', \tau, \tau_0, x_0, f, \bar{f}] \Delta f(\tau', v[\tau', \tau_0, x_0, f])\, d\tau' \tag{3.49}$$

where $\Delta f = \bar{f} - f$. The elements of $L$ have continuity of type II. Bliss takes this equation as the definition of the difference function: the emphasis is on finding the

[29] "In order that $\Delta\phi$ shall approach zero, therefore, it is necessary, according to this definition, that the *derivatives* of the elements of $\Delta f$ with respect to the elements of $x$ shall approach zero as well as the elements of $\Delta\tau$, $\Delta\tau_0$, $\Delta x_0$, $\Delta f$. Furthermore, the uniformity of the continuity is with respect to the arguments $\tau$, $\tau_0$, $x_0$ but not with respect to $f$, as is indicated by the fact that the constant $\delta'$ depends upon $f$ as well as upon $\epsilon'$ [p. 84].

[30] The assumptions are the continuous differentiability of the $f$ and the uniform boundedness of $|\frac{\partial f^i}{\partial x_j}|$ by $\kappa$ on $C_\epsilon$.

matrix $L$. To get the integral expression, Bliss will integrate a differential equation for $\Delta v$:

$$\frac{d\Delta v}{d\tau} = \bar{f}(\tau, v + \Delta v) - f(\tau, v) = \bar{f}(\tau, v + \Delta v) - \bar{f}(\tau, v) + \Delta f(\tau, v) \tag{3.50}$$

and to the next to the last term is applied Taylor's theorem with integral form of the remainder to get

$$\frac{d\Delta v}{d\tau} = A\Delta v + \Delta f(\tau, v) \tag{3.51}$$

where $A$ is the matrix of terms

$$A = \int_0^1 \bar{f}_x(\tau, v + \alpha\Delta v)\, d\alpha \tag{3.52}$$

(Note that this equation is not linear in $\Delta v$ due to the presence of $\Delta v$ terms in $A$.)

This (3.77) is the differential equation to be integrated, but this cannot be done as it stands. The technique is to find an "integrating factor matrix" $L$ which will serve as the matrix in the definition of the difference function. To this end, Bliss introduces another "associated" set of differential equations

$$\frac{dl}{d\tau} = -lA \tag{3.53}$$

where $l$ is a p-partite row. An argument is given to prove the existence of a solution matrix $L$ whose rows are solutions to this equation satisfying certain initial conditions. Upon multiplying both sides of 6.8 by $L$, one obtains

$$L\frac{d}{d\tau}\Delta v = -\frac{dL}{d\tau}\Delta v + L\Delta f(\tau, v) \tag{3.54}$$

which upon integration gives the desired difference relation 6.6.

Bliss then proceeds to derive similar difference functions with respect to each of the other arguments $\tau$, $\tau_0$, $x_0$ of $v$, and an overall difference function expression for a change in all four arguments simultaneously. But before discussing this, let us note the return of a familiar concept in disguise: the matrix $L$ of solutions of (3.79) acts in a manner similar to the adjoint in the ballistic work of Bliss, as an integrating factor. In fact its definition resembles that of the adjoint of a linear system.

Returning to the idea of difference functions, Bliss notes that for the remaining arguments of $v$, one can let

$$\Delta_1 v = v[\bar{\tau}, \bar{\tau_0}, \bar{x_0}, \bar{f}] - v[\tau, \tau_0, x_0, \bar{f}] \tag{3.55}$$

$$\Delta_2 v = v[\tau, \tau_0, x_0, \bar{f}] - v[\tau, \tau_0, x_0, f] \tag{3.56}$$

$$\Delta v = \Delta_1 v + \Delta_2 v \tag{3.57}$$

and one lets $\Delta\tau = \bar{\tau} - \tau$ and similarly for $\Delta\tau_0$, $\Delta x_0$ and $\Delta f$. Then applying Taylor's theorem as before yields

$$\Delta v = a\Delta\tau + b\Delta\tau_0 + C\Delta x_0 + \int_{\tau_0}^{\tau} L\Delta f d\tau' \tag{3.58}$$

where $a$, $b$ and $C$ are integrals, analogous to 6.9, of $v_\tau$, $v_{\tau_0}$, and $v_{x_0}$, respectively. This expression is then the difference function for $\Delta v = v[\bar{\tau}, \bar{\tau_0}, \bar{x_0}, \bar{f}] - v[\tau, \tau_0, x_0, f]$.

From these considerations of the difference function follows quickly the existence of the differential ultimately desired, in fact, it is a differential in all arguments simultaneously, "in a sense analogous to that defined by Fréchet in the theory of the functions of lines." Setting $\bar{\tau} = \tau$, $\bar{\tau_0} = \tau_0$, $\bar{x}_0 = x_0$ and $\bar{f} = f$ reduces the integral coefficients to $a = v_\tau$, $b = v_{\tau_0}$, $C = v_{x_0}$; if one sets $L'$ equal to the value of $L$ at these same arguments, and one defines

$$\delta v = v_\tau\Delta\tau + v_{\tau_0}\Delta\tau_0 + v_{x_0}\Delta x_0 + \int_{\tau_0}^{\tau} L'\Delta f\, d\tau' \tag{3.59}$$

then one has

$$\Delta v = \delta v + \mu h[\Delta\tau, \Delta\tau_0, \Delta x_0, \Delta f] \tag{3.60}$$

where $\mu$ is the maximum of $|\Delta\tau|$, $|\Delta\tau_0|$, $\text{mod}\Delta x_0$, $\text{mod}\Delta f$, and $\text{mod}(\Delta f)_x$ in $C_\epsilon$, and $h$ is a multipartite function whose modulus approaches zero with $\mu$.[31] The equation (3.85) is called the differential of $v$ at $[\tau, \tau_0, x_0, f]$; "It gives the value of $\Delta v$ to terms of the first order in the variables $\Delta\tau$, $\Delta\tau_0$, $\Delta x_0$, $\Delta f$ as described above [718, p. 89]." It is interesting that no additional smoothness hypotheses are necessary to achieve this existence.

This closes the discussion of the existence of differentials; details left out may of course be filled in by a perusal of the paper. Before moving on to the last section, a few comments should be made. Bliss' references to the literature of differentials include [784, p. 139], and a look at that page is interesting. Fréchet's definition of a linear functional with argument $f$, denoted $U_f$, takes place in the context of the domain ("champ") of functions continuous on an interval; this domain is denoted $C$. Frèchet begins by assuming that a linear functional is "distributive" and continuous. Continuity is defined relative to the topology of uniform convergence on $(a, b)$, though this language is not used. Distributivity is defined by the equation $U_{f_1+f_2} = U_{f_1} + U_{f_2}$. "We say that a functional $U_f$ defined on the domain $C$ has a differential for the argument $f_0$ if there exists a linear functional $T_{\Delta f}$ of the increment $\Delta f$ of the argument, which differs from the argument of the functional $U_f$ only by a quantity infinitely small with respect to the maximum of the absolute value of the argument

[31] THe expression for $h$ is given explicitly in the paper on p. 89.

$\Delta f(x)$ in $(a, b)$. Calling $m\Delta f$ this maximum, one has

$$U_{f+\Delta f} = U_f + T_{\Delta f} + \epsilon m\Delta f \tag{3.61}$$

$\epsilon$ tending to zero with $m\Delta f$ [ibid, p. 139]. Bliss' adaptation of this concept is straightforward as shown in (3.86). Bliss mentions neither the linearity of $\delta v$ nor its continuity, perhaps assuming both to be obvious.[32]

On the practical side, one may ask about the relation of this work to Bliss' unpublished technical report regarding antiaircraft fire, treated in "Introduction". There are clear similarities in the manipulations, especially in the use of an adjoint-like expression to make a system of differential equations exact. The relation between the two papers would make an interesting study.

In the final section of DECAF, Bliss proves an implicit function theorem and, unusually for such theorems, provides rules for differentiation of quantities. The statement of the theorem reads very much like a result in classical analysis from Bolza: Bliss considers an equation of the form $\phi[\tau, \tau_0, x_0, f] = 0$ having the particular solution $[\tau', \tau_0', x_0', f']$ and $[\tau', \tau_0', x_0', f']_\epsilon$ is the "totality of elements" $[\tau, \tau_0, x_0, f]$ for which

$$|\tau - \tau'|,\ |\tau_0 - \tau_0'|,\ \mathrm{mod}(x_0 - x_0'),\ \mathrm{mod}(f - f') < \epsilon \tag{3.62}$$

Then if the functions $\phi$ and $\phi_\tau$ have continuity of type I in $[\tau', \tau_0', x_0', f']_\epsilon$ and if $\phi[\tau', \tau_0', x_0', f'] = 0$ and $\phi_\tau[\tau', \tau_0', x_0', f'] \neq 0$, one has that

(1) the constant $\epsilon$ can be chosen so that in $[\tau', \tau_0', x_0', f']_\epsilon$ no two solutions $[\tau_1, \tau_0, x_0, f]$ and $[\tau_2, \tau_0, x_0, f]$ of the equation $\phi = 0$ exist with $\tau_1 \neq \tau_2$
(2) there is a neighborhood $[\tau', \tau_0', f']_\delta$ with every element of which there is associated a solution $[\tau, \tau_0, x_0, f]$ in $[\tau', \tau_0', x_0', f']_\epsilon$
(3) the function $\tau[\tau_0, x_0, f]$ so defined has continuity of type I in $[\tau_0', x_0', f']_\delta$
(4) One then also has $\tau' = \tau[\tau_0', x_0', f']$.

The proof will not be given here, but it is similar to the arguments presented in [738, pp. 79–80] for the existence of a field discussed earlier. Thus Bliss' first implicit function theorem in the domain of functions of a line has a proof following the path of a "classical" result for the existence of a field. He however also stated [p. 80] that his results here "are special cases of some much more general ones in [E. H. Moore's] General Analysis," though without citing those theorems.

The most important part of the theorem for application is the final remark, a rule given for the differential of the composition of two differential expressions. This is

[32] For completeness, we give the modern definition of the differential: following [807, p. 42], in the most abstract backdrop: suppose $G$ is a function on an open set $Y_0$ of a normed vector space $Y$ with values in a second normed vector space $B$. Then $G$ is said to be differentiable at a point $y_1$ in $Y_0$ if there is a continuous linear function $L$ mapping $Y$ into $B$ such that $\frac{|G(y)-G(y_1)-L(y-y_1)|}{|y-y_1|}$ tends to zero with the denominator. The function $L$ is called the differential of $G$ at $y_1$.

used in the ballistic considerations of the follow-up paper. Bliss lets $\phi$ and $\psi$ be particular functions of the set $v$ in the preceding section. He then "solves" $\phi = 0$, by the use of his implicit function theorem, to get $\tau = \tau[\tau_0, x_0, f]$ and substitutes this into $\psi$ to get $\psi[\tau_0, x_0, f] = \psi[\tau[\tau_0, x_0, f], \tau_0, x_0, f]$ thus $\psi$ is now a function of three variables $\tau_0$, $x_0$, and $f$. By substituting the difference function for $\tau = \tau[\tau_0, x_0, f]$ into the difference function for $\psi$, he states that " the differential $\delta_1\psi$ of $\psi$ [the composition] with respect to the arguments $\tau$, $\tau_0$, $f$ after the substitution $\tau = \tau[\tau_0, x_0, f]$ is evidently

$$\delta_1\psi = -\frac{\psi_\tau}{\phi_\tau}(\phi_{\tau_0}\Delta\tau_0+\phi_{x_0}\Delta x_0+\int_{\tau_0}^{\tau} l'\Delta f\, d\tau')+(\psi_{\tau_0}\Delta\tau_0+\psi_{x_0}\Delta x_0+\int_{\tau_0}^{\tau} \bar{l}'\Delta f\, d\tau') \tag{3.63}$$

where the arguments of the derivatives of $\phi$, $\psi$ are $\tau[\tau_0, x_0, f]$, $\tau_0$, $x_0$, and $f$" [p. 92]. Also, $l$ corresponds to $\phi$ and $\bar{l}$ to $\psi$. Implicit in this conclusion is the existence of a differential for $\tau = \tau[\tau_0, x_0, f]$, an analogue of the differentiability of the explicit solution in the classical case. The complication of the differential formula perhaps obscures its formal agreement with the simplest scalar analogue: given the relation $f(x, y) = 0$ and a second function $g(x, y)$, if one solves the first for $y = y(x)$, giving $f(x, y(x)) = 0$, then differentiation of this formula with respect to $x$ gives, upon substitution into $g$,

$$dg(x, y(x)) = -\frac{g_y(x, y(x))}{f_y(x, y(x))}[f_x(x, y(x))dx] + [g_x(x, y(x))dx] = -\frac{g_y}{f_y}df + dg \tag{3.64}$$

The fact that one can in this sense "calculate" with functionals in a calculus-like manner is an example of Hadamard's description of these operations as the "functional calculus." Bliss' interest in the implicit function theorem here stemmed from its to-be-demonstrated practical use but drew on his knowledge of such theorems from their frequent use in the calculus of variations. Their use for related but more abstract problems has already been mentioned in 1913s [776] and continued with the 1920 publication of [875]. In this work, by K. W. Lamson, also a student of Bliss', implicit function theorems "in the theory of Functions of a Line" [p. 143] are presented for two types of equations, originally considered by Goursat [799] (his 1903 paper mentioned above) in the context of functions of a finite number of variables. Lamson's goal is to present a rigorous demonstration of an idea for such a proof suggested by Volterra in 1913 [980]. The author's theorems include Goursat's as special cases and use his iterative method as well. The paper was a stepping stone along the way to the most general of functional-analytic implicit function theorems produced by the Chicago school, to be considered in a later section.

E. J. McShane was responsible for further interest in DECAF. In a paper of 1950 [895], and in the 1953 volume [898], he addressed the issues surrounding the differential introduced by Bliss.[33] (Recall that in his review of [898], the proof of the existence of the differentials was one of the few positive remarks made by Clifford Truesdell [974].) The paper shows an intriguing combination of mathematical sophistication and a willingness to consider different measures of smoothness of variation of the trajectory variables, depending on how well the measures fit the realities of the effects of external forces on flight [p. 116].

Though his major contributions in [895] involved an alteration of Bliss' mathematical assumptions in his treatment of the ballistics problem, his insistence on the importance of proving the existence of the first differential, regardless of which norms were used in its definition, is in evidence. Speaking of a new norm which he introduces, he states: "But the existence of a differential in this sense has not been before established. To the best of my knowledge, the only adequate mathematical discussion of the differential effects is due to G. A. Bliss," and our two papers are then cited. And again: "Since the practical importance of being able to compute differential effects has been established beyond dispute, it is logically important to prove that differential effects exist. [p. 471]"[34]

New proofs of Bliss' continuity and differentiability results are given which use a Gronwall-type inequality and are phrased in the crisp language of spaces and mappings. A lead-up theorem to Bliss' continuity result (stating a differential result would lead to lengthy complexity) illustrates the new style. McShane explains how the equations of motion, including functions like wind and temperature as functions of position, can be put into the form

$$\frac{dq^i}{dt} = F^i(\phi^j(t,q)),\ i = 1\ldots n,\ j = 1\ldots m \tag{3.65}$$

with initial conditions

$$q^i(t_1) = q_0^i \tag{3.66}$$

Using McShane's terminology exclusively, in the continuity theorem, the $F^i$ are assumed continuous on a bounded open set $R'$ in $R^m$, the $\phi^j$ are assumed continuous in $(t, q)$ where $q$ the $q$ lie in a bounded open set $R_0$ in $R^n$, except possibly for a finite number of values of $t$ where $\phi^j$ may have simple jump discontinuities. One has $t_1 \leq t \leq t_2$. A collection $\Phi_0$ is defined to be the system of "points" $[q_0, \phi, F]$ where $q_0$ is interior to $R_0$, and the solution to (3.65), (3.66) corresponding to the bracketed point is defined and in $R_0$ for $t_1 \leq t \leq t_2$. Now the equations (3.65),

---

[33] McShane worked in the Ballistic Research Laboratory at Aberdeen Proving Grounds, founded in 1938, during part of the second World War; this accounts for his interest in this material.

[34] This remark brings to mind Moulton's insistence that his method of numerical integration may very well have been introduced by any number of other people, but that his proof that the method converges is of a higher mathematical order. Similarly, it would seem evident that variations in a trajectory exist, but to prove mathematically that solutions of systems of differential equations possess a certain well-defined differential is higher-order matter.

(3.66) "map $\Phi_0$ into the space $C_n$ of $n$-tuples of real-valued functions defined and continuous on $[t_1, t_2]$." [p. 119] Defining $N[q] = \sup(|q(t)| : t_1 \le t \le t_2)$ as a norm on $C_n$, he lets "the analogous norm" $N_0[q_0, \phi, F]$ be defined by the greatest of the three numbers

$$|q_0| \tag{3.67}$$

$$\sup(|\phi(t, q)| : t_1 \le s \le t_2,\ q \in R_0) \tag{3.68}$$

$$\sup(|F(r)| : r \in R') \tag{3.69}$$

The theorem then reads: when the spaces $\Phi_0$ and $C_n$ are metrized in accordance with the respective norms $N_0$ and $N$, the mapping of $\Phi_0$ into $C_n$ defined by (3.65) and (3.66) is continuous.[35] Bliss' theorem on the differential is given a similar framing, and the bulk of the paper is devoted to alternative norms, one-branched trajectories, different parametrizations, different co-ordinate systems, uniqueness of solutions to the equations of motion under alternative hypotheses, and other largely mathematical topics.[36]

In [898], the notions of differential in the function space context is writ large in Chapter VIII where practical and theoretical issues run side by side. This exposition can be considered as a descendent of Bliss' second *Transactions* paper, to which we now turn.

## 3.6 Bliss 1920—FOLIB

In the second of the *Transactions* papers, which followed DECAF sequentially in the volume, Bliss attempted an altered exposition of the material of differentials directed more toward the ballistics problems to which it gave rise. Entitled *Functions of Lines in Ballistics*, hereafter referred to as FOLIB, its purpose was simply stated on the second page: noting that his adjoint method had already been published "... with

[35] The wording here omits a superset $\Phi$ of $\Phi_0$ which is not necessary for the statement of the theorem, see p. 119.

[36] It is interesting that Bliss refers to the *solutions* $v$ of the differential equations having a differential $\delta v$ at an element $[\tau, \tau_0, x_0, f]$, though Frèchet's definition refers to the *mapping* from one space to another as possessing a differential at a point of the domain space. The domain space here is the space of data specifying the system of differential equations, the range space the space of solutions to the equation, and the mapping the assignment of solutions to differential equations. McShane sometimes uses the latter phrasing. Thus one writes $dG(y; \delta y)$ for the value of the differential of $G$ at input $y$ and the input increment $\delta y$. The difference is comparable to the simplest notational case $y = y(x)$, where the two expressions for a change of output corresponding to a change of input $dx$ might be $dy|(x = x_o, dx)$ with the quantities in the right of the bar perhaps not even present and $df(x_0)dx$. The same comment applies to the continuity of the solutions $v$ with respect to $[\tau, \tau_0, x_0, f]$.

such mathematical details as are necessary for the practical ballistician" [i.e., his JUSA/CAJ article] "the purpose of the present paper is to show the connection between the ballistic theory and the theory of functions of lines" [719, p. 94]. He remarks that the formulas produced are applications of his previous paper to the equations of motion in ballistics. He aims at a level of exposition more general that of the JUSA/CAJ report but less abstract than DECAF, making the proofs as independent as possible from those in DECAF.

Bliss begins with three second-order differential equations, generalizations of the equations of motion:

$$\begin{aligned} x'' &= f(t, x, y, z, x', x', z') \\ y'' &= g(t, x, y, z, x', x', z') \\ z'' &= h(t, x, y, z, x', x', z') \end{aligned} \tag{3.70}$$

having initial solution

$$x = x(t), \; y = y(t), \; z = z(t) \tag{3.71}$$

for $t_1 \leq t \leq t_2$, for which the defining functions are $C^2$, having initial "element" of conditions $(t_0, x_0, y_0, z_0, x_0', y_0', z_0')$. In a neighborhood of the solution curve the functions $f$, $g$, and $h$ are also $C^2$. These three defining equations are then replaced by $\bar{f}, \bar{g}, \bar{h}$, also $C^2$, and a new set of equations

$$\begin{aligned} \bar{x}'' &= \bar{f}(t, \bar{x}, \bar{y}, \bar{z}, \bar{x}', \bar{y}', \bar{z}') \\ \bar{y}'' &= \bar{g}(t, \bar{x}, \bar{y}, \bar{z}, \bar{x}', \bar{y}', \bar{z}') \\ \bar{z}'' &= \bar{h}(t, \bar{x}, \bar{y}, \bar{z}, \bar{x}', \bar{y}', \bar{z}') \end{aligned} \tag{3.72}$$

is introduced, where a unique solution determined by the element $(t_0, x_0, y_0, z_0, \bar{x}_0', \bar{y}_0', \bar{z}_0')$ also exists. (In ballistics the initial position of the object is constant, and the initial time of fire does not change.) If the first four entries of the element are kept fixed, one then has $\bar{x}$ as a functional $\bar{x}[\bar{x}_0', \bar{y}_0', \bar{z}_0', \bar{f}, \bar{g}, \bar{h}]$ with similar expressions for $\bar{y}$ and $\bar{z}$. Replacing 7.1 with the equivalent first order system

$$\begin{aligned} &\frac{dx}{dt} = x', && \frac{dy}{dt} = y', && \frac{dz}{dt} = z', \\ &\frac{dx'}{dt} = f, && \frac{dy'}{dt} = g, && \frac{dz'}{dt} = h \end{aligned} \tag{3.73}$$

Bliss refers to DECAF to point out that its results, applied to the first-order system, indicate that one could expect the solutions to (3.90) to approach the solutions to (3.88) as the element for (3.90) approaches the element for (3.88).

By subtracting (3.88) from (3.89), applying Taylor's theorem, and writing the result as a first order system, one gets

$$\begin{aligned}
\tfrac{d}{dt}\Delta x &= \Delta x' \\
\tfrac{d}{dt}\Delta y &= \Delta y' \\
\tfrac{d}{dt}\Delta z &= \Delta z' \\
\tfrac{d}{dt}\Delta x' &= f_x\Delta x + f_y\Delta y + f_z\Delta z + f_{x'}\Delta x' + f_{y'}\Delta y' + f_{z'}\Delta z' + \Delta f + \phi \\
\tfrac{d}{dt}\Delta y' &= g_x\Delta x + g_y\Delta y + g_z\Delta z + g_{x'}\Delta x' + g_{y'}\Delta y' + g_{z'}\Delta z' + \Delta g + \psi \\
\tfrac{d}{dt}\Delta z' &= h_x\Delta x + h_y\Delta y + h_z\Delta z + h_{x'}\Delta x' + h_{y'}\Delta y' + h_{z'}\Delta z' + \Delta h + \chi
\end{aligned} \tag{3.74}$$

In this system, one has $\Delta x = \bar{x} - x$ and similarly for the other five variables. Also, each of $\phi$, $\psi$, and $\chi$ are of second order in the 15 differences $\Delta x, \ldots, \Delta x'$, $\ldots \Delta f, \ldots, \Delta f_x, \ldots$, and $\Delta f_{x'}, \ldots$.

In what is by now a familiar move, the adjoint system to the system just given is written in the variables $\lambda$, $\mu$, $\nu$, $\rho$, $\sigma$, and $\tau$, and the integrated version [from $t_0$ to $t$] of the earlier "fundamental relation" between the original and adjoint system is given:

$$\begin{aligned}
&[\lambda\Delta x + \mu\Delta y + \nu\Delta z + \rho\Delta x' + \sigma\Delta y' + \tau\Delta z']|t \\
&= [\lambda\Delta x + \mu\Delta y + \nu\Delta z + \rho\Delta x' + \sigma\Delta y' + \tau\Delta z']|t_0 \\
&+ \int_{t_0}^{t} (\rho\Delta f + \sigma\Delta g + \tau\Delta h)\, dt + \int_{t_0}^{t} (\rho\phi + \sigma\psi + \tau\xi)\, dt
\end{aligned} \tag{3.75}$$

An argument then follows from which Bliss deduces that if the higher-order terms contained in the last integral on the right are dropped, what is left can be set equal to the sum on the left, in which each $\Delta$ is replaced by the differential $\delta$. Hence

$$\begin{aligned}
&[\lambda\delta x + \mu\delta y + \nu\delta z + \rho\delta x' + \sigma\delta y' + \tau\delta z']|t \\
&= \rho\Delta x_0' + \sigma\Delta y_0' + \tau\Delta z_0']|t_0 + \int_{t_0}^{t} (\rho\Delta f + \sigma\Delta g + \tau\Delta h)\, dt
\end{aligned} \tag{3.76}$$

It is also pointed out that the existence of the six differentials on the left follows from the results in DECAF. This equation "gives the first-order terms with respect to the variables $\Delta x_0'$, $\Delta y_0'$, $\Delta z_0'$, $\Delta f$, $\Delta g$, $\Delta h$ of every expression of the form

$$\lambda\Delta x + \mu\Delta y + \nu\Delta z + \rho\Delta x' + \sigma\Delta y' + \tau\Delta z' \tag{3.77}$$

noindent for which $(\lambda, \mu, \nu, \rho, \sigma, \tau)$ is a solution to the adjoint system, and it may be called the first differential of such an expression with respect to the variables $x_0'$, $y_0'$, $z_0'$, and the functions $f, g, h$." [ibid., p. 97] The reader is already familiar with how this relation is used, for example, to obtain an expression for the range differential corresponding to changes in initial conditions or the presence of a wind, though the generality here allows of any possible source of variation. A use given by Bliss has also been encountered in the antiaircraft worksheet: "…if $t$ is a variable, then six linearly independent solutions of [the adjoint] give six linearly independent equations of the type (3.94) from which the six differentials $\delta x$, $\delta y$, $\delta z$, $\delta x'$, $\delta y'$, $\delta z'$ can be calculated." [p. 97]

Thus it ends the first section. Given what has already been presented in "Introduction", this is a rather straightforward, if congested, generalization, with the exception of the notion of the differential. In the second section, however, there is an elegant surprise. The equations of motion for a projectile are given and the associated functions defined and discussed. Bliss then begins to derive the equations for the range and deflection variations. It is assumed that a trajectory under normal firing conditions has been computed:

$$x = x(t), \ y = y(t), \ z = 0 \tag{3.78}$$

for $0 \leq t \leq T$, with initial conditions $x = y = z = 0$, $x' = x_0'$, $y' = y_0'$, $z' = 0$, at $t = 0$. As before, under normal firing conditions, the projectile remains in the plane of fire ($z = 0$). Then the time of flight is $T$ and the range is denoted by $X$. A second trajectory, caused by disturbing factors in the initial conditions and by changes in the defining functions $f + \Delta f$, $g + \Delta g$, $h + \Delta h$, is assumed, where $f$, $g$, and $h$ are those on the right-hand side of the equation of motion.

Focusing first on the range variation $\Delta X$, the old derivation for

$$\Delta X = [\Delta x + \cot(\omega)\Delta y]|(t = T) \tag{3.79}$$

where $\omega$ is the angle of fall, $(x + \Delta x, y + \Delta y, \Delta z)$ are the co-ordinates of the projectile on the disturbed trajectory, is presented; see "Introduction". This derivation is called "intuitive" [p. 99]. Then the $\Delta x$ and $\Delta y$ are replaced by $\delta x$ and $\delta y$ to get the differential result

$$\delta X = [\delta x + \cot(\omega)\delta y]|(t = T) \tag{3.80}$$

One would assume that this would be followed by a replacement of the increments in the integral portion of (3.94) by specific ballistic expressions, and indeed this does occur in a later section, but instead attention is refocused on the left-hand side. Bliss' goal is to reproduce (3.98) *by calculation alone*, using the implicit function theorem from DECAF. No references to a diagram are needed. Bliss introduces the process with the remark: "A more satisfactory method of determining $\delta X$ from the standpoint of the mathematical analyst is, however, to apply the theorems on implicit functions of the paper referred to above" [p. 99].

In presenting Bliss' argument, we shall here present the exposition in [727, Section 36], since FOLIB omits much of the detail and thus does not give a sense of how easily the calculus of functionals works. The volume [727] was authored by Bliss[37] and meant to be used as an undergraduate text; minimal amounts of definition and proof appear, but for justification of the argument, the reader is referred to FOLIB and DECAF. The notation has changed slightly: the time of fall on a trajectory is given by

$$t = T(x_0', y_0', z_0', W, W_z, K, \gamma) \tag{3.81}$$

where $W,\ W_z$ are the arcs (in the sense of functions of a line) of horizontal and cross winds $w(y)$, $w_z(y)$; $K$ is the arc of the function $\kappa(y)$, where $1+\kappa(y)$ multiplies the atmospheric density function $H(y)$ to account for a $100\kappa$ percent variation in it; $\gamma$ a constant which attaches the same way to the ballistic coefficient.

The time of fall $T$ satisfies the equation

$$y(T, x_0', y_0', z_0', W, W_z, K, \gamma) = 0 \tag{3.82}$$

For simplicity let $B = (x_0', y_0', z_0', W, W_z, K, \gamma)$ so that $y(T, B) = 0$ defines $T$ implicitly as a function of $B$; $y(T(B), B) = 0$. Taking differentials gives

$$y'(T(B), B)\delta T(B) + \delta y(B) = 0 \tag{3.83}$$

giving

$$\delta T(B) = -\frac{\delta y(B)}{y'(T(B), B)} \tag{3.84}$$

Now substituting $t = T(B)$ into $x(t, B)$ gives the range function

$$X(B) = x(T(B), B) \tag{3.85}$$

Taking differentials again gives

$$\delta X(B) = x'(T(B), B)\delta T(B) + \delta X(B) \tag{3.86}$$

or, upon solving for $\delta X$ after substituting for $\delta T$, and dropping the arguments,

$$\delta X = -(\frac{x'}{y'})\delta y + \delta x = \delta x + \cot(\omega)\delta y \tag{3.87}$$

The strength of the technique lies not only in its production of the known range variation expression but its yield of similar expressions for other quantities of interest. For example, though not noted by Bliss, one can use this technique to find the differential of the time of fall. This was addressed by Gronwall in [817,

[37] Goldstine played an authorship role which will be discussed in the last section of this chapter.

p. 514], who found an alternative derivation, but also footnoted DECAF "For the general theorem of which this determination [of the variation of the time of fall] is a special application." Expressions obtained in a similar manner for differentials of time of maximum ordinate and angle of fall are given in [727, pp 85–87], where it is remarked that "The corrections for which formulas are to be computed ... are interesting illustrations of the elasticity of the theory of differential variations as explained in the preceding pages ... It will not be possible to give in detail the theorems concerning functions of lines and their differentials which underlie the theory of differentials as presented here ... A more complete treatment has been published elsewhere by the author" [727, p. 63]. This substantiates Wiener's reference to Bliss "brilliant use of functionals." With the two *Transactions* papers, Bliss' abstract methods yielded solid contributions to the ballistic theory comparable to those of Moulton. Moulton's insistence on rigor resulted in the mathematical treatment given in his *New Methods*, a clear step beyond the work of the Army ballisticians, while Bliss' use of the new functions of a line, a generalization of classical analysis ideas, resulted in material which extended the scope of Moulton's work, much to Bliss' own surprise.[38] Moulton does not appear to have been interested in the more abstract structures and ideas advocated by his colleagues at the University of Chicago, preferring to adhere to classical methods. Each made his own contribution to the advancement of the subject according to his own interests.

In a similar vein, it is interesting to compare Bliss' rigorous introduction of an extension to functionals of an implicit function theorem and differentiation by chain rule to Franklin and Hoar's attempt to do a calculus with their $\delta$ operator. The latter was intentionally kept as closely tied to the physical aspects of the ballistic trajectory as possible. Rules corresponding to those of conventional calculus were developed and proved within a certain framework and used with as much explanation as was needed to convince the students of the Ordnance School that some logical process underlay the resulting formulas. Often appeals to the geometry of the ballistic trajectory was the basis of such a proof. Bliss' theorems came from pure mathematics, beginning as an extension of the existence theorems for solutions to differential equations. He was convinced that the generality of his results allowed them to be used elsewhere: "In the following pages the results attained are of a more general character than is necessary for the ballistic problem, and it seems probable that they will be useful in many other connections" [718, p. 79].

The rest of the paper consists of deriving the expressions for range and deflection using the integral and adjoint variables we have already seen, plus a few remarks on the antiaircraft material. The specifics are undemanding and so will be omitted.

We close this section with some remarks, which could have been made in Bliss' time as well, on the material in FOLIB as expanded upon by the exuberant

[38] It is well to remember his comments in the 1927 report on the applications of pure science [722, pp. 314–315]: "Now the theory of functions of a line is on the farther boundary of the explored mathematical domain of today, and I for one should never have expected it to have an application in ballistics. Its methods were effective, however..."

commentator E. J. McShane in [898]. Having had the differential idea available for several decades, his exposition has more of a textbook style to it. On the replacement of actual differences in ballistic quantities by approximations:

> In the simpler case of disturbances depending on a single number, we found the use of differential effects convenient; the closeness of the approximation between differential and difference permitted the substitution of the differential for the difference with little error ...provided the disturbance remained within small enough bounds, and the linearity of the differential effects made it easier to work with them than with the actual differences. All of those remarks apply to this more complicated case of functionals, and in fact in an intensified form. It is still permissible to replace the difference by the differential effect with only a small error, if the disturbance (or its norm) remains within small enough bounds; and the gain in simplicity is even more marked, because the differences are essentially more complicated, being functionals, than they were in the simpler case ... Therefore it is important, first, to show that the differentials exist; second, to investigate methods by which they can be found and made available to the using services. ... The proof of the existence will automatically furnish us with one feasible method of computing the differential effects; more practical methods will be exhibited in the next chapter. [898, pp. 470–471]

A typical discussion of magnitude of errors made in an approximation:

> The equations with which we finished are the same as we would have obtained had we simply expanded everything involved in the equations by Taylor's theorem, stopping with the linear terms. But this process would have been logically inadequate. It would have left us without information as to the magnitude of the errors that might result. Such a difficulty can occur even in the study of functions of two real variables, which are much simpler than the functions which we have been investigating. For example, it might be thought that if a function has a directional derivative at the origin, it must necessarily be closely approximated by a constant on some region of the origin. But this is false; functions with such a directional derivatives can nevertheless be discontinuous at the origin. Expansions of our equations to terms linear in [a ballistic quantity], are analogous to directional derivatives, and are incapable of furnishing adequate information about the behavior of the changes in [another quantity]. [898, pp. 483–484]

The theory and practice of functional use in ballistics appear completed here.

## 3.7 Hadamard's *Leçons sur le calcul des variations*, Volterra's *Leçons sur les Fonctions des Lignes*, Some Student Work

In the remaining sections of this chapter, we consider questions of origin, context, and subsequent mathematical development. Bliss' ideas did not occur in isolation. So it is natural, for example, to ask about the sources of his use of the adjoint, a principal idea both in theory and in practice. How did the idea of functions of lines become familiar to him, and what, if any, other use of this material did he make in his work? How typical an idea was it to "show the connection between the ballistic theory and the theory of functions of lines [719, p. 94]," that is, were there other physical theories to which functions of lines was being applied? Since the latter has been considered a predecessor to the body of work known as functional analysis, did any portion of this later development stem from or contribute to the work of

Bliss and his colleagues and students? An attempt to answer these questions is now begun, ending with Bliss' final words on ballistics and "analysis of functionals" [39] as contained in his 1944 work [727].

Some answers can be found by considering Hadamard's *Leçons sur le calcul des variations* [821], published a year after Bolza's *Vorlesung*, in 1910. It has already been mentioned that these twin volumes, along with Kneser's work, constituted the three principal references for the calculus of variations in the first decades of the twentieth century. Hadamard's *Leçons* contains in the first six sections of its first book (of three) material on the first variation, extremals, and the parametric from of problems in the calculus of variation, geodesics, and all of the elementary topics reviewed in this chapter. The remaining books treat, as does Bolza, the Mayer problem, a much-investigated generalization of the standard calculus of variation problem which will be discussed in the next section and many other topics.

Reviews of the book were glowing. Constantin Carathéodory published a review [743] in 1911 which praised the book in the highest terms: "In publishing the lectures of the Calculus of Variations, which he delivered at the Collège de France, M. Hadamard has not only filled a gap strongly felt in France, but above all, written a book which, it seems to me, marks a date in the history of this chapter of Analysis" [743, p. 124]. A delayed review in 1914 in the *Bulletin of the American Mathematical Society* states: "Hadamard at once commanded respectful attention everywhere; an immediate review would not have increased nor diminished the desire for instant examination of the work. The reception the book received throughout the world fully justifies this sweeping statement" [834, p. 30].

What precipitated these words of praise was the perspective from which the book was written. This view was summarized succinctly by the author in the first sentence, widely quoted at the time and since, of the Forward: " The Calculus of Variations is nothing more than a first chapter of the doctrine that one calls today the Functional Calculus [le Calcul Fonctionnel] and whose development will doubtlessly be one of the first imposing tasks of the analysis of the future" [821, p. vii]. Much of the remainder of the Forward is devoted to elaborating this statement. The phrase "functional calculus" was Hadamard's, meant to continue the work of Volterra's functions of a line[40] Chapter VII of the first book devotes roughly 20 pages to an introduction to this material from his own point of view as well as that of Volterra and Fréchet. Hadamard wrote extensively elsewhere on the functional calculus, tracing its history and significance, and explaining the contributions of its founders [823, 824]. We shall in the next few sections use the *Leçons* as a point of departure for responses to some of the questions raised above. Our present section is devoted to Bliss' use of functions of a line, in particular the derivative

[39] The latter is not Bliss' usage. It is not proper at this point to speak of functional analysis per se, as that term refers to the analysis of linear functionals and operators on general spaces; its formal appearance could be said to date from the early 1930s. The "analysis of functionals" better describes the stage of developments discussed here.

[40] For a history of the contributions of Volterra, Hadamard, and Fréchet and many others to the subject now called functional analysis; see [700] and [764].

and differential notions developed by Volterra and Fréchet. Each of the remaining sections will relate in some way to functional analysis.

It is difficult to date Bliss' familiarity with Volterra's work since the latter began publishing on functions of a line in 1887, but surely inclusion of this material in Hadamard's *Leçons* would guarantee its coming to Bliss' attention. Whatever the original source, Bliss was highly taken with these ideas, as is evidenced by his review of Volterra's 1913 *Leçons sur les Fonctions des Lignes* [711]. His review is a thorough chapter-by-chapter examination of the book; several points relevant to the present work should be emphasized. A function of a line is a function $F$ defined on a collection of functions $f$, the latter defined on a real interval $[a, b]$. The usual notation is

$$F|[f(x)]| \tag{3.88}$$

and this is considered as a generalization of a function $F(y_1, y_2, \ldots y_n)$ of a finite number of variables.[41] "This process of passing from a conception involving a finite number to a conception involving an infinite number of mathematical symbols or operations is one of the most striking features of mathematical thought of our time" [711, p. 346]. Among other such examples Bliss cites Volerra's treatment of "linear integral equations, as generalizations from a finite to an infinite number of linear equations ..." [ibid., p. 346]. For Bliss, though not stated, the functionals of greatest interest are those associated with the calculus of variation, such as the familiar expression taking $y(x)$ as input:

$$\int_a^b G(x, y, y')\,dx \tag{3.89}$$

As mentioned earlier, the minimizing of this functional might begin with its "derivative," in some sense involving the input curves as variables, being set equal to zero. Bliss reviews Volterra's definition of the derivative of a function of a line (an obvious candidate for such a calculus of variations derivative), an important formula for a differential of $F$, and a Taylor series-like expansion of such a function, to be given below. He then summarizes Volterra's discussion of the consequences of a derivative vanishing along a domain curve, pointing out that in the calculus of variations, this would lead to the Euler equations, but that other outcomes are possible for other functionals. A chapter is devoted to implicit function theory, a source of the generalizations given by Bliss and some of his students. Quite interesting in what might be considered a domain of pure mathematics is the inclusion of several chapters devoted to applications of the theory to such topics as

[41] Occasionally one sees an $a$ below the $[f(x)]$ and a $b$ above to emphasize the interval.

the theories of elasticity and electromagnetism;[42] perhaps Bliss felt, as the opening quote of this section suggests, that his *Transactions* papers followed this tradition.

We begin with Volterra's derivative concept. For the definition of the derivative of a function of a line, we will follow Bliss' account in [710]: for $F$ a function of a line, let $\xi$ be a point in $[a, b]$, $\phi$ and $\psi$ functions defined on $[a, b]$, $\delta > 0$ and $h > 0$ small positive numbers. Suppose further that $\psi$ is supported on an interval of length $h$ centered at $\xi$, with $\psi$ not changing sign on the interval, and

$$|\psi(x)| < \delta \tag{3.90}$$

for $a \leq x \leq b$. Then if

$$|\Delta F| = |F[\phi(x) + \psi(x)] - F[\phi]| \tag{3.91}$$

the derivative of $F$ at $\xi$ is defined by

$$F'[\phi(x), \xi] = lim\frac{\Delta F}{\sigma} \tag{3.92}$$

as $\delta$ and $h$ approach zero, where

$$\sigma = \int_a^b \psi(x)\, dx \tag{3.93}$$

Higher-order derivatives are also defined by Volterra: if, for example, $F'[\phi, \xi_1]$ is a function of $\phi$ only, where $\xi_1$ is fixed, one may calculate its derivative $F''[\phi, \xi_1, \xi_2]$ at $\xi_2$. In an attempt to generalize the idea of a differential from multivariable calculus, it is pointed out that if $f$ were a function of $n$ variables $y_1, \ldots y_n$, then one has

$$\delta F = \sum_1^n F_i' \delta y_i \tag{3.94}$$

so one has by analogy

$$\delta F = \int_a^b F'[\phi(x), \xi]\delta\phi(\xi)\, d\xi \tag{3.95}$$

The $\delta\phi(x)$ in the integral is not rigorously defined. This expression was termed "so important in the calculus of variations" by Bliss in his review of the *Leçons* [p.

[42] Hadamard's *Leçons* also contain applications to Green's functions and the Dirichlet problem [821, pp. 303–312].

347]; note the similarity to (3.32). In what we will take as the last calculus analogy, Volterra develops a Taylor-like series for $F$:

$$F[\phi(x) + \psi x)] = F[\phi(x)] + \int_a^b F'[\phi(x), \xi_1]\psi(\xi_1)\, d\xi_1 + \tag{3.96}$$

$$(1/2) \int_a^b [\int_a^b F''[\phi(x), \xi_1, \xi_2]\psi(\xi_1)\psi(\xi_2)\, d\xi_1]\, d\xi_2 + \dots$$

In the *Leçons*, Volterra does not provide conditions under which these definitions make sense but defers proofs to cited papers. These same ideas are also in the first few pages of Hadamard's section on functions of a line, [821, pp. 281–284].

In [710], Bliss addressed the adequacy of these definitions to deal with the problems of the calculus of variations. "It is important that these considerations should apply to the integrals of the calculus of variations in terms of which the line functions cited above by way of illustration, with many others, are expressible. Such integrals in general are not continuous, do not possess derivatives, and do not satisfy the conditions ... in the forms specified by Volterra" [710, p. 174]. To clarify and remedy this situation, Bliss first illustrates the deficiency of the derivative definition by returning to

$$F[\phi(x)] = \int_a^b f(x, \phi(x), \phi'(x))\, dx \tag{3.97}$$

Bliss shows that it is possible to construct a family of functions $\psi$ such that, for arbitrary $f$ and $\phi$ in class $C^2$, the Volterra derivative fails to exist. (The exact statement occurs on pp. 175–176.) But he also shows that if

$$|\psi(x)| < \delta \tag{3.98}$$

$$|\psi'(x)| < \delta \tag{3.99}$$

$$|\psi''(x)| < \delta \tag{3.100}$$

for all $x$ in $[a, b]$, that the derivative does exist and has the value $[f_y - \frac{d}{dx} f_{y'}]$. This assumes that all of Volterra's other conditions for the existence of the derivative hold as well. This is the goal Bliss wanted to achieve: to find a functional analytic derivative of his functional which yielded under appropriate conditions the Euler quantity; setting this derivative equation to zero gives the Euler equation for an extremal.

In the last paragraph of [710], Bliss cites a generalization of these results due to Charles Fischer [776], from whose paper the implicit function theorem was quoted in Sect. 3.5. This generalization involves an integral of the form

$$f[\phi(x)] = \int_a^b f(x, \phi, \phi', \dots, \phi^{(n)})\, dx \tag{3.101}$$

If $\phi$ is of class $C^{2n}$ and $f$ of class $C^{n+1}$, then "f has a derivative at any value $x = \xi$ which is approached with order $2n$, that is, if $\psi$ does not change sign and vanishes except on an interval of length less than $h$ including $x = \xi$ and if furthermore"

$$|\psi^i(x)| < \delta,\ i = 1, \ldots, 2n \tag{3.102}$$

then the limit in (3.119) exists, it has a limiting value in the sense of (3.120).[43] Fischer's work contains even further generalizations. Its publication date of 1913 demonstrates how early was Bliss' interest in this material.

More relevant for an appreciation of the ballistics works of 1920 is the thesis from that year of another student, Elizabeth Le Stourgeon, [879], who essentially considered the same question as Fischer but using the Fréchet differential. As a backdrop to this work, one can quote the first paragraph of the work of Fréchet [784, p. 90] cited by Bliss in DECAF: "After studying the particular case of the function of lines, it appears to me useful to show that the definition of the differential that I propose is inspired by general considerations applicable to an arbitrary functional ... These authors [Volterra and Hadamard] were occupied uniquely with the derivative of functionals that one knows from the calculus of variations." Le Stourgeon's work attempts to address necessary and sufficient conditions on a functional $F[\lambda(x)]$ that it assume a minimum at an arc $\lambda_0$, conditions expressed using a modification of the Fréchet differential.

Le Stourgeon considers an arc $\lambda_0$ of class $C^n$ on an interval $[a, b]$ and defines a neighborhood $(\lambda_0)^n_\delta$ of order $n$ as the set of arcs of class $C^n$ satisfying

$$|\lambda^{(i)}(x) - \lambda_0^{(i)}(x)| < \delta,\ i = 0, \ldots, n \tag{3.103}$$

on the interval. A functional is said to have continuity of order $n$ at $\lambda_0$ if for every $\epsilon$ there exists a $\delta$ such that

$$|F(\lambda) - F(\lambda_0)| < \epsilon \tag{3.104}$$

[43] The notion of "order of approach," also used by Bliss in his 1915 paper and implicitly in his work on the Fréchet differential in DECAF, and by his student Le Stourgeon as described below, is easily related to the idea of a general neighborhood as presented, for example, by Hadamard in his *Leçons*, p. 49: "One says two functions $\psi$ and $\Psi$ have between them (in the interval where one considers them) a neighborhood of order $p$ defined by the positive number $\epsilon$ if one can establish a unique and reciprocal correspondence between the two numerical variables $x'$, $x''$, such that one has: $|\psi(x') - \psi(x'')| < \epsilon, \ldots, |\frac{d^p\psi(x')}{d(x')^p} - \frac{d^p\psi(x'')}{d(x'')^p}| < \epsilon$ with $|x' - x''| < \delta$." In a later publication of 1923 [822, p. 35], Hadamard states that defining such neighborhoods is "now classic in the Calculus of Variations and its recent generalization, the Functional Calculus," in a footnote, he credits the idea as originating in the 1894 dissertation of Ernst Zermelo.

whenever $\lambda$ is in $(\lambda_0)^n_\delta$. (Note the similarity to Fischer's work.) A functional $L(\lambda)$ is said to be linear if

$$L(c_1\lambda_1 + c_2\lambda_2) = c_1L(\lambda_1) + c_2L(\lambda_2) \tag{3.105}$$

and

$$|L(\lambda)| < AM(\lambda) \tag{3.106}$$

for some constant A and all $\lambda$. Thus linearity of a functional entails boundedness in this definition.

A variation of the recently proved Riesz representation theorem for functionals of $C[a, b]$ is stated: If $L$ satisfies (3.132) on the class of all $C^1$ arcs on $[a, b]$ and is continuous of order 1, it is always expressible, in fact in an infinite number of ways, as the sum of two Stieltjes integrals

$$L(\lambda) = \int_a^b \lambda(x)\,du(x) + \int_a^b \lambda'(x)\,du_1(x) \tag{3.107}$$

where $u$ and $u_1$ are functions of bounded variation.[44] (This theorem was proved first by Charles Fischer [777].)

Then follows a definition of a variation of Fréchet's differential: for $\lambda_0$ an arc of class $C^1$, and $F(\lambda)$ a functional defined at least in $(\lambda_0)^1_\delta$, then $F(\lambda)$ is said to have a differential at $\lambda_0$ if there exists a linear functional $L(\Delta\lambda)$ with continuity of order 1 such that for all arcs $\lambda_0 + \Delta\lambda$ in $(\lambda_0)^1_\delta$ one has

$$F(\lambda_0 + \Delta\lambda) - F(\lambda_0) = L(\Delta\lambda) + \epsilon(\Delta\lambda)M_1(\Delta\lambda) \tag{3.108}$$

where $M_1(\Delta\lambda)$ is the maximum of the values of $|\Delta\lambda|$ and $|\Delta\lambda'|$ on $[a, b]$ and $\epsilon(\Delta\lambda)$ is functional which vanishes with $M_1(\Delta\lambda)$. Then $L(\Delta\lambda)$ is expressible in the form (3.134). The important point to note here is that this variation of the Fréchet differential is analogous to that used by Bliss in DECAF in its requirement that $L$ possess continuity of order 1 and that $M_1(\Delta\lambda)$ be the maximum of both $\Delta\lambda$ and $\Delta\lambda'$.

To address the question of a minimum of a functional[45] $F(\lambda)$ Le Stourgeon considers a $C^1$ arc $\lambda_0$ on an interval $[x_1, x_2]$, $F$ defined on a $(\lambda_0)^1_\delta$ having a differential in the restricted sense just defined, and defines $F$ as having a minimum at $\lambda_0$ in the obvious way. A necessary condition follows quickly as in the calculus of variations: defining a one-parameter family of arcs

$$y = \lambda_0(x) + \alpha\eta(x) \tag{3.109}$$

[44] Le Stourgeon uses the phrase "limited variation."

[45] "It is proposed in this section to study the properties of the first variation, in other words, the first differential, of a functional..." [879, p. 361].

where $\eta(x_1) = \eta(x_2) = 0$. It follows immediately from the definition of $L$ and the representation theorem for $L$ that, using this family, if $F(\lambda_0)$ is a minimum, then

$$L(\eta) = \int_{x_1}^{x_2} \eta \, du + \int_{x_1}^{x_2} \eta' \, du_1 \tag{3.110}$$

vanishes for all $\eta$ of class $C^1$ such that $\eta(x_1) = \eta(x_2) = 0$, where $u$ and $u_1$ are the functions used in the representation of $L$. It is now of interest to find equivalent conditions for this vanishing. A necessary condition achieved for this purpose is that the vanishing occurs if there is a relation between $u$ and $u_1$, namely,

$$u_1(x) - \int_{x_1}^{x_2} u(x) \, dx = kx + l \tag{3.111}$$

must hold on $[x_1, x_2]$ for some constants $k$ and $l$. For our purpose, this is the result of interest. A similar treatment of the second variation is also given.

In the last section, Le Stourgeon "interprets" her results in this context: a consideration of the standard functional

$$F(\lambda) = \int_{x_1}^{x_2} f(x, \lambda, \lambda') \, dx \tag{3.112}$$

and the two functions

$$u(x) = \int_{x_1}^{x_2} f_\lambda \, dx, \; u_1(x) = \int_{x_1}^{x_2} f_{\lambda'} \, dx \tag{3.113}$$

leads to the conclusion that (3.137) "assumes the familiar form"

$$L(\eta) = \int_{x_1}^{x_2} (f_\lambda \eta + f_{\lambda'} \eta') \, dx \tag{3.114}$$

and (3.138) yields, after two differentiations, the Euler equation. Thus Le Stourgeon has linked the Fréchet differential with classical expression for $\delta J$ given by Bolza and earlier writers. The thesis contains much more material.

The appearance of Le Stourgeon's thesis in the same year as Bliss' *Transactions* papers shows the close connection in the latter's thinking between the two problems,[46] and constituted, along with Fischer's earlier work, a continuum of concern with these new tools for the study of variation. The question of a derivative

[46] Thus the recurrent comment from [729, p. 601]: [when beginning his service as a technical expert] "...the theory and application of the so-called 'differential corrections' of trajectories was still very imperfect ...The calculation of the differentials of functionals is a principal problem of my special field, the calculus of variations. I found that the methods of that calculus were effective..."

concept useful for both the simplest integral expression of the calculus of variations and the variation problem in ballistics found a solution in a modified form of the Fréchet differential.[47] Fréchet's idea connected the two fields for Bliss as had the existence, smoothness, and implicit function theorems noted earlier. The notion of a higher "order of approach" was the necessary key in the work of Bliss and these two students. For their work, together with the contributions of Griffith Evans (a student of Bôcher's), Bliss and Fischer received a mention as two of the (regrettably) few "devotees" of "le calcul Fonctionnel" in a 1921 survey of research interests of American mathematicians [869, p. 544]. Bliss' two *Transactions* papers were explicitly mentioned. A zenith of a sort for this abstract account of the calculus of variations was reached in [972], a 1920s text mentioned in [720], where the subject was presented entirely in terms of functionals.

It appears debatable how effective this method was in the long run, however. In the opinion of Morris Kline in [871, p. 1981], writing from a vantage point of roughly 50 years after these publications, "Though the abstract approach was pursued subsequently by many men, the progress made in the application of the theory of functionals to the calculus of variations has not been great." But this opinion does not diminish the interesting results which were achieved, nor does it neutralize the debate over the classical versus the modern view of this material. One example of this debate will be presented in a later section devoted to Herman Goldstine's contributions to the problem of Bolza.

We conclude this section with a remark about the General Analysis of E. H. Moore. Moore was a central figure in the founding and development of mathematics at the University of Chicago, and his research interests included an ongoing program, begun around 1906, of an abstract treatment of analysis under the above name. His efforts were noted, for example, by Hadamard, and the following quote, from a talk Moore gave at the International Congress of Mathematicians in Rome in 1908 [904, p. 98], was often repeated at that time and used as a rallying cry for advocates of abstraction:

> The existence of analogies between central features of various theories implies the existence of a more fundamental general theory embracing the special theories as particular instances and unifying them as to those central features.

Bliss was quite familiar with Moore's work; he wrote Moore's technical obituary [706] in 1934. It is natural to ask about the possible influence of Moore's work on Bliss' research efforts. But it appears that there was little as far as the ballistics work was concerned. A few references to Moore's work have been given in this chapter,

[47] Fréchet, who was a student of Hadamard, considered himself as working in the framework of Volterra's functions of a line, though his name is most likely associated in the reader's mind with his later concepts of linear topological spaces with completeness properties. In his work prior to 1914, he had indeed introduced the notion of a metric space and "distance in an infinite-dimensional function space" [700, p. 271], but these concepts played no explicit role in the work of Bliss and Le Stourgeon, central as they are to today's conception of functional analysis. They referred to "a family of arcs" or a "collection of curves" with no underlying structure, though DECAF essentially used distance measures without referring to them as such.

and Bliss' implicit function theorem in DECAF was an acknowledged special case of Moore's; however, Bliss' proof is self-contained. But it is fair to say that none of Bliss' results depends on Moore's ideas, though some hypotheses may be phrased in Moore's terminology. There was, on the other hand, an acknowledgment of Moore's work in the later efforts of Lawrence Graves to evolve a general implicit function theorem in function spaces; this will be treated in a later section of this chapter. An evaluation of General Analysis from a modern viewpoint can be found in [960]. A 1918 publication, *Functionals and Their Applications* [770], an entry in the American Mathematical Society Colloquium Series, gives an interesting exposition of the ideas of Volterra and Fréchet, especially their views of differentiation, using some of Moore's Analysis.

## 3.8 Adjoints and Multiplier Rules

Bliss' method of computing differential variations came to be known as the "Bliss adjoint method" for its reliance on the adjoint. One may ask how he came to this procedure: what was the status of the adjoint idea at this time? How did it fit into the broader mathematical research picture? Was it an isolated piece of knowledge, as the ballisticians who criticized its use felt? How novel was his use of it? In this section, we briefly review the history of the adjoint; find that Bliss did not have to look far to encounter the germ of his method, again in the realm of the calculus of variations; and note a pioneering attempt to introduce the adjoint into a the functional calculus.

In the eighth and early nineteenth centuries, there were two types of adjoints which came to be employed in the study of ordinary differential equations.[48] Each of these was introduced in an attempt to lower the order of the equation or system of equations being considered. The first was due to Lagrange, who used it in a paper which appeared roughly at the same time as his first paper in the calculus of variations [874]. The definition and properties, as they appeared in contemporary treatises, can be found, for example, in [800, pp. 115 ff.].

Letting

$$L[u] = p_0 \frac{d^n u}{dx^n} + p_1 \frac{d^{n-1} u}{dx^{n-1}} + \ldots + p_{n-1} \frac{du}{dx} + p_n u \tag{3.115}$$

one defines

$$M[v] = (-1)^n \frac{d^n(p_0 v)}{dx^n} + (-1)^{n-1} \frac{d^{n-1}(p_1 v)}{dx^{n-1}} + \ldots - \frac{d(p_{n-1} v)}{dx} + p_n v \tag{3.116}$$

[48] The adjoint came to play a major role in partial differential equations and potential theory, as will be discussed in the next section.

Then $M$ is called the "adjoint" operator corresponding to $L$[49] One has

$$vL[u] - uM[v] = \frac{d}{dx}P(u, v) \tag{3.117}$$

where $P(u, v)$ is linear and homogeneous in $u, u', \ldots, u^{n-1}, v, v', \ldots, v^{n-1}$. This had come to be known as Lagrange's Identity. Now in order to find an integrating factor for $L[u] = 0$ one attempts to solve $M[v] = 0$. if $v$ is a solution to the adjoint, then the equation $L[u] = 0$ reduces to a linear equation of order $n - 1$, namely $P(u, v) = C$, for $C$ an arbitrary constant. The process can be continued if more specific solutions of the adjoint are found.[50]

For systems of linear equations, the adjoint was introduced by Jacobi in a lengthy 1837[51] paper [865] on solutions to ordinary differential equation.[52]Again, the adjoint system was a standard topic in many of the analysis treatises of Bliss' time. One may find it also in [800, pp. 156–157]. The properties analogous to Lagrange's adjoint are straightforward. In modern notation, if $A = [a_{ij}]$ is an $n$ by $n$ matrix of functions and $A^* = [a_{ji}]$ its transpose, $Y = (y_i)$, $Z = (z_i)$ columns of functions, then the system

$$Y' = AY \tag{3.118}$$

is said to have adjoint system

$$Z' = -A^*Z \tag{3.119}$$

Alternatively, the adjoint system may be written as

$$-Z' = ZA \tag{3.120}$$

if $Z$ is regarded as a row vector.

As in Bliss' usage, the first system allowed a nonhomogeneous term. One has that if $(y_i)$ and $(z_i)$ are particular solutions of (3.154) and (3.155), then $y_1z_1 + \ldots + y_nz_n = C$, for $C$ constant. Thus if one knows a particular solution $(z_i)$ of the adjoint, then a "first integral" of the original system is provided by this relation. This first integral allows the original system to be reduced in order by one; further integrals decrease the order more.

---

[49] Lagrange did not use the term "adjoint," which first appeared in an 1873 paper of L. Fuchs [785]. Fuchs called equations related as above "adjungirte Differentialgleichungen" [p. 421].

[50] An operator equal to its own adjoint is called "self-adjoint"; Jacobi's differential operator $\Psi$ (3.21) is self-adjoint, a fact which Bolza uses to analyze the second variation; see [738, pp. 52–54].

[51] This was the same year as his paper on the second variation considered above.

[52] One finds on page 404: "Aequationibus (1) alterum systema aequationum differentialium linearum coniugatum est . . ." introducing the adjoint system.

It has already been pointed out that Gronwall's discovery of a first integral of the adjoint system to the equations of variation is an example of the first property.[53] There were many physical applications of the adjoint of a linear system at the time of Bliss' work on ballistics; one such will be described in the next section. Self-adjoint systems, that is, systems equal to their own adjoints, were singled out by Goursat by example; the solution of such a system reduces to integration of Ricatti equations [800, p. 79, p. 170].

There is a connection between the two types of adjoints; an account of it can be found, for example, in an 1889 treatise on differential equations by a professor at Johns Hopkins University [753, pp. 17–18]. If a linear homogeneous differential equation of order $n$, say $L[y] = 0$ is given, it can be replaced in the standard way by an equivalent first order-order system $Y' = AY$. If one then takes the adjoint of this system, it is possible through a systematic elimination process applied to the adjoint to produce the Lagrangian adjoint of the original $n^{th}$ order equation.[54] Thus in a sense the system adjoint is a generalization of Lagrange's adjoint.[55]

With no obvious connection to the adjoint topic, we now turn to the problems of the calculus of variations involving constraints. Until now the questions considered have involved the minimizing of integral expressions of a function and its derivatives with no further conditions which must be satisfied by the functions. But some of the classical problems in the subject involved restricting the solutions in some way. The isoperimetric problem is of this sort: finding, for a closed curve of given length $l$ the maximum area contained by the curve. Analytically, find an arc with equations $(x(t), y(t)),\ t_1 \leq t \leq t_2$ with $x(t_1) = x(t_2),\ y(t_1) = y(t_2)$ which maximizes

$$(1/2)\int_{t_1}^{t_2} (xy' - x'y)\,dt \tag{3.121}$$

subject to the constraint

$$\int_{t_1}^{t_2} ((x')^2 + (y')^2)^{1/2}\,dt = l \tag{3.122}$$

Those problems had been considered simultaneously with the others in our survey and with a similar history: special cases first, often with incomplete or unrigorous proofs, and the gradual introduction of careful consideration, with Weierstrass as a chief figure in the latter. Thus Chapter VI of Bolza's *Lectures*

[53] Moulton made a reference to this fact in his deflating reaction to Hoar's great praise of Gronwall's first integral: " ...it is a direct consequence in this particular problem of general principles which have been well known for at least fifty years" [915, p. 249].

[54] Craig used the word "adjunct" as a translation of the German "adjungierte," being unaware of the usage of "adjoint."

[55] "Adjoint" operations were proliferating in other mathematical contexts as well at this time; for example, integral equation had adjoint equations, and Moore's General Analysis gave rise to other attempts to introduce them.

contains a treatment by Weierstrass of a general isoperimetric problem: among all curves joining two points 0 and 1[56] for which the definite integral

$$\int_{t_1}^{t_2} G(x, y, x', y')\, dt \tag{3.123}$$

has a fixed value $l$, determine the one which minimizes another definite integral

$$\int_{t_1}^{t_2} F(x, y, x', y')\, dt \tag{3.124}$$

As in the unconstrained problems, the first goal is to find an extremal; that is, a function or set of functions for which some derivative expression, like the Euler equation, must necessarily equal zero. For the problem just stated, the necessary condition bears a strong resemblance to the Lagrange multiplier problem in multivariable calculus. In order that $(x(t), y(t))$ be a solution, there must exist a constant $\lambda$ for which $H = F + \lambda G$ satisfies the two Euler-type equations

$$H_x - \frac{d}{dt}H_{x'} = 0,\; H_y - \frac{d}{dt}H_{y'} = 0, \tag{3.125}$$

The similarity to the multivariable calculus problem leads to the interesting fact that both problems were considered by Lagrange, who formulated the familiar necessary condition for the constrained optimization of a point function (see [796, p. 149]) and also gave a similar approach to the isoperimetric problem (see [796, pp. 149–150]) for details and references.)[57] Further generalizations of Lagrange's work have been subsumed under the name the "Problem of Lagrange." At this point, a new figure appears: Adolf Mayer, a professor of mathematics at Heidelberg who wrote extensively on the calculus of variations from the late 1860s through the early years of the new century. In an 1886 paper [892], he stated and attempted to treat the Problem of Lagrange as it was given at the this time: among all functions $y_1, y_2, \ldots, y_n$ of $x$ which satisfy the constraints

$$\phi_k(x, y_1, \ldots, y_n, y_1', \ldots, y_n') = 0,\; k = 1, 2, \ldots m \tag{3.126}$$

[56] Here is another example of the practice of labeling a point in the plane with an integer.

[57] It is a remarkable fact that even the theory of maxima and minima for functions of several variables was not considered to be on a completely rigorous basis until after 1900. Harris Hancock, an American who wrote an early text on the calculus of variations [829], published in 1903 a University of Cincinnati Bulletin on maxima and minima, based on lectures given by Weierstrass, as a companion volume. It was expanded and published in 1917 as [830]. The introduction explains the status of the deficiencies in existing expositions. Chapter VI is devoted to constrained optimization.

where $m < n$, and have given values at the endpoints of $[x_0, x_1]$, find those which minimize

$$\int_{x_0}^{x_1} f(x, y_1, \dots, y_n, y_1', \dots, y_n')\, dx \tag{3.127}$$

His paper is quite involved (see [796, pp. 282–287]), but among his conclusions is the fact that, in certain cases, one has that a necessary condition for a minimum is that there are constants $\lambda_1, \dots, \lambda_m$ such that if $F = \lambda_1\phi_1 + \dots + \lambda_m\phi_m$, then

$$\frac{\partial F}{\partial y_i} - \frac{d}{dx}\frac{\partial F}{\partial y_i'} = 0,\ i = 1, \dots n \tag{3.128}$$

This type of result had come to be known as a "multiplier rule" for the type of problem with constraints exemplified by Lagrange. Mayer's proofs had gaps which were not fixed until many years later, but the Problem of Lagrange held interest for quite some time. Bliss devoted much attention to necessary and sufficient conditions for its solution and published a long essay on alternative methods as late as 1930 [724].

Mayer himself in 1878 proposed a problem which he considered "the most general problem of the calculus of variations of a single independent variable," to cite the title of his first paper on this subject [891]. We will take the statement of it given in Hadamard's *Leçons* [821, p. 223], whose solution involves the adjoint idea. He considers functions $y_0, y_1, \dots, y_n$ subject to $p + 1$ differential equations

$$g_i(y_0', y_1', \dots, y_n', y_0, \dots, y_n, x) = 0,\ i = 1, \dots, p \tag{3.129}$$

where $p < n$, and the values of the $y_i$ are all prescribed at the endpoints of the interval $[x^0, x^1]$ with one exception: the value $y_0(x^1)$ is left free. The problem is to find the functions $y_0, y_1, \dots, y_n$ which minimize $y_0(x^1)$. This has come to be known as the Problem of Mayer, again the subject of much attention, by Bliss, among others [714]. Bolza gives a treatment of the problem in his *Vorlesungen* [pp. 572–574], but it is only mentioned in a footnote in his *Lectures*, on page 1, as an example of how the calculus of variations has been "extended beyond the domain of definite integrals." This statement indicates a shift; the Mayer problem, stated with the constraining equations first, almost appears to be about differential equations exclusively, concealing its inspiration from optimizing integral expressions.[58]

He gives a long, leisurely discussion of his proof and relies on material developed in earlier chapters. It will therefore be possible to give a fairly complete picture of

[58] Hadamard notes briefly that the Mayer problem is a true generalization of the Lagrange problem by setting $g_0 := y_0' - f(y_1', \dots, y_n', y_1, \dots, y_n, x)$ and $g_h := g_h(y_1', \dots, y_n', y_1, \dots, y_n, x)$, $h = 1, 2, .., p$ and considering $(y_0)_{x^1} = (y_0)_{x^0} + \int_{x^0}^{x^1} f(y_1', \dots, y_n, x)\, dx$.

his procedure only up to a certain point, highlighting the role of the adjoint, after which the earlier results must be used. His theorem should be stated first:

> In order that a curve $(y_0, y_1, \ldots, y_n)$ be a solution to the Mayer problem, it is necessary that one can determine a system of functions $l_0, l_1, \ldots, l_p$, not all identically zero, defined on $[x^0, x^1]$, such that if $f = l_0 g_0 + \ldots + l_p g_p$, then $F_1(x) = \ldots. = F_n(x) = 0$ for all $x$, where $F_i = f_{y_i} - \frac{d}{dx} f'_{y_i}$, $i = 1, \ldots, n$.

Note that the $l_k$ here are *functions*, they are called the multipliers, and the solution curve is called an extremal.

Hadamard assumes that his solution curve is part of a one-parameter family $y_k = y_k(x, \alpha)$ where $\alpha = 0$ corresponds to the solution curve and the variation of $y_k$ is defined as $\eta_k = \frac{\partial y_k}{\partial \alpha}|(\alpha = 0)$; a variation of Bolza's form with $\epsilon$ replaced by $\alpha$ would suffice. Substituting $y_k(x, \alpha)$ into (3.164), differentiating with respect to $\alpha$ and setting $\alpha = 0$ yields what we have called the equations of variation for the $g_i$, namely,

$$\sum_{i=0}^{n} (\frac{\partial g_h}{\partial y'_i}\eta'_i + \frac{\partial g_h}{\partial y_i}\eta_i) = 0, \ h = 0, \ldots, p \tag{3.130}$$

At this point, no conditions are imposed on the $\eta_i$, but Hadamard now requires $\eta_i(x^0) = 0$ for all $i$, and $\eta_i(x^1) = 0, i \neq 0$; the latter to deal with the variability of $y_0(x^1)$. The equations of variation are then written in a different way, namely, as a nonhomogeneous system in $\eta_0, \ldots, \eta_p, \eta'_0, \ldots, \eta'_p$:

$$(\frac{\partial g_i}{\partial y'_0}\eta'_0 + \ldots + \frac{\partial g_i}{\partial y'_p}\eta'_p) + (\frac{\partial g_i}{\partial y_0}\eta_0 + \ldots + \frac{\partial g_i}{\partial y_p}\eta_p)$$

$$= - \sum_{k=p+1}^{n} (\frac{\partial g_i}{\partial y'_k}\eta'_k + \frac{\partial g_i}{\partial y_k}\eta_k), \ i = 0, \ldots, p \tag{3.131}$$

Note that this system is linear in the $\eta_k$ and $\eta'_k$; Hadamard wishes to ignore the nonhomogeneous terms on the right and get his multipliers $l_k$ from the adjoint of the resulting system. Since the left-hand side contains $\eta'_k$ terms, the adjoint must be more general as well. It is

$$\frac{d}{dx}[l_0\frac{\partial g_0}{\partial y'_j} + \ldots + l_p\frac{\partial g_p}{\partial y'_j}] = l_0\frac{\partial g_0}{\partial y_j} + \ldots + l_p\frac{\partial g_p}{\partial y_j}, \ j = 0, \ldots, p \tag{3.132}$$

If one takes $\frac{\partial g_r}{\partial y'_s} = 0$ for $r = s$ and 0 otherwise, one gets the familiar adjoint in the variable $l_k$.

Now, using a procedure which would be repeated in Dunham Jackson's ordnance textbook, Bliss' JUSA/CAJ article, and other research publications, Hadamard bids us to multiply the first of the equations of variation (3.166) by $l_0$, the second by $l_1$, and so on until the last is multiplied by $l_p$, add both sides of the equations, and use the adjoint to replace the left-hand side to get

$$\frac{d}{dx}[l_0(\frac{\partial g_0}{\partial y_0'}\eta_0+\ldots+\frac{\partial g_0}{\partial y_p'}\eta_p)+\ldots+l_p(\frac{\partial g_p}{\partial y_0'}\eta_0+\ldots+\frac{\partial g_p}{\partial y_p'}\eta_p)]$$

$$=-\sum_{h=0}^{p}\sum_{i=p+1}^{n} l_h(\frac{\partial g_h}{\partial y_i'}\eta_i'+\frac{\partial g_h}{\partial y_i}\eta_i) \tag{3.133}$$

An integration of both sides of this gives the version of what had earlier been called the fundamental relation between the original system and the adjoint. It is assumed that the variations have $\eta_k(x^0)=0, k=0,\ldots,p$, so integration yields

$$[l_0(\frac{\partial g_0}{\partial y_0'}\eta_0+\ldots+\frac{\partial g_0}{\partial y_p'}\eta_p)+\ldots+l_p(\frac{\partial g_p}{\partial y_0'}\eta_0+\ldots+\frac{\partial g_p}{\partial y_p'}\eta_p)]|x^1$$

$$=\int_{x^0}^{x^1}(-\sum_{h=0}^{p}\sum_{i=p+1}^{n} l_h(\frac{\partial g_h}{\partial y_i'}\eta_i'+\frac{\partial g_h}{\partial y_i}\eta_i))\,dx \tag{3.134}$$

Both the left- and right-hand sides of this fundamental relation can be rewritten to give

$$(l_0\frac{\partial g_0}{\partial y_0'}+\ldots+l_p\frac{\partial g_p}{\partial y_0'})\eta_0|x^1+\ldots+(l_0\frac{\partial g_0}{\partial y_p'}+\ldots+l_p\frac{\partial g_p}{\partial y_p'})\eta_p|x_1$$

$$=\int_{x^0}^{x^1}(\sum_{i=p+1}^{n} F_i\eta_i)\,dx \tag{3.135}$$

provided that $\eta_i(x^0)=\eta_i(x^1)=0,\ i=p+1,\ldots,n$.

(Under the conditions described so far, one can additionally say that, given arbitrary constants $q_0, q_1, \ldots, q_p$, there is one and only one solution to the adjoint system (3.168) for which

$$(l_0\frac{\partial g_0}{\partial y_i'}+\ldots+l_p\frac{\partial g_p}{\partial y_i'})|x^1=q_i \tag{3.136}$$

for $i=0,\ldots,p$. Bliss, and Fischer in the next example, use a similar choice of adjoint solution where the $q_k$ are the components of a unit vector.)

By an appropriate choice of $q_k$ and $l_k$ it follows, through an extended argument relying on here unstated previous results, that $F_i=0,\ i=0,\ldots,p$ and

$$0=\int_{x^0}^{x^1}(\sum_{i=p+1}^{n} F_i\eta_i)\,dx \tag{3.137}$$

for all variations $\eta_i$ for which $\eta_i(x^0) = \eta_i(x^1) = 0, i = p+1, .., n$. Endpoint conditions on the remaining $\eta_k$ are also discussed in the process. This variability is enough to yield, by the fundamental lemma of the calculus of variations, that $F_i = 0$ for $i = p+1, \ldots, n$ as well.

Although occurring in a context different indeed from that of ballistics, the basic manipulations are the same. Hadamard wrote the equations of variation as a nonhomogeneous square linear system and invoked the adjoint of this system, properly defined, to obtain a fundamental integral relation between the variations and the solutions to the adjoint equations. *The adjoint solutions are the Lagrange multipliers for the Mayer problem.* This resulted in a situation in which the fundamental lemma of the calculus of variations could be called on to achieve Euler equation-like results from the integral in the fundamental relation.[59] Bliss' insight was to realize that the same procedure could be used to achieve a fundamental relation which could be used in the calculation of the variations once the adjoint system had been solved.

Bliss' acquaintance with this argument might be suspect, especially in view of the absence in any of his publications (to the date of his ballistics work) of a reference to it, but one piece of evidence dispels any doubts. Fischer's 1913 paper [776], which will be mined one last time, contains a special case of the argument with a specific citation to Hadamard's treatment of the Mayer problem. It is buried in the proof of a theorem [p. 379] which itself is a technical lemma in support of his result on a generalized Volterra derivative. He considers a set of differential equations

$$z_i' = g_i(x, y, y', z_1, \ldots, z_m), \ i = 1 \ldots m \tag{3.138}$$

He wishes to consider curves $y = y(x)$ joining points $(x_1, y_1)$ and $(x_2, y_2)$ in the plane which, when substituted into the previous equation, give solutions $z_i$ to the system with prescribed values at both $x_1$ and $x_2$. With such a $y$ given, associated with these solutions are the usual equations of variation

$$\zeta_i' = \frac{\partial g_i}{\partial y}\eta + \frac{\partial g_i}{\partial y'}\eta' + \sum_j \frac{\partial g_i}{\partial z_j}\zeta_j \tag{3.139}$$

and the adjoint system (Hadamard's definition in the context of the Mayer Problem is cited here.)

$$\lambda_i' + \sum_j \frac{\partial g_j}{\partial z_i}\lambda_j = 0 \tag{3.140}$$

[59] Use of the adjoint is not necessary to prove the multiplier rule for Mayer's problem: Bolza's treatment, for example, makes no use of it. It also appears that the adjoint system was not widely used in other contexts in the calculus of variations. In addition, in a comprehensive bibliography of the literature of the calculus of variations from 1850 to 1913 [876, note 550], one finds the comments "J. Hadamard studied particularly the Mayer Problem, only speaking subsidiarily of the Lagrange problem. O. Bolza ... operates otherwise. Hadamard's manner seems better."

Fischer then lets $[\lambda_{h1}, \lambda_{h2}, \ldots, \lambda_{hm}]$ be the solutions of the adjoint which have the initial values $[\lambda_{11}, \lambda_{12}, \ldots, \lambda_{1m}] = [1, 0, \ldots, 0]; \ldots; [\lambda_{m1}, \lambda_{m2}, \ldots, \lambda_{mm}] = [0, 0, \ldots, 1]$. (Hadamard's treatment is also cited here.) The goal of this portion of the technical lemma is: "By means of the solutions $[\lambda_{h1}, \lambda_{h2}, \ldots, \lambda_{hm}]$ of the adjoint equations the values $\zeta_i(x_2)$ of the solutions of [the equations of variation] which vanish at $x = x_1$ can readily be expressed in terms of the functions $\eta_j(x)$." Writing $Z$ as the column $\zeta = [\zeta_j],\ i = 1, \ldots, m$, $G$ the $m$ by $m$ matrix $G = [\frac{\partial g_i}{\partial z_j}]$ and $y$ the column $Y = \frac{\partial g_i}{\partial y}\eta + \frac{\partial g_i}{\partial y'}\eta', i = 1, \ldots, m$ and finally $L$ the $m$ by $m$ matrix of adjoint row solutions $[\lambda_{h1}, \lambda_{h2}, \ldots, \lambda_{hm}],\ h = 1, \ldots, m$, the equations of variation can be given as[60]

$$Z' = GZ + Y \tag{3.141}$$

It follows that

$$LZ' = LGZ + LY \tag{3.142}$$

$$LZ' = -L'Z + LY \tag{3.143}$$

$$LZ' + L'Z = LY \tag{3.144}$$

$$\frac{d}{dx}(LZ) = LY \tag{3.145}$$

whereupon, integrating from $x_1$ to $x_2$ one obtains

$$Z(x_2) = \int_{x_1}^{x_2} LY\,dx \tag{3.146}$$

which is the desired expression. (Compare to (1.60)–(1.66) in Chap. 1, which follows Bliss' antiaircraft work.)

Not much can be said about the significance of this result itself, but the plucking of the technique from Hadamard's work is striking. It cements the connection between Hadamard's and Bliss' use of the adjoint, with Bliss' familiarity of the idea established prior to his World War work. His claim that the methods of the calculus of variations proved sufficient to deal with the problems set him is then broad enough to include the adjoint technique as well as the use of functionals arising from an abstract consideration of smoothness of solutions to systems of differential equations.

We close this section with an admittedly unrelated but natural aspect of the adjoint process. The word "adjoint" conjures up, in, for example, Hilbert space

60 Fischer does not give the following equations; they are a translation of his verbal description of steps to be taken.

theory, an operator $A : H \to H$ and its "adjoining" operator $A^* : H \to H$ defined by $< x, A^*y > = < Ax, y >$ for $x$, $y$ in $H$, where $<, >$ is the inner product operation on $H$. Of course, these ideas were yet to be evolved at the time considered here. It is natural to ask if any notion of this nature was extant for the Lagrange or Jacobi version of the adjoint in the differential equation sense. Certainly the defining equation above is true of the transpose matrix $A^*$, where $x$ and $y$ are vectors.

An interesting answer of sorts was given in the work of a lesser known enthusiast of the functional calculus, Salvatore Pincherle, an Italian mathematician who taught at the University of Bologna.[61] Pincherle, whose work was mentioned by Hadamard, wrote several papers on the topic of a general adjoint, the first [927] in 1897, always with Lagrange's operator in mind. A paper of 1912 [928] shows a more developed treatment. Pincherle considers two "spaces" (essentially vector spaces) $S$ and $S'$, two elements $\mathbf{a}$ in $S$ and $\alpha$ in $S'$, and an operation called $P$ applied to the pair, denoted by $(\mathbf{a}, \alpha)$, taking values in a third space $S''$. The operator $P$ has multilinearity and othogonality properties. Then given an operator $A : S \to S$, he defines the adjoint $\bar{A} : S' \to S'$ as

$$(A\mathbf{a}, \alpha) = (\mathbf{a}, \bar{A}(\alpha)) \tag{3.147}$$

Properties of the adjoint thus defined are proved. Three examples are then provided. The first is essentially the transpose of a square matrix in $R^n$, though a generalization to infinite dimensions is mentioned, "supposing that the operation $A$ is defined on a 'limited' matrix in the sense of Hilbert" [928, p. 148]. The second example takes place in $L^2[0, 1]$ (not denoted as such), where the operator $P$ is given by

$$(\alpha, \beta) = \int_0^1 \alpha(x)\beta(x)\,dx \tag{3.148}$$

The operation $A$ is integration against a kernel:

$$A(\alpha) = \int_0^1 K(x, y)\alpha(y)\,dy \tag{3.149}$$

and its adjoint is shown to be the expected integral equations adjoint

$$\bar{A} = \int_0^1 K(x, y)\beta(x)\,dx \tag{3.150}$$

The final example uses functions of sufficient differentiability in $L^2[0, 1]$ with the same inner product as the second example and shows that, assuming that the functions satisfy appropriate boundary conditions, the adjoint of the linear form

[61] See [700, p. 265] for a summary of Pincherle's work.

$\alpha_0\phi + \alpha_1 D\phi + \ldots + \alpha_n D^n\phi$ is "the classical adjoint of Lagrange," [ibid., p. 150]. In these examples continuity of the operations is not considered.

Though much more could be said and cited about this aspect of the adjoint during this period, we will limit ourselves to this small contribution whose influence is frankly unknown. We shall return to the subject of multiplier rules in a later section, when their solution is approached using the fully developed tools of functional analysis from the 1930s and 1940s.

## 3.9 Partial Differential Equations, Maxime Bôcher, Max Mason, and "Applied Mathematics"

The opening sentence of the review of Bliss' *Mathematics for Exterior Ballistics* of 1944 [727] reads: "This book treats exterior ballistics as a problem in particle dynamics, that is, rotational effects are ignored and resistance is assumed to act along the tangent to the trajectory." This remark points out that the problems of exterior ballistics, including Bliss' variability theory, can in their most simplified form be considered as problems in the "applied mathematics" of the day[62] and as such share features with other physical problems of interest at the time, be they from potential theory, the theory of elasticity, or electromagnetic phenomena. While Bolza's *Lectures* lay "entirely on the theoretical side," Hadamard's book was praised not only for its basis in the functional calculus but for its inclusion of many applied topics; thus the review in the *Bulletin of the American Mathematical Society* claims "Especial emphasis is laid on practical applications, too often much neglected ... profuse individual problems characterize the entire book as one of more than usual practical bearing" [834, p. 30]. Bliss himself showed an interest in many such problems, especially those utilizing partial differential equations in their solution, almost to his own embarrassment: in [722, p. 317], his 1927 talk on applied mathematics and engineering, he states: "Examples of problems of this type are the determination of the potential functions for a gravitational field, the determination of temperature variations in a field when temperatures and the boundary are assigned, vibration of strings and membranes, the determination of potential functions for electrostatic and magnetic fields, and many others. I must admit that these problems

[62] The quotations around the phrase are meant to convey the looseness of this term, especially as it was understood at that time in the United States. The work of Bliss and Moulton, with its theory, calculation through numerical integration, computation sheets, entering of results into range firing tables, and testing of theory against practice, earns the title applied mathematics. The labeling as a piece of applied mathematics and the solution of a partial differential equation for the deflection of a supported rectangular plate by an infinite series of trigonometric and hyperbolic functions also occurred [884, p. 493]; there is a difference. As another example, taking at random the New Publications list under Applied Mathematics of the January 1917 *Bulletin of the American Mathematical Society*, one finds a book on particle physics by an Italian professor of physics, a translation of Joukowski's work on aerodynamics, as well as volumes on mechanical drawing and surveying. But the commonality of dealing with physical phenomena suffices to bring them under the same umbrella.

have a great fascination for the mathematician, irrespective of their applicability in the industries or in neighboring sciences, and I am willing to make this admission in spite of the fact that the yielding to such fascination is sometimes implied to be a fault rather than a virtue." Given these considerations, it is natural to ask how Bliss' methods in the "problem of particle dynamics" fit in with solutions of other applied problems. In particular, one could ask if and how the adjoint appeared in solutions to other applied problems in partial differential equations.[63]

A review of all the uses of the adjoint idea in applied problems of this time would be a large undertaking. We shall limit this discussion to the work of the American mathematicians Maxime Bôcher and his students, and Max Mason.[64] The work of Hadamard, whose continued efforts to treat applications dealt with generalizing the techniques of earlier mathematicians such as Riemann and Volterra, is central and will be treated first. Our conclusions regarding the influence of these men on Bliss' ideas will be speculative in that, absent any explicit mention by Bliss, one can only hope to demonstrate that he may have become familiar with current research in partial differential equations by association with Mason or by routine absorption of contemporary work. To omit the contemporary discussions of these equations would be to neglect parallel developments in which uses of the adjoint operator were many.[65]

Early in the nineteenth century, attention was focused on the potential equation

$$\Delta U = \frac{\partial^2 U}{\partial x^2} + \frac{\partial^2 U}{\partial y^2} + \frac{\partial^2 U}{\partial z^2} = 0 \tag{3.151}$$

where $U$ is a function of $x$, $y$ and $z$ in three-dimensional space; the middle term is the familiar Laplacian operator. This equation arose in problems of heat transfer, gravitational attraction, and electromagnetic phenomena and was studied by Fourier, Laplace, Poisson, and many others. In time, a function satisfying this equation was called harmonic. A key investigator in this group was George Green, whose work on this equation in the context of electricity and magnetism was eventually published in the 1850s [809]. Among Green's results were the following identity, in which $U$ and $V$ are again functions in $R^3$, $\Omega$ is a body and $S$ its boundary;[66] partials with respect to $n$ are the normal derivatives:

$$\int\int_{\Omega}\int (U\Delta V - V\Delta U)\,dv = \int_S\int (U\frac{\partial V}{\partial n} - V\frac{\partial U}{\partial n})\,d\sigma \tag{3.152}$$

[63] For an overview of partial differential equations in the nineteenth century by an enthusiast of the subject, see [871, Chapter 28].

[64] Recall that during this Mason and Bliss were friends and collaborators.

[65] From the autobiographical account [729], one finds that Bliss' experience with applied mathematics includes the attendance of a lecture series on "Mechanik der Kontinua" by Hilbert during the former's 1902–1903 stay in Göttingen, attendance of lectures on "Dynamical Theory of Gases" while at Princeton, and the reading of a book on electromagnetic theory by James Jeans.

[66] In the plane $\Omega$ will be a region and $S$ its curvilinear boundary.

This eventually became known as Green's theorem and was referred to as such as late as the second decade of the twentieth century.[67] In the statement of this theorem and others immediately to follow differentiability and continuity conditions for the validity of equations will not be given but were present in rigorous statements of theorems at the time.

Another key result of Green's was the following: If $\Omega$ and $S$ are as above, and $U$ is harmonic on a region containing $\Omega$, then its values in $\Omega$ can be represented by an integral expression involving the values of $U(X)$ and $G = G(X, Y)$ for $X, Y$ in $R^3$, $X$ on $S$, $Y$ a fixed point of $\Omega$. Here $G$ is another harmonic function in $X$ satisfying properties to be listed below. The representation formula is the following:

$$u(Y) = -\frac{1}{2\pi} \int_S \int U(X) \frac{\partial G(X, Y)}{\partial n} \, d\sigma \tag{3.153}$$

Here $\frac{\partial G}{\partial n}$ is the normal derivative of $G$. The function $G$ must, in addition to being harmonic in $X$, have value zero for $X$ on $S$ and arbitrary $Y$ in $\Omega$, and, at a fixed $Y$ in $\Omega$ must have a singularity of the form $\frac{1}{|X-Y|}$ for $X$ in $\Omega$.[68] The function $G$ came in time to be known as a Green's function for the domain $\Omega$. The existence of such a function is not obvious, and much work was expended either proving its existence or finding explicit formulae for it. Green inferred its existence on physical grounds; see [869, p. 277]. One hopes that $G$ can be found for a particular $\Omega$ more easily than $U$ itself. If it can, the representation theorem can be considered as a solution to the boundary value problem consisting of the potential equation together with values of $U$ on $S$.

Of great importance for our narrative is the work of Riemann [934] in 1860 which investigates the so-called hyperbolic partial differential equation

$$F[u] = \frac{\partial^2 u}{\partial x \partial y} + A\frac{\partial u}{\partial x} + B\frac{\partial u}{\partial y} + Cu = f \tag{3.154}$$

where $u$, $A$, $B$, $C$, and $f$ are functions in the plane. This equation arose in the study of the propagation of sound waves. Our account follows that given in [822], a 1923 volume rich in history and references dedicated to generalizations of Riemann's method and their use in physical problems. The problem is to find the value of the solution function to (3.154) at a point $a(x_0, y_0)$ in the plane, when $u$ and $\frac{\partial u}{\partial n}$ are known on a monotone arc lying above $a$; see Fig. 3.2.

---

67 This is not the identity usually referred to by that name today, but (3.172) is easily derived from the modern namesake; see, for example, [778, pp. 61–62]. Terminology changed and (3.172) became one of three "Green's identities"; see [869] as a book from the 1920s on potential theory by an American mathematician in which such a reference is made.

68 In $R^2$ the singularity is as $ln|X - Y|$.

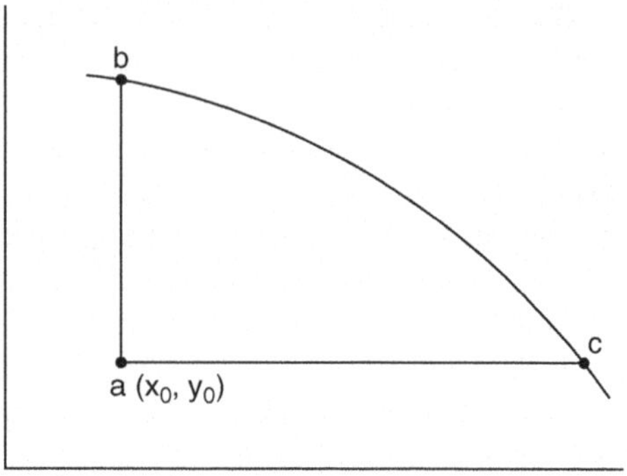


**Fig. 3.2** Rendering of diagram from [822], p. 66

In Hadamard's account, the adjoint operator $G[v] = \frac{\partial^2 v}{\partial x \partial y} - \frac{\partial(Av)}{\partial x} - \frac{\partial(Bv)}{\partial y} + Cv$ is introduced and an analogue of Green's theorem is also deduced;

$$\int\int (vF[u] - uG[v])\, dA = \int (Pdy - Qdx) \tag{3.155}$$

where

$$P = \frac{1}{2}(v\frac{\partial u}{\partial y} - u\frac{\partial v}{\partial y}) + Auv \tag{3.156}$$

and

$$Q = \frac{1}{2}(v\frac{\partial u}{\partial x} - u\frac{\partial v}{\partial x}) + Buv \tag{3.157}$$

This version of Green's theorem called by Hadamard the Fundamental Formula.[69] The Fundamental Formula is applied to $u$, the solution of the given boundary value problem, and $v_0 = v_0(x, y; x_0, y_0)$, a solution to the adjoint problem $G[v] = 0$ which satisfies certain carefully chosen boundary value conditions. The domain of application is the triangular region $T$ shown in Fig. 3.2, bounded by an arc $A$ from $b$ to $c$ and the segments connecting these points to $a$. The upshot is the integral representation formula:

$$u_a = (1/2)(uv_0)_b + (1/2)(uv_0)_c + \int_A (Pdy - Qdx) - \int\int f\, dA \tag{3.158}$$

where $P$ and $Q$ are as above with $v = v_0$.

[69] In Riemann's original account (pp. 158–161) neither the word "adjoint" nor the phrase "Green's theorem" is used; these are labels of later expositors. In each case a reference to integration by parts is made.

Thus the representation theorem expresses $u$ at a point outside a curve in terms of $u$ and its partials *as well as* $v_0$*, the solution to an adjoint boundary value problem, and its partials*, on $S$ itself.

The significance of this result is clear. First, it can be taken as a generalization of the ideas of Green to partial differential operators other than the Laplacian and shows how, provided the adjoint is introduced, analogues of Green's theorem and representation theorem are possible. The Laplacian is self-adjoint, so Green's results can be considered as a special case of this idea if Riemann's method can be extended beyond the hyperbolic case. Second, it provides a method for solution of a class of boundary value problems provided that the adjoint can be found. Of the generalizations of Riemann's method which followed, to be described next, Hadamard noted "This formula [the generalization of Riemann's Fundamental Formula obtained by Hadamard for the hyperbolic operator in $n$ variables] is, as we have said, the basis of any research concerning linear partial differential equations of the second order and especially of the above quoted investigations. $u$ habitually denotes the unknown function of the problem; $v$ is an auxiliary arbitrary function, precisely in the choice of which the whole skill of the operator lies. It will be most generally chosen so as to satisfy the adjoint equation $G[v] = 0$" [822, p. 64].[70] In one case considered by Hadamard $v_0$ involved a Bessel function of order zero. (One might at this point begin to consider the similarity of this scenario with Bliss' use of the adjoint system: using the solution of the adjoint to compute solutions of the differential equations for the differential variations.)

There followed the desired generalizations of Riemann's method to other classes of partial differential operators, culminating in a publication by Paul du Bois-Reymond [769] which appeared in 1889, the year of his death. In this work, the author was able to obtain results for the general linear partial differential operator

$$L[u] = A\frac{\partial^2 u}{\partial x^2} + 2B\frac{\partial^2 u}{\partial x \partial y} + C\frac{\partial^2 u}{\partial y^2} + D\frac{\partial u}{\partial x} + E\frac{\partial u}{\partial y} + Fu \tag{3.159}$$

and its adjoint

$$M[v] = \frac{\partial^2 (Av)}{\partial x^2} + 2\frac{\partial^2 (Bv)}{\partial x \partial y} + \frac{\partial^2 (Cv)}{\partial y^2} - \frac{\partial (Dv)}{\partial x} - \frac{\partial (Ev)}{\partial y} + Fv \tag{3.160}$$

The general Green's theorem appeared as

$$\int_\Omega \int vL[u] - uM[v]\,dA = \int_S Pdy - Qdx \tag{3.161}$$

where $P$ and $Q$ were expressions analogous to those used by Riemann, $\Omega$ and $S$ as before. From this it would be possible to obtain integral representation theorems like

[70] The studies mentioned by Hadamard were investigations into optics.

those of Green and Riemann.[71] du Bois-Reymond humbly acknowledged Riemann as the predecessor of this paper, saying that what was presented was following the path opened by him [769, p. 282].

If Hadamard is to be believed, Riemann's method became quite well known by the turn of the twentieth century: a mention of it by du Bois Reymond in 1883 preceded his 1889 work, "...and it was finally brought to the knowledge of every mathematician, in its most general form, by the classic *Leçons sur la theorie des surfaces* of Darboux.[72] Since that time, Riemann's method may be considered as universally known..." [822, p. 57]. Hadamard himself mentioned it in his 1903 work on the propagation of waves [820]. Max Mason referred to it briefly in his New Haven Mathematical Colloquium lectures, published in 1910 [890, p. 182]. And Hadamard's volume [822] is a book-length set of variations on the theme of Riemann's method which includes other related materials.

This having been said, the connection between this material and Bliss' use of the adjoint system, though mathematically explicable, must, as previously noted, be considered speculative from the point of view of influence. Bliss used the adjoint to a system of ordinary linear differential equations for the variations in the ballistic trajectory as a computational method in which one solution of the adjoint provided many solutions for the variations themselves. In this respect, there is a similarity of purpose in solving the adjoint equation and connecting it to the solution of the original system. Gronwall's paper [817] presents the use of the solutions of the adjoint to get qualitative information about the ballistic trajectory. Curiously, Gronwall published a paper in July of 1919 [814] in which Riemann's method was used to help in the solution of a problem in differential geometry set by a former Princeton colleague; Gronwall was still employed by Moulton's working group at that time. Bliss was familiar with Darboux's treatise, having cited it several times in papers as early as 1906 [705], he was, after all, one of the collection "every mathematician" referred to by Hadamard. But lacking any explicit mention by Bliss of Riemann's method, one can only suggest that the idea was in the air at this time and may have served Bliss as a parallel development for his use of the adjoint.

While Rowe's review of Moulton's ballistics book decried the adjoint as a topic too esoteric for the practicing ballistician, it was not so for American research mathematicians of the era. We now consider the contributions of Bôcher and Mason to the material on the adjoint. Each can be classed as a mathematician having an applied bent, if by that one includes the authoring of papers devoted to differential equations with moderate attention to their use.[73] Now and then one encounters a

[71] The classification of linear partial differential equations into hyperbolic, elliptic and parabolic was first presented in this paper.

[72] The 1889 edition of this multivolume series [757] contains the comment "The adjoint equation to a given one [equation] was presented for the first time in a memoir of Riemann on the propagation of sound" [757, Volume II, book IV, p. 75].

[73] Mason was engaged in genuine applied mathematics in the World War era; he was involved in submarine sound ranging work.

justification of this practice: Bôcher himself explained while explaining his attention to such a problem: "Why is [the problem at hand] an important one? The most obvious answer is that it is one of which special cases come up constantly in applied mathematics; that even its special cases are of sufficient difficulty to have demanded the serious attention of the best mathematicians for nearly two hundred years..." [736, p. 164]. The stream of papers to be described next culminates in the showcasing of the adjoint used by Bliss (in fact its generalization presented nearly simultaneously by Hadamard) arrived at by a circuitous route.[74]

Bôcher received his doctorate in Göttingen in 1891 under the direction of Felix Klein. Mason, the younger man by 10 years, arrived in Göttingen in 1900 and earned his doctorate under David Hilbert in 1903. Mason's work involved Hilbert's new view of the calculus of variations which linked that subject with boundary value problems and integral equations; see [796, p. 362 ff.]. Bôcher's involvement stemmed from his efforts to appropriate the notions of Green's theorem and Green's functions from partial differential equations and apply them to ordinary differential equations. In 1901, on a return visit to Göttingen, Bôcher issued a note [733] which offered a solution method for the general $nth$ order linear equation

$$L[y] = \frac{d^n y}{dx^n} + p_1 \frac{d^{n-1} y}{dx^{n-1}} + \ldots + p_n y = p \tag{3.162}$$

where $p_k$ are functions of $x$ defined on an interval $[a, b]$. He introduced what he called the Green's function $G(x, \xi)$ for this solution. This is a function of two real variables, each an element of $[a, b]$ with $\xi$ fixed, satisfying various differentiability and continuity properties and some others reminiscent of their counterparts. The function must satisfy $L[G(x, \xi)] = 0$ in $x$ for a fixed $\xi$, and at each endpoint of the interval, it must satisfy a *system* of linearly independent homogeneous relations. In certain cases, this produces the vanishing of $G(a, \xi)$and $G(b, \xi)$.

With the use of these functions Bôcher announced the result that the solution of the equation $L[y] = p$ which satisfies the same condition as does $G(x, \xi)$ at $a$ and $b$ is given by the formula

$$y = -\int_a^b G(x, \xi)\, p(\xi)\, d\xi \tag{3.163}$$

One notes the similarity to the integral representation theorem for harmonic functions.

In 1904, Mason [886] published a paper based on his thesis which shows evidence of the influence of Bôcher's idea. In particular on page 529, one sees the

[74] The distance between this notion of applied mathematics and the numerical thrust of Moulton's efforts can be gauged by a footnote on the first page of Bôcher's 32-page survey from which the above quote is taken: "The question of finding effective means for computing the solutions in question is also one which might well be considered here." No such consideration is given, however.

equation $L[y] = \frac{d^2y}{dx^2} + \lambda A(x)y = f(x)$ with a set of linear relations among $y$ and $\frac{dy}{dx}$ at the endpoints of an interval $[a, b]$. On the next page, a Green's theorem analogue is given for the operator $L[y] = \frac{d^2y}{dx^2} + \lambda A(x)y$, namely,

$$\int_{x_0}^{x_1} (vL[y] - yL[v])dx = [v\frac{dy}{dx} - y\frac{dv}{dx}]_{x_0}^{x_1} \tag{3.164}$$

though it is not identified as such. This identity is applied to a solution $y$ of the above differential equation and $v$ the one-dimensional Green's function for $\frac{d^2y}{dx^2} = 0$;[75] this $G(x, \xi)$ turns out to be a tent function vanishing at $a$ and $b$ with a maximum at $\xi$. The result after some manipulation is that $y$ satisfies

$$y(\xi) = F(\xi) + \lambda \int_a^b G(x, \xi)\, A(x)\, y(x)\, dx \tag{3.165}$$

where $F(\xi)$ involves an integral of the Bôcher type with $f$. Thus in this case, Bôcher's ideas are used by Mason to change a differential equation into an integral equation, a technique with an interesting history which will be referred to again in what follows. The rest of the paper deals with Hilbert's techniques for relating this equation to a problem in calculus of variations. In a paper of 1906 [888], Mason considered an arbitrary second-order linear nonhomogeneous differential equation with boundary conditions explicitly expressed as linear equations in $y$ and $\frac{dy}{dx}$ evaluated at the endpoints of the interval of definition; extra restrictions on these conditions were involved. This came to be known as a "linear boundary value problem." At some point, either during Bôcher's stay in Göttingen or after Mason's return to America, Mason began to consider systems of partial differential equations of the elliptic type, that is, of the form

$$\Delta u + a\frac{\partial u}{\partial x} + b\frac{\partial u}{\partial y} + cu = f \tag{3.166}$$

where $u$, $a$, $b$, $c$, and $f$ are all functions of $x$ and $y$ and $\Delta u$ is the Laplacian of u. His goal was to extend Green's theorem and functions to *systems* of this type of equation, a suggestion apparently made by Bôcher (see [887, p. 221 footnote]).[76]

[75] This equation was sometimes referred to as the "one-dimensional Laplace's equation."

[76] Mason says that Bôcher pointed out the idea of adjoints of systems of ordinary differential equations as presented in Jordan's *Cours d'Analyse*; however Bôcher mistakenly identified Jordan as the originator of the idea. Bôcher's work seems to have been limited to ordinary differential equations, but he encouraged his students and others like Mason to consider analogous results for partial differential equations.

To this end, Mason published [887], which also appeared in 1904, in which he described the system

$$L_i(u_1, u_2, \ldots, u_n) = \Delta u_i + \sum_{j=1}^{n}[a_{ij}\frac{\partial u_j}{\partial x} + b_{ij}\frac{\partial u_j}{\partial y} + c_{ij}u_j],\ i = 1, 2, \ldots, n \tag{3.167}$$

We note here that this system is more general than that mentioned in Jordan since the derivatives of all the $u_j$ are included in each equation, as in Hadamard's system. He then introduces and adjoint system as did Hadamard:

$$M_i(v_1, v_2, \ldots, v_n) = \Delta v_i - \sum_{j=1}^{n}[\frac{\partial(a_{ji}v_j)}{\partial x} + \frac{\partial(b_{ji}v_j)}{\partial y} - c_{ji}v_j],\ i = 1, 2, \ldots, n \tag{3.168}$$

Again one notes the "adjointness" of this expression both with respect to the individual operators and the reflection around the diagonal elements of the system. These equations are connected by a Green's theorem:

$$\sum_{i=1}^{n}\int_{\Omega}\int v_i L_i(u_1, \ldots, u_n) - u_i M(v_1, \ldots, v_n)\, dA = \int_S (Qdy - Pdx) \tag{3.169}$$

where $S$ is the boundary of the region $\Omega$, and $P$ and $Q$ are expressions involving $u_{ij}$, $v_{ij}$ and their partials.

Arrays of $n$ sets of $n$ functions $G_{ki}(x, y, \xi, \eta)$ and $H_{ki}(x, y, \xi, \eta)$ called Green's functions for $L_i$ and $M_i$ are then introduced. These functions have properties similar to those in the harmonic case, for example, each $G_{ki}$, $i = 1, \ldots, n$ is a solution to $L_i(u_1, \ldots, u_n) = 0$, in $\Omega$ with $G_{ii}$ containing a logarithmic singularity at a fixed $(\xi, \eta)$, and all vanishing identically in $(\xi, \eta)$ on the boundary $S$. The conclusion is an integral representation formula for the solutions $u$ to $L_i(u_1, \ldots, u_n) = f_i, i = 1, \ldots, n$ involving sums of integrals over $S$ and $\Omega$ of $u$ and $f$ respectively against the $H_{ki}$.

This work seems to have been a one-off for Mason, since he never followed up on the ideas presented in it. Yet it represents a continuation of the generalizations of the adjoint operation in an arc from Riemann through du Bois-Raymond's results and those of Bôcher and his students. Thus, for example, in [861] and [697] from 1908 and 1910, respectively, we have efforts to generalize Mason's results to the most general partial differential equation of order two in $n$ variables with accompanying Green's functions and Green's theorem; invariants and covariants are investigated and conditions for existence of solutions are given. References to Mason's paper and du Bois-Raymond's 1889 results are made. On a different note, G. D. Birkhoff contributed to the literature in 1908 [701] by writing on arbitrary $nth$ order linear boundary value problem, with the addition of an adjoint equation and *adjoint boundary conditions* defined appropriately.

In a final trio of papers by Bôcher, [735] which provides proofs of the assertions of his 1901 note which were omitted generalizes some of Birkhoff's results. In an invited address before the fifth International Congress of Mathematicians in Cambridge England in 1912 [736], from which several quotes have already been taken, a survey of results to that date is presented. Lastly, in [737], he presents a necessary and sufficient condition that a linear boundary value problem for a general $nth$ order ordinary equation have a solution; Mason's [888] is cited here.

In a development that is the goal of presenting this series of papers, Bôcher here defines the $nth$ order linear ordinary differential operator $L[u]$ and its adjoint $M[v]$ as usual, writes the Lagrange identity

$$vL[u] - uM[v] = \frac{d}{dx}P(u, v) \tag{3.170}$$

where $P(u, v)$ is a bilinear form in $u$, $v$, and their derivatives. He then integrates both sides of the identity on the interval of definition and labels the result Green's theorem.[77] In his final section, he extends his results to a system of $m$ linear $nth$ order ordinary equations in $m$ unknowns and its adjoint and establishes a Green's theorem here again by integrating a Lagrange-type identity. When specialized to a system of order one equations this Green's theorem becomes essentially the Fundamental Relation used by Bliss in his method for computing differential variations.[78] Our account can be concluded with the assertion that *The Fundamental Relation of Bliss' first order system and adjoint can be seen as a version of Green's Theorem.*[79]

Again, how much of this came to Bliss' attention is a matter of speculation. Some of these ideas may have been discussed by Bliss and Mason during their time at Göttingen, and their collaboration in the 1900s may have resulted in sharing information on other current projects. Bliss himself, a few years after the war, wrote a theoretical paper on linear boundary value problems for systems [721], though no connection to his ballistics work is mentioned. Even if no explicit connection can be documented, this contemporary coincidental appearance of the adjoint system in a potential-theoretic vein is quite interesting.

[77] This had been done previously in [697].

[78] Actually a system and Fundamental Relation like Hadamard's is achieved. Credit is also given to one Bounitzky who in 1909 also published on this topic.

[79] One can see this appellation in some modern textbooks, for example, [832] notes this connection.

## 3.10 Existence Theorems Again and Integral Equations

In this section, we mention some "pure" aspects of the equations dealt with in the previous section, specifically, solution existence theorems, some of which have already been referenced. Bôcher was well aware of the limitations of dwelling on these matters at the expense of studying the nature of the solutions; in his address before the International Congress he remarked:

> It must not be inferred from what I have said so far that the theory of boundary problems consists wholly, or even chiefly, in establishing existence-theorems or in proving by the exact methods of modern analysis facts which a hundred years ago would have seemed self-evident to any mathematician. Some applied mathematicians make it a reproach to pure mathematics that it has come now to a state where it is interested solely in questions of this sort. ... The good old-fashioned view that it is the main object of mathematics to discover essentially new facts is, however, hardly in danger of becoming obsolete in a generation which has just witnessed the splendid achievements of Poincaré. ... In the subject of boundary problems, while we need as a foundation the existence-theorems, and exact proofs of facts which in themselves are quite to be expected, they are *only* a foundation. [735, p. 171]

Yet he contributed his share of existence theorems, as well as showing interest in more applied matters.

The wave of Weierstrassian rigor which so informed Moulton's work and inspired Bolza to produce a more rigorous treatment of the calculus of variations had its effect on partial differential equations as well. Hadamard in particular wrote on the difficulties in ensuring that a problem is accurately posed and praised Bôcher for his insistence on defining precisely what is meant by a solution [822, p. 32]. In Riemann's method, while replacing the original equation for $u$ by an equation in the adjoint variable $v$ leads to a search for a solution to a possibly easier problem, the existence of such a solution may not be at all obvious. Darboux made some comments on this problem in his exposition [757, pp. 96–106]. And it has already been noted that the existence of a Green's function for a region $\Omega$ was not obvious.[80]

In Mason's thesis-derived paper cited above, he employed a Green's theorem analogue to transform a differential equation into an integral equation. This is an example of a technique of dealing with existence problems for both ordinary and partial differential equations alike as the subject of integral equations came into its own at the turn of the twentieth century. A common strategy was to show that the existence of a solution to a boundary value problem, for example, was equivalent to the existence of the solution to a corresponding integral equation and then focus on theorems which guaranteed the latter.

[80] "Green himself argued that such a function existed from the physical evidence. Of course the static charge on $S$ exists! We have here an excellent example of the value and danger of intuitional reasoning." remarked Kellogg wryly [869, p. 237]. The story of the Dirichlet principle provides another example of intuition successfully employed but not rigorously justified [869, pp. 279–280], one in which Weierstrass himself played a role.

Among the first works in English devoted to integral equations was the 1909 volume by Bôcher [734] mentioned above. This volume and Bliss' favorable review [707] of it may have encouraged Bolza to give the set of lectures on the subject in the summer of 1913 [740] [81] which was eventually released in modified book form (*Linear Integral Equations*) in 1924 by an auditor [885]. (Mason, while not devoting a book to the subject, used it in a major way in a paper of 1908 on partial differential equations [889].) These works begin with an existence proof for the so-called Fredholm's equation:

$$u(x) = f(x) + \lambda \int_a^b K(x,t)\,u(t)\,dt \tag{3.171}$$

where $\lambda$ is a parameter. It is easily seen that Mason's equation (10.15) is of this type. The solution strategy is the "method of successive substitution," used by Bliss in his dealing with solution derivatives in [712] (see Section 5 above). *Linear Integral Equations* outlines the procedure by pointing out that one can substitute into the right-hand side of the equation the equation itself to obtain a sum of three terms; repetition of this produces a finite series of iterated integrals and powers of $\lambda$. Analysis requiring the usual convergence criteria for an infinite series leads to a solution for $u(x)$ in the form of a power series in $\lambda$ whose coefficients are iterated integral expressions involving $K(x,t)f(t)$; the series converges for restricted values of $\lambda$ by comparison to a geometric series. Bôcher gives the same proof for $\lambda = 1$.[82] (It is possible that Bliss' use of the method was influenced by his knowledge of integral equations.)

For the theorem just cited Bolza introduces the function

$$R[u] = \int_a^b K(x,t)\,u(t)\,dt \tag{3.172}$$

which he proves is a linear "operator," though no notion of continuity of the operator with respect to domain or range spaces is discussed. Fredholm's equation is then expressed as

$$u = f + \lambda R(u) \tag{3.173}$$

and the successive substitution procedure produces the mercifully more briefly expressed infinite series

$$u = f + \lambda R(f) + \lambda^2 R^2(f) + \dots \tag{3.174}$$

which is proved convergent as before.

[81] The notes which constitute this reference are a fine example of the art of calligraphy.

[82] The method of Picard which, as noted above, differs from this, was often called "successive approximation"; the two terms were used interchangeably by Bolza in his lecture notes and by Mason. *Linear Integral Equations* notes this distinction [p. 15].

Bolza was likely influenced in this choice of presentation by Mason's 1908 paper [889], which gives a very general application of the method of successive substitution to the solution of hyperbolic partial differential equations. After mentioning successive approximation and Riemann's method as popular existence proof methods, he introduces a linear integral operator corresponding to the hyperbolic expression. It is more complicated than that involving the Fredholm equation as is the convergence proof but the idea is the same.[83] Mason also points out how several results of Picard and others are included in this single theorem.

An even stronger move in the direction of generality was displayed by Mason in his *New Haven Mathematical Colloquium* lectures published in 1910, a collection of some of his earlier papers on partial differential equations worked into book form. Here he considers the equation $f = g + Sf$, where $g$ is a given continuous function in an arbitrary region $R$ (domain not specified), $S$ is identified as a linear operator, and $S$ called "convergent" if

(1) $S\phi$ is continuous in $R$ if $\phi$ is
(2) $\phi + S\phi + S^2\phi + \ldots$ converges in $R$ to a continuous function in $R$, and
(3) "The result of the operation $S$ on the function represented by this series is equal to the result of the term by term operation on the series."

He quickly proves the theorem that for S a convergent operator, the equation $f = g + Sf$ has a unique solution given by $f = g + Sg + S^2g + S^3g + \ldots$ From this general statement, Mason is able to deduce existence of solutions to both partial and ordinary differential equations of many types, including Green's original representation theorem for harmonic functions, by converting them to integral equation problems to which his theorem is applicable. Remarkable in all this is the conception of a large collection of existence problems solved in the framework of an operator with more general domain and range and supplied with properties easily related to continuity, a trend toward functional analytic thinking which will culminate in the next section.

## 3.11 General Implicit Function Theory, General Multiplier Rules, and More Student Work

Given the number of implicit function and embedding theorems used by Bliss and Bolza in their work in the calculus of variations, one naturally wonders if there is some general theorem from which these could emerge as special cases, somewhat in the same spirit of Mason's theorem mentioned in the last paragraph. In this section, we trace the development of such a theorem, a general implicit function theorem, which had its origins in Goursat's elegant proof in 1903 of the

[83] This paper was cited positively by Bôcher, [734, p. 13].

elementary implicit function theorem for functions on Euclidean space.[84] Goursat's work passed through the hands of Volterra, who attempted generalizations of it which culminated in a sketch of a quite extensive result which would allow for the existence of an explicit solution to an implicitly defined functional relation between two functions of a line. This sketch was in turn fleshed out by Kenneth Lamson, a student of Bliss, who provided a rigorous argument for Volterra's suggestions using some of the concepts of E. H. Moore's general analysis. In a final series of papers, Lawrence Graves, another student of Bliss, gave several versions of a general implicit function theorem, the last ones stated in the language of modern functional analysis. During part of this evolution, Bliss' 1905 and 1917 papers on existence of solutions to ordinary differential equations, as well as DECAF and FOLIB, were cited as motivating influences. The final versions of the theorem came from a point of view more comprehensive than Mason's, a "general point of view, [in which] algebraic equations, differential equations, integral equations, and many other types are all gathered under one head" [803, p. 654], to quote from a 1935 paper of Graves. In reviewing these results, one thus sees not only the changing treatment of implicitly defined relations but the abstract theories in which these treatments occurred: functions of a line, functional calculus, general analysis, and Banach's functional analysis. An application of the general implicit function theorem outside of embedding theory is mentioned: we discuss the early work of Herman Goldstine in proving multiplier rules for Banach spaces of functions, results which yielded at one stroke the existence of such a rule for many different specific cases. Thus we return to Bliss' ongoing concern with classical multiplier theorems.

Goursat proves his result in two stages.[85] First, he begins with a continuous function $F(x, y)$ of two real variables defined in a neighborhood of a fixed point $(x_0, y_0)$ in the plane; here $F(x_0, y_0) = 0$, and $F$ is a contraction in $y$. Without loss of generality, he takes $(x_0, y_0) = (0, 0)$ and proves that the equation $y = F(x, y)$ admits a unique solution, where $x$ is fixed independently of $y$. To this end, he sets up the sequence of iterates $y_1 = F(x, 0)$, $y_n = F(x, y_{n-1})$ and proves in the familiar way that this sequence remains in the rectangle of definition of $F$ about $(x_0, y_0)$ and converges to a limit by consideration of the series $y_1 + (y_2 - y_1) + (y_3 - y_2) + \ldots$. The limiting value $Y$ is proved to be the unique solution of $y = F(x, y)$ and $(x, Y)$ tends toward zero with $x$. Then follows the main result: if $G(x, y)$ is a continuous function of two variables $x$, $y$ in the neighborhood of $(x_0, y_0)$ at which $G(x_0, y_0) = 0$, and $G$ admits a continuous partial derivative $\frac{\partial G}{\partial y}$ such that $m = \frac{\partial G(x_0, y_0)}{\partial y} \neq 0$, then the equation $G(x, y) = 0$ has, for each $x$ a unique solution y tending towards $y_0$ as $x$ tends towards $x_0$. The proof follows from the observation that the preceding equation is equivalent to $y - y_0 = y - y_0 - \frac{G(x,y)}{m}$, wherein the right-hand side is set equal to his $F(x, y)$ and the preceding argument is applied.

---

[84] It has already been noted that Goursat's proof was inspired by the success of Picard's iteration method for existence proofs for differential equations.

[85] In each implicit function theorems which follow we note that the authors proved or referred to the existence of *local* explicit solutions to implicitly defined relations. To avoid complication the domains and ranges of these solutions are not given.

This is the format for the argument in higher-dimensional Euclidean space. The result is stated as follows: let $G_i(x_1, \ldots, x_n; y_1, \ldots, y_p)$, $i = 1, .., p$ be $p$ continuous functions with continuous partial derivatives $\frac{\partial G_i}{\partial y_k}$ in a neighborhood of $x_1^0, \ldots, x_n^0; y_1^0 \ldots, y_p^0$, suppose that $G_i(x_1^0, \ldots, x_n^0; y_1^0, \ldots, y_p^0) = 0$ for all $i$ and $\frac{\partial G_i}{\partial y_k}$ are all continuous in a neighborhood of $(x_1^0, \ldots, x_n^0; y_1^0, \ldots, y_p^0)$. Then if the "functional determinant" $\frac{D(G_1, \ldots, G_p)}{D(y_1, \ldots y_p)} \neq 0$ at $(x_1^0, \ldots, x_n^0; y_1^0, \ldots, y_p^0)$ the system $G_i(x_1, \ldots, x_n; y_1, \ldots, y_p) = 0$, $i = 1, .., p$ has a unique solution $y_1 = \phi_1(x_1, \ldots, x_n), \ldots, y_p = \phi_p(x_1, \ldots, x_n)$ where the $\phi_i$ are continuous functions which tend toward $y_1^0, \ldots, y_p^0$ as $x_1, \ldots, x_n$ tends toward $x_1^0, \ldots, x_n^0$. The proof uses an auxiliary function set $F_i, i = 1, \ldots, p$ as in the single variable case, on which the sequence of iterates is defined.

A decade later, Volterra, in Chapter IV of [980], attempted the first generalization of this technique beyond Euclidean space in the context of his theory of the functions of a line. Recall that here a real valued function $f$ defined on an interval $[a, b]$ could be used as the input to a real-valued "function of a line" $F|[f(x)_a^b]|$; one could have, for example $F|[f(x)_a^b]| = \int_a^b f(x)\phi(x)\,dx$, for a fixed $\phi$. As noted previously Volterra considered continuity and derivatives $F'|[f(x)_a^b, \xi]|$ of his functions; he also introduced a differential $\delta F = \int_a^b F'|[f(x)_a^b, \xi]|\,\delta f(\xi)\,d\xi$ corresponding to $\delta f(\xi)$.

Before turning to Volterra's exposition on general implicit function theory, it is important to note that his *Leçons* must be considered a source of the idea of considering solutions to differential equations as dependent on the functions which define the equation, a point of view which provided the gateway for Bliss. Volterra explicitly mentions this as an example of a function of a line. In Chap. 1, he notes that, for example, in the second-order linear differential equation

$$\frac{d^2y}{dx^2} + p(x)\frac{dy}{dx} + q(x)y = 0 \qquad (3.175)$$

if $X$ is a fixed point on the abscissa, then the solution to an initial value problem $y(X)$ can be considered as a function of a line of $p(x)$ and/or $q(x)$ on the interval $(0, X)$. He calculates the derivative $y'|[q(x)_0^X]|$ using the familiar variation of parameters technique; many more examples of such derivatives for higher-order nonlinear equations are calculated. Integral equations of the type dealt with above also can be considered functions of a line.

Chapter IV is devoted to implicit functions of a line; the general problem of "resolution" of the equation $F|[f(x)_a^b, \phi(x)_a^b, \xi]| = 0$ with respect to $f(x)$ for a given $\phi$ fixed. This problem is considered as a limiting case of the implicit function theorem just treated. He deals first with the special case of an integral equation

$$f(x) + \int_a^b \phi(\xi)\,K(x, \xi)\,d\xi = \phi(x) \qquad (3.176)$$

where $\phi(x)$ and $K(x, \xi)$ are given. His solution method falls back on an analogous finite problem; rather than discuss the method, which was given to emphasize the connection between functions of a line as limits of related finite problems, it will serve our purposes to state the results in terms of a presentation, given in Bôcher's book [734, pp. 19–24], which is itself based on Volterra's earlier expositions of this solution. Either way, "Volterra's elegant solution" [734, p. 21] takes the form

$$\phi(x) + \int_a^b f(\xi)\, k(x, \xi)\, d\xi = f(x) \tag{3.177}$$

where $k(x, \xi)$ is a kernel defined as a sum of iterates involving $K(x, \xi)$, and the relations

$$K(x, y) + k(x, y) = \int_a^b K(x, \xi)\, k(\xi, y)\, d\xi = \int_a^b k(x, \xi)\, K(\xi, y)\, d\xi \tag{3.178}$$

hold. This is the defining condition that $K(x, y)$ and $k(x, y)$ be called *reciprocals*.

Volterra finally outlines ["Nous sommes maintenant en mesure de traiter le cas general...] [980, p. 71] a solution for the general case which is dependent on the integral equation result just given. Volterra assumes that the function of a line $F|[f(x)_a^b, \phi(x)_a^b]|$ has an expansion as a Taylor series in integral expressions involving $f$ and $\phi$. Again the reduction is made to the case in which $F|[f(x)_a^b, \phi(x)_a^b, \xi]| = 0$ is satisfied by $f = 0$, and $\phi = 0$. With $\phi$ then given, he notes that

$$F|[f(x)_a^b, \phi(x)_a^b, \xi]| = 0 \tag{3.179}$$

is equivalent to finding a solution to

$$\begin{aligned} f(\xi) &+ \int_a^b K(\xi, \xi_1)\, f(\xi_1)\, d\xi_1 \\ &= f(\xi) + \int_a^b K(\xi, \xi_1)\, f(\xi_1)\, d\xi_1 - F|[f(x)_a^b, \phi(x)_a^b, \xi]| \end{aligned} \tag{3.180}$$

(note the similarity to Goursat's argument), where $K$ is a kernel related to the expansion of $F$ in its series. Calling the right-hand side of this equation $\Theta|[f(x)_a^b, \phi_a^b, \xi]|$ one then "resolves" the equation

$$f(\xi) + \int_a^b K(\xi, \xi_1)\, f(\xi_1)\, d\xi_1 = \Theta|[f(x)_a^b, \phi(x)_a^b, \xi]| \tag{3.181}$$

by his previously developed method of dealing with integral equations. He sets

$$f(\xi) = \Omega|[f(x)_a^b, \phi(x)_a^b, \xi]| \tag{3.182}$$

"$\Omega$ being a quantity enjoying the same property as $\Theta$" in the resolution. This equation is resolved by iteration in a functional domain of $\phi = 0$: $f_1(\xi) = \Omega|[0, \phi(x)_a^b, \xi]|, \ldots, f_n(\xi) = \Omega|[f_{n-1}(x)_a^b, \phi_a^b(x), \xi]|, \ldots$ "of which one can study the convergence [to $f$] by posing certain conditions."

Two concluding remarks [980, p. 73] "The method we have indicated is none other than the method of successive approximations of M. Picard, which M. Goursat has utilized in his proof of the implicit function theorem." After one final calculation relating differentials of $f$ and $\phi$ Volterra concludes; "The reader has already perceived the profound parallel between this last calculation and that of the derivatives of ordinary implicit functions of which it merely the extension."

In view of his contemporary work on embedding theorems and his positive review of Volterra's *Leçons*, Bliss must have seen Volterra's speculations on a general implicit function theorem as an invitation to explore a rigorously proved result of this nature. Following his earlier pattern, he involved a student in this foray into the functional calculus. The result was contained in the thesis of Kenneth Lamson, appearing in 1920.[86] He begins by citing the results of both Goursat and Volterra and then sets himself the task of moving these results to a general milieu by solving the equations $y(p) = F[y(q), z(q);\ p],\ G[y(q), z(q);\ p] = 0$.

Here $p$ and $q$ are elements of a set $B$, which, for example, could be $1, 2, \ldots, n$ or $[a, b]$;[87] $y$ and $z$ real-valued functions from a fixed class $M$; $F$ and $G$ are called functionals after Fréchet's term "functionelle." Elements of $M$ satisfy postulates which later became those defining a Banach space but are merely listed in the paper.

To solve the first equation Lamson fixes $y_0$ and $z_0$ in $M$ such that $F[y_0, z_0;\ p] = y_0(p)$ for all $p$ in $B$; $F$ is assumed to be a contraction in $y$ for a fixed $z$ in a prescribed set, and the successive approximations $y_1 = F[y_0, z;\ p],\ y_{i+1} = F[y_i, z;\ p]$ are the defined. By the usual Banach space arguments, a limiting value $y$ is obtained which is the unique continuous solution to $y = F[y, z;\ p]$. Thus an implicit function theorem is proved for the first of the two equations by methods which are the straightforward generalizations of Goursat's and Volterra's and resemble in all important respects a modern argument.[88]

For the equation $G[y(q), z(q);\ p] = 0$ Lamson deals with only a simpler form $G[y;\ p] = z(p)$, to be solved for $y$, given that $G[y_0;\ p] = z_0(p)$. A quantity which plays the role of the Jacobian must be introduced at this point; this quantity is a functional $A[y_1, y_2, \eta;\ p]$. $A$ satisfies a list of properties evidently assigned to make the proof work: membership in $M$ in the variable $p$, linearity and boundedness as an operator in $\eta$, and uniform continuity in $(y_1, y_2)$ at $(y_0, y_0)$ when $||\eta|| = 1$.

---

86 In his Vita accompanying the thesis Lamson states that he finished his studies at the University of Chicago in 1917, but the thesis is a reprint of his 1920 article [875] in the *American Journal of Mathematics*. Bliss' military commitments no doubt contributed to this gap.

87 The set $B$ is described as the "general range" due to E. H. Moore, "not defined and on which no postulates bear explicitly" [875, p. 244]. This notion seems to simply play the role of the domain of the functions $y$ and $z$.

88 Lamson states that his method "is similar to that given by Goursat" [p. 246].

This $A$ is associated with the functional $G$ to impart to it a kind of differentiability through the equation

$$G[y_1; p] - G[y_2; p] = A[y_1, y_2, y_1 - y_2; p]. \tag{3.183}$$

When $y_1 = y_2 = y_0$, the functional is referred to as the differential of $G$ at $y_0$. Invertibility of A, not identified as such, is suggested by the requirement that $A$ has a "reciprocal" for $y_1 = y_2 = y_0$, namely a bounded linear functional $\bar{A}[\eta; p]$ such that $\bar{A}[A[y_0, y_0, \eta; q]; p] = \eta(p)$; it vanishes in $p$ if and only if $\eta$ is identically 0.

With these definitions and properties, Lamson, following Goursat and Volterra, defines

$$F[y, z; p] = y - \bar{A}[G[y; p] - z; p] \tag{3.184}$$

and, as his predecessors, arrives at the conclusion that the equation solution to $y = F[y, z; p]$ satisfies uniquely the equation $G[y; p] = z(p)$; $G$ is proved to be a continuous functional of $z$. Again, as was the case with Bliss' earlier students, a result in the calculus of variations is obtained as a special case of the implicit function theorem; it includes the results of Goursat and Volterra for the special form of the second equation.

The final form of a general implicit function theorem was given by another student of Bliss', Lawrence Graves, who, together with fellow University of Chicago PhD T. H. Hildebrandt, described it in a paper of 1927 [808].[89] Graves received his doctorate under Bliss in 1924 for a thesis on a topic in the calculus of variations. At this time he apparently developed an interest in implicit function problems; his two lengthy 1927 papers [808] and [802] give extensive coverage to this topic. The theorem remained of interest to him over the decades following these efforts; he gave expositions on it at an invited address of the American Mathematical Society in 1935 [803], he wrote a survey article in the *American Mathematical Monthly* in 1948 [804], and his contributed paper to the volume *Studies in Real and Complex Analysis* in 1965 [807] has already been noted. He even used Goursat's proof of the theorem in Euclidean space in his textbook [806]. His papers show a considerable knowledge of the existing literature,[90] and ample acknowledgment of Bliss' work is made: DECAF is cited all of the references mentioned, FOLIB in [804] and [807], and even Bliss and Mason's third collaborative paper [732] is referenced in [802]. "These various theories suggest the desirability of a general unifying theory" [808, p. 128].

[89] Hildebrandt received his doctorate under E. H Moore in 1910; he made many contributions to functional analysis. We choose to focus on Graves in the following account due to his multiple publications on implicit function theory.

[90] A contribution by Moulton is mentioned: see [808, p. 127].

Lamson's effort at generality has several insufficiencies: first, his result discusses conditions under which $G(x, y) = 0$ can be solved explicitly only for $G$ of a special form. Second, the quantity playing the role of the Jacobian is defined axiomatically to have properties needed to make the argument work and bears no inherent relation to the functions involved, as does, for example, the "functional determinant" in the Euclidean case. Third, and explicitly mentioned by Graves, is the lack of information about the differentiability of the solution function. (We remember that a similar concern with the nature of the solution drew Bliss initially to improvements of embedding theorems.) Graves addresses these problems in two lengthy papers, in [808], the implicit function theorem is given, and in [802], a host of examples of its use is presented. A summary of the contents of [808] follows.[91]

The first result proved is the standard existence locally of unique solutions $Y = Y(x)$ to the equation $y = F(x, y)$ where $F : X \times Y \to Y$ for $X, Y$ metric spaces with $Y$ complete, given that $y_0 = F(x_0, y_0)$, and that $F$ is a contraction in $y$; the $Y(x)$ is locally continuous if $F$ is.[92]

If now $X_0$ is a subdomain of $X$ and the function $F : X_0 \to Y$, the definition of a differential as given by Fréchet is stated: $F$ is said to have a differential at $x_0$ if there exists a bounded linear function $dF$ on $X$ such that the function $R$ on $X_0$ defined by

$$F(x_1) - F(x_0) - dF(x_1 - x_0) = R(x_1)\,\|x_1 - x_0\| \tag{3.185}$$

is continuous in $x_1$ at $x_0$. One writes $dF = dF(dx)$. A class $C^n$ for the function $F$ is defined by requiring various continuity properties of $dF$ and its higher-order differentials, and if $X$ is a Cartesian product of spaces, partial differentials with respect to each co-ordinate can be defined.

Invertibility is then addressed: for $Y$ a Banach space and $K : Y \to Y$ bounded and linear, $K$ is said to have a "reciprocal" if there is a bounded linear function $L : Y \to Y$ such that $K(L(y)) = L(K(y)) = y$ for $y$ in $Y$. The present reciprocal is explicitly connected to the reciprocal as used in linear integral equations: if $M_0$ is a subregion of $M$ a normed linear space, $G : M_0 \times Y \to Y$ bounded and linear on $Y$ for each fixed $w$ of $M_0$ with $\|(G, w)\| < 1$ then for each $w$ in $M_0$ there is a reciprocal for $y - G(w, y)$. The proof is by successive substitution: $G_1(w, y) = G(w, y)$,

[91] The authors spend roughly five initial pages listing the axioms and definitions of their abstract spaces, identified as metric spaces and linear spaces. Completeness is defined as are the usual topological notions of connectedness and boundary. The summary herein presented is stated in terms of Banach spaces, an easy step from those in Graves and Hildebrandt, a step which was taken by Graves himself in [804] and [807]. The notion of a complete metric space in the older work coincides with present usage.

[92] This result is connected with a recently appearing paper [702] by Birkhoff and Kellogg: "We note that the hypotheses... imply that for each $x$ of the [domain] the function $F(x, y)$ transforms the neighborhood $(y_0)_a$ into a part of itself. For each $x$, the solution $Y(x)$ is the unique invariant point of this transformation. This point of view has been developed by Birkhoff and Kellogg..." [808, p. 134]. Graves' view of implicit function theorems does not emphasize fixed points of transformations, as do typical textbook accounts of differential equation existence theorems which follow as corollaries of the Banach Fixed Point Theorem, for example.

$G_n(w, y) = G(w, G_{n-1}(w, y))$, and one has that the reciprocal of $y - G(w, y)$ is

$$y - H(w, y) = y + \sum_{1}^{\infty} G_n(w, y) \tag{3.186}$$

where $H$ is introduced to display the relations

$$H(w, y) + G(w, y) = G(w, H(w, y)) - H(w, G(w, y)) \tag{3.187}$$

"which are reminiscent of the reciprocal relations of linear integral equations" [808, p. 146] (see (12.13) above.) Note the similarity of this result to Mason's generalized procedure for solving partial and ordinary boundary value problems, given nearly two decades prior.

The general implicit function theorem follows as in its predecessors. If $G$ has the property that $G(x, y_1) - G(x, y_2)$ is approximable in norm by $K(y_1 - y_2)$ uniformly for $x$ in $X_0$, where $K$ is an invertible bounded linear function, then one defines $F(x, y) = y - L(G(x, y))$ where $L$ is the reciprocal of $K$, then solutions to $G(x, y) = 0$ follow from solutions to $y = F(x, y)$. Then taking $d_y G(x_0, y_0; dy) = K_0(dy)$ gives the final result, which is given in its entirety in an Appendix at the end of the next chapter.

None of this will come as a surprise to the reader who has followed this account, but the connections made by the authors to the earlier subject of integral equations and the parallel developments of the fixed-point perspective, as well as the still-developing space axioms and multitudinous examples in the follow-up paper [802], give a far more accurate view of the context in which the theorem emerged than one encounters in typical modern presentations. These connections can be made after the implicit function theorem is presented in isolation, and the theorem remains "a *cornerstone* of nonlinear analysis" [984, p. 226], to quote an excellent modern textbook.[93] Also noteworthy in our story is the roller-coaster-like path from Picard's papers on existence theorems for differential equations to Goursat's adaptation of the iteration scheme to an implicit function theorem for Euclidean functions, which in turn was picked up by Volterra in his attempt to provide the ultimate general version of such a theorem for implicitly defined functions, an attempt not to be successfully achieved in totality until the work of Graves and Hildebrandt, with Lamson's thesis as an intermediate step.

Examples of the use of the theorem were given by Graves in later papers. In his expository account [804, p. 470], he applies the result to a nonlinear integral equation

$$y(x) = \int_a^b K(x, t, y(t))\, dt + z(t) \tag{3.188}$$

[93] On p. 226 of this text a comprehensive chart gives the connections between the theorem and other topics in analysis.

If this equation has an initial solution $(y_0(x), z_0(x))$ at which the linear differential equation

$$\eta(x) = \int_a^b K_y(x, t, y_0(t))\,\eta(t)\,dt + z_0(t) \tag{3.189}$$

has a unique solution $\eta(x)$, then the previous equation has a unique solution $y(x)$ near $y_0(x)$ for each $z(x)$near $z_0(x)$. In [803, p. 656], one finds a discussion of the equation $y'' - h(t, y, y') = z(t)$ with boundary conditions and its application to a problem in the calculus of variations. In his final account of the material [807, pp. 50–51], a straightforward classical embedding theorem for the single first-order initial value problem $f[r, v'(r), v(r)] = 0,\ \ v(a) = x$ on $[a, b]$ is presented.

In this way, an arc is traced out: it begins with the interesting similarity of implicit function theorems for Euclidean functions with embedding theorems for systems of differential equations as pointed out by Bliss in his 1909 colloquium lectures and ends with a demonstration of their kinship through the general implicit function theorem of Graves. Algebraic, differential, and integral equations have indeed all been gathered under one head, as Graves pointed out, and Moore's dictum regarding similarities of theories has been validated.

The generality of the theorem suggests the possibility of its use in proofs of theorems in Banach spaces from which known and new results in the classical mode follow as special cases. This avenue was investigated by the last of the mathematicians to be mentioned in this chapter, Herman H. Goldstine. Goldstine, already encountered, worked closely with Bliss on several projects, including [727] and [728]. His doctorate was taken under Graves.[94] He went on to publish several papers on the topic of Lagrange multiplier rules in an abstract setting. We summarize the main result of one of his multiplier papers, [793] from 1938, as a conclusion to this section. (It has been mentioned that Hadamard's work on a multiplier rule for the Mayer problem was a possible source of Bliss' for the use of the adjoint of a linear system of differential equations in dealing with variations.) The topic of multiplier rules in abstract spaces still appears in present-day books on applied analysis; see again [984, pp. 270–271].

Goldstine begins by acknowledging previous work on this topic by Graves and himself,[95] and announces that his present result subsumes this work as well as the problem of Bolza[96] "and numerous other examples are included as special cases" [793, p. 388]. Mention is made of the central role played by the implicit function theorem of Hildebrandt and Graves, and that his proof lies "along lines which are

[94] Goldstine stated in an interview [797] that although Graves was listed as his advisor, Graves was absent during the year the thesis was written; Bliss informally followed Goldstine's progress.

[95] His own work stemmed from his thesis, in which Le Stourgeon's work with Bliss was cited as a predecessor.

[96] This problem, first stated by Bolza in 1913 and receiving its name from Bliss in 1932, is a generalization of the Mayer problem; see [796, Section 7.10] for an exposition and references. A statement of the problem is given in the Appendix of this book.

essentially generalizations of the methods of the calculus of variations" [793, p. 388]. What follows here is a slightly simplified version of the paper's main result: given first are domain Banach spaces $X$, $M$, and $N$ with subregions $X_0$, $M_0$, and $N_0$, respectively, and range Banach spaces $U$ and the real line $R$. A functional $f : X_0 \times M_0 \to R$ is given whose minimum is examined. A set of constraint functionals $\psi_k : X_0 \times M_0 \to R, k = 1, \ldots, q$ is given along with a vector-valued constraint function $\phi_1 : X_0 \times M_0 \times N_0 \to U$. An auxiliary function $n : X_0 \times M_0 \to N_0$ is given such that $\phi_1[x_0, \mu_0, n(x_0, \mu_0)] = 0$ where $(x_0, \mu_0)$ is the point of minimization; certain other technical conditions on $n$ are required. Another function and functional relation guaranteeing the needed invertibility condition is given but will not be reproduced here. Goldstine uses these spaces to create a framework for a general minimization problem.

The functional $f$ to be minimized is subject to the constraints $\phi_1[x, \mu, n(x, \mu)] = 0$ and $\psi_k(x, \mu) = 0, k = 1, \ldots, q$. Goldstine's multiplier result then states that if the constrained minimum occurs at $(x_0, \mu_0)$, then there exist constants $l_0, c_1, \ldots, c_q$, not all zero, and a continuous linear functional $J : U \to R$ such that

$$l_0 d_\mu f(x_0, \mu_0; \mu) + \sum_{k=1}^{q} c_k d_\mu \psi_k(x_0, \mu_0; \mu) - J(\Phi_1[0, \mu, N(0, \mu)]) = 0 \qquad (3.190)$$

for all $\mu$ in $M$ and

$$l_0 d_x f(x_0, \mu_0; x) + \sum_{k=1}^{q} c_k d_x \psi_k(x_0, \mu_0; x) - J(\Phi_1[x, 0, N(x, 0)] = 0 \qquad (3.191)$$

for all $x$ in $X$ where $\Phi_1 : X_0 \times M_0 \times N_0 \to U$ is given by $d\phi_1(x_0, \mu_0, n(x_0, \mu_0); x, \mu, \nu)$, and $N : X_0 \times M_0 \to N_0$ is given by $N(x, \mu) = dn(x_0, \mu_0; x, \mu)$. The implicit function theorem is used to obtain a needed embedding theorem.

To show the scope of this theorem, Goldstine gives as special cases the simplest and most advanced minimization scenarios: he first obtains the familiar Lagrange multiplier condition for the constrained minimum of a Euclidean function of several variables by taking $M$, $N$ and $U$ to be the real line and $X$ to be $m$-dimensional Euclidean space. This is followed by establishing the multiplier rule for the problem of Bolza, in which a functional

$$I[x_1, x_2, y, y'] = g[x_1, y(x_1), x_2, y(x_2)] + \int_{x_1}^{x_2} h[x, y(x), y'(x)]\, dx \qquad (3.192)$$

is minimized over the class of arcs $y(x) = (y_1(x), \ldots, y_s(x))$ defined on an interval $[x_1, x_2]$ satisfying side conditions $\phi_\alpha[x, y(x), y'(x)] = 0, \alpha = 1, \ldots, r < s$ and end conditions $\psi_k[x_1, y(x_1), x_2, y(x_2)] = 0,\ k = 1, \ldots, q$. For this latter case, the functional $J$ is defined by making reference to an identification of the dual space of continuous functions on an interval with the operation of integration against a function of bounded variation.

While Goldstine's result is daunting, its power is easy to appreciate, and he remarks in the opening paragraphs of his thesis that this type of result is highly desirable:

> A study of the extremizing properties of numerically valued functions is of fundamental importance to a theory of such functions and has been treated at great length in the calculus of variations. However, the functionals which are discussed in this branch of analysis are somewhat special in character; hence their treatment by no means exhausts the theory of the minima of functionals. With the inception of the general theory of functionals considerable impetus was given to this problem with the result that a number of attempts were made to solve it. The most complete of these was made by Le Stourgeon who obtained two conditions which are necessary for a minimum; however, the generality of her results is marred by the explicit assumption of certain stringent hypotheses and the tacit use of others" [792, p. 1].

Yet Bliss had his reservations about this line of inquiry. In a 1936 expository article on the problems in the calculus of variations, he concludes with the following opinion:

> Interesting studies of the determination of maxima and minima for functionals of less restricted types than those of the calculus of variations, have been made by Le Stourgeon... and within the past year by Goldstine... What has been gained in generality seems to have been counter-balanced in each of these cases, however, by a loss of some properties of the functionals of the calculus of variations upon which the most interesting parts of the theory depend. For the present the problem of Bolza is one of the most satisfactory of the general problems in which these properties are preserved. Its location as a chapter in a more inclusive theory of maxima and minima of functionals is one of the interesting problems of the future [726, pp, 607–608].

## 3.12 BLISS 1944

Sometime in the late 1930s, Goldstine became a research associate of and assistant to Bliss, who was then the chairman of the mathematics department at University of Chicago. This close association resulted in collaborative efforts which eventually brought about Bliss' major book on the calculus of variations [728] and his final contribution to exterior ballistics, [727], 1944s *Mathematics for Exterior Ballistics*. The latter was initiated by Bliss out of a feeling of duty, since it seemed by this time that another world war was on the horizon. He had earlier announced a course in exterior ballistics, as had Moulton decades previously, which Goldstine was allowed to teach. (Goldstine was also asked to instruct other of Bliss' offerings.) Bliss' declining health made Goldstine's assistance quite welcome, and their process consisted of daily visits by Goldstine to Bliss' office during which the former acted as instructor at the board to the latter [798]. Changes would be decided upon, and then Goldstine would spend the evening drafting the next section of whatever volume was in progress.

*Mathematics for Exterior Ballistics*, conceived as an upper level undergraduate text for use at the University of Chicago [798], covered in a little over one hundred pages all the materials with which Bliss was familiar from his own work as well as some recent developments; for example, the differential analyzer was introduced and discussed. In the fifth chapter, one finds material on differential corrections, and it will be worth noting, as a conclusion to the present chapter, some aspects of the presentation. In particular, some influence due to the new ideas of functional analysis can be detected if one reads between the lines. "It will not be possible to give in detail the theorems concerning functions of lines and their differentials which underlie the theory of differential corrections as presented here." The text states at the beginning of the chapter, and references to DECAF, FOLIB, and [803] are offered. But assurance is given that after a few definitions and computing methods are presented, "the analysis takes on a relatively elementary form..." [727, p. 63]. The present summary is not comprehensive but is intended to highlight the aforementioned influences.[97]

Bliss begins by stating Moulton's equations of motion using all three co-ordinate functions, along with the corresponding six initial conditions. Allowances are then made for following and cross wind functions $w(y)$ and $w_z(y)$, respectively, multiplying factors of $(1+\kappa(y))$ and $(1+\gamma)$ are introduced to account for percentage variation $100\kappa$ and $100\gamma$ in air density and ballistic coefficient respectively. The differential equations for a trajectory disturbed by these factors alone are then derived, using the same initial conditions. It is then pointed out that solutions to this disturbed trajectory are functions of the form

$$x(t, x_0', y_0', z_0', W, W_z, K, \gamma) \tag{3.193}$$

where the second, third, and fourth entries are the components of the initial velocity and the next three entries are the arcs defined by $w(y)$,$w_z(y)$, and $\kappa(y)$; similar expressions for $y$ and $z$ are given. The three solution components are described as functions of a line, and an explanatory section follows:

Bliss considers a single function $x(t, W)$ of one scalar and one arc. Then $W$, $W + \delta W$, $\delta W$ are arcs corresponding to functions $w(y)$, $w(y) + \delta w(y)$, and $\delta w(y)$ defined on an interval $[y_1, y_2]$. He then states that the function $x(t, W)$ is "said to have a linear expression

$$dx = adt + \int_{y_1}^{y_2} A(y)\,\delta w(y)\,dy \tag{3.194}$$

[97] We have already cited in Sect. 3.7 of this chapter the presentation in this book of the use of Bliss' implicit function theorem to calculate the range differential. Some of the material which follows is a repetition of Sect. 3.7.

as its differential at the set $(t, W)$" if the usual defining condition for a differential is met. The calculation of this differential "follows readily" from the definition as

$$dx = lim_{\alpha \to 0} \frac{x(t + \alpha dt, W + \alpha \delta W) - x(t, W)}{a} \tag{3.195}$$

in an analogue to the directional derivative.[98] This novel evaluation is likely introduced with the knowledge of a theorem of Graves [808, p. 136–137] which states that, regardless of the form of a differential of mapping from one Banach space to another, it may be evaluated as a limit of difference quotients as shown; this result is more plausible if differential is of the type given by Bliss. Additionally, the fact that the differential is the sum of a scalar multiplication and an integral against a fixed function can be linked to foreknowledge that this is the form which the ballistic variations will take, but also that recent results on the nature of continuous linear functionals give them as integrals of this type, perhaps against functions of bounded variation, as in the previous section.[99]

The adjoint of a system of differential equations is then defined, and the goal of reaching the "fundamental formula" for differential variations, the familiar expression involving the integral and solutions of the adjoint system, is announced. In a switch of notation, $\delta x$ now denotes the differential of (3.193) with respect to all variables except $t$ at $(t, x_0', y_0', z_0', 0, 0, 0, 0)$ with similar expressions for $y$ and $z$. Our account will end with the derivation of the "equations of variation" for these differentials, since the technique of obtaining the "fundamental formula" from them is the same as previously given.

The initial value problem for the disturbed trajectory is written as

$$\frac{d^2x}{dt^2} = f(x, y, z, x', y', z', \alpha w, \alpha w_z, \alpha \kappa, \alpha \gamma) \tag{3.196}$$

$$\frac{d^2y}{dt^2} = g(x, y, z, x', y', z', \alpha w, \alpha w_z, \alpha \kappa, \alpha \gamma) \tag{3.197}$$

$$\frac{d^2z}{dt^2} = h(x, y, z, x', y', z', \alpha w, \alpha w_z, \alpha \kappa, \alpha \gamma) \tag{3.198}$$

---

[98] The expression on the right became known as the Gateaux derivative or differential, after R. Gateaux, a young French mathematician who used it in [788, p. 84]. Graves cited this work in [803, p. 651]. One also remembers Bolza's definition, going back to Lagrange, of the first variation of a function $\phi(x, y, y', y'' \ldots)$ with respect to an increment $\eta$ as essentially its directional derivative in the "direction" $\eta$ [738, pp. 16–17].

[99] Goldstine was probably responsible for this innovation due to his involvement with the later versions of the functional calculus. Bliss still referred to functionals in Volterra's locution as "functions of a line"; his lack of involvement in Goldstine's multiplier rule work and the absence of the functional calculus in his own later work suggest an indifference to this new version of abstract analysis.

with

$$x(0,\alpha) = y(0,\alpha) = z(0,\alpha) = 0 \tag{3.199}$$

and

$$x'(0,\alpha) = x_0' + \alpha dx_0',\ y'(0,\alpha) = y_0' + \alpha dy_0',\ z'(0,\alpha) = z_0' + \alpha dz_0' \tag{3.200}$$

It is noted that setting $\alpha = 0$ in this initial value problem reduces it to that of the original trajectory. The differentials can be evaluated by directional derivatives; hence, for example, $\delta x = x_\alpha(t, 0)$, where the subscript denotes differentiation with respect to $\alpha$ of

$$x(t,\alpha) = x(x_0' + \alpha dx_0', y_0' + \alpha dy_0', z_0' + \alpha dz_0', \alpha w, \alpha w_z, \alpha\kappa, \alpha\gamma) \tag{3.201}$$

the solution to the disturbed initial value problem.

When (3.201) and analogous expressions are substituted into (3.196)–(3.198) and both sides are differentiated with respect to $\alpha$, then $\alpha$ set equal to zero, one obtains the six equations of variation, which are given as

$$\frac{d\delta x}{dt} = \delta x' \tag{3.202}$$

and similarly for $y$ and $z$,

$$\frac{d\delta x'}{dt} = f_x\delta x + f_y\delta y + f_z\delta z + f_{x'}\delta x' + \ldots + f_w w + f_{w_z} w_z + f_\kappa \kappa + f_\gamma \gamma \tag{3.203}$$

and similarly for $y$ and $z$.

The differentials are then the unique solution of this system with initial conditions

$$\delta x = \delta y = \delta z = 0,\ \delta x' = dx_0',\ \delta y' = dy_0',\ \delta z' = dx_0' \tag{3.204}$$

As a final comment, we note that this form for the equations of variation is dependent on a reversal of elements $\delta$ and $d$, that is, $\delta d = d\delta$, a reminder of the early issue as it presented itself in the earlier calculus of variations and in Hoar's extensive commentary.

This ends our account of the last exposition of the adjoint method executed by Bliss himself, with the assistance of Goldstine. If one compares it to the spare presentations in the technical reports from Aberdeen and their polished versions in Bliss' JUSA/JUSA papers and Dunham Jackson's ordnance textbook, Gronwall's rigorous papers, Hitchckock's guided presentation in the range firing table construction book of 1934, and Bliss' own theoretical papers, one sees the great variety of mathematical ideas and styles in an evolution over nearly three decades. Yet even this was not the last version of the idea, as will be evident in the concluding chapter.

## Archival Materials (Abbreviations)

[Veblen Papers] Oswald Veblen Papers, Archives of the Library of Congress, Library of Congress, Washington, D.C.
[NARA Aberdeen] Record Group 156–866, National Archives and Records Administration, College Park, Maryland
[NARA Washington] Record Group 156.7.18, National Archives and Records Administration, College Park, Maryland
[NARA Ordnance School] Record Group 156–1512A, Box 2, National Archives and Records Administration, Philadelphia, Pennsylvania
[NARA ENIAC] Record Group 156–1512A, Box 9, National Archives and Records Administration, Philadelphia, Pennsylvania
[Columbia University] University Archives and Columbiana Library, Columbia University in the City of New York
[Princeton University] University Archives Project, Seeley G. Mudd Manuscript Library, Princeton University, Princeton, New Jersey
[MIT] MIT Institute Archives and Special Collections, Massachusetts Institute of Technology, Cambridge, Massachusetts
[University of Chicago] Special Collections Research Center, University of Chicago Library, Chicago, Illinois
[Concord] Hoar Family Papers, Concord Free Public Library, Concord, Massachusetts
[Harvard] Office of Harvard War Records, Harvard University, Cambridge, Massachusetts

# Chapter 4
# Conclusion of the Narrative

## 4.1 Generalities

We begin by returning to the point made in the Introduction: mathematical advances arising from the first World War have not been as much in evidence as those arising from the War beginning two decades later. Veblen's mass call to mathematicians to contribute to the War effort has received attention in current times thanks to works like [1003] as well as [1140, 1142], but knowledge of the resulting mathematics seems quite limited. The list of four specific topics of interest mentioned in Chap. 1 likely seems unimpressive measured against those mentioned in the Introduction. The entré of numerical integration for the solution of differential equations is gaining increased awareness due to its connection with the ENIAC,[1] which was built largely for the quick solution of these equations. But we have seen that this introduction was greatly downplayed by its initiator, Forest Ray Moulton. Moulton impatiently claimed that any number of people could have done what he did, though he pointedly emphasized the importance of his proof of the convergence of his method. All the powers in the War employed similar techniques; Moulton perhaps felt that the United States was merely catching up with existing practice.

This obscurity can be attributed in part to the fact that, at least in the view of Moulton and others, the introduction of a rigorous version of science itself was new to the War. We have seen how, under his influence, ballistic discourse changed from European-dependent calculus-based treatments given by experienced artillery officers to a vision of an ongoing scientific effort[2] backed by the latest mathematical

[1] Electronic Numerical Integrator and Computer, built during the Second World War years at the University of Pennsylvania.

[2] Of course the word "scientific" begs a consideration of context and definition. Hamilton and Ingalls no doubt thought of their efforts as scientific, expressed in current mathematical garb and accompanied by the limited experimental efforts available. Euler had always been acknowledged as having provided the first truly mathematical treatment of the ballistic trajectory and together



A. Gluchoff, *From Fréchet Differentials to Firing Tables*,
https://doi.org/10.1007/978-3-031-68267-4_4

theory and co-ordinated with experimental verification at the newly constructed Aberdeen Proving Grounds. "Rough and ready" rule-of-thumb ideas were to be discarded in favor of mathematically described and tested theories. These goals were articulated in the concluding recommendations of his *History* [1246, pp. 85–87]:

> In the last four years the methods of warfare have been entirely changed. . . . If the German Army had had in 1914 the material and methods it had in 1918, it would have swept over Paris and on to the sea; if the French army had had in 1914 the equipment and methods it had in 1919, it would have thrown the Germans back across the Rhine. The methods of four years ago are largely obsolete today, and failure to recognize this fact will be dangerous. *The changes in warfare in the last four years have almost entirely been due to the applications of science* [emphasis added]. The scientific men of the most advanced nations of the world have applied themselves to these problems under the spur of the keenest necessity. Their accomplishments have been amazing.

This observation implicitly contains a contrast with an earlier attitude, discussed in the present author's paper [1119] and elsewhere that the ingenuity of American inventors (the "wizardry" of Thomas Edison was often cited in this regard) would be sufficient to meet the needs of the war effort. But contemporary opinion pointed out the inadequacy of this view,[3] an inadequacy based in part on the realities of existing ordnance. It was Moulton's opinion that the sheer magnitude of types of guns ammunition required by the war effort ruled out range table construction by firing alone; scientific ballistic theory was an economic necessity. Moulton stressed this need in his recommendations, as noted in Chap. 1:

> The feature of ordnance work most striking to the civilian scientist is the almost complete lack of exact knowledge of its fundamentals. Consider cannons and projectiles and the hundreds and even thousands of millions of dollars spent for them! It is doubtful whether in the history of the world any other expenditures comparable to these have ever been made on so slender a basis either in well-conducted experiments or sound theory. . . . Considered in this way, the development of fundamental scientific principles is the most important and the cheapest war preparation that can be made. [1246, p. 86]

Seen in this light, the problem of the variability of the ballistic trajectory stands as one typical of its time. The realities of the variety of weapons and ammunition resulted in a need to, for example, consider mathematically the variations in ammunition used for a particular weapon. More accurate scientific measurements of wind, air density, and other meteorological conditions made their incorporation as varying factors possible. Variability was mentioned in the same breath as trajectory calculation itself by Moulton: "These problems [new conditions of warfare] and

---

with Robins formed a scientific team; Francis Bashforth was accorded the honor of having raised ballistics to the level of a science in an 1885 exterior ballistics text: "The methods of solving problems in Ballistics here given are substantially due to professors Francis Bashforth and W. D. Niven; to the former, by his admirable series of experiments and methods of reducing the same, belongs the honor of having raised Exterior Ballistics to the dignity of a science" [1234, Preface]. New in Moulton's time, however, was the large-scale institutionalization of scientific development for military purposes in the form of, for example, the Aberdeen Proving Grounds and the introduction of new schools and programs for instruction in ballistics.

[3] See, for example, [1184].

related ones which need not be enumerated, led Major Moulton, in May, 1918, to formulate new processes for treating the problems of computing trajectories and making allowances for abnormal conditions. These processes are characterized by their mathematical rigor, their generality, and their simplicity and directness of numerical applications" [1246, p. 12]. He added "Improvements were made on the basis of practical experience." It is hoped that the previous chapters have supported the contention that this focus on the variability problem results in its being considered a major mathematical achievement of the World War.

Moulton's views were shared by other mathematicians and physicists who expressed opinions in *Science*, the publication of the American Association for the Advancement of Science. J. S. Ames, then vice-president of the American Physical Society, took the opportunity to point out the change in attitude of the American public toward men of science:

> [The mobilization of science in service of the war effort by Germany] was a matter of great surprise to most Americans, and illustrates clearly the comparatively insignificant position held by scientific men in the minds of our people. ... When this country entered the war, it is true beyond a doubt that the American people had great expectations, nay a conviction, that with our so-called inventive genius we would seriously influence the war, perhaps stop it, by the epoch-making inventions which our professional, highly advertised, inventors would quickly make. ... There was a great disappointment, almost a shock, as days went by, the periods promised for great accomplishment passed, and certain names almost disappeared from the public press. ... The point I wish to emphasize is that the ability and knowledge required in waging this war successfully are not those possessed by any body of men except those with a profound knowledge of science and of scientific method. The problems are too complicated. [1000, p. 403] ... The scientific men in America have suffered greatly at the hands of the people. [ibid., p. 410].

Veblen also appreciated the scientific aspect of his war work; in his address cited earlier [1318, p. 338], he described experiments made by members of his team on modifying the form of projectiles, noting that "In one case, ... by a very slight modification - cutting off a little copper from the rotating band - the range of the projectile was increased by over a mile and its accuracy made several times as great." In Gordon Hull's opinion piece, also cited earlier, the issue of the rise of the scientific man is framed in terms of a territorial dispute between physics and mathematics, whose opening salvo was cited in Chap. 1:[4] "For two hundred years or more the subject of exterior ballistics has been regarded as belonging to pure mathematics. But into this realm physicists at times intruded. ..." [1184, p. 222]. There follows a lengthy discussion which addresses the forces acting on a projectile and how its trajectory is dependent on its shape.[5]

---

[4] This conflict is interesting in light of the fact that Hull, as previously noted, was a member of Moulton's original Washington working group.

[5] This discussion can be considered as a reply to the nagging question which rises in evaluating the work of Moulton and the other mathematicians: How much faith can be put in purely mathematical expositions of exterior ballistics when the projectile is considered a particle, yet Veblen's own group showed that a tremendous change in range follows from a minor modification of it?

To give a sample:

> ...the system of forces under which a projectile moves is not the simple one implied by the equations just given [Moulton's equations]. For a projectile is a body spinning rapidly about an axis probably nearly identical with its geometrical axis. It emerges from the gun either with a small yaw, or with a static rate of change of yaw, or both. ...As in the case of a top, precessional motion results. ...In short, the projectile tumbles. Loss of range and great dispersion are the results. [p. 225] ...the retardation must be not only a function of the velocity but also of the yaw [p. 227].[6]

The conclusion: "The work belongs to physics, notwithstanding that one of these civilian officers [Moulton, in charge of his ballistics group] was and is a professor of mathematics of the purest quality" [p. 229]. And: "Enough has been said to show that the fundamental problem is the projectile is not one of mathematics. There are various mathematical methods of handling the problem" [p. 231]. But Hull's paper is written from the point of view of a physicist with no doubt that his field contributed mightily to winning the war.

To conclude, this opinion may be compared with an account [1325] of the role of mathematics in aerodynamics given in 1916 by Edwin Bidwell Wilson, a professor of mathematics and physics at MIT and an early enthusiast of the use of mathematics in the design of aircraft. In discussing the stages of the development of a branch of engineering, after the first era of "bold inventors," for example, the Wright brothers, there comes the mathematics and physics stage, whose practitioners "construct a theory to correspond to the branch of engineering and by the indications of that theory aid materially in hastening the development...they assist in codifying the fundamental scientific principles which underlie the subject..." [1325, p. 292]. There follows of list of mathematicians and physicists, all English, who contributed to this effort, and later it is noted that the most appropriate type of mathematics for this work is "the good old traditional Cambridge, England type, the mathematics of Newton, of Green, of Kelvin, of Rayleigh, a type of mathematician which in this country, owing perhaps to our preponderance of German-trained mathematicians, has all too little prestige" [1325, p. 297]. Interesting also is a comparison of his description of the analysis of airplane motion to the problem of the ballistic trajectory:

> Owing partly to the divergence between theory and experiment in fluid motion in simple cases and partly to the impossibility of obtaining a theoretical hydrodynamic solution for the motion of so complicated an object as an airship or airplane, it is necessary in the main to treat the ship or plane as a rigid body to which are applied certain forces...determined by wind tunnel experiments upon models. [Analogous to treating the projectile as a particle.] The problem of the motion of the machine then becomes one in the dynamics of a rigid-body free to move in space...subject to known forces. ...the forces represented by empirical equations [note similarity to the resistance function] are too difficult to integrate without approximation...By considering the motion as made up of small departures from uniform flight in straight lines, the differential equations of the motion become linear equations with constant coefficients, and their solution is therefore a matter of well-known, though sometimes very tedious, routine. ...the art and science of getting a sufficiently accurate

[6] The issue of spin will be taken up later in this chapter.

numerical solution of a numerical problem [e.g., by numerical integration] are of equal importance and dignity with general theory in all applications of mathematics. [1325, pp. 295–296]

Wilson, when contrasted with Moulton, provides alternative example of a contemporary mathematical practitioner's approach to an applied mathematical problem. Though writing a dissertation on a strictly mathematical topic, Wilson's close association with Josiah Gibbs made him aware of its use in physics, a parallel to Moulton's relation to celestial mechanics and Poincaré. However, Wilson was not entranced with rigor: in his 1912 book [1324], which some regard as the first book on Advanced Calculus authored by an American, he stated "With the large necessities of the physicist and the growing requirements of the engineer, it is inevitable that the great majority of our students of calculus should need to use their mathematics readily and vigorously rather than with hesitation and rigor. Hence, although due attention has been paid to the modern questions of rigor, the chief desire has been to confirm and to extend the student's working knowledge of those great algorisms of of mathematics which are naturally associated with the calculus" [1324, p. iv]. This attitude may be compared with Moulton's comments in his volume on differential equations. As in Moulton's case, Wilson instructed a course in his chosen area of military applied mathematics at his home institution, in this instance during the War, but eventually strayed from the topic to follow other interests, as Moulton eventually gave up active involvement in ballistics by 1926.

## 4.2 Individual Responses

In assessing the impact of the variations work, it is instructive to consider the responses of several individuals immediately affected by its introduction. These men were chosen as something of a cross section of involved parties in the years immediately following the War.

To Bliss we owe what may be the first application in American mathematics of abstract material at the frontiers of its contemporary research to a military problem. What is most striking about this application is the seamlessness with which it follows his research interests at the time. The papers FOLIB and DECAF fit in so smoothly with his earlier work on the smoothness of solutions to systems of differential equations that they appear as obvious extensions. He was already familiar at the beginning of his time at Aberdeen with the work of Volterra and Fréchet; he had set some of these students to work on problems using these ideas. In turn he used the idea of Fréchet differentials to express his results in variability theory. Bliss showed surprise tempered with characteristic modesty that his area of expertise was applicable to ballistic work; we are reminded of his earlier comment: "This [the differential corrections] was my particular field of study, and I was astonished to find that the most advanced mathematics which I knew was needed and effective. ... The calculation of the differentials of such functionals [as wind corrections] is a principle problem of my special field, the calculus of variations" [1050, p. 601]. This was his only brush with applications; he was in no way an applied mathematician nor did his prior mathematical work impart to him an overall vision of the role of mathematics in practical problems, as Moulton's had. The differential variations remained for him an interesting chapter of a mathematical life largely devoted to furthering research in the calculus of variations. His contributions affirm the old saw that one can never anticipate the possible uses to which results in pure mathematics can be put. At the University of Chicago, Bliss' mathematical descendants were kept busy furthering implicit function theory and the calculus of variations in Banach spaces, branches stemming from his war-time investigations.

Enough has already been said about Moulton's view of this material that little can be added. His devotion to high-level mathematics, with special emphasis on "Weierstrassian rigor," is evident throughout his *History* and elsewhere; one comes across the phrases "solid logical basis," "mathematical rigor," "solid foundation," and their opposites "rough and ready manner" and "rough methods" on page after page. His care in addressing the variations issue was such that he devoted a paper to exploring the error committed by replacing the actual deviation by its first-order approximation. He was unafraid to challenge established ballisticians of the status of Charbonnier on this topic; at one point his analysis of the latter's results leads to the statement "The [results of the] rough methods heretofore in use, such as given in Charbonnier's *Traité de Balistique*, are sometimes even of the wrong sign" [1246, p. 25]. In his dispute with Hoar, one sees an example of how he viewed the gap between his methods and those of earlier practitioners as he remarks that he dispelled the "fogs" that surrounded the matter of the variations with the introduction of the differential equations. His work used the results of modern analysis, such as the idea of a basis of solutions for a linear system, without introducing corresponding research of his own, as Bliss had done. But this use set the tone for exposition of this material in the mathematical community.

With Roger Sherman Hoar, we come to the issue of transmission of the new material to a military audience, to education in the new methods. Where Moulton and Bliss put forth textbooks appropriate for university students of mathematics, military students would not generally be expected to receive this material at the level it was created, if indeed they needed to absorb it at all. Hoar himself was an odd candidate for an instructor of this mathematics to army officers: a man with degree and career in law, inventor of various ballistic devices, and instructor of elementary mathematics who found himself at a crossroads in career choices at the end of the War. But with the help of Phillip Franklin, he enthusiastically threw himself into the job of bridging the gap between high-level mathematics and the needs and abilities of his students by writing an exterior ballistics textbook and instructing a course using it. The resulting book featured his own interpretations of the variational symbols as he attempted explanation of equations using physical ideas. How successful his approach was is not known, and his book came in for some sharp criticism from Moulton, but he faced the fact that the uninspiring job of data collection, ballistic calculation, and tabulation of results had to be accomplished if Moulton's "New Methods" were to have any impact. His computation sheets for the trajectories as well as the variations show an engagement in the material at a lower but essential level than that of, say, the academic mathematician Dunham Jackson's comparable volume. Hoar also showed somewhat of a continued interest in the variational topics nearly a decade after his departure from the army, when in response to the controversy over his review of Moulton's book he organized an informal seminar at Marquette College to discuss these issues. His contributions as a mathematical amateur of great enthusiasm are a lighter side of this story. He was of a type referred to in the introduction of the recent study by David L. Roberts [1277] of figures in American mathematics: one who "had [a] relatively brief encounter[s] with mathematics and then moved on to other activities. This trajectory becomes increasingly rare in recent years, reflecting the growing professionalization of the subject. A part-time or temporary affiliation with mathematics has become unusual [1277, Introduction]." Philip Schwartz was a career military man, receiver of the new material, and unabashed critic of some of it. While the mathematical basis of the calculations for a range firing table would likely be a matter of indifference to many men in his position, Schwartz, with an undergraduate degree in mathematics, took it upon himself to stir up some interest among his fellow officers by contributing technical reports and articles to military journals. He reported on experiments at Aberdeen intended to establish the validity of the new variational laws and wrote notes describing other new developments in ballistic matters. In his internal essay on the technical reports of Bliss, Gronwall, and Moulton, he was concerned about the high level of mathematics used, presumably because of the clear gap between this level and the effort needed to make the material usable. He pointed out that the older methods, like the long-used Siacci theory, were likely to be used for quite some time in spite of their limitations. Whereas Bliss, Hoar, and Moulton eventually left military service, Schwartz elected to stay and in doing so found his place as a writer critical of what he perceived as the weakness of military mathematical instruction, at least throughout the 1920s. His involvement in the controversial topic of probable error displays his critical talents at their best. Schwartz's contributions as well as his correspondence with Veblen show an awareness of the need for mathematical analysis in military affairs ranging from ballistics to sampling theory, along with a frustration at the lack of educational opportunities to remedy ignorance on these matters.

Two reviews of books produced by the second wave reformers are of final interest here. Both were written by J. E. Rowe, whose criticism of the unnaturalness of Bliss' adjoint system was quoted in Chap. 1. A 1910 PhD in mathematics from Johns Hopkins, Rowe held several posts during the War years and beyond; he taught a course in ballistics at Aberdeen using Hoar's book, was a research associate at Pennsylvania State College, and was a member of the National Advisory Committee for Aeronautics. Thus he had a position somewhat between research mathematician and military member. In his 1923 review of Hoar's work [1280], many points were raised. The system developed by Franklin and Hoar did not find favor; he preferred the original derivation of Moulton. Bliss' adjoint was not criticized here, though again Rowe's preference for a traditional derivation like Dunham Jackson's was stated. Other particular criticisms were leveled, but production of the tables by these methods was given high praise, as were the tables themselves:

> The methods developed in this book, with some variations and additions, have been used by Dr. A. A. Bennett in the construction of new ballistic table. It was my privilege to use these in the course art Aberdeen. They are easy to understand, easy to use, and from every standpoint are all that could be desired at the present time by the practical ballistician. [1280, p. 279]

A review of Moulton's book, given in 1928, began with a cautionary note:

> There has been a failure on the part of some properly to appraise the work done in this country during the War by our mathematical men who were interested in ballistics. The contributions that they made to the subject were advertised in most complimentary terms at meetings of the national mathematical organizations and elsewhere. No one was disposed to question them; no one was in a position to question them. It came to be usual to say certain things about these researches in ballistics.... [After a quote from a positive review of Moulton's book he remarks] There is little doubt that the above quotation is an expression of confirmed opinion on the part of the author. Also it must be said that it embodies the attitude that has become all too prevalent regarding this work and which has not been corrected.[7] [1281, p. 230]

The heaviest criticism of Moulton's book was directed at the use of numerical integration as a panacea of sorts; its imitations as Rowe sees them need not be listed here. Additionally, "The subject of differential variation as presented in Chapter IV will be interesting to the mathematician, but is not elementary, and in my opinion, is not given in a sufficiently elementary and practical manner" [1281, p. 231]. However, Moulton's work on rotating projectiles, to be described later in this chapter, earned top marks:

> This study of the rotating projectile is of fundamental importance in ballistics and especially in projectile design. ... Although this [Moulton's theoretical exposition] has been added to since the War, it is the best presentation of the subject in our language, or in any language, that I have seen. The study of projectile design, and in general the problems of the rotating projectile, may be studied through the mathematics of this chapter as a tool, provided that experimentation can be conducted as it should be conducted. In this sense Professor Moulton as put the subject of ballistics upon such a basis that no essential change will have to be made in the mathematical method of attacking such problems for some time to come.[8]

The review ends on a positive note, with remaining inadequacies in ballistic practice being ascribed to lack of physical research:

---

[7] These comments were preceded by the following remark, having a mild degree of sarcasm: "Toward the close of the World War, and even during the period of hostilities, every person who made any pretense of being interested in contributions to mathematics longed to secure a book on ballistics from which be could compute a trajectory; not by the old antiquated methods which, he had heard, had been entirely discarded, but by the method that had been devised by American mathematicians who had patriotically devoted their talents to the solution of these important problems" [1281, p. 229].

[8] Rowe's assessment of Moulton's work should be considered in light of the account given below, which stresses the importance the [1108], a British work on spinning projectiles.

> Placing ballistics research upon a sound mathematical basis is also a matter of great importance. Credit for this must be given to Professor Moulton and his associates. Also, they must have credit for giving an impetus to mathematical research in the Army and Navy that has had excellent results. ... Greater accuracy in such work from the practical standpoint will depend upon more accurate physical data rather than upon new methods of mathematical attack. Although Professor Moulton's book is by far the best book on modern ballistics published in this country, it is for use of the mathematician rather than for the practical ballistician. [1281, p. 232]

With the dual concerns of, on the one hand, the needs practical ballistics and, on the other, the overemphasis on the role the contributions of pure mathematicians to what is an applied subject, one is led to ask about the existence of what would naturally be described in modern terms as an applied mathematician. Would Rowe qualify as one? Again, in this account, A. A. Bennett and L. S. Dederick, the latter introduced below, would come closest to matching this description, if only because they were pure mathematicians whose interests involved "applications" which were undeniably real. The issue of what constitutes applied mathematics and who its practitioners were at this time deserves a closer look.

## 4.3 Community Responses

Two communities may be thought of as having collective responses to the new ballistic developments. The academic mathematical community occasionally regarded itself as changed in a minor way its response the War effort, and the military as a whole underwent something of a transformation in its reaction to the work of the mathematicians. We will present some observations on these challenging topics.

Most of the mathematicians left Aberdeen at the end of the War and returned to their former careers. Their service to their country was perhaps mentioned in personal memoirs and obituaries, but, with the possible exception of figures of such stature as Oswald Veblen and Forest Ray Moulton, their military experience was not considered by them to be individually transformative. A few PhDs in mathematics did choose a career associated with ballistics in the military. Yet commentators like Norbert Wiener insisted that the community underwent something of an increase in prestige, having demonstrated that they were capable of being more than mere fumblers with symbols. They could have in fact contributed something of practical value. One consequence of this new self-image was the perception of greater possibility for funding their efforts; results of two such efforts are recounted in [1102].

The validity of this broad claim of increased prestige is hard to verify, but the theme does recur in addresses by some mathematicians given between the wars. Mention has already been made of Bliss' 1927 presentation of the uses of mathematics in industry, in which his own work in the differential variations was stressed. We will quote again from Dunham Jackson's remarks on his retirement as president of the Mathematical Association of America in 1927 [1197]. Jackson attempted his version of the view that research in mathematics can ultimately be useful. After self-ironically pointing out his contribution to the "enlargement of human knowledge" by the machine calculation of the product of two large integers, he continued, "Collectively and in the long run we are finding out things that the world wants to know, however ephemeral most of our individual efforts may be. To mention specific instances, the methods in exterior ballistics developed by Moulton and Bliss during the war depend on this Picard process of successive approximations and the introduction of the adjoint system of differential equations, familiar products of modern mathematical research" [1197, p. 410]. This was mentioned in a list which included relativity and the "forbiddingly abstruse" nature of statistics as topics illuminated by mathematics. "Discoveries and inventions are no longer 'cut and try' methods, but rather the application of the laws of mathematics," according to a 1935 article in National Mathematics Magazine by an instructor at a small college in Pennsylvania [1301]. On a personal level, the short New York Times obituary of Gronwall, who died in obscurity in 1932, contained a sentence noting that he was a ballistician during the World War; this would be demonstrative of a practical contribution on the part of the Columbia University mathematician. In later years Bliss' war-time work was referred to both in a popular account of the history of computing by Goldstine [1123, p. 132] and in a contribution by Graves to a Mathematical Association of America volume on *Studies In Real and Complex Analysis* [1137, p. 35].

It would be foolish to talk about the response of the "military community" to as specialized a topic as the differential variations. There were many communities within the military, and recorded responses are not plentiful; the existing ones ran the gamut from articles praising the revolutionary work of Moulton and his associates to annoyance with the tediousness of the newly introduced numerical integration. But a few observations might prove illuminating. The most obvious change was to the nature and use of the range firing tables; it has been mentioned that provisional tables were released within Moulton's tenure in Washington. The new tables had material devoted to firing under nonstandard conditions, the work of Bliss and Moulton figured here. In addition, elaborate new protocols of firing at the commander's battery were established. The preparation for fire included systematic readings of wind aloft by balloons; temperatures and barometric pressures were also recorded, and adjustments to fire were made accordingly. Some new instruments, based on a design by J. W. Alexander, were created to assist in this process. Of course planned procedures would not necessarily be implemented as desired, and the new instruments would have to be tested, but these initial steps at least speak to a change in thinking about fire control.

Among Moulton's recommendations at the end of his *History* are some which speak to the explosion of knowledge applied to the War, on which he has already been quoted.

> The most obvious lesson from the application of science in the war is that specialists are required. ... In the days when the only weapons were swords and spears and men smote one another hip and thigh, a man could be an expert in everything. But those days are past, not only in war but in other things. ... [The universities] have abandoned the mental picture

> of the universal genius. What probability is there, therefore, that West Point or Annapolis will make a universal expert out of a man who would not think of being one in civil life? ... The first thing that is implied by these statements is graduate work. West Point and Annapolis compare with high grade colleges. ... There should be a settled policy of sending officers to Universities or Technical Schools for special work in such subjects as Chemistry, Mathematics, and Physics. [1246, pp. 86–87]

In addition, Moulton planted the seed of what became the ongoing concern with experimentation and mathematical analysis of ballistics questions. "Rapid progress has been made during the year in the applications of mathematics to ballistics problems. The field is not exhausted; in fact, there is no reason to believe it is more than well begun. The services of a man with a good familiarity with modern mathematics and a taste for the applications of mathematics should be constantly employed by the Technical Staff [Moulton's office in Washington]" [1246, p. 89]. Though as noted above most of the war-time mathematicians chose not to stay at Aberdeen and Washington, the following decades saw the gradual accumulation of mathematical talent. L. S. Dederick, a 1909 Harvard Ph. D. in mathematics, chose the military as his domain of research in the decades between the wars.[9] He was involved in many facets of ballistics during his career, some of which will be cited in what follows. Other names can be found in [1123, pp. 81–83]. The founding of the Ballistics Research Laboratory at Aberdeen over the period from 1935 to 1938 was the embodiment of Moulton's suggestion; it was a continuation of an existing working group. Dederick was an original associate director; see [1123, pp. 127–128] for more on the establishment of the laboratory.

One product of all this attention to ballistic matters is a volume which became highly regarded after World War II, 1953 *Exterior Ballistics*, by McShane, Reno, and Kelley. Remarkable for our purposes is the material in Chapter 8, devoted to differential effects. This material was likely the writing of E. J. McShane, who may be regarded as a mathematical heir of Bliss' World War I work. What is unusual in this chapter is the depth in which the subject matter is discussed, a short course on functionals and their differentials. Practical aspects of using Bliss' methods and alternatives are also discussed.

In contrast to this massively researched chapter, one sees a continued interest in computation of differential variations by mathematical people of all stations. Such a suggestion was given in a 1934 issue of the *Journal of the Franklin Institute*.[10] Written by a Navy man of unknown education, it offered a method which bypassed the need for numerical integration, "which is a very laborious process" [1321, p. 729].

A later figure straddling both mathematical and military communities was Harry Polachek (1913–2002). Polachek earned a Ph.D. in mathematics from

---

[9] One of Dederick's publications in pure mathematics from 1913 discussed the implicit function theorem for points on the boundary of the domain of definition.

[10] In the years before the establishment of the Society for Industrial and Applied Mathematics and its journal, the Franklin Institute's journal served as an outlet for many scientific and technological papers.

Columbia University in 1947;[11] earlier, in 1941, he had left a career in high school teaching of mathematics to join the Ballistic Research Laboratory, participating in what was then the production of range firing tables by mechanical calculators. The majority of his career was, however, spent with the US Navy, beginning with his having been chosen in 1952 to form the first Navy Laboratory charged with using electronic computers for the solution of military problems.

Polachek's PhD thesis, published in 1949 [1266], involved the solution of the differential equations of motion of a projectile subject to what he termed a "quasi-Newtonian resistance" or a resistance law close to Newtonian (where the resistance is proportional to the square of the velocity) in a precisely defined sense. The solution is effected by using the differential variations "developed during the first world war by G. A. Bliss, F. R. Moulton and T. H Gronwall..." [1266, p. 277]. [Punctuation in the following quotes is Polachek's.] Both the theories of Moulton and Bliss were used, and much numerical computation was done.

In a 1943 publication [1265] Polachek attempted an account, for a mathematical audience (requiring only a knowledge of what a differential equation is), of the familiar exterior ballistics of a particle. He gave a presentation which was both mathematically nontrivial, but accessible, and militarily informed by the citation of specific weapons and the trajectory of their shells. His introduction stressed the importance of the work of computing accurate tables by citing the contemporary action of MacArthur in Bataan:

> In the first days of his heroic stand against the Japanese invaders of the Bataan Peninsula General Douglas MacArthur was able, by accurate aiming of his guns to disperse a large fleet of invading troops assembling in Manila Bay and thus thwart an early Japanese victory. Much credit for that is due to the skill and valor of MacArthur's men, but just as much credit is due to the genius of the scientists and mathematicians back home, more than five thousand miles away from the scene of the battle, whose careful calculations resulted in the precision of MacArthur's guns. [1265, p. 27]

Later in the article [p. 31] mention is made of the need for tables with corrections for nonstandard conditions. Two methods are given for the computation of trajectories corresponding to these conditions; the first a computation of a new trajectory incorporating the new conditions via the Differential Analyzer (discussed in the next section), the other "...a set of equations which would give directly the effects of nonstandard conditions. Such a system of equations has been discovered for use by the US Army by three American mathematicians, Moulton, Bliss, and Gronwall, each of whom made an important contribution in its development" [1265, p. 31].

The article concludes with a statement regarding the precision of the guns, the improvement of which is the "job of the ballistician" [1265, p. 31]:

> Recently, the ranges given in a firing table for a medium sized howitzer were questioned. Extended range firings were conducted at seven different elevations...Corrections were made for wind, density of the air, muzzle velocity, temperature of the air and weight of

[11] His advisor was Edward Kasner.

> projectile. In five of the seven groups fired the corrected ranges obtained, which were several miles in magnitude, differed from the firing range tables by less than 7 yards. In the remaining groups the difference was slightly greater." [1265, p. 32]

The pride in the efforts of the ballisticians is evident as is the patriotic attitude of the author.

One final issue should be stressed. Both the mathematical (including physicists) and military communities emphasized the importance of effectiveness of fire and accuracy in prediction of the trajectory in starkly financial terms. Again and again one encounters arguments which point to monetary gains to be made by obtaining better information about the behavior of projectiles. In the present, account what began as an off-hand remark by Alston Hamilton, in the pre-War years, to the effect that making a firing table by theory alone was economical became with Moulton a rallying cry for a revision of ballistic theory with obvious financial benefits, and a declaration that enormous expenditures were made during the War on artillery matters with a slim knowledge base. Bliss was able to quote the cost of "reduc[ing] by one the number of shots fired for adjustment from a fourteen inch gun in order to attain a target" [1043, p. 314] to emphasize the importance of his variations work. Bennett, a mathematician with close ties to the military, hoped that one day the general ballistic tables wit their increased accuracy would offset expensive costs of range firing. R. H. Kent will be quoted below on the financial importance of "projectile design." Gordon Hull:

> In America we often measure things by money. The total amount of money authorized for artillery, including motor equipment, was 3,188,000,000 and for machine guns 1,102,600,000 [dollars]. Judged by the money expended for them, guns are of importance. It is essential that we get as effective guns as possible and that we know how to use them. Aircraft, and anti-aircraft warfare, barrage firing, long-range guns - all of these call for a very complete and accurate knowledge concerning the motion of a projectile and the energy required to carry it to a certain place and to cause it there to explode at a chosen time. Exterior and interior ballistics are thus matters of great importance. [1184, p. 222]

General Hagood's opinions of 1924, given in the Preface, on the importance of preparation of fire could also be construed financially; dumping the huge amount of ammunition in the ocean would be quite costly.

## 4.4 On Aerodynamic Forces, Shape of Projectiles, Spinning, and Tumbling

A key assumption made by the ballisticians considered in this book is the reduction of a projectile to a point mass subject to various forces-"particle ballistics," in Polachek's words. The gap between this assumption and reality is perhaps the most serious obstacle to accepting their analyses, and one might well wonder at what point the shape of a projectile and the resulting aerodynamic forces on it began to be taken into account.[12] In this section, an attempt is made to address that question. Such an account could be the subject of a book on its own, but mention needs to be made of these issues for several reasons:

(1) They were the subject of some of Moulton's researches of a theoretical nature, and his reports on the experimental efforts both planned and executed by his associates to address these matters form a large part of his *History*. They also figure prominently in his *New Methods* in its long final section.
(2) They help complete the story of the evolution of the equations of motion, as set down by Moulton, at the hands of American ballisticians after the World War by their inclusion of terms which account for the aerodynamic forces.[13]
(3) This section returns to the beginning of the era treated in the first chapter, the post-American Civil War, and retraces the development of a relevant ballistic problem from its early empirical study to its attempted analytical resolution, effected for our purposes by a key paper of 1920, and highlights the larger picture of extra-American ballistic activity and its influence on American work.
(4) Phenomena which affected a projectile's trajectory, such as those associated with its spin, were often experimentally observed well before they were theoretically accounted for and capable of mathematical expression. They were often spoken of informally as quantities which "varied" the trajectory from some imagined normal in which they were absent. In this context, the word "variation" has a different meaning than the change of a well-known quantity like temperature from an accepted "normal" value, as in the case of the differential variations of Moulton and Bliss. Of course, once constants were introduced to account for these forces in the equations of motion, their variation could be studied by the earlier methods. This was in fact done in [1095] in a different context, to be discussed in Sect. 4.6.
(5) The real impact of these issues was felt more in the design of projectiles than in their use in fire control, and design became increasingly important after the World War.

A beginning can be made by considering several pages of the historical appendix of the 1953 volume just mentioned in the previous section [1233, pp. 790–792]. There one learns that

> The first need for treating the projectile as a body subject to aerodynamic forces other than the drag arose with the employment of rifled artillery in the latter part of the nineteenth century. Some information on the character of the general aerodynamic force system acting on a projectile had long been available - physicists from the time of da Vince had recognized

[12] Images of the shapes of various shells through history can be found in [1013, p. 39].

[13] There are other examples of post-War ballistic activities. One such is the continuing attention paid to numerical integration, as described in [1120]. Here one finds a proliferation of new methods and comparisons of efficiency among them, as well as an account of how long it took before this now-standard topic became a legitimate object of study for mathematicians in general. It is interesting to note that the time after the War saw declining investments in all such research activities.

> the aerodynamic lift and overturning moment on elongated yawing bodies moving in fluids. ... The Magnus force acting to swerve a spinning cannon ball from the path of a particle acted upon only by drag and gravity was recognized early in the nineteenth century. However, aerodynamic forces other than the drag were first considered in ballistics with the advent of the spinning shell in gunnery.

Two phenomena of spinning shells are then discussed: the first is tumbling, an occurrence which induced a huge drag, shortening the ranges of the projectiles. Investigations of the causes led to a so-called first stability condition, involving various moments and air speed. "This relation was initially found by considerations analogous to those employed in the theory of tops: it was probably known to George Greenhill in England and K[C]arl Cranz in Germany early in the the twentieth century" [1233, p. 791]. The second was a certain right deflection of a shell fired from a weapon with a right-hand twist of rifling. "This effect was found to be independent of other known perturbing, or 'abnormal' conditions, such as cross wind... which also affected deflection" [1233, p. 791]. This deflection was given the technical term "drift." The authors note that drift was first theoretically treated in the 1920 British paper [1108], a source often referred to in aerodynamic matters at this time; it was of particular interest to the American ballisticians, as will be described below.[14]

The design of aircraft at the beginning of the twentieth century provided a boost to the study of aerodynamic forces of a projectile, since wind tunnel studies for the former could easily be adapted to the latter. This development was also an influence for the writing of [1108]; and solutions to equations from the paper were found by R. H. Kent and A. P. Hitchcock, at Aberdeen, in the years between the World Wars.

The three pages whose contents were just summarized barely scratch the surface of the complexity inherent in these matters. The classic 1921 volume [1083][15] gives a much fuller picture. On pages 70–82, one finds a description of the effects of the shape of the head of a projectile on its air resistance, including worked mathematical examples and a use of the calculus of variations to design a shell which minimizes drag. Chapter X, roughly 50 pages in length, is devoted to an analysis of the drift just mentioned, including a graphical solution by the author, dating from 1898 [1083, pp. 345–350]. In each chapter, numerous experiments by ballisticians across different eras and nations are given, with critical analyses.

In order to reduce this mass of material to manageable size and relevance, we now limit ourselves to contemporary American sources, most have which have already been cited earlier in this book.

The earliest American gunnery volume mentioned in the first chapter, the 1867 volume by Benton [1017], begins its section on the "Science of Gunnery" with

[14] The authors also assert that "If all the aerodynamic coefficients of the shell were known, the drift could be predicted in advance of range firings by methods given in Chapter XII of this book" [1233, p. 791].

[15] One of the authors, Carl Cranz (1886–1941), had a status in Germany comparable to Charbonnier in France; he contributed much to the literature during his career. The volume is a 1921 translation into English, published in London, of the second edition of the original German work.

brief historical remarks, noting Benjamin Robins' 1765 introduction of the ballistic pendulum. "He also discovered that the rotary motion which spherical projectiles generally assume around their centres of gravity, will cause them to deviate from their true direction" [1017, p. 383]. Regarding experiments to determine the drag law, he mentions Charles Hutton. He cites firings at Metz in France (1839–1840) using projectiles of large size with high velocities. One striking aspect of Benton's account is the amount of attention paid to the way in which the firings were conducted, as well as the machinery of measurement, which was in a state of constant development. In particular, devices to measure initial velocity of a projectile were presented, with several pages devoted to West Point's own "ballistic machine."[16] Elementary laws of resistance of projectiles in the air are given, justified by rough physical assumptions; their alteration by comparison with firing data, for example, those of Guillaume Piobert (1791–1871),[17] are included. The experiments of Charles Hutton and Jean-Charles de Borda (1733–1799), another French investigator, regarding the effects of the shape of the projectile, are listed; the heads were spherical, hemispherical, conical, prismoidal, and elliptical. Benton's conclusions include remarks to the effect that theory and practice do not yet agree, rounded and pointed solids are affected less by resistance, the sharpest points do not always encounter the least resistance, and that "where the front surfaces were the same, the resistance was least with those in which the posterior surfaces were the flatest" [1017, p. 411]. "Now that oblong projectiles are used in all kinds of fire-arms, it is important to determine that form which will be least affected by resistance of air" [1017, p. 411]. Also, "Several distinguished mathematicians have obtained expressions which approximate to the true results, but the expressions are generally too complicated to be of much practical value" [1017, p. 412].

In a section entitled "Deviation of Projectiles," Benton considers "normal trajectory" as one where the projectile moves only under the forces of gravity and tangential resistance and discusses the familiar sources of deviation from it but also discusses rotation for spherical and oblong shells. No mathematics is involved, but attention is paid to a device which measures the position of the center of inertia of the shell,[18] and, more important, the device of "Professor Magnus of Berlin" to demonstrate the Magnus effect (1852): spinning object moving through a medium is deflected from its path of no spin by the difference in the pressures of the medium on the sides of the object. From this principle, Benton lists conclusions relevant to projectiles.

From these rough beginnings emerged the works of the first wave of ballistic reformers, Ingalls prominent among them. In his 1886 volume [1188], pressure on the surface of revolution forming the head of a projectile is discussed mathematically. He references the work of Jean Victor Poncelet (1788–1867), a French

[16] Not noted by Benton is the fact that the ballistic pendulum was replaced by electric screens suggested by Charles Wheatstone in 1840 [1011, p. 16].

[17] Piobert was a French general who participated in the Metz firings.

[18] The device is called an eccentrometer.

engineer most closely associated with the École Polytechnique, who collected empirical formulae for normal pressure on the shell head. These formulae were used by Ingalls to derive a general mathematical expression for total pressure. Declaring that it fit the experimental data better than any other, Ingalls used this expression extensively; in particular, it was applied to hemispherical and ogival heads. As with Benton, at least as much space was devoted to the descriptions of measuring devices: ballistic pendulums and chronographs. As the title of the book (Exterior Ballistics in the Plane of Fire) might suggest, deviations from the plane are not covered; that the level and amount of mathematics used in the book in general was greater than the earlier volumes has already been discussed.

By the 1890s, it was admitted by the author of an American text on ordnance and gunnery, Henry Metcalfe [1235], that the subject had grown so enormous due to recent advances that a comprehensive revision of the West Point course established by Benton was needed. Part of this set of advances was attributed to the eclipse of spherical projectiles by those of newer shapes [1235, p. 5]. Consequently, material relating to such projectiles began quickly, on page 8, where the exposition was devoted exclusively to experiments of Francis Bashforth for the British government from 1865 to 1880.[19] Conclusions of Bashforth's studies are given, most prominent among them being that the shape of the head is less important than "the form of the head, near its junction with the cylindrical body of the shot" [1235, p. 10]. Contemporary experiments done by Krupp (place and time unnoted) are also reported, with a single sentence conclusion: "...resistance varies also with the shapes of the sides and rear of the projectile" [1235, p. 10]. Attention was paid to the mathematics of data analysis. No mention is made of rotating shells. Some study of drift must have occurred during this time, however, as the quantity would appear in some range tables of the 1890s; for example, a table for the 10 inch B. L. steel rifle was presented without comment in JUSA/CAJ in 1892 [989], where a column for the drift associated with range is given, assuming no wind. No mention of how these numbers were attained is given.

Perhaps these entries were a result of efforts by the other prominent first-wave reformer, Alston Hamilton, who did some original work of the drift of rifled projectiles. This appeared first as a paper in JUSA/CAJ in 1907 but was also included as an appendix of the second volume of his major work [1159, p. 150 ff.]. In the paper, Hamilton developed "a drift formula of simple form independent of any explicit hypothesis as to the law of retardation or as to the character of fire ... the formula to be generally applicable to practical problems of fire" [1159, p. 150]. He cites the work of Charbonnier in this connection, referring the reader to this material for the "analytical treatment" and deals with the "facts of the case" in the remainder [1159, p. 150]. Thus as in the case of Ingalls, who presented Siacci's theory, the reliance on European sources was explicit. This represents the first serious attack of the problem by an American ballistician and earned Hamilton one of his two citations in Cranz and Becker's list of references on drift [1083, p. 154].[20]

[19] Metcalfe, citing Robins as a predecessor, referred to the latter as the "Father of Gunnery."

[20] Here Hamilton is erroneously identified as "E. Hamilton."

Events in the run-up to the World War are summarized in a post-War paper by R. H. Kent [1203]. Citing a "lethargy from which projectile design had been suffering for a period of over fifty years" [1203, p. 204] prior to 1911, he drew attention to a series of experiments by the French taking place that year regarding shape. He claimed that their results started to have an effect of the design of shells at the beginning of the War, when "the shape of the projectiles used by all belligerents ... was to all intents the same as the shape of the projectiles manufactured in the early 1860s when the use of rifled cannon first became general." His attempt to find the source of the lethargy led him to the conclusion that, in spite of the presence of engineering and scientific knowledge, "there was at hand during this period no adequate theory of motion of projectiles" [1203, p. 204]. He found existing efforts inadequate.[21]

A significant change in this state of affairs in America occurred with the advent of the World War. The establishment of the Aberdeen Proving Ground and the construction of wind tunnel facilities made it possible for the United States to devote attention to its own firing studies and secure the services of mathematical personnel to address more difficult design issues. Moulton's *History* confirms work on the first of these activities. Before addressing this account, mention should be made of the economic motive behind better understanding of shape, as described by Kent at the opening of his paper:

> During the World War, the Ordnance Department expended approximately four billion dollars. Of this sum approximately two-thirds was expended for artillery ammunition. In view of the vastness of these sums it would seem to be a matter of considerable importance that this ammunition should be of the best possible design. A large portion of this ammunition would have been used for barrage work or against definite objectives in which case the value of the ammunition would probably be approximately proportional to its accuracy. It is therefore a fair statement to say that an improvement in the design of these projectiles would have increased the value of this ammunition by an amount which would be measured in the hundreds of millions of dollars.

Returning to the *History*, Section VI contains a description of "boat-tailing" experiments conducted by Veblen "with enthusiasm" [1246, p. 44]. Taking a cue from their French allies, Veblen's team at Aberdeen altered the shape of several existing types of shells by boat-tailing or streamlining them (no further details of this process are discussed). Summaries of the results are given, including tables of changes in ranges of projectiles boat-tailed with differing numerical dimensions [1246, p. 49].

Later, Section X describes "Form of Projectile Experiments":

> Conclusions from experiments with different types of projectiles are exceedingly costly because of the time and expense attached to making experimental shell, and they are limited in their scope and certainty because one can never be absolutely certain as to what feature, or combination of features, of a projectile the results are due. [An example of this difficulty is

[21] The remainder of the paper described the experiments at Aberdeen and the theory (from a 1922 British paper) on which they were based, to be given below, in detail.

given in detail.] Consequently the problem of determining experimentally, without essential aid from general principles, the best form for a projectile seems hopeless. [1246, p. 62]

This passage is remarkable for its straightforward admission of the difficulty of the problem, its acknowledgment of the need for experiments and the care necessary for their interpretation, and expression of the desire for explanations based on "general principles." Plans to meet these difficulties were then formulated by A. A. Bennett and executed by R. H. Kent and H. P. Hitchcock: special shells made up of four parts were to be manufactured, each of the four parts having several shapes and lengths available for use. Each individual part was to account for a certain aerodynamic feature whose effect was changed according to which of the available forms of the part was used in the overall shell. Guns with different riflings were also constructed, some with uniform and others with increased rifling toward the opening of the barrel. The shells were then to be fired and resulting ranges noted. This would provide data for analysis and hopefully, eventually, a theory to aid in projectile design.

As for drift due to rotation of the projectile, there is no mention of this in the *History*, though the section on the differential effects of the rotation of the earth on the "deflection" from the plane of fire is derived [1246, pp. 24–25]. Moulton mentions a "careful discussion," done by Ritt, of the effect of the rotation of the earth on the trajectory of a particular projectile, a study which includes "the great importance of the indirect effects of the earth's atmosphere near the end of the trajectory" [1246, p. 25]. In this context, there occurs a criticism of the "rough methods heretofore in use, such as given in Charbonnier's Traité de Balistique [and which formed the basis of Hamilton's piece]" [1246, p.25], which sometimes gave results of the wrong sign! As in the case of the first wave of reformers, some work of an indeterminate nature must have been done to include a column of drift data among the entries in the Provisional Range Table of 1918 issued by Moulton's group [1268, p. 16]. These figures would not represent the effects due to the earth's rotation, since that effect requires a latitude value, missing from the table.

This design initiative begun at Aberdeen was helped mathematically by the appearance in 1920 of [1108], a paper by four British Technical Officers of the Munitions Inventions Department.[22] In this paper, both theoretical and experimental results were reported, the firings having taken place on the grounds of the H.M.S. "Excellent" in Portsmouth. The subject of the paper "is the motion of a spinning shell through air at velocities both greater and less than the velocity of sound" [1108, p. 296]. The problem had hitherto been attacked by both Charbonnier and Cranz and a host of others mentioned in the introductory pages. In what follows, this paper will be referred to as FGLR; it had a significant impact on shell design in the 1920s in America and was cited repeatedly in different publications.

Philip Schwartz, acting in his self-appointed role of reporter of ballistic innovations at the Aberdeen Proving Ground, described in a note from 1924 of use of FGLR by its American counterparts:

[22] No further affiliation of this department within the British military establishment is given.

> The mathematical theory of the motion of a modern projectile considered as a particle is relatively simple, compared with the theory of the projectile considered as a rigid body. ... Improvements have been made ... by obtaining a retardation law which is suitable for all velocities and which considers the shape of the projectile as a factor. ... However, it has only been since the past war that the United States and other governments have actually taken up the more difficult but complete problem of the projectile considered as a rigid body. [1289, p. 340]

Schwartz notes that procedures for approaching this problem were developed by the British during the World War and are now being followed at Aberdeen by Kent, whose expressed need for a theory like that contained in FGLR was evidently answered. The experimental aspect is described:

> [Projectiles are fired] through cardboard screens placed at successive positions in front of the guns and along the expected path of the projectile. Examination of the holes made in the cardboard by the projectile, permits determining the direction in which the axis of the projectile was pointing and the angle which the axis was making with the path of the center of gravity of the projectile at the time when it went through each cardboard. [1289, p. 340]

Schwartz then presents "a graphical figure of the motion of every part of the shell for about 1200 feet from the muzzle," citing the use of the theory from FGLR together with the results of a particular firing to determine constants for use in producing the image.[23] The diagram is whirl of spirals giving the "true motion of the nose of a projectile." There is at least one lesson to be drawn from a study of the diagram: "Before the World War it was customary to teach in the United States that the nose of a projectile precessed around the tangent to the trajectory less than once in a whole trajectory. In the case chosen, the nose of the projectile has turned five times around the tangent in the first 1200 feet." ([1289, p. 431] for all quotes in this paragraph.)

The paper FGLR caused a bit of consternation for Moulton, who in the meantime had worked out a preliminary theory of his own. It is not known when this was begun, but apparently the issues described in his *History* regarding the aerodynamic experiments tempted him to undertake the mathematical explanations "based on general principles" he referred to in his report. Nothing of this work appears in the *History*. The results of his efforts, which he regarded as his most important contribution to the subject of ballistics, and which reportedly cost him at least three times as much effort as the rest, appeared as the last chapter of his *New Methods*. "It is my opinion that all essential future improvements in projectile design and firing control will be guided by such a theory."[24]

The contents of this chapter are impossible to summarize here but basically consist of the solution of a set of differential equations "for a rigid body in three dimensions without restraints. The form of the differential equations for the problem have been well-known since the time of Euler ... " [1249, p. 172]. The equations are laid out, each component explained geometrically, and solved for various cases, with

[23] No draftsman is credited for this figure.

[24] Moulton's comments in the last two sentences are from [1250, p. 247].

masses of approximations and techniques of integration. The last few sections are however devoted to the use of the material in design, addressing such problems as stability:

> The practical problem in every special case is so to control $S$ through the shape and distribution of mass of the projectile and through the pitch of the rifling of the gun from which it is to be fired, that it shall, in the first place, be stable and, in the second place, not so stable that is will fail to trail[25] satisfactorily. Since the solution depends upon the numerical values of $\mu$, $\nu$ and $p$, it can be reached practically only with the aid of such experiments as those that have been carried out extensively at the Aberdeen Proving Ground by R. H. Kent and H. P. Hitchcock... but the experiments could not lead to the optimum design unless their interpretation were guided by a complete theory, such as given here, that shows the character of the dependence upon all quantities that are subject to control. [1249, p. 254]

The variables quoted are part of Moulton's theory. Applications to drift are also given.

While this work was in progress, FGLR appeared. Moulton considered it "noteworthy" but declared that "The methods used by these authors are entirely different from those which I employed" [1250, p. 247–248]. These remarks were taken from the dispute between Moulton and Roger Sherman Hoar detailed in Chap. 1 and were part of the former's reply to the latter upon Hoar's declaration that Moulton's chapter "does not appear to bear much resemblance to, or take in to account, the very complete theory built up by Capt. R. H. Kent, as a result of painstaking research at Aberdeen from 1919 to date" [1180, p. 396]. It has already been noted that Moulton took Hoar to task for not dealing with these issues in his own ordnance text and essentially pronounced Hoar incapable of understanding either effort.

Another, perhaps more serious, criticism was leveled by Kent himself, who in [1204] gently chided Moulton's often-stated attitude of independence from the work of his predecessors by pointing out that this resulted in some unfortunate consequences for the last chapter of *New Methods*. Kent claimed that Moulton's lack of attention to the contents of FGLR resulted in formulae which, as they stood, could not be used in application. He also remarked that "The results of Fowler [i.e., FGLR] have been confirmed in the experiments conducted under my direction at Aberdeen Proving Ground," where Moulton had hoped his theory would be tested. But overall the tone was respectful and hopes for a revision were expressed. Moulton's reply to this was extensive but similarly respectful, as we have previously explained.

The contents of FGLR moved into the realm of firing table discourse when they were used as the basis of a treatment of drift in Hitchcock's 1934 volume [1171]. The amount of work done on the drift problem between the end of World War and the decades of the 1920s and 1930s must have been considerable when one contemplates here, as elsewhere, the number of new variables introduced into ballistic discourse, such as $N$, the spin of the shell in radians per second, $\delta$, the yaw in radians, couple coefficients, cross-wind force coefficients, overturning

[25] The technical meaning of this term is given on pp. 251 ff.

moments, and so on. We will discuss briefly Hitchcock's exposition of drift, the term mentioned in a technical way above in [1233].

> The drift is the resultant of three effects: gyroscopic, Magnus, and Poisson; given in the order of their magnitude. The gyroscopic action causes the point of the shell to turn to the right of the plane of fire, ... so that the air resistance has a component which compels the center of gravity to turn to the right. At the same time, the viscosity of the air causes a thin layer of it to rotate with the shell, resulting in a greater pressure on the right and a consequent deflection to the left; this is the Magnus effect. The fall of the shell causes a "cushion" of air to accumulate below it, over which it tends to "roll" to the right. This is the Poisson effect; but it is negligible in comparison with the others and will not be considered further. [1171, p. 120]

The Magnus effect is stated to be of consequence only when the yaw is large; the assumption, for "practical purposes," of a yaw of less than ten degrees is made, and thus this effect is ignored.

In the next several pages, more constants are introduced, some coming from FGLR, which is explicitly acknowledged. Through suitable approximations, assumptions, and a borrowing from tables of data from British three-inch gun firings, a new "drift constant" $K$ is obtained and related to the "rate of change of direction of motion" $\psi'$ through a first-order differential equation. After a section of material dealing, interestingly, with Moulton's original theory of differential variations, a sample trajectory with such variations is worked out by numerical integration and accompanied by a small "Computation of Drift" table produced by numerically integrating the equation just mentioned. The table lists the change in $z$ co-ordinate of the center of gravity of the shell in yards over the first six seconds of the flight [1171, p. 136].

The final contribution in the sequence considered here is the 1938 volume [1166], the last of the West Point ordnance textbooks issued prior to the beginning of the second World War. Quite a difference is indeed felt when turning the pages of the chapter devoted to exterior ballistics. Slightly more that three pages alone is devoted to a list of symbols needed to express the new equivalent of Moulton's equations of motion, which are derived from pages 397 through 433. The issues of drift, yaw, and twist of rifling are all accounted for by variables in the list. The work of Bliss and Gronwall is noted in a single sentence [1166, p. 462]. But perhaps most salient is the set of references given at the end of the chapter, which includes not only FGLR but Dunham Jackson's 1919 numerical integration volume, a table by Ingalls, Moulton's book, and works by Charbonnier and Cranz. This study of exterior ballistics for the students at least acknowledges the vast theoretical work done nationally and internationally in the period between the wars. As always, citing a reference is no guarantee that its contents will be digested, and even less that it has a practical influence on gunnery, but the present section attests to the interest taken by the mathematicians, physicists, and engineers in what was the "next" problem of variation of the trajectory.

We conclude by noting that Bliss' 1944 book [1048], being confined to "particle ballistics," was silent on the issues discussed in this section. Hoar's 1921 book [1176, pp. 70–75]however has an instructive chapter devoted to deflection formulae.

Four sources are identified: cross-wind, lateral jump (the side jump of the gun upon firing), drift = gyroscopic action, and cant (when the two cylindrical supporting pieces on which a gun lies are of unequal height). A simple formula for the deflection due to cross wind is developed, as is one for the cant, derived from basic trigonometry. A formula for lateral jump is similarly given. The drift is treated as "wholly empirical," that is, as entered in the table it is the measured deviation from the firing adjusted algebraically by the other three quantities. This bare-bones approach was sufficient for its time, data dependent on a particular firing, and can be contrasted with the process used roughly ten years later by Hitchcock. On the other hand, one notes that cant and lateral jump, two intrusively realistic factors of genuine consequence, do not figure in the writings developed of Moulton or Bliss. Again Hoar's is positioned more closely to the facts of firing.

## 4.5 Computational Issues

The "very laborious process" of numerical integration has been documented to some extent in this book as well as in others.[26] There are several aspects to the computational issues involved. David Grier has discussed the communities of "human computers" necessary to execute the trajectory computations in [1142]. The present account has emphasized the interplay between the mathematics and computation; the authors of the technical reports not only set aside sections of their papers outlining computational procedures to effect their work but were at great pains to introduce mathematical simplifications to reduce the computational labor. Thus, for example, Gronwall's reduction of the order of the adjoint system was a significant achievement at the time. The creation of computation sheets by Hoar was a necessary, if somewhat clerical, effort. The tools computation, beginning with slide rules, logarithmic tables, and mechanical calculators, changed over the years between the wars. The topic of numerical integration itself, given higher visibility due to its introduction during the war, remained of limited interest to the American mathematical community for several decades, as the present author has argued; see [1120].

Extensive volumes devoted to the calculation of trajectories appeared in the years after the war. Dederick contributed such a tome, and the 1934 *Report* of A. P. Hitchcock mentioned earlier is typical [1171]. Large fold-out sheets display the data tabulated during range firings; computation sheets for trajectories, descendants

[26] In a short contribution to the Moulton-Hoar debate, Dederick wanted to go on record in this matter: "The fact that numerical integration is extremely laborious is not brought out in the review [by Hoar], although of course well known to the reviewer. The individual steps are easy to perform and easy to understand but are very numerous. It would appear that this fact should be stressed in any general estimate of the value of the method, as it constitute the outstanding objection to its use" [1087, p. 667].

of those of Hoar, are given, and the methods of calculating differential variations according to both Moulton and Bliss are described.

A new element in the history of trajectory computation was the introduction of the Differential Analyzer. This analog device is described in Chapter 10 of [1123], and its intermediate role between hand calculation and electronic computation of trajectories is documented there. More can be found in [1016, pp. 7–13], an extensive account of the machine written by A. A. Bennett. Briefly, initial discussions between the machine's inventor, Vannevar Bush of MIT, and representatives of the Proving Grounds, including Dederick, took place at a Christmas meeting of the American Mathematical Society in 1929. A demonstration of the Analyzer's calculation of a trajectory was given on September of 1931 at MIT; there followed more extensive trials in the summer of 1932. By December of 1933, the Proving Ground was authorized to procure a differential analyzer similar to that used at MIT. Two machines, one for use at Aberdeen and another for use at the Moore School of Engineering at the University of Pennsylvania in Philadelphia, were eventually built. On September 30, 1935, the first complete mechanically calculated trajectory was run at the Proving Grounds.[27] Dederick was exhaustively familiar with Aberdeen's device and was a major participant in the negotiations to get the two machines constructed.[28]

The variations in the ballistic trajectory figured in two contrasting ways with respect to the Analyzer.[29] On one hand, they were used as in the earlier era in the "reduction of observed range firings," that is, the determination of the ballistic coefficient $C$ for trajectories computed under standard conditions by using data from actual range firings, which will always differ from standard conditions. Three "corrections" are used: observed values for wind aloft, temperature, and air pressure yield corrections to the observed trajectory. These three corrected values are then used with several possible candidate values of $C$ and the computed ranges and times of flight examined to yield the best candidate for $C$ by various criteria; specifics are given in [1016, p. 57]. As in Hoar's time, one might say that the process is itself subject to "variation." The use of a correction required an adjustment to relevant parts of the machinery.

On the other hand, the introduction of the Differential Analyzer made the use of the Bliss adjoint method for producing range firing table entries obsolete at this time.[30] In a 1940 article by Dederick in the *American Mathematical Monthly* [1088], the author explains:

> With the use of mechanical integration this method is not practical. For mechanical integration provides for the rapid finding of numerous solutions of one system of differential

[27] The Differential Analyzer is mentioned briefly in Hitchcock's volume, [1171, pp. 49–51].

[28] Bennett notes that the Philadelphia machine was built with the use of C. W. A. labor [p. 11].

[29] Moulton's equations of motion appear on p. 52 of the report.

[30] Hitchcock [1171, p. 274] states somewhat unclearly: "Although the differential equations for the variations from standard conditions cannot be solved with the analyzer, the results can be obtained by solving trajectories with non-standard conditions and applying the appropriate formulas."

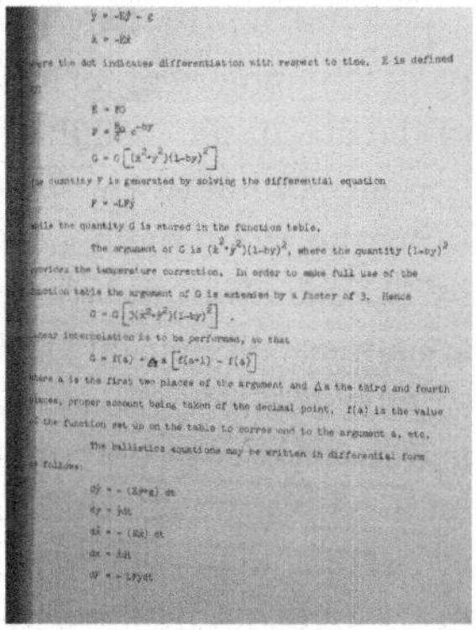

$\ddot{y} = -E\dot{y} - g$

$\ddot{x} = -E\dot{x}$

...re the dot indicates differentiation with respect to time. E is defined

$E = FG$

$F = \frac{h_0}{C} e^{-by}$

$G = G\left[(\dot{x}^2+\dot{y}^2)(1-by)^2\right]$

...e quantity F is generated by solving the differential equation

$\dot{F} = -LF\dot{y}$

...ile the quantity G is stored in the function table.

The argument of G is $(\dot{x}^2+\dot{y}^2)(1-by)^2$, where the quantity $(1-by)^2$ provides the temperature correction. In order to make full use of the ...ction table the argument of G is extended by a factor of 3. Hence

$G = G\left[3(\dot{x}^2+\dot{y}^2)(1-by)^2\right]$ .

...near interpolation is to be performed, so that

$G = f(a) + \Delta a\left[f(a+1) - f(a)\right]$

...ere a is the first two places of the argument and $\Delta a$ the third and fourth places, proper account being taken of the decimal point. f(a) is the value ... the function set up on the table to correspond to the argument a, etc.

The ballistics equations may be written in differential form ... follows:

$d\dot{y} = -(E\dot{y}+g)\,dt$

$dy = \dot{y}dt$

$d\dot{x} = -(E\dot{x})\,dt$

$dx = \dot{x}dt$

$dF = -LF\dot{y}dt$

**Fig. 4.1** Moulton's equations as they appear in the ENIAC progress report from 1943. [NARA ENIAC], p. XIV (4)

equations, but not for rapid work when there is a change in the differential equations themselves. Now the coefficients in the adjoint system depend on the particular solution of the trajectory equations. Apparently then the use of the adjoint would have no value in connection with mechanical integration, where the variations are obtained by running independent trajectories and noting actual differences. [1088, p. 633][31]

The article continues by outlining a possible theoretical use of Bliss' method, due to Dunham Jackson, in the derivation of equations which suggest that the "effect of change of density can be expressed in terms of those of velocity and angle. The practical result anticipated is that density runs as well as wind runs may be omitted in preparing a firing table and the data derived from the effect of variation in angle and muzzle velocity" [1088, p. 634]. This reduction in labor could be effective "even with mechanical integration." Evidently savings of effort in table creation were still being sought.

By far the most well-known outcome of attempting to find faster methods of computation for the range firing tables was the war-time development of the ENIAC. (See Fig. 4.1.) Many accounts of this historic effort exist; [1123] was written by one of the participants, Herman Goldstine. The story needs not be repeated here, but two aspects relevant to differential variations can be noted. Goldstine was involved with both the Differential Analyzer and ENIAC, and in a 2004 interview with the present author [1127], Goldstine mentioned that he used the adjoint method to "speed up" the calculations done on each device. Due to his advanced age, however, details of this use were not forthcoming, and this remark must remain a tantalizing but unsubstantiated utterance. A more well-documented use occurred in an unlikely place. The use of any device to effect calculation introduces errors; such errors in the Differential Analyzer had to be considered before it was even thought suitable for use in trajectory computations. The same was true of the ENIAC, and Hans Rademacher, a professor at the University of Pennsylvania, was given the task of estimating the magnitude of such errors on the digital machine. The result was a paper presented at a two-month course at that institution in 1946 discussing all

[31] The use of the Differential Analyzer in this capacity was also noted in [1265, p. 31].

aspects of the ENIAC design and function to a limited audience [1239]. By dint of much numerical analysis, some bounds were achieved, assuring that the results of the electronic calculation could be trusted. Among the tools used in the analysis was the adjoint to a system of linear differential equations in the error introduced by using the Heun method of numerical integration [1239, pp. 7–8]. Error introduced by the use of digital rounding off was also treated; in this case, "adjoint difference equations" were used [1239, p. 13].

One can see by this chronology that the relations between variation theory and methods of computation are manifold. In the next and last sections our story ends with a transformation of the differential variations material into a different environment in which the use of electronic computers enables their application, one which simultaneously involves the evolution of the calculus of variations into a successor discipline.

## 4.6 From Projectiles to Missiles

In April of 1939,[32] after having received his Ph. D. in theoretical physics at the University of Vienna the previous month, Rudolf Drenick (1914–2010) came to the United States in order to escape the Nazi threat. He took a teaching position at Villanova College, where he taught mathematics, physics, and navigation in the Navy's V 12 program, an accelerated curriculum for Navy recruits. In 1945, he was drafted as a private but through some connections of the college president was placed in the Technical Attachment to the Ordnance Department at the Aberdeen Proving Grounds. There he, along with many volunteers, was presented with a warehouse full of crates containing documents captured by the Allies at Peenemünde, the German proving grounds for the new V-2 ballistic missile. The documents, when organized and translated, would reveal the workings of this new weapon. Drenick spent about a year on this project and then was discharged from the Army and returned to civilian life. The immersion in the V-2 material had changed his career interests from physics to engineering, and he began working for various engineering companies, among them RCA, the Radio Corporation of America.

In 1951 (the year of Bliss' death), while employed at RCA, he published a paper [1095] which drew on his experience with the V-2 documents. He referred to the differential variation theory of Moulton and Bliss as "the calculus of perturbations" and proposed to use it for control of "rocket missiles." He pointed out that this calculus could be used in rocket design much as it had been in projectile design, to see how the weapon's performance would be changed by alterations. But more important was its use in cutoff and guidance. The ballistic missile experienced thrust for only a limited portion of its flight; after cutoff of thrust the V-2, for example, acted as a massive projectile for the rest of its flight. Other missiles

[32] See [1096] for the following biographical information regarding Drenick.

could be guided after cutoff from radio stations. The paper considered the following scenario: suppose a missile had a flight under "standard" conditions-similar to those conditions assumed earlier for a projectile-including a cutoff time for thrust, $t_c$, and the range to target denoted by $X$. If the conditions of fire are now changed, how can the cutoff time, for example, be changed so that the same target is hit? More generally, how can the missile be guided after the cutoff to achieve the same target? Drenick states at the outset that his "treatment is strictly theoretical in all cases and no effort is made ... to take account of many of the practical difficulties which might be encountered in realizing the solutions. The purpose of this study is not to present eminently practicable control principles but to show the usefulness of the analytical model [1095, p. 424]." We provide a summary of this account.

Following Moulton and Bliss, whose books he cited, Drenick set out equations of motion for the missile trajectory in the form

$$\begin{aligned} \frac{d}{dt}x &= v_x \\ \frac{d}{dt}y &= v_y \\ \frac{d}{dt}v_x &= F(y, v_x, v_y, \alpha, t) \\ \frac{d}{dt}v_y &= G(y, v_x, v_y, \alpha, t) \end{aligned} \tag{4.1}$$

where $F$ and $G$ are functions containing a list of constant and time-dependent expressions descriptive of the rocket: drag, lift, thrust, structural weight, and so on. The variable $\alpha$, which "constitutes the chief departure of guided missile ballistics from conventional ballistics," is the angle of attack (not precisely defined); it can be controlled. The burn time, or time before cutoff, is the other controllable quantity. The simplifying assumptions for the use of these equations are the same as Moulton's for a projectile; the missile is treated as a point mass with no inertia. A standard trajectory is a solution to this system under given conditions including a cutoff time $t_c$ and a "guidance equation" which gives $\bar{\alpha}$ as a function of time and is given by

$$x = \bar{x}, \quad y = \bar{y}, \quad v_x = \bar{v_x}, \quad v_y = \bar{v_y}, \quad \alpha = \bar{\alpha} \tag{4.2}$$

(Note the reversal of notation as used by Bliss.) Drenick then assumes that $F$ and $G$ contain a parameter $\mu$ and that the standard trajectory has been determined for the value $\mu = \bar{\mu}$. The parameter $\mu$ might be, for example, the drag or lift coefficient, structural weight, or specific impulse. A new trajectory results if $\bar{\mu}$ is changed to $\mu = \bar{\mu} + \delta\mu$, and each of the quantities appearing in the previous equation will be changed to $x = \bar{x} + \delta x$, and similarly for the remaining variables. The equations of variation are then given as

$$\begin{aligned}
\frac{d}{dt}(\delta x) &= \delta v_x \\
\frac{d}{dt}(\delta y) &= \delta v_y \\
\frac{d}{dt}(\delta v_x) &= a_1\delta y + a_2\delta v_x + a_3\delta v_y + a_4\delta\alpha + a_5\delta\mu \\
\frac{d}{dt}(\delta v_y) &= b_1\delta y + b_2\delta v_x + b_3\delta v_y + b_4\delta\alpha + b_5\delta\mu
\end{aligned} \tag{4.3}$$

The coefficients, as earlier, are the appropriate partial derivatives given at the values of the standard trajectory; their expressions are listed in the paper. The adjoint system is then introduced (Bliss is acknowledged here):

$$\begin{aligned}
-\frac{d}{dt}\lambda_1 &= 0 \\
-\frac{d}{dt}\lambda_2 &= \qquad +a_1\lambda_3 + b_1\lambda_4 \\
-\frac{d}{dt}\lambda_3 &= \lambda_1 \; +a_2\lambda_3 + b_2\lambda_4 \\
-\frac{d}{dt}\lambda_4 &= \lambda_2 \; +a_3\lambda_3 + b_3\lambda_4
\end{aligned} \tag{4.4}$$

and the fundamental relation

$$[\lambda_1\delta x + \lambda_2\delta y + \lambda_3\delta v_x + \lambda_4\delta v_y]_{t=t_2} = [\lambda_1\delta x + \lambda_2\delta y + \lambda_3\delta v_x + \lambda_4\delta v_y]_{t=t_1}$$
$$+ \int_{t_1}^{t_2} [(\lambda_3 a_4 + \lambda_4 b_4)\delta\alpha + (\lambda_3 a_5 + \lambda_4 b_5)\delta\mu]\, dt \tag{4.5}$$

is given for $t_1 < t_2$.

The cutoff criteria is then discussed. The "path variables" have at the standard cutoff time $t_c$ the values $\bar{x}_c,\ \bar{y}_c,\ \bar{v}_{xc},\ \bar{v}_{yc}$; along the altered trajectory, these will have values $x_c,\ y_c,\ v_{xc},\ v_{yc}$ which differ from the first set by $\delta x_c,\ \delta y_c,\ \delta v_{xc},\ \delta v_{yc}$ respectively. One wishes to change the cutoff point so that despite the changes, the range remains the same. Drenick reasons that the range $X = X(x_c, y_c, v_{xc}, v_{yc})$ and so the range error is to terms of first-order $\delta X = \lambda_1\delta x_c + \lambda_2\delta y_c + \lambda_3\delta v_{xc} + \lambda_4\delta v_{yc}$, as Franklin had observed, with the values of $\lambda_k$ being obtained by integrating the adjoint system backward from the standard time of impact $T$ to the standard cutoff time $t_c$ using Bliss' initial conditions. Thus in accordance with the usual practice, the $\lambda_k$ are found as soon as the standard trajectory is known. If one wants zero change in range, then one must have $0 = \delta X = \lambda_1\delta x_c + \lambda_2\delta y_c + \lambda_3\delta v_{xc} + \lambda_4\delta v_{yc}$. In order of effect this, record the path variables $x_c,\ y_c,\ v_{xc},\ v_{yc}$ during the power portion of the flight and keep track of the quantity $J = \lambda_1 x + \lambda_2 y + \lambda_3 v_x + \lambda_4 v_y$, which is continuously compared to the corresponding precomputed quantity $\bar{J} = \lambda_1\bar{x}_c + \lambda_2\bar{y}_c + \lambda_3\bar{v}_{xc} + \lambda_4\bar{v}_{yc}$. Then the new cutoff time occurs when $J = \bar{J}$.

The mathematics having been disposed of, the practical problem of keeping a running record of the path variables on the nonstandard flight remains. There were two plans proposed for measurements, the first involving writing the quantity $J$ in polar co-ordinate form $J = \mu_1 r + \mu_2\phi + \mu_3\frac{d}{dt}r + \mu_4\frac{d}{dt}\phi$ with origin at a radio station which continually tracked the four new path variables. A second involved the placement of two integrating accelerometers, devices for determining velocity and position from measured acceleration. Drenick gives some discussion of how this theoretical plan compared with what was actually done in the case of the V-2.

In the second part of the paper, a "guidance equation" is developed along similar lines. Roughly speaking, the idea is that, for each $t$ two quantities are involved: $J = \lambda_1 x + \lambda_2 y + \lambda_3 v_x + \lambda_4 v_y$ and $\bar{J} = \lambda_1\bar{x} + \lambda_2\bar{y} + \lambda_3\bar{v}_x + \lambda_4\bar{v}_y$. This time the $\lambda_k$ are now "precomputed from the standard trajectory by the adjoint system" as known *functions of time*. The path variables on the standard trajectory are similarly known at each $t$, hence $\bar{J} = \bar{J}(t)$ is a known function. In $J$, the path variables will be continually measurable during flight, and the condition necessary that no range error occur is that the missile be guided (by adjusting the variable $\alpha$ = angle of attack) in such a way that $J = \bar{J}$ for all t. This is the required "guidance equation." Once again the scheme requires measuring devices. Drenick proposes the establishment of three radio tracking stations, each of which measures the distances from the station to the missile; these distances are called the "slant ranges." The three slant ranges form a co-ordinate system which is to be used instead of the Cartesian system. These ranges, $r_1$, $r_2$ and $r_3$, functions of $t$, are then differentiated by an "automatic computer" and the guidance equation rewritten as $\bar{J} = \nu_1 r_1 + \nu_2 r_2 + \nu_3 r_3 + \nu_4\frac{d}{dt}r_1 + \nu_5\frac{d}{dt}r_2 + \nu_6\frac{d}{dt}r_3$. Deflection errors are handled simultaneously in a similar way.

The V-2 played a role in the second World War somewhat similar to that of the Paris Canon in the first: a frightening new war machine whose effectiveness demanded a response. From the opposite side of the country, in California, such a response was achieved in 1944. The "Private A," a ballistic missile, was developed at what was soon to be called the Jet Propulsion Laboratory, under the direction of H. S. Tsien; test flights took place in December of that year. Tsien[33] had come to the United States from China in 1935 and took a master's degree at MIT before moving to California to study with Theodore von Kármán at Caltech. There Tsien earned a doctorate under von Kármán's direction in 1939. After teaching at both MIT and Caltech and a period of war-time service, Tsien, upon being accused of having Communist sympathies, entered a period of effective detention in 1951, not being allowed to leave Los Angelos County.

[33] The story of Tsien's life and work, including details of the following account, has been told in [1068].

Under these difficult circumstances Tsien continued to work, in particular, Drenick's paper came to his attention. In response, Tsien and two colleagues published [1315] in 1952, from which the following quote is taken:

> The present theory of control and stability is built almost exclusively upon the theory of [linear] differential equations with constant coefficients. Therefore to study the disturbed motion of rockets, new methods [other than those used with non-constant coefficients] have to be used. R. Drenick in a recent paper [reference given] demonstrated the usefulness of ballistic and guidance problem of ballistic trajectories described by equations with time-varying coefficients. His theory is based upon the method of adjoint functions, first introduced by G. A. Bliss [reference given]during World War I. The purpose of the present paper is to make Drenick's theory more complete and definite and to apply it to the problem of automatic navigation of a long-range winged rocket vehicle. ... The main objective here is not, however, to give the final design of such an automatic navigation system, but rather to show the power of the ballistic disturbance theory for solving such problems and the various elements necessary to such a navigational system. [1315, p. 192]

The ideas of this paper were incorporated into Tsien's *Engineering Cybernetics*, eventually published in 1954 [1316]. This volume was highly regarded in certain engineering circles; it was a synthesis of various studies and ideas having as a rough backdrop Norbert Wiener's 1948 book *Cybernetics*.

The goal of Tsien's work was "to study those parts of the broad science of cybernetics which have direct engineering applications in designing controlled or guided systems." Tsien stressed that his work was an "engineering science" - "predominated by theoretical analysis and very often us[ing] the tool of advanced mathematics." "No gadget [a favorite word of Wiener's] is mentioned [1316, Introduction]." In Chapter 13 of this book, entitled "Control Design By Perturbation Theory," one finds a nearly 20-page enlargement of the 1952 paper. The Bliss adjoint system is again the centerpiece, and Tsien repeated his assertion that his effort "more complete" than Drenick's, "includ[ing] the problem of the automatic guidance of such vehicles [1316, p. 178]." The chapter goes into more detail on the operation of the rocket, containing descriptions of propellant release and elevator motion, for example. It also contains a flowchart of the "feedback" process [the word used by Tsien] inherent in Drenick's plan. The last section is devoted to a discussion of "Control Computers"; again, no "gadget" is mentioned, but a four-page discussion of digital and analog computers follows. It is apparently from this chapter that the ideas in Drenick's original paper received a wider hearing, as later references invariably cite it.[34]

---

[34] This technical material was cited in [1068], whose contents are claimed nontechnical by the author. After his period of confinement in Los Angeles Tsien returned to China to become a leader of that country's missile program. Following a few failed attempts at producing working missiles, he and his colleagues, using ideas from *Engineering Cybernetics*, pressed ahead; Chang claims that "Particularly important was Tsien's interpretation of the Bliss formula, based on the work of a mathematician best known for this work on the calculus of variations and for the application of calculus to the field of ballistics during World War II [sic]. In *Engineering Cybernetics*, Tsien applied Bliss' work to the science of guided missiles, and his application was later exploited by Chinese engineers to design a simple guidance system for the new, improved DF2 [a followup to a

Several features of this sequence of events can be noted. First, and most obvious, a new spin was given on Bliss' original theory. The differential variations as considered by Bliss and Moulton were quantities describing an altered trajectory of a projectile, that is, an object which would be passively observed after launch and the effect of various factors merely noted and compared to theoretical predictions. The only possible element of control of the path would be the initial adjustment of fire needed to make the projectile compensate for nonstandard conditions, if desired. In Drenick and Tsien's work, the path of the object is subject to continual control. This control is determined by constant measurement of variations from a standard and instantaneous adjustment using the adjoint variables to conform to the same range as the standard. Second, the control was to be effected using technology nonexistent in the First World War-accurate radio tracking-and what were coming to be known as "high-speed" computers to perform integrations and differentiations on the fly. Thus evolving technology enabled one to imagine such a scenario. Third, the work of both authors was self-admittedly theoretical. Neither Tsien nor Drenick dwelt on the specifics of practical aspects of their proposals, which, at least in the case of Tsien, who envisioned the computers as being carried by the missile, would be formidable at the time of writing. The technical problems did not prevent them from proposing their ideas. Fourth, as pointed out at the end of the previous section, this application took place more than a decade after Dederick announced that the differential variations were no longer calculated due to the ability of the new computing devices to quickly deliver trajectories under varied conditions. One might have expected the topic to no longer be of any great interest. Instead, the existence of the devices breathed new life into the matter. Last, by placing his scheme in the context of a work with ties to other control topics, Tsien made it possible to see this idea as part of larger bodies of knowledge, be it design methodologies, cybernetics, or engineering mathematics. Drenick and Tsien also removed Bliss' ideas from the province of ordnance and ballistics textbooks, though not from a military environment.

## 4.7 From the Calculus of Variations to Control Theory

It is generally acknowledged that the classical calculus of variations, the source of Bliss' work on implicit function theorems, embedding theorems, existence theorems, and the functional calculus, all of which proved so useful to his ballistics work, came to somewhat of a standstill by the mid-twentieth century. Bliss himself was considered the master of this classical material. Indeed, it has been noted that

failed first version]. Chinese aeronautical engineers claim today that Tsien's book helped provide a sound theoretical foundation for an entire generation of Chinese rockets from the DF2 to the DF5" [1068, pp. 220–221]. The author based the latter comment on first-hand interviews with these engineers.

there is an arc extending from Weierstrass to Bolza, who attended the former's lectures on the subject in 1879, to Bliss. Bliss himself was a tree with many branches: from Bliss to Graves, whose work included the consideration of abstract spaces, from Graves to Goldstine, who attempted an extension of the classical calculus of variations material to abstract spaces. Another branch from Bliss leads to E. J. McShane, whose further exploration of Bliss' ballistics work has been noted. Yet another branch from Bliss leads to his student Magnus Hestenes, author of 1966s *The Calculus of Variations and Optimal Control Theory* [1169]. As the title suggests, Hestenes' work started in the calculus of variations but ended in a different but related discipline, optimal control theory. The same is true of E. J. McShane. It is here that our story concludes.

The transition from the calculus of variations to optimal control was discussed by E.J. McShane in 1978 [1232] with typical good humor.[35] After noting that a trio of his 1940 papers on the problem of Bolza was received in the mathematical world "with all the éclat of a butterfly's hiccough," he claimed that the "the whole subject [calculus of variations] was introverted," continuing that "Its [the problem of Bolza] mastery gave us the power to answer many deep and complicated questions that no one was asking." This outcome was contrasted with the theory of optimal control, as advanced by Pontryagin in the Soviet Union and introduced to the United States in the late 1950s. "Pontryagin and his students encountered some problems in engineering and in economics that urgently asked for answers. They answered the questions, and in the process they incidentally introduced new and important ideas into the calculus of variations." McShane attributes the stagnation of the subject in America to the much-discussed long-standing separation of theory and application that characterized the establishment of the mathematical community: "Like most mathematicians in the United States, I was not paying attention to the problems of engineers [1232, p. 921]." In order to make a transition from the classical subject to optimal control theory, McShane introduces a new way of formulating extremum problems which includes those of the older theory and the newer subject, stating that it was due to Hestenes in a RAND report in 1950; he also cited a 1933 result of Graves which anticipated the control theory format. After stating some theorems of control theory, he takes four classical problems in the calculus of variations and translates them into the language of the new discipline and uses the theorems stated to solve them. There follow some suggestions on introducing control theory to the undergraduate curriculum.

The relevance of this evolution to our subject is that Bliss' adjoint technique reemerged as a tool to attack the military problems of a new era, problems stemming from the work of Drenick and Tsien but having nontrivial intersection with control theory. An example of this use is given in a series of technical reports and papers by Frank D. Faulkner and his students at the Naval Postgraduate School, Monterey, California, in the 1960s. One notes in the works cited that the goal is the guided hitting of a target under nonstandard conditions, but no quantity involved is subject

[35] Recall McShane's description of the calculus of variations at the beginning of Bliss' career as suffering from "elephantiasis."

to minimization in the process. In Faulkner's first research paper [1099] of 1961, an optimization issue is addressed. What follows is a descriptive summary of its contents.[36]

Faulkner sets the problem of interception of a target following a given path above the atmosphere, specifically, obtaining the trajectory of a missile which achieves the interception point $(x_f, y_f)$ with, for example, minimum fuel expenditure. The equations of motion for the interceptor are given as

$$\frac{d^2x}{dt^2} = g_1 + a\cos(p),$$
$$\frac{d^2y}{dt^2} = g_2 + a\sin(p) \tag{4.6}$$

where $(x, y)$ is the position, $(g_1, g_2)$ is the gravitational vector, which now is a function of $(x, y)$, and $(a\cos(p), a\sin(p))$ is the acceleration vector. If $(u_i, v_i)$ are two particular solutions to the adjoint of the variational equations for the system just given, then Faulkner obtains after simple manipulations

$$\delta x(T) = -[a(u_1\cos(p) + v_1\sin(p)]_{t_1^-}^{t_1^+}\delta t_1 + \tag{4.7}$$
$$\int_0^T [\delta a(u_1\cos(p) + v_1\sin(p)) + a\delta p(-u_1\sin(p) + v_1\cos(p))]\,dt,$$
$$\delta y(T) = -[a(u_2\cos(p) + v_2\sin(p)]_{t_1^-}^{t_1^+}\delta t_1 +$$
$$\int_0^T [\delta a(u_2\cos(p) + v_2\sin(p)) + a\delta p(-u_2\sin(p) + v_2\cos(p))]\,dt$$

where $t_1$ is a time at which the acceleration is discontinuous. These are the fundamental variation and control equations; they express the effects on the terminal values of $(x, y)$ of small changes in the "control variables" $a$ and $p$. Thus the adjoint variables now play a role in expressing the variation and control equations.

An iterative scheme for the simultaneous solutions of $(u, v)$ and $(x, y)$ is then set up using these equations, after initial guesses for the variables are specified; the description of the procedure spans several pages. Faulkner then argues that if the routine as described actually does converge, one has at least a "stationary value" for the fuel consumption, which must be a minimum.

Again, several comments are in order. First, the work of Bliss is acknowledged on the first page of the document and several other times in the body; for example, in the final section, "The procedure seems to furnish a rather simple extension of the procedures which Bliss introduced into Ballistics, for calculating differentials" [1099, p. 10]. The papers DECAF and FOLIB are cited as is Bliss' JUSA/CAJ

[36] The paper was also issued as a report for the Boeing Airplane Company in Seattle, Washington.

article. Second, there is no proof offered that the procedure as described actually does converge. Third, actual computation on a digital computer was done, using a modified Runge-Kutta method, and the author reports that "No convergence problems have been encountered in the ballistic missile interception problem." Thus, in a sense, this new application, from the point of view of rigorous mathematics, was at a pre-Moulton stage, an experimental stage, yet for the first time in the present account the computation envisioned by Drenick and Tsien was implemented. Fourth, the author notes an additional benefit to the procedures "due to Bliss," namely, that "should the actual trajectory deviate in some way from the planned trajectory, the same equations for the variations can be used to determine a corrective thrust schedule" [1099, pp. 11–12]. Connections to the new Pontryagin maximal principle are made, and the paper concludes with the comment "It seems to the author that one of the most important features in recent control theory rests on the use and interpretation of the adjoint system of differential equations. As remarked earlier, the foundations were laid by Bliss, with more recent applications and contributions by Drenick, Tsien,..., among others. The mathematical contribution to the calculus of variations is in the numerical procedures for solving two- (and more) point boundary-value problems on high speed computers [1099, p. 13]." Faulkner wrote several more papers along these lines in the following years. In 1961, he advised a master's thesis in which the adjoint system was employed in the treatment of the so-called Bang-Bang problem in control theory [1228]. The same strategy was used in problems of optimal ship and submarine routing; in these cases, the specifics of the particular digital computer and number of iterations were given. In some instances, the papers appeared in journals; others remained in Technical Report status with lengthy distribution lists. One interesting Research Paper [1100] involved a comparison of two computational methods of arriving at optimum solutions in the problem of Bolza with no specific application in mind.[37] In this manner, a problem in the calculus of variations stated in 1913, to which Bliss and others devoted so much theoretical effort, entered the digital age.

One wonders what Forest Ray Moulton, could he have become accustomed to in a world with radio-guided missiles and "high-speed" computers; would he have thought of these investigations? Would he have been concerned about the lack of labeling of the functions involved as continuous, differentiable, or analytic? Would he have been appalled by the conviction of the author that his computational schemes would always converge, let alone to an extreme for the problem? Would he have cast about for the highest-level analysis volumes relevant to the problem, coming across, say Hestenes' [1169] or Pontyagin's material? One hopes that he would have seen these papers as opening a door to further work in spite of their purely mathematical limitations.[38]

---

[37] In some of these papers what was called the Fundamental Equation by Bliss, the equation linking system and adjoint variables, is called Green's theorem.

[38] These limitations were addressed in Faulkner's 1969 PhD thesis from the University of Michigan [1101].

Similarly, what would Bliss have thought of computerized searches for solutions to the Bolza problem? The Mayer problem, whose solution by Hadamard was likely the source of Bliss' adjoint method, was a special case. Would Bliss have found useful or even accepted the methodology of an computer-aided iterative search for a Mayer problem solution, while appreciating the broadening of his method to control and optimization problems?

It would be unwise to infer from this limited collection of reports from a single military institution that Bliss' idea had a great presence in the mathematical literature of the Cold War era or beyond. There is evidence that it was considered one of several "methods of adjoint systems," to cite the title of a 1964 NASA Technical Note [1185], being collected for use in applications to engineering problems; Bliss' *Mathematics for Exterior Ballistics* was listed in the bibliography. Faulkner's work formed a chapter of a 1962 book from Academic Press on optimization methods [1212]. But further examination of the history of this type of application would be the beginning of a new story, the story of the reinvention of a technique and its gradual regeneration, its passing from one era into another, and its initiation into a new form of life. That might be a subject of a new story, but our present story is ended.

## Archival Materials (Abbreviations)

[Veblen Papers] Oswald Veblen Papers, Archives of the Library of Congress, Library of Congress, Washington, D.C.

[NARA Aberdeen] Record Group 156–866, National Archives and Records Administration, College Park, Maryland

[NARA Washington] Record Group 156.7.18, National Archives and Records Administration, College Park, Maryland

[NARA Ordnance School] Record Group 156–1512A, Box 2, National Archives and Records Administration, Philadelphia, Pennsylvania

[NARA ENIAC] Record Group 156–1512A, Box 9, National Archives and Records Administration, Philadelphia, Pennsylvania

[Columbia University] University Archives and Columbiana Library, Columbia University in the City of New York

[Princeton University] University Archives Project, Seeley G. Mudd Manuscript Library, Princeton University, Princeton, New Jersey

[MIT] MIT Institute Archives and Special Collections, Massachusetts Institute of Technology, Cambridge, Massachusetts

[University of Chicago] Special Collections Research Center, University of Chicago Library, Chicago, Illinois

[Concord] Hoar Family Papers, Concord Free Public Library, Concord, Massachusetts

[Harvard] Office of Harvard War Records, Harvard University, Cambridge, Massachusetts

# Chapter 5
# Broader Historical Issues

## 5.1 Introduction

The preceding four chapters have given an account of the technical aspects of the differential variation work of Forest Ray Moulton, Gilbert Ames Bliss, and their followers during and after the World War. Aspects of its reception in the military and mathematical communities have also been considered. It remains to pull back from the particular case of these concepts and their reception and ask if one can distill any lessons which can be examined in the larger context of the history of mathematics in America during these years. In this chapter, such a focus is attempted.

In particular, we look at the attitudes toward the application of mathematical ideas in two contemporary engineering communities in America, the mechanical and civil engineering groups. This allows a comparison of the military absorption of ballistic ideas with similar confrontations with mathematics in parallel settings. The ongoing research programs of Bliss in calculus of variations and the functional calculus are put in the context of other such programs in the American mathematical community in the post war period, as described by Karen Parshall in [1643]. The idea of the Aberdeen/Washington ballistics group as a school and a research center are investigated. Comparisons with accounts in the important volume [1349] are made.

The attention paid to the engineering material may seem out of place, but one can note that the military itself had engineering needs; needs were addressed during the War by the Corps of Engineers. These men were recruited from West Point and leading civilian military colleges where one imagines that new developments in the disciplines were transmitted to all. The engineers were detailed to many tasks, some of which involved, for example, assembly and maintenance of railroad equipment and transport and assembly of military bridging strong enough to support



A. Gluchoff, *From Fréchet Differentials to Firing Tables*,
https://doi.org/10.1007/978-3-031-68267-4_5

rail traffic.[1] It is thus arguable that accounts of mathematics within these disciplines can stand side by side with accounts of mathematics in ballistics, trends in one being reflected in the other.

## 5.2 Mechanical Engineering and Mathematics in America, 1870–1910

> Most commentators agreed that the technical graduate [of a mechanical engineering program] generally had a higher level of mathematical sophistication and skill than the uneducated mechanic in the shop, but there was sharp disagreement on whether he was able profitably to apply what he knew to practical problems in the shop. The mechanics' magazines of the 1870 to 1910 period were full of folksy stories about the smart-aleck young technical graduate who spouted mathematical terms and berated practical mechanics for their lack of theoretical knowledge, but who could not himself perform the simplest of mechanical jobs. Frequently, but not always, it was implied that in addition to being ignorant of practice the college-graduated engineer had his mathematics and theory garbled. Always, he was rescued (just before blowing up an engine or ruining a machine) by a grisly, greasy old mechanic who solved the problem with the direct application of a basic scientific law which he had, of course, learned the hard way - by experience.
> —Monte A. Calvert - The Mechanical Engineer in America, 1830–1910 [1418, p. 60]

> The infantry and cavalry sharpshooter, firing at the target range, has ammunition and a rear sight which has been prepared for him by the expert in that line of business. He does not have to take a book of logarithms and of tables and compute the value of C for any variation of barometer or thermometer, or other changes in any of its factors. He learns by *practical experience* that, under certain conditions of temperature, light, position or what not, he must vary his elevation from day to day to make good shooting. In the course of a time of greater or less length he learns to almost instinctively give somewhere near the proper allowance for the first or trial shot, and to make the proper alteration to correct any inaccuracies observed in this trial shot, and once on the target he can stay there indefinitely. Admitting at once that heavy gun practice does not agree exactly with small arms firing, particularly in the cost of ammunition, and that for the latter reason if no other every shot should be made to tell if possible, the question arises how to make this a possibility? ... logarithms and ballistic tables have about as much place in action in my opinion as glass houses. I think that in action the computation that must be made should be done by reference to tables that have been computed and compiled by the expert long before; that they contain any and all of the possible conditions so arranged that they can be readily referred to...
> —First Lieutenant C. D. Parkhurst, 4th Artillery, quoted in "The Theoretical Instruction of Gunners," James Chester, [1431, p. 200], 1892

These sketches show similarities between two settings in which mathematics became more prominent in the years given. The first shows how raising the status of mathematics in the mechanical engineering community, corresponding in years to the first wave of ballistic reform, was received in some circles. This account appears in the highly regarded study [1418], a work which, among other debates,

[1] See [1695] for some photographs of World War era military bridging by the British forces.

highlights a conflict in educational methods between an older, more entrenched "shop culture" of engineers and the new "school culture" of technically educated students of engineering in America. The second sketch reminds us of the first wave of ballistic reform discussed in the Preface and first chapter of the present work, where a debate regarding on-the-fly calculation of ballistic quantities during battle was highlighted. Too often in the literature of World War era ballistics history one never hears of this wave, the first in America in which mathematics began to play a significant role, however "clumsily" presented. It appears only as a defective theory, to be superseded by those of the professional mathematician. It is not appreciated that the first wave itself was considered an installation of "experts" of its own, to whom respect was paid, even if predictable reservations were felt.

Each scenario related the difficulty of applying mathematics (or science, whose language is mathematics) in a given instance of urgency for which it is deemed an appropriate tool-firing heavy artillery against a target or preventing an imminent mechanical failure. In neither case is the need for these tools ultimately denied. In the first case, the author later notes that the decision facing mechanical engineers of the given era was one of choosing whether education in the sciences should be the province of the experienced shop man or the institutions promoted by the "school" movement. In the second case, the lieutenant is willing to concede that firing the heavy guns cannot be a matter of experience only but that the procedures and tables need to be made up ahead of time by an "expert." But in each case, one senses a mild discomfort with mathematics, be it in the form of a wildly inappropriate depiction of its use by a sharpshooter or in the comic blusterings of a school graduate.

It may seem odd to compare developments in exterior ballistics with those of mechanical engineering. It would seem that this subject could more easily be described as belonging to physics, as was indeed claimed by some of the critics of Moulton's work. But a case can be made for the parallel treatment in these early years. In browsing the journal articles in the professional publications of the mechanical engineers, one notices an increasing use of calculus in the decades delineated above, an increase mirrored in the textbooks and professional journals such as CAJ/JUSA. In some sense, perhaps impossible to define explicitly, the engineering articles evince a certain style, reminiscent of the of the "milieu" referred to in the first wave literature, with their somewhat casual use of average values and other approximations. Furthermore, the early expositions of exterior ballistics such and Benton's, from West Point, began by devoting significant space to the task of measuring the ballistic quantities, such as the velocity of projectiles, by machines like the gun pendulum and electro-ballistic devices. Six pages [1365] (pp. 390–395) are devoted to the explanation of the West Point Ballistic Machine, a device created by the author. Even as late as 1915, chapters were dedicated to measurement of velocity, time, and quantities related to interior ballistics [1594]. Ingalls taught mathematics at several institutions prior to taking his post at the Coast Guard Artillery School where he also taught engineering in 1988. In one odd case, an important paper in the first issue of JUSA (described below), on the effect of wind on the ballistic trajectory, found its preliminary version appearing in the School of Mines Quarterly, a publication of the Columbia University's School of Mines. (No military journal existed at that time.)

Calvert traces the beginnings of the shop culture to independent shops in the early 1800s which made parts for the steam engine, the symbol of the industrial revolution. "The large, relatively specialized machine shop came into its own in the 1840 to 1860 period." (p. 5) He notes that contrary to what one might expect, the shop culture was one of information-sharing, resulting in a "vast, mutually owned store of knowledge and experience closely akin to a body of scientific knowledge." (p.7) Engineers emerging from this system retained the knowledge of the technicians with whom his experience was shared. The apprentices in such shops often came from "upper-class and elite groups in American society" (p. 12). The introduction of railroads also increased the need for mechanical engineers, and the use of steam engines in naval vessels forged a close connection between these engineers and this branch of the military. The gradual entrance of scientific knowledge into the shop culture is described by the careers of several key figures. "Not only did scientific studies liberalize the mind but they could be brought to bear upon the solution of otherwise difficult or impossible problems." (p. 35) ... "[One such figure] devoted himself night and day to mathematics, natural philosophy, and astronomy, particularly to mathematics, which he regarded as all-important. ... [He felt that] mechanics were put at a disadvantage by their neglect of mathematics." (p. 35) Thus as early as the 1830s, the time described, the stage was being set for the conflicting attitudes of engineers toward mathematics. "Stressing science was one way of expressing concern over the status of the mechanic, and those who did so were interested in raising that status." (p. 37) Among the institutions which turned out scientifically trained mechanical engineers was the Naval Academy.

Robert H. Thurston, a member of a wealthy New England family who graduated from Brown University in 1856, and who also served in the Navy during the Civil War, was one of the early figures in what came to be known as the "school" culture. He became an instructor in natural philosophy at the Naval Academy and, with only limited early exposure to the shop culture, he "rapidly became convinced that a thorough technical education was the first prerequisite to becoming a successful mechanical engineer." (p. 46) Calvert cites the Morrill Land Grant Colleges Act of 1862 as a push to the school movement, allowing curricula in mechanical engineering to be set up using funds from governmental grants. Some of these programs fared well, but Thurston's importance stems from his involvement in the Stevens Institute of Technology, a privately funded technical school in Hoboken, New Jersey, founded in 1870. He was wooed away from the Naval Academy to teach there, and "from 1871 to 1885, Thurston devoted his life to creating a mechanical engineering curriculum that could serve as a model for the pure technical school or professional school with no admixture of general cultural education. The establishment of Stevens Institute of Technology, especially devoted to the training of mechanical engineers, added greatly to the rationale for such programs at other universities, colleges,and technical institutes" (p. 49).[2] A list of

[2] Calvert notes the importance of the Navy to these new institutions: "The success or failure of many schools in mechanical engineering depended on the naval engineer assigned to them." (p. 50).

universities, colleges, and technical schools which followed in the wake of Stevens by inaugurating similar programs is then given.

The following passage from Calvert's study has some points of contact with exterior ballistics in America:

> Mechanical Engineering emerged in Europe before it did in the United States, and it was also in Europe that the theoretical basis was created, much of which was, of course, not specifically mechanical engineering, but basic physics. ... Thermodynamics, or the science of heat and energy, was first explored by French and German engineers and scientists. Strength of materials was another subject of direct concern to the mechanical engineer, and here the basic work was done by Navier, a Frenchman. ... A Scottish engineering professor, Rankine, studied fatigue in metals, and the basic science involved in the workings of the steam engine. America had no scientist-engineers to rival these pioneers until late in the nineteenth century. ... American engineers recognized this debt to Europe for basic principles in their science. *Engineer* in 1860 complained that the United States had no first-class theorists in the field of mechanical engineering. ... Even the published works of the European scientist-engineers were not used extensively in America until the educational institutions forced them to the attention of a new generation of college-trained engineers. [1418, p. 53]

One notes that in exterior ballistics as well, the theories prior to and during the first wave of reform were in debt to the masters of European study. It was not until the appearance of Ingalls' book, which introduced the Siacci theory, that attention of the officers was forced on these formulations. Hamilton's books, appearing in 1908, were the first stirrings of America's equivalent of independent theoretical and practical exterior ballistics. Furthermore, as Calvert notes–

> There was, however, one tool which was necessary if one was to read and understand these basic works. Higher mathematics - algebra, geometry, and, increasingly throughout the latter half of the nineteenth century, calculus - was required to cope with engineering literature. Rankine's *The Steam Engine and other Prime Movers* (1859), the first systematic synthesis of steam engine theory, demanded of the potential reader a relatively high mathematical sophistication. This is particularly significant since mathematics was, from the very first, the only area of engineering knowledge and skill which the technical school could claim as its special province. ... Although shop culture maintained a high level of sophisticated experience in the actual building of machines and engines, mathematics, other than simple addition, was something that the shop could not or did not wish to take responsibility for in the training of mechanical engineers. As calculus and other areas of higher mathematics became increasingly indispensable to the practice of mechanical engineering, the educational complex gained in prestige. It was the common denominator of all serious efforts in technical education in mechanical engineering. As *Technologist* noted in 1871: "In all institutions which afford a scientific preparation for the different departments of engineering, the differential and integral calculus appears as one of the studies of the course."... *American Engineer's* editors thought in 1881 that higher mathematics was indispensable to the practical engineer. Those who thought otherwise would, in the editor's opinion, set up a "real barrier to progress." There was, however, no universal feeling that the necessity for mathematics was a good thing for mechanical engineering. [1418, pp 54–55]

There follows a description of a debate on the merits of mathematics in the mechanical engineering community, a debate whose contents closely resemble those of the military personnel in the mathematical ballistics debate.[3] If one recalls that the founders of the Coast Artillery School regarded the institution as a post-graduate facility, West Point being required for entrance, one can conceive of it as playing an analogous role to the "school culture" of the mechanical engineering debate, with some evident differences. The "school culture" was attempting to create a demand for its educational services, especially in mathematics, which could not be supplied by the shop culture; the Coast Artillery School was not in competition with West Point in this regard. The West Point texts did in some cases assume an elementary knowledge of calculus, but the Coast Artillery publications increased the sophistication of this material in their presentations of the Siacci theory as well as in articles. Thus in the sense of the level of sophistication of mathematics used, the higher level ultimately desired by the school culture of the 1870–1910 period is comparable to that in the first-wave reformers' literature.

This can be seen in the 1914 *Report of the Committee on the Teaching of Mathematics to Students of Engineering* [1696], issued by the Society for the Promotion of Engineering Education, a group founded in 1907. Among its members were Philip Alger, a professor of mathematics at the Naval Academy,[4] the president of Worcester Polytechnic Institute, several professors of mechanical engineering from universities, Charles Steinmetz, the outstanding electrical engineer from General Electric, Edward V. Huntington and William Fogg Osgood, mathematicians from Harvard, and several civil engineers. Its contents begin with algebra and range through calculus of one variable. Though accompanied by a healthy discussion of its merits, again typical of the engineering publication format, it stands as the agreed-upon goal of its school-oriented authors.[5]

Benton uses calculus in his exposition of the ballistic trajectory, first in a complete treatment of the motion in a vacuum and then in an elementary derivation of the elements of the trajectory assuming a polynomial resistance law. Ingalls' book, by contrast, uses calculus quite liberally, for example, by deriving an integral

[3] It should be noted that the calculus referred to here as well as that in the ballistic context is one-variable calculus only, as an examination of the journals of mechanical engineering shows. That of the writers of the first and second wave as well restricts itself to one-variable calculus, although the latter deals with systems of ordinary differential equations. As a side comment, in his excellent account of the activities of Oswald Veblen in the World War, Grier [1502] mistakenly refers to the Siacci theory as a subtopic in partial differential equations.

[4] Alger wrote a naval gunnery textbook for the Academy [1342] which included a computation of ballistic tables "Revised and Extended to Include the Formulae and Methods of Colonel James Ingalls, U. S. Army."

[5] For an example of a publication by a "school" engineer, one can cite [1456] in which the author, H. T. Eddy, a 1872 Ph.D in mathematics from Cornell and a graduate of Yale's Sheffield Scientific School, used one-variable differential calculus to determine the most economical use of a unit of steam in a steam engine in terms of ratios of compression and expansion. This result was referred to as "Eddy's theorem" by some sources. For more on Eddy, see [1641, pp. 317–318]. Also, it is not to be inferred from above that the shop culture eschewed the calculus. George B. Grant, the "Father of the American Gear Cutting Industry," with shops in Lexington, New York and Philadelpia, used it in his books on the theory and practice of gearing and in several publications.

formula for the pressure on a shell of arbitrary shape, explaining the electric fence mechanism used to determine a more accurate resistance law, and extensively in the Siacci theory. The contemporary Navy volume [1613] shows a similarly intensive usage but sprinkles in an occasional MacLaurin series. By contrast, a 1915 West Point text [1594] retains the modest use of the tool. The full flowering of mathematics in ballistics by the first wave occurred with Hamilton's books, as previously noted.

None of the works mentioned uses calculus to discuss differential variations. A most interesting paper in this regard, by John W. Ruckman, is [1667], the first paper of the first issue of the JUSA. This work was an extension of the previous effort by Lieutenant Ruckman from 1890 appearing in the School of Mines Quarterly. (Ruckman was an editor of JUSA.) The contents treat the effect of wind on the ballistic trajectory, and it appears that this is the first original American military approach to the problem of a differential variation. (The method used in [1613], for example, is borrowed from Hélie at l'École d'Artillerie de la Marine in an 1874 pamphlet.) A fine example of a first wave effort, it features a geometric method which is extensively described; calculus is also used. The method incorporates other factors such as atmospheric data, and tables of variations are computed. Typical of these men, as previously noted, is the familiarity they have with their weapons, which is nearly always used in examples. Ruckman states: "The preceding discussion is perfectly general and therefore applies to all projectiles. The illustrations are taken from the Springfield Rifle for two reasons. (1) When the discussion was developed two years ago our practical knowledge of this gun was better than that of any other. (2) At that time it was impossible to obtain any reliable data relative to any other gun in our service" [1667, p. 12].

In conclusion, it is hoped that the parallel examples of "school" culture in mechanical engineering and the first wave of ballistic reform demonstrate the American ambivalence and ultimate acceptance of mathematics in the period preceding the work of Moulton and Bliss. Even as basic a tool as calculus of one variable had a slow embrace during this era. Under these conditions, the work of the professional mathematicians, with their use of general systems of differential equations, and contact with weaponry confined to numbers used for calculation of trajectories by numerical integration, must have appeared quite foreign. Ultimately, in both settings, the use of higher mathematics became a permanent fixture. In a final nod to the school movement, the second wave of ballistic reformers contributed a mathematical solution to a mechanical engineering problem—T. H. Gronwall published in 1919 an analysis of stress distributions in a keyed cylindrical shaft [1508]. The technique of analysis involved potential theory and complex variables and the use of the Saint Venant theory of torsion. Details can be found in [1481], where it is noted that in 1909 a report was issued from the University of Illinois Engineering Experiment Station describing torsional tests to which various keyed columns were subjected.

## 5.3 Empirical and Rational Knowledge, Column Formulae

The profession of civil engineering is teeming with problems awaiting the solutions of a St. Venant, a Laplace, a Newton. ... The profession of building is the oldest in the world and yet how little we have progressed in the analysis of the stresses in the frame itself. The knowledge of stresses in a column remains practically in the state in which Euler left it. There are, indeed, some empirical column formulas, but an empirical formula is always a sad apology for lack of scientific knowledge. It may be said that empirical formulas have sufficed in the past for design, but the failures of structures designed on that basis most effectively answers that. The failure of the largest bridge in the world–the Quebec bridge–was due to lack of knowledge of the action of large compression members (columns) and the only path open to engineers was that of experimentation on larger size test pieces: mathematicians had failed them. ... When we come to the analysis of structures proper, there are either empirical formulas or faulty analysis. The subject of moving bodies with their resultant impacts and vibrations is at present hopeless–even the empirical formulas fail us here–and we must make up for lack of knowledge by excessive and wasteful strength. ... It may be visionary dreams of the wildest type, but the day may come when, from the knowledge of the chemical constituents of a body, its crystallographic history and the temperature and space element, its stress and strain relations shall at once be known and the effects of external loadings, static or otherwise, at once interpreted.

—George Paaswell, C. E., "An Appeal to Producing Mathematicians," 1911 [1639]

Formerly, the comparative inexpensiveness of artillery firing, together with the large element of guesswork, which professional sentiment then permitted, caused this subject [the effect of the wind on the ballistic trajectory] to be entirely neglected. Now however, when every discharge of a modern sea-coast gun costs from one hundred to one thousand dollars, we cannot afford to waste our ammunition, and every effort must be made to remove all causes which introduce inaccuracy of fire, or to make allowances for them. The wind problem having already assumed such importance in our work today, it is perfectly clear to all observing minds that it must become more important with the increasing demand for higher efficiency. ... In dealing with it some have restricted themselves to theoretical considerations, and upon various assumptions have founded their formulas. Others, turning their attention to the practical side of the problem, have endeavored to obtain sufficient data for firing tables. None within the writer's knowledge have thoroughly combined these two methods, which in all matters of investigation are so essential to each other, and from which combination only, the best results can be obtained. ... In attempting the solution practically, the difficulties are almost insurmountable. The want of adaptability of what few instruments can be obtained for the experiments; the difficulty of placing them where they will give reliable results; and the many other discordant considerations make the problem almost impossible of solution in this manner. ... In order to obtain the necessary data an independent series of experiments must be had for each gun in the service, and under all the different conditions of fire. In this manner the solution of the problem would be well-nigh impossible. The laws of deviation, or rather the law of generation of the deviating force, if once discovered and expressed in analytical form, could soon be verified with reasonable accuracy by experiment. However, so long as the observed deviations do not agree with the theoretical ones, we can come to but one conclusion - that the theory is not a correct one.

—Second Lieutenant John W. Ruckman, "The Wind Problem in Gunnery," 1890 [1666]

The two passages above, one from the civil engineering profession and the other from the military, strike common themes. In each we find the concern with two types of knowledge: that gained from experiment, which we shall call empirical, and that based on logical reasoning from first scientific principles, which we shall call rational.[6] In the plea to mathematicians made by the civil engineer, one sees an extreme disparaging of empirical explanations and a desperate call for the current crop of research-oriented mathematicians to turn their efforts toward rational explanations of civil engineering problems. By contrast, the lieutenant acknowledges that both empirical rational explanations are necessary but laments the near-impossibility of the former in the wind problem in gunnery. Ruckman's paper was followed by the 1892 article in JUSA mentioned above, in which his graphical method was proposed to obtain some degree of theoretical insight into the problem. In each case, a need for efficiency in one form or another is voiced: wastefulness in design on columns is to be avoided and economy of design for structures is desired, and the previously documented costliness of firing a sea-coast gun is repeated. The civil engineer even expresses a dream of determining the stresses and strains in a column based on limited knowledge of the piece alone, recalling Moulton's vision of firing "a few shots," after which no more are necessary to determine the performance of the gun.

These themes will be discussed in this section in another parallel treatment of engineering and artillery fire control in an attempt to bring out a contemporary dichotomy affecting each effort, the roles of empirical and rational explanation.[7] One explanation for the playout of these two philosophical modes in America is the association of empiricism with "American Baconianism," pragmatism, and practical science, with rational formulae having been tied with more prestigious European science, which Americans were attempting to absorb and copy. The American character may be said to contain a "self-created identity of being a practical people who made machines that worked and conquered the wilderness" (Amy Ackerberg-Hastings, email to present author, 12/29/2020) These distinctions were relatively new in early nineteenth century America but became more sharply delineated as the century progressed. Paaswell, for example, by the early years of the twentieth century, was clearly dismissive of the overreliance on experiment and desirous of rational treatment of column strength. Though the roots of this dichotomy go back to at least the late eighth-century Scotland, when John Playfair, for example, dealt with it, and the work of Scottish engineer Rankine was affected by it, the predominance of American empiricism came under question in the later era, when engineering became professionalized, as discussed in the previous section. It will be seen that, far from remaining philosophical terms, "empiricism" and "rational knowledge" were

[6] These definitions are made in the spirit in which one encounters them in the literature herein considered. The purely philosophical definitions and the history of their development are out of place here, though occasionally addressed by professors of engineering or science. One example is [1717], by Edwin B. Wilson, mentioned in connection with aircraft design in the previous chapter. He discusses in general terms the role of experience of the senses, *a priori* innate conceptions, idealization, and the place of mathematics in a lengthy 1926 address. The context is the use of statistics in the design of experiments.

[7] The author is indebted to communications with Amy Ackerberg-Hastings and Ellen Abrams for the contents of this paragraph.

explicitly used in the civil engineering literature, though not so much in discussions of artillery firings. The era of industrialization, with its demands for efficiency, put a sharp edge on these modes of knowledge, where experimentation and mass testing of column strength were costly.

Turning first to our topic of differential corrections, to which Ruckman's 1890 comments are addressed, one can first recall from the previous chapter the involved back-and-forth movement between empiricism and rational explanation in the history of the French ballisticians at the Gâvre. Nothing as drawn out or explicitly identified as such occurred in American institutions, which were younger; the establishment of the Coast Artillery School and Ingalls' installation as organizer of the ballistic program may be said to be the first attempt at organized rational discourse on the subject. Given the difficulty of experimentation on differential variation, one might expect a rational treatment of the subject to dominate, with greater sophistication as the century progressed, and this is borne out by another examination of the texts. The empirical nature of the treatment of differential variations by Benton has already been noted; no mathematics is present here. Ingalls' book has no rational discourse on the variations but, for example, uses an empirical formula due to Poncelet [1559, p. 9] as the basis of his calculus-based formula for pressure on the surface of a shell. He does mention the difficulties "both experimental and analytical" [1559, p. 26] in determining the resistance law, difficulties which plagued all ballisticians including Moulton. The early first-wave naval effort [1613] is similarly devoid of the use of mathematics in the problem of deviations.[8] It was not until Hamilton's 1908 volume, subtitled "Analytical Exterior Ballistics" [1525], that a calculus-based treatment of the variations was presented by the first-wave reformers, though Ruckman's 1892 graphical effort [1667] probably marks the first attempt at a rational method.

With the arrival of Moulton on the scene, we have not only a solid endorsement of the combination of theory and experiment but is implementation. Enough has been said about Moulton's bolstering of rigorous analytic argument and disparagement of slipshod reasoning, but it might seem that his endorsement of empirical processes would be hesitant. One can recall his disgust on finding that "...range tables [of his predecessors] were [sometimes] based on large numbers of experimental firings at various angles of elevation and were largely independent of any mathematical theory. ...The ballistic tables became as a matter of fact not much more than interpolation formulas for supplying results in the intervals" [1628, p. 10].[9] Also, it

[8] One finds, however, the following rationale for their study and application: "In these days of high-powered guns and high speed of modern ships, much valuable time would be lost in trying to find the amount of the corrections necessary to apply to the range and direction of flight by trial shots alone, especially if the guns and target are in motion and if a ship's battery is composed of guns of different calibre and class; and it is believed that with expensive ammunition, of which but a limited supply can be carried, ...a great saving would result if approximate corrections were applied to the first trial shot..." [1613, p. 53].

[9] In the 1925 published version [1677] of his 1920 technical report [1670] Philip Schwartz's assertion that this practice was still taking place at Aberdeen remains.

is to be cited that some of the reviewers of his *New Methods* were complimentary about his discarding of certain empirically based results. On the other hand, when viewed in its entirety, his embrace of experiment is more than evident, as is the measured presentation of rational and empirical formulae. In [1631], his discussion of the resistance law, for example, is notable for its "rational" discussion of the form that must be taken by this function before his admission that "The theoretical difficulties are so serious that at present [the function] can be found only by experiments" [1631, p. 36]. Similarly, the exponential density function is supported by a "rational" argument, but the next paragraph admits that it "is subject to certain errors resulting from the fact that the assumptions under which it was derived are not exactly fulfilled in experience" [1631, pp. 49–50]. As for the differential variations, a telling comment made in response to Hoar's review of *New Methods* can be repeated: "In my first paper, written shortly after I entered the service I seized the problem of computing differential variations and pulled it out of the fogs in which it was involved and placed it on a solid foundation" [1633, p. 321]. This remark implies that, in spite of the mixture of empirical and rational knowledge which went in to the equations of motion themselves, he believed his derivation of the corrections was strictly logical. The winds aloft were also incorporated into his model, awaiting certain on-the-spot readings for use. Government funds made possible the testing of some of these formulae, as reported in 1924s report by Schwartz [1673].

Perhaps the most ambitious experimental program initiated by Moulton's group were the wind tunnel studies of the motion of projectiles and its dependence on the shape of the shell, some of which is described in [1346, p. 246]. One could easily add to that account one of the objects of study which is analogous to the consideration of differential variations: the attempt to determine how each property of the shape determines the range:

> The range which a projectile will have with a given muzzle velocity can, of course, be determined by firing. The results obtained, however, are the integrated effects of all losses of energy. Moreover, the experimental firings give no direct information regarding what feature of the projectile causes the principal loss of energy. For these reasons it is important to have some means of determining the instantaneous, instead of the integrated, effects, and of finding what properties of the projectile are involved in the losses of energy. [1628, p. 66]

These investigations involved enlisting the General Electric Company of Lynn, Massachusetts, to provide time on their commercial blowers. "It is believed that fundamental facts can be established in this way much more cheaply and certainly than by firing projectiles by guns" [1628, p. 67].

In spite of these examples, it is perhaps not surprising to find evidence that purely empirically determined results were still used in the services and mathematics criticized. For example, in an article [1713] in the first volume the the journal *Army Ordnance*, dating from 1920, one finds a report on "Long Range Small Arms Firing," the outcomes of range firings at Borden Brook Reservoir in Massachusetts. The entire first page [1713, p. 773] of this piece is taken up with a cautionary tale which might very well have been aimed Moulton's new ideas in ballistics by questioning the numerical methods of celestial mechanics. The author begins by stating that the discovery of the planet Neptune by Leverrier in Paris and Adams in England may be regarded as the "greatest triumph of mathematical astronomy." He continues:

> However, wonderful as this mathematical achievement is, it is marred by an unfortunate error which occurred in the computations of both men. In addition to determining the planet's position in the sky, they also had calculated the elements of its orbit and a value for its mass, which were greatly in error. [A reason was given for this error, an inappropriate assumption having been made.] Their supposition, however, was justified by the available data at that time ... Nevertheless the computed elements were so much in error that several authorities have contended that the discovery of Neptune was a "happy accident." ... Another conspicuous example of the wonders of mathematics is in its application to the determination of the path which a bullet travels through the air. Mathematics is an exact science which is sometimes erroneously held to be infallible and of unlimited application without regards to facts. ... Some enthusiasts of no mean mathematical ability let their confidence get the better of their judgment and proceed on a policy which might be stated as follows: "When an experiment disagrees with mathematics, repeat the experiment until it does agree."

This suspicion of the use of mathematics is given as a preamble to the simple statement that for larger guns the "assumptions" needed to appropriately use mathematics are present in the form of experimental determination of the resistance law whereas for small arms firing no such law is available. This justifies the use of firing data only for the determination of the ranges.

Let us turn now to the "Appeal to Producing Mathematicians," a note which appeared in the *Bulletin of the American Mathematical Society* in 1914. Exploring the topics listed in this plea leads to some interesting material produced by the American civil engineering community during the years from 1880 to the first decades of the twentieth century. Motivated to some extent by the catastrophic collapse of the Quebec Bridge in 1907, the author scolds the mathematical research community for its lack of attention to problems of construction. In the process, he calls attention to various means of determining the strength of columns and focuses his ire on the reliance of empirical results, in this case formulae which arise as a result of testing the crushing and bending strengths at experimental facilities. He asserts that mathematicians had failed to provide adequate scientifically based formulae to aid in understanding the nature of column stresses.

It is surely true that this subject was of little interest to most of the addressees of the indictment, but the topic was in fact vigorously mathematically explored by the engineers in the era in question. To be sure, the level of mathematics involved was once again limited to calculus of one variable, and nothing approaching the theories of elasticity of Saint Venant, for example, was pressed into service. But a look at the papers published is enlightening for its revelation of the balance between empirical and rational explanation and its parallels to ballistic issues. Throughout this discussion, we shall use *Columns* [1668], a 1921 volume by Salmon, as a contemporary reference.[10]

[10] This work was the author's thesis for the Doctor of Science in Engineering at the University of London and is a comprehensive account of the literature up to the years immediately preceding its publication.

If the introductions to American treatises on the subject are any indication, an interest in rational explanation for the strength of columns was beginning to make itself felt at least by the early 1880s. Thus in Cousins' Beams and Columns [1436], whose author had been a professor of mathematics at the Virginia Military Institute, we find–

> During the present century much time and means have been expended in attempts to solve, experimentally, the problems that have engaged the attention of the mathematicians and as a result of their labors we find such experimentors as Hodgkinson, Fairbairn, and others, whose names are household words in the literature of the subject, adopting *empirical* rules for the strength of Beams and Columns rather than *rational* formulas deduced by the scientists. [1436, pp. iii-iv]

The author aims to give a rational account of the subject.

> The great practical benefits to be derived from the correct theory of the strength of Beams and Columns will be evident, when we consider the countless tons of metal that have been made into railroad rails, rolled beams, and the other various shapes, and that the manufacturers were without knowledge of the work to be performed by the different parts of the beam or column in sustaining the load that it was intended to carry. ... The correct theory will enable them to foretell the strength of any untried shape, and the reason for the strength of those that have been long in use, which is the "true object of the theory."

Again one sees the often repeated vision of prediction of behavior by theory alone. Another author, a professor of civil engineering at Columbia University, presents a text [1415] in which the first part consists of a "Rational" discourse, stating "The writer would esteem himself fortunate if the mathematical portion of the book should find favor with such individuals and be useful to them" [1415, p. iv]. Variation among materials tested is noted: "But it has been the aim of the author ... to show not only the variation in products of different mills but the variation in different products of the same mill; to exhibit the variations due to the difference in size, shape, relative dimensions and conditions of specimens; to show that specimens apparently identically the same may even give considerable diversity in results..." [1415, p. v].

The strength of columns has a long history and an extensive literature; the latter up to 1920 occupies a section of Salmon. A short version can be found in [1587, pp. 324–333], an 1885 textbook by a professor of theoretical and applied mechanics at the Massachusetts Institute of Technology and in the 1886 paper [1574] by a civil engineer. The modern mathematical aspect begins, as in the case of exterior ballistics, with Euler, who in 1757 gave a rational formula for the strength of long columns, often called the "Euler crushing formula." It gave the "crushing load" of a column in terms of certain measurable features of the column. This proved of limited use to engineers and to be based on questionable hypotheses as well; the builders were thus forced to rely on experiment. Euler's formula, however, continued to be of interest; indeed it is still presented in modern engineering courses. A key set of experiments was made by Eaton Hodgkinson in 1840 for the Conway and Brittania Bridges; the columns used were iron, and again the maximum load before failure was the quantity of interest. From the data obtained, he was able to fit two empirical formulae for the load in terms of simple characteristics of the columns. The engineer Gordon looked at the data from these trials and, making certain assumptions, derived a failure formula which met, within reason, all of Hodgkinson data. Rankine modified Gordon's formula, which thus became known as the Rankine-Gordon formula. According to [1587, p. 325], as of 1885 "The formulae most frequently used in American practice are those of Gordon." As in the case of Euler's contribution, this result remains of current interest. (Some sources refer to the Rankine-Gordon formula as simply Gordon's formula, others as Rankine's. References for the above works can be found in Salmon's book.) However, according to Johnson [1574, p. 519] of the same year, "But engineers have gradually come to distrust all the formulas, and to believe either that the experiments have not been properly interpreted, or that iron has one mode of behavior in small test pieces, and another different one in full sized members as used in structures. To this feeling of doubt and uncertainly we are indebted for the large number of experiments upon full-sized columns which have been made in recent years..." Such experiments were performed in places as varied as the Cincinnati Southern Railway facilities and the Watertown Arsenal in Watertown, New York. It can also be noted that in [1587], following the column formulae, one finds sections on "Strength of Shafting," the mechanical engineering counterpart of column strength.

Given the importance of trusting the Rankine-Gordon formula, a great deal of attention was paid to determining its accuracy by both experimental and rational means. Indeed, the resulting efforts turned into a minor conflict of deciding once and for all whether the formula was empirical or rational, but the definition of these terms changed. The debate now centered on the relation of the constants in the formula to known, measurable, or otherwise determinable characteristics of the column in question. If such a connection could be made by arguments from first principles, the formula would be deemed rational; otherwise it would be called empirical. The size of the literature on this topic can be judged by a remark in Salmon: "It [Rankine-Gordon] has been the subject of almost as much controversy as Euler's formula. On the one hand its claims as a rational formula have been attacked, on the other its merits as an empirical expression of experimental results have been decried" [1668, p. 133]. Salmon himself concluded that the formula was not rational, or, rather, he analyzed the extent to which it could claim to be rational. His conclusion: up to a certain point in its derivation from first principles, it was indeed rational, however, past that point one of the variables in the expression "could not be expressed in terms of known factors of the problem." "If a quantity depends on factors which are not merely unknown, but which cannot from the nature of the case be determined, then it cannot be expressed in terms of known factors, and by no amount of mathematical ingenuity can it be evaluated" [1668, p. 134]. There follows a critique of all efforts to circumvent this difficulty, including attempts by American civil engineers.

The journal and book literature on column formulae by American engineers is extensive and clearly indicates a sharpening of the focus on scientific and rational discourse. A relatively brief summary of some items follows. (Again the reader is asked to indulge in this topic, whose connections with ballistic matters will be made in due course.) In 1881, there appeared in *Engineering News* a note [1414] on "Gordon's" formula, in which the author, a professor of rational and technical mechanics at Renssalaer Polytechnic Institute in Troy, New York, simply asserted that the formula had an empirical constant, proposed a method of finding it from experiment, and, using data from existing experiments, laid out explicit forms for use with different columns on the market. What strikes the reader is the number of possibilities: there are four shapes of columns considered, each having two conditions on the end pieces. Clearly more types could have been included. Among the earliest theoretical papers is an 1882 effort [1614] by a professor at Lehigh University in Bethlehem, Pennsylvania, then a large steel-producing area. Here the goal appears to be to discredit the formula both by deriving it and, in the process, proving that a constant must be empirical and then stating that the formula really only works for certain columns. In view of this alleged inadequacy, it is not surprising that alternative formulae were proposed; such was the goal of an 1894 text [1571] on framed structures. Here yet another derivation of Gordon-Rankine is proposed, a general intermediate formula achieved, which is claimed to be "strictly rational" [p. 145] (this phrase is replaced by "nearly rational" in the next edition of the book) and then reduced to the desired form. Though found "satisfactory," a simpler replacement formula which also fits some experimental data is presented. Examples of the formula are given for what is a different variety of columns, this time listed by material: wrought-iron, mild steel, cast-iron, and so on.

Throughout this list of works, one notices a common theme: they are all authored by professors of the subject "rational and technical mechanics" or of civil engineering. This is reminiscent of the "school culture" considered in the previous section, though some of these men had colleagues who were involved in actual construction projects: one of the co-authors of [1571], for example, was a bridge engineer working in Delaware.

We conclude with two papers which are of interest because they lean toward a mathematical validation of a new Rankine-Gordon-like expression. The first, [1416], from 1898, was written by William Cain, a professor of mathematics at the University of North Carolina. He was the author of a calculus textbook [1417] and had held a teaching post at South Carolina Military Academy. His goal is a new rational derivation of a modification of Euler's formula,[11] but of equal interest is a response by another civil engineering professor, this time from the University of Iowa. This response consists of a lengthy derivation of a formula for the breaking point of an eccentrically loaded column, that is, a column loaded on a point at a distance from the center of area of the end cross section: "...the writer is not aware that any correct analysis of the theory of the ideal column eccentrically loaded has

[11] In a response to a 1908 paper on steel columns [1718], in which the author developed his own theory of loading and produced new formula by largely rational means, Cain opined that the Gordon-Rankine formula is "irrational, and it is surprising that it should still be used in such problems..." [1718, p. 208].

ever been made, though it is quite possible that this may have been done" [1416, p. 108]. He defends his formula but then points out its difficulty in application and in fact "does not advocate its use" [1416, p. 112]. His presentation ends with the revelation: "Since the above was prepared the writer has accidentally learned that Mr. Carl G. L. Barth, of Scranton, [Pennsylvania], has made a rigid analysis of columns eccentrically loaded. The writer, having addressed a letter of inquiry to Mr. Barth, learns that this analysis leads to the same formula as the writer's" [1416, p. 113].

Carl Barth was perhaps the most interesting of the figures in the column debate. A native of Norway, he studied mechanical engineering at a technical school in his native country and learned one variable calculus there so well that he taught it at the same school upon graduation. Immigrating to the United States in 1880, he found employment with Frederick Winslow Taylor, the father of Scientific Management and an operator of a "shop culture" steel tool company in Philadelphia. Bath eventually became the "mathematician" associated with that group of managers.[12] On a sojourn to the Midwest in which he did some consulting on the Emery Testing Machine, a machine for testing the strength of steel samples, he met one of the authors of [1571], who induced him to study the Gordon-Rankine formula. He did so before moving to Scranton, where he accepted a position as a professor in Applied Mechanics at the International Correspondence School.[13] The result was the 1898 paper [1356]. "When I for the first time became acquainted with the development of the Rankine-Gordon formula for the strength of columns, I was much surprised to find it resting on a very unsatisfactory analysis. I accordingly set to work on an independent investigation, and soon received sufficient encouragement to go ahead, with reasonable assurance of accomplishing something. As a result, it is the purpose of this paper to submit a new and purely rational formula for the proper working load on a column, which rests on arguments to which it is thought that no exception can be taken by anyone who will try to follow them in an unbiased spirit" [1356, p. 21]. There follows a beautifully detailed mathematical argument to achieve the goal.[14]

This section must end on a note that defends the amount of attention given to the column literature. One justification is the need for a view of a larger context in which mathematics was practiced in America at this time; such a view was also provided in the first section. Connections between engineering mathematicians, industry, and universities are also evident, though not deeply analyzed. Taken as a whole, the debate shows a remarkable variety of professorial interests and concerns for rational explanation, highlighting the increasing concern for scientific discourse

[12] For more on Barth's contributions to Scientific Management, see [1575]. He attempted to organize the machine works at the Watertown Arsenal in New York in 1911.

[13] Barth had been offered an instructorship in machine design at State College, Pennsylvania, and was on good terms with professors at Lehigh University; see [1599].

[14] This effort won no great notice in the mechanical engineering journals, which by then were used to Barth's submissions on more applied subjects. One such [1330] merely states: "This is an elaborate mathematical investigation in a well trodden field."

in the equivalent of the "school" culture of mechanical engineers.[15] In the process of attempting to discredit and replace the Gordon-Rankine formula, the critical faculties brought to bear are considerable, conducted at the same level as that of the second wave of ballistic reformers. Parallels with solutions to ballistic problems are evident; once again recall Moulton's remark that "... the methods of determining the effects of such disturbing factors as winds, abnormal air densities, and the rotation of the earth were imperfect in the extreme, and in some cases even gave results of the One of 5 mentions sign" [1628, p. 10]. Experiments in the matter of differential variations were in the 1890s considered unfeasible, and the rational discourse on them, beginning with Ruckman's paper and culminating in Moulton and Bliss' work, therefore had to dominate until government spending allowed them to a modest degree at Aberdeen. By contrast, experimental data on column strength were available from the outset of the Gordon-Rankine debate, making the accuracy of competing formulae easier to judge.[16]

One may consider the list of requirements of any column formula as given by Moncreiff's publication of 1901 [1619] and compare them to the corresponding case of equations of motion of a projectile or the differential variations solutions:

(1) The theory should be based on correct principles and the formulas should be of the correct form, without introducing refinements which have but little practical value or influence on the results. [Such refinements may have been the use of elliptic functions by Greenhill to determine the equations of motion of a projectile using a cubic law of resistance, as remarked upon by Charbonnier, or the determination by Drach of all resistance functions which allow the equations of motion to be reduced to quadratures [1452].]

(2) The theory and formulas should have a wide range of application, to cover the conditions met in engineering practice, and should not be limited to the case of columns under presumably central loads. [One notes efforts by ballisticians to have equations of motion suitable for any type of shell and weapon, and the use of the ballistic coefficient

[15] To give a balanced perspective, Calvert's book also reports criticism of this culture based on the remoteness from experience and in some cases incompetence of some of the instruction.

[16] To give the experimental side in ballistics its due, we mention two articles. The first, concerning the first wave of reform, is [1479] from 1891, in which the author compares the predicted range of a projectile using both the first and second powers of the velocity as a retardation law with the actual range firings from two different weapons. The second, from the second wave, is by Gordon Hull; recall that Hull challenged the mathematical treatment of ballistic problems by claiming them for the domain of physics. He justified his assertion by publishing in 1921 an article [1555], addressed to the Mechanical and Engineering Section of the Franklin Institute in Philadelphia. Two of his four topics for discussion involved the forces acting on a projectile and the retardation law. "When we consider the vast amount of money which is expended in guns and projectiles, we feel justified in devoting time and energy to a study of scientific problems connected with their functions" [p. 32]. The experiments formulated with Moulton's input are described. In what might be considered a note of caution addressed to the mathematicians' efforts, he declared that the resistance law used to compute all firing range tables for the Army and Navy used the method of firing through screens, "... a method which has not been carried to great precision, and which as conducted in the past can lead to very incomplete results." [p. 328] Kent's work on projectile design is emphasized.

to attempt this. Also differential variations are an attempt to cover all conditions of fire.]

(3) The reasoning and the formulas should be sufficiently simple to be understood by more than expert mathematicians, and should be suitable for use in everyday engineering practice. [This can be said of Franklin's efforts to rederive the equations of variation using "elementary" calculus of variation ideas, rather than the adjoint method.]

(4) The theory must show agreement with, and be a reasonable explanation of, the results of practical experiment. Moulton's group was the first to achieve this.]

(5) The formulas should in practice be applicable generally to different materials, by simply introducing the values of the ordinary physical constants of strength and stiffness. [Again this recalls the ballistic coefficient, which could if necessary be computed from known data, except for the coefficient of form, for which some guidance by illustration was available; see Fig. 1.1.]

(6) Empirical factors should be reduced to a minimum. [Moulton attempted this, but was cognizant of the difficulties of it.] [1619, pp. 334–335]

It might be added that the most theoretical treatments of the differential variations allow for arbitrary laws of retardation and atmospheric density as well as the ballistic coefficient, and then allow for an examination of the change in range due to these changes simultaneously. This can be considered an acknowledgment that these quantities will never be known to a certainty, but at least provision can be made for any changes that occur in the state of existing knowledge of them.

We conclude with the mention of another paper by T. H. Gronwall [1507], a 1919 two-page effort in which a solid of infinite extent with a finite spherical cavity is subjected to tension and compression. The stress components at the surface of the cavity are computed "by the method of Lord Kelvin"; the material is applicable to columns. As in the example given at the end of the preceding section, the method required much advanced knowledge of "applied mathematics" and would be beyond the scope of the efforts described here.

## 5.4 Variety and Efficiency

Outside of the empirical/rational debate just described, one encounters two obvious themes when comparing the strength of columns and ballistic trajectory problems. The first is the theme of variety; surely the word has appeared frequently enough in the column material quoted above that it is evidently a factor in considering the application of these formulae. (See 2 in the above list.) The columns treated were primarily iron and steel, but the different types of iron as well as the manufacturer's design of shapes needed to be taken into account; in [1571] different woods were also discussed. Long and short columns behaved differently when loaded. The first note quoted above gave the Rankine-Gordon formula with constants adjusted for a dozen or so types of columns; the expression had to be able to handle many different types, and the theory valid for any new product, not to mention variation between individual columns manufactured by the same company. In some cases, new formulae introduced as alternatives had to be tested as well and would work better in some cases than existing ones. Would all varieties and samples have the same degree of agreement with the predicted crushing loads? Could the expressions be relied upon to predict failure with the same accuracy? Could there be unknown factors not accounted for in a new type of column?

An analogous situation confronted the ballisticians when the great variety of weapons began to appear: recall the 1907 textbook introduction "The weapons of attack and defense have become more numerous, more complicated, and vastly more efficient... The science of gunnery constantly requires of the officer greater knowledge and higher attainments." Moulton's *History* lists, for its one year of reporting, 17 types of guns for which range tables were prepared, [1628, p. 42], and 11 guns for which complete trajectories were computed [1628, pp 74–76]. This greater variety of weapons, in addition to the attempts to take into account a greater number of secondary factors, made determination of range by firing prohibitively costly; hence theory became more important. The trajectory and differential variations might be computable by formulae, which would then have to be elastic enough to be usable on all weapons with great accuracy. (Recall the difficulty of finding a range firing table for the long-range small arm pieces, for which no resistance law was available.) But, for example, it is clear that the equations of motion as set down by Moulton left out factors such as the shape of the projectile, leaving these to be dealt with much later. The phenomenon of variety was a result of the increasing industrialization of America in this era, when the scene was flooded with new types of mechanical and building products as well as weaponry.

The second theme is efficiency. Again, this word appears in many places in the quoted material as does its opposite, waste. The efficiency of a gun battery or ship in battle was closely tied to the ready availability of information regarding the behavior of a gun under different battle conditions, the cost of firing again making itself felt. Thus tables or plaques with the computed information attached to the gun were a must. In the design of columns, waste in the incorporation of unnecessary safety factors was an issue. Also illuminating is a comment on a 1901 paper on eccentric loading of columns; the comment was made by a civil engineer familiar with the construction of tall steel-skeleton buildings in cities: "This is an extremely practical subject for engineers who have many columns to dimension or design. In the Astoria Hotel there are about 3600 separate columns, each of which had to be especially considered. Ten minutes given to each would require 600 hours for all of them. In the speaker's practice every column is analyzed; if there are eccentric loads the exact effects must be determined, and it is therefore important that the methods should be such as will secure the greatest amount of work in the least possible time" [1619, p. 435]. Again, industrial manufacture made efficiency a byword of the times.[17]

[17] The word "efficiency" was of course associated with Frederick W. Taylor's Scientific Management movement, but Taylor was concerned primarily with the efficiency in the management of factories, as in his steel tool-making concerns of Philadelphia. This would sometimes entail the study of the most efficient way to operate machinery, as illustrated by the well-known 1906 study "The Art of Cutting Metals." But Taylor is only the most obvious proponent of this idea.

## 5.5 Mathematics and War

> It will not be profitable to make further specific recommendations [for topics of ballistic research] which must be carried out by others if they are carried out at all. Plenty of others spring to mind. They all have elements of scientific interest, and some are indeed of great interest. The sad thing is that whether we regard them as having for their purpose the destruction of our fellow beings, or adopt the more comforting theory that they are being solved to deter man from engaging in war, they are nevertheless important only because of unworthy human ambitions and the wickedness of the human heart.
>
> —Concluding remark of Moulton from his *History*.

> These results and savings were important for the Range Firing Section because of the large number of computers engaged in the construction of range tables and the constant flow of data into the Section concerning new experimental firings and new types of guns. I am happy to say that there are many applications other than in ballistics of the method which I devised. Some of them I have pursued to my own satisfaction.
>
> —Bliss in *Autobiographical Notes* [1400, p. 601]

These two remarks, each in the characteristic tone of its author, return us to the fact that these developments took place in the late stages and period after the World War. Though their sentiments do not approach those of, for example, Bliss' mentor Oskar Bolza, who abandoned mathematics for lengthy period after the War and pursued religious matters, they give pause to consider the War context.[18] Debates on mathematics and war have of course a long history, and one might well ask what contributions the present study can make toward it. The following is a list of related themes touched upon herein:

(1) Placing the innovations in military historical context: the observation that the contributions of Moulton and Bliss took place as a second wave of reform, following a first wave that is usually not acknowledged except for its "inadequacies" as perceived by the research mathematicians. The latter took their standards of operation from the rigorization of mathematics which occurred in the second half of the nineteenth century, in which the names of Picard and Weierstrass were prominent. The first wave used mathematics in the style of the engineering community.
(2) The effect of societal trends as a motivating force: the processes of industrialization brought about two special forces, the variety of weapons produced, and the need for efficient operation of military ventures, including the production of the range firing tables themselves. Variety: as reflected in other sectors of society (production of columns), one need only to recall the remarks of Moulton regarding General Ballistic Tables, "But in view of the great variety of guns and projectiles in modern artillery practice, it is desirable to have general ballistic tables from which all special range tables can be interpolated,

[18] In [1346, p. 238] it is reported that while at the University of Chicago, Bolza took an antiallied stance which resulted in his feeling quite isolated. He greatly preferred to reject the comforts of his life to enlist in the German cause.

or otherwise determined, by a relatively small amount of work." "Efficiency" and "waste" have occurred so often that nothing more need be said. It can be affirmed that the differential variations enterprise itself was an effort to streamline the production of tables. Moulton introduced them to reduce the number of complete trajectories that would need to be computed, Bliss' method reduced the number of integrations executed, and Gronwall's contribution had a similar effect.

(3) Critical appraisals by fellow practitioners: This study has included criticisms of the innovations by such men as Gordon Hull, Philip Schwartz, and J. E. Rowe, who called into question what they felt to be exaggerated claims of the second-wave reformers. It is important to include such reactions in a realistic account of their introduction.

(4) Advocacy and resistance within the military: it is not surprising that any novel approach will have both its advocates and resistors within the community which receives it. These have been documented; the grumblings and suspicions of the new methods found their way into the CAJ/JUSA in the form of side remarks in technical articles. On the other hand, the advocacy of Roger Sherman Hoar in his "New Ballistics" series, the confidence expressed by the student Guion in his review of Moulton's book ("Although it is probable that the theories set forth in this book will be adequate for all practical requirements for a long time to come, Professor Moulton concludes his treatise by indicating how the methods that he has developed are capable of being easily extended to meet any needs that may ever arise" [1515, p. 155].), and the concluding remark of Bennett's Army Ordnance piece [1361, p. 40] may be cited again.[19]

(5) Long-term consequences of the innovations: Although the previous chapter concluded with the adaptation of Bliss' contribution to rocketry in the work of a single naval instructor in the early 1960s, it is not difficult to produce similar examples from the internet. The papers [1474, 1490, 1505, 1598], all dating from the late 1950s and early 1960s, each cite explicitly the adjoint system of Bliss. The sponsors are telling: one is from the Lawrence Livermore Radiation Laboratory, one a technical report from NASA on guidance systems for travel from Earth to Mars, and another a Soviet piece on low-thrust space flight translated for NASA and the National Science Foundation. This argues in favor of Bliss' work outlasting its period of introduction and surviving into a new era.

[19] These statements read: "It is to be expected, however, that something like astronomical accuracy may be expected by their [the new methods] use, provided that corresponding improvement in material is sustained. ... If a more or less permanent peace becomes probable, and target practice with major caliber artillery continues to increase in cost, the trajectory of a large projectile may become as rare but as dependable a phenomenon as a reappearance of Halley's Comet." Here the accurate prediction of the return of the comet in 1910, which resulted in tables of visibility schedules for popular use (see [1699, p. 11]), is adduced as evidence of the reliability of ballistic tables under the new regime.

(6) Stratification in the dispersion of the innovations: It has been considered important to provide an account not only of the introduction of the new ballistic ideas by the mathematicians but of attempts to understand and convey them in educational settings such as the Coast Artillery School. In such attempts, modifications were sometimes felt necessary, and these required work by mathematically knowledgeable nonresearchers such as R. S. Hoar. Indeed, even among the mathematicians, Bliss was more purely research oriented than Moulton; descriptions of such stratification give a more comprehensive picture of the realities of the reception of the new ideas. The critical work of Schwartz may also be put into this list of strata, a mathematically degreed member of the military offering cogent analyses.
(7) The unusual ease with which the innovation fit with one contributor's research agenda: Bliss' work with the nature of solutions to systems of differential equations and his familiarity with Hadamard's use of the adjoint system in the solution of Mayer's problem provided the basis for his method.

In this section, some of the literature on war and mathematics is reviewed, and points of contact with the present study are attempted. It must be said that some deeper sociological and political themes raised do not find counterparts in the present study, so the comments which follow may seem somewhat limited.

In the mid-1960s and early 1970s, comments of a historical nature on the subject of this study by men who were associated with Bliss appeared; each has been previously noted. Lawrence Graves' article on the general implicit function theory written for the MAA series "MAA Studies in Mathematics" ("directed generally to readers at the advanced undergraduate and first-year graduate level") dating from 1965 had an introductory section devoted to the idea of a functional, and Bliss' work applying these notions to "problems of calculating the effect of wind and other factors on the trajectories of projectiles" is mentioned [1499, p. 35]. This remark applies to the highest theoretical level with which this study deals, in functional analysis; as has been shown, it was not necessary to adduce these results in order to proceed to application. In 1972, the highly regarded *The Computer from Pascal to Von Neumann* by Goldstine was published. This sprawling account of the history of the computer from a key participant in the development of the ENIAC contains sketches of Moulton and Bliss as well as a host of other figures. It is the single richest source of material relevant to the present study available to the historian. The importance of Moulton's numerical integration method is explained, though his insistence on the importance of his convergence proof is not. The pathway from this method to the ENIAC is given in detail. Bliss' adjoint system is described as a "perturbational or variational technique" [1485, p. 132] and explained flawlessly.[20] In a minor quibble, Gronwall is described as a "Swedish scientist" rather than a practicing mathematical analyst.

[20] The words in quotes have been avoided in this study in an attempt to prevent the association of Bliss' work with any other material described as "perturbation theory" or "variational methods." These phrases did not occur in any of the historical references used.

According to David Aubin and Catherine Goldstein in [1349, p. 5], it was only in the 1990s that historians of mathematics started to pay attention to the World War. In 2001, computer historian David Grier published [1502] a short and informative study of Veblen's time with the military, geared toward a mathematical audience. Issue has been taken in the present study with Grier's assertion that the range tables were primarily used as engineering tools and his downplaying of their role in action. But it is clear from the early debates in the Coast Artillery that their batteries should be equipped with charts containing information arrived at from the range firing tables. In his account of the Siacci method, no mention is made of the first wave of reformers who introduced the method into American ballistics, though his presentation of the Siacci equations [p. 925] is exactly as it appears in Ingall's book on page 117, and his example of finding the ballistic coefficient resembles an example done in that volume on page 58 (Example 3). His description is accurate, but the lack of acknowledgment of men like Ingalls leaves an incomplete picture of the transition from the older to the newer methods. Conversely, the vivid descriptions of the mathematicians using the Siacci method to complete a range table for an actual weapon provide a refreshing counterpart to the dreaded "mathematics professors at their tables" deplored by the French pacifist quoted in [1349, p. 1]; these men literally got their hands dirty. Though not the main subject of the piece, Moulton is mentioned as the director of the theoretical ballistics program in Washington. The description of Moulton's contributions, necessarily brief, underplays his near complete dissatisfaction with every aspect of the first wave techniques and his devotion to the ideas of mathematical rigor. Bliss and his contributions are mentioned in a sentence, which gives only two differential effects: wind and rotation of the earth. But surely no more is necessary given the goals of the work. Then again, he points out the proving ground was "a part of the industrial complex that was preparing weapons for the expeditionary forces," [p. 928], that is, that there were other aspects to the grounds than the theoretical work. He creates the analogy that the mathematicians were "manufacturing numbers" by numerical integration. The bulk of this account is taken up by the description of the computers' activities, the natural focus for a computer historian and author of [1504]. The present study fills in the story of the reception and attitudes of mathematicians and ballisticians toward the method with further comments in [1482]. The piece concludes with a description of offices held by all participants after the War.

At the turn of the twenty-first century came renewed interest in the general topic of mathematics and war, as [1412] appeared. This essay offered a call for reconsideration in light of the recent Kosovo war and a conference organized to discuss related questions. Focusing mostly on post-World War II activities, the paper made some interesting points about the first World War. It is pointed out that prior to the latter, mathematics served as a "toolbox" from which the military among others could draw as needed, "... but the involvement of mathematics as a general endeavor within the military institution was not very intimate, and specific military applications had no independent shaping force for mathematics" [1412, p. 13]. However, this situation changed around World War, and the authors cite "two important new military technologies [which] depended on *mathematics in the making* [emphasis in the original]: sonar and aerodynamics" [1412, p. 14].

In the United States, aerodynamics was not developed at this time, but submarine detection was investigated by Max Mason, among others. The difficulty in applying this assertion to the American situation lies in defining the phrase "mathematics in the making"; what activities are involved in this, and what new mathematics was produced? Were genuinely new concepts evolved, new methods for solving, say, partial differential equations? One could argue that Moulton's solution of the system of ordinary differential equations by finding a basis for the solution space was an application of relatively new mathematical ideas, though no new ideas were introduced. Fréchet introduced the idea of the differential of a functional only seven years prior to its use by Bliss. It would be plausible to say that Bliss made use of "tools" of more recent acquisition in the toolbox when he refashioned Hadamard's 1910 use of the adjoint system in the solution of the Mayer problem for the differential variations and cited the differential in the ballistic context. Since the "Bliss' adjoint method" found use much later in rocketry and was repeatedly cited as such, is this an example of making new mathematics? However one chooses to describe it, the authors' assertion that the role of mathematicians among other pure scientists was "...to implement theoretical knowledge and make it function in practice" [1412, p. 15] stands as an accurate depiction of the differential variations work, in contrast to that of the first wave reformers, not to mention their predecessors.

"Nobody will claim that mathematics was in any way decisive for the outcome of the war, nor that WWI applications of mathematics left important traces in the world..." [1412, p.15]. The timeline of the reports of the Aberdeen and Washington investigators makes the first claim undeniable; the second depends again on the meaning of the phrase "important traces." The authors are surely aware of the pathway from Moulton's numerical methods to the ENIAC described in [1485], though the period of development spanned decades and was hardly direct. Does the adaptation of the adjoint method to rocketry count as a "trace"? Much would depend on proof of actual application in particular cases, a topic for further investigation. Its appearance as a theoretical tool is evident.

Three other comments made by the authors regarding the use of mathematics from the military perspective have been touched on in this study. First, they note that "mathematics has become an integral and even an essential part of modern warfare. (This does not mean that mathematics has become a major *expense* [italics in the original] of the military apparatus - mathematics is a cheap way to use costly resources more efficiently" [1412, p. 18]). This is an affirmation of the onset of efficiency as an advantage offered by mathematical analysis. Second, "the strategic planning of the possible use of weapon systems depends on the mathematical calculation..." [1412, p. 18], which in their context refers to the highly sophisticated missilIt's appee systems of the modern era, brings us back to the general ballistic tables and their alleged ability to predict the firepower and any new weapon. Even earlier were the uses of the Siacci theory to do the same and arrange for harbor defense, as noted in JUSA/CAJ articles. Third, the authors make a point which if anything has become even more relevant at the present time than when it was written:

> Ideologically, the waging of war is made more acceptable to the public by the presentation of warfare as precise and hence "more rational and clean"...the increased precision of weapons (which is real) depends essentially on the application of mathematics. ...mathematics presents modern warfare as "fast by avionics, precise by GPS, safe by optimized operations planning." [1412, p. 18]

Adjusting for the changes in technology, one can detect the beginnings of these attitudes in the shift from first- to second-wave reformers. The former, spearheaded by Ingalls and Hamilton, were sometimes war veterans but at the least were familiar with ordnance and perhaps even witnesses to the horrors of war; the latter were "professors at their writing tables," able to concern themselves with "increased precision of weapons." This was for them indeed war at a distance.

In 2003 came Reinhard Siegmund-Schultze's effort [1686], an essay which attempted an international perspective on military work in mathematics in the period 1914–1945. This was an ambitious attempt to find themes and explore connections among them in war-related mathematical activity in many countries in the first half of the twentieth century. This lengthy work does not slight the first World War and contains many open questions, case studies, outlines of specific work and responsible personnel, surveys, and speculations. Several of the points made in here have already been addressed and need not be repeated, but one can comment on some points. In his case study of "USA", Siegmund-Schultze points out the continuities from the First World War to the second in state funded military work at Aberdeen. He describes Moulton's algorithm as a "finite difference method," though not citing his initial work on differential variations, and casts Bliss' contribution as the introduction of "higher methods of the calculus of variations...which lay 'on the farther boundary of the explored mathematical domain of today'" [1686, p. 40], quoting Bliss' own phrase from his 1927 *Science* article, and citing Goldstine's book as a reference. While, as in the case of Grier's work, one cannot expect a more detailed description of the development of Bliss' ideas, one can see here what has become the time-honored short explanation: it comes down to the calculus of variations. Siegmund-Schultze cites the use of "line functions" but does not elaborate that this subject itself was not necessary to the military application, as the present study has shown. One might in this manner lead to the conclusion that either Volterra'a functions of a line or the subject matter of the calculus of variations precipitated the whole use of the adjoint, which is not the case. Hopefully the present study provides a disentanglement of these notions. In his chart containing extra notes on fields of application, methods, and mathematicians/appliers involved for the USA, he lists the above information in short entries but cites as "sources" Moulton's *New Methods* and Bliss' article in *Science*. Though not particularly consequential, it is asymmetric in that the former contains all of Moulton's technical work but the latter is only a one-line self-description meant for a general audience. Taken as a whole, these lapses are surely understandable in a work of this magnitude, in which the present subject matter plays a small role.

A century after the onset of the World War, the volume *The War of Guns and Mathematics* [1349] was published, a collection of essays by historians of mathematics subtitled *Mathematical Practices and Communities in France and its Western Allies around World War I.* We shall consider two essays from this work: the introductory "Placing World War I in the History of Mathematics" and the view of American participation, "A Mobilized Community: Mathematicians in the United States during the First World War." Naturally, only certain parts of each of these have relevance to the present study.

In the first work, the authors explain the long absence of the War from historical mathematical literature, proposing various explanations, concluding that one could receive the impression from certain studies that "War seemed to leave little imprint, if at all, on the development of mathematics either as a discipline or as a body of knowledge." [p. 7] They cite a 1994s landmark work [1641] which they paraphrase as saying that "war was a 'furlough' away from pure research for most American mathematicians after which they, by and large, had 'little inclination' to pursue applications." [p. 6] While the latter portion of this statement is likely valid, this study has emphasized the exceptional nature of Bliss' trajectory, in which his pure studies in the smoothness of solutions to systems of differential equations passed "seamlessly" into his variations work. The existence of the differential for the appropriate system was the next logical step in these studies and yet was a theoretical part of the work which confronted him at Aberdeen. Thus he, perhaps uniquely among his colleagues, experienced no furlough.[21]

Two particular activities to which mathematicians were able to contribute are discussed: sound-ranging and exterior ballistics. The section on ballistics follows the standard historical accounts up to and including the Siacci method, noting that "Despite strong opposition, the mathematical sophistication of exterior ballistics was slowly gaining acceptance from artillery officers, especially in France where many officers were trained at the École Polytechnique which had a strong mathematical tradition. ... exterior ballistics was in 1914 strongly pursued by military research centers in several European countries and developed into an autonomous branch of military science" [p. 40]. Moulton's comments on the inadequacies of the method are quoted, but the description of the American history is inaccurate: "There were strong contrasts between countries, like France, where parts of the military possessed a strong mathematical culture instilled by polytechnical establishments and those, like the US, where since the mathematical skill of army officers was much lower mathematicians played a more active role in introducing Siacci's methods and adapting them to the requirements of the Great War" [p. 41]. In fact it was the military first wave which introduced the Siacci method in the early 1880s and developed it; the mathematicians had limited use for it in their War work.

The discussion on ballistics includes a frank evaluation of the adequacy of mathematics alone to deal with these issues, and the citation from Hull's remarks

[21] The authors' comments most likely refer to European experiences, but their query "How was WWI a disruption, if at all, in the mathematical activities of its contemporaries?" [p. 10] can apply to Americans as well.

and plea for more experimental work is given [p. 42]. The following challenging quote is given:

> For the most part, the needs for true mathematical innovation were limited in those fields [sound-ranging and exterior ballistics]. What was required mostly consisted in efficient methods to carry out length computations and mathematical analyses of physical problems. ...for the most part they [mathematicians] proved no more useful in that task than mathematically trained physicists and engineers, while lagging behind in practical matters. By and large, mathematicians were not the best people to address the mathematical problems raised by modern warfare. [p. 43]

In spite of the elaborate story of Moulton and Bliss herein presented, it remains a fact that the range-firing tables they helped create needed to be tested against actual firings to be of any genuine use and were subject to the vagaries of empirical data such as the resistance function and the ballistic coefficient. The authors' conclusion is supported by the writings of the critics listed in the present study. Hull's technical remarks in support of this position can be found in his Franklin Institute address [1555]. Schwartz's 1925 critique [1677] remains the most succinct assessment of the limitations of the second wave of reform on range-firing tables. On the other hand, the mathematicians Moulton, Bliss, and Gronwall, who introduced the "efficient methods to carry out lengthy computations" of the range firing tables, did so with considerable ingenuity in the case of differential variations, allowing these factors in range tables in a comprehensive way. Moulton proposed many other practical uses for them, pointing out, for example, that changes in a resistance law would not necessarily result in massive reworking of the tables, merely an adjustment based on variation theory. Even if we agree with Schwartz that the tables were only one among many tools needed for fire control, tables were needed and expected. The efforts of these men appear as a natural extension of the activities of their predecessors in the first wave. It has also been shown that Bliss' contribution had a second, purely mathematical component: his study on the nature of solutions to differential equations, elements of theory of functionals, and implicit function theory was a part of his work. The theory had an independent life in the research publications of Bliss and Gronwall. The real "mathematical analyses of physical problems" involving projectiles did indeed occur at the hands of physicists and engineers, in the period after the departure of the mathematicians. The work of the British team on rotating projectiles began to be absorbed, and Kent's pioneering consideration of the shape of the projectile was begun. Moulton had of course made an attempt at addressing these issues.

An acknowledgment of the importance of mathematical analysis (in the technical sense) including "perturbation theory," ordinary and partial differential equations, and conformal mapping, which was applied to the problems, concludes this section.[22]

[22] Although Bliss appropriately is not mentioned in this introduction, his work in "functions of lines" appears in another essay, as does one of Gronwall's papers, as "notable contributions to ballistics that raised interesting problems at a mathematical level." [p. 166]

This introductory chapter concludes with the observation that "... many mathematicians of the war generation extended the limits of their profession beyond academic research." [p. 46]. This to a certain extent describes Moulton. Although primarily as celestial mechanician, he had published purely mathematical papers on the solutions of systems of differential equations, but he surely extended his limits beyond pure research into the messy realm of ballistics.

Another work from this volume is the account of the American response to the War given in [1346], to which we have referred from time to time already. Some familiar criticisms might be leveled at this comprehensive account; again the importance of them must be taken in context of the entire piece. The adaptation of the Siacci method by Ingalls and Hamilton prior to its use by Veblen is ignored, as in the description of the differential variations work at Aberdeen [pp. 241–244], while only Veblen's efforts are described. Similarly the description of Moulton's group in Washington omits any reference to research in the secondary factors [pp. 244–246], though it is later mentioned as a the topic of a chapter in Moulton's *New Methods* [p. 260]. Yet again, Bliss is one of the exceptions to the statement that "The work [of the mathematicians] clearly had very little to do with the research specialties of the participants, at least in many cases." [p. 247] Although mathematical education and training for the military are discussed, and programs given in universities, YMCAs, and other settings are detailed, an opportunity to relate the ballistics material produced by the mathematicians for military use is missed. Lester Ford's book [1468], discussed in the introductory piece, does not appear here. Although Dunham Jackson appears as the recipient of a post-War offer to teach at Harvard, and his experiences with Moulton's group are given, no reference to his excellent ballistics pamphlet [1567] is made. Bennett's ballistics book [1360] is absent, though as noted the contents of Moulton's book are detailed. The most interesting revelation is that Hoar's treatment is figured in a thesis produced by a Harvard graduate in December 1920. [pp. 262–263] Hoar himself was considered an instructor in ballistics at that institution. A nice summary of post-war accounts of wartime mathematical activity is given [pp. 257–258] in which Bliss and Gronwall are listed among those who improved and extended Moulton's numerical integration method, though variability is again not discussed. In the authors' review of Moulton's book, an appreciation for the magnitude of the effort undertaken by Moulton is given its due, in the previously quoted line "Moulton appears to have taken it on himself to wholly recast ballistics, in particular noting the importance of combining theory and experiment." [p.260] Robert Kent, whose work on the motion of projectiles has been discussed in the previous chapter, emerges as "almost surely, the country's only 'complete' ballistician" and "the outstanding carrier of the torch of science in the Army" [p. 267], quoting sources which can be found in the volume.

Although it has no parallel in the American setting, the piece [1350] from *The War of Guns and Mathematics* describes to some extent an instance of stratification in France which is in a sense the reverse of the American one. Here the French military, with its own attitudes toward mathematics, calls on French mathematicians for assistance in war-related work. The comment which forms part of the title, "I'm just a Mathematician," was made by Jules Haag, whose work was noted in an earlier chapter of this study, a "young professor of rational mechanics from the University of Clermont-Ferrand" [1350, p. 307], who was commenting on his status relative to the military. The entire quote is "In their eyes I am just a mathematician, without practical use other than serving as a computing machine when the occasion arises." Here the military is the entrenched authority in whose eyes mere mathematicians would be of limited use. In America, the military was in some sense asked to step aside for the mathematicians, whose strata was perceived by some parties as higher. One gets a sense of the French situation from the observation by the author that "From the minute description of the work done at Gâvre, one gets a rich impression of what it meant to use the mathematicians' skills in the First World War. That impression is far from the romantic vision of genius solving a problem that had frustrated ignorant militarymen for ages. On the contrary, ballisticians at Gâvre had enough mathematical sophistication to be able to see the added value mathematicians were susceptible to bringing to their trade. It also seems clear that mathematicians were not 'well-prepared for their new role' [source given] and that, at first, their experience at Gâvre may have been rather humbling" [1350, p.337]. Here the contrast with the American situation is evident. Moulton in the eyes of some was seen as having unapologetically recast all of American ballistics regardless of existing traditions. Indeed, one of his students thought that he may very well have solved the most important issues completely, as has been mentioned.

## 5.6 Mathematical Research Programs, Schools, and Communities

> The officers, privates, and civilians of the Ballistics Branch came from widely separated parts of the country-from Maine to Texas and California, yet it would be difficult to gather in any way an equal number of individuals who would have more in common and whose relations would be more harmonious. The unfailing courtesy and the evidence of mutual helpfulness which were manifested in numerous ways were inspired not alone by military customs and the proprieties of the situation but much more by sincere mutual respect and real personal regard. ... This spirit of mutual helpfulness and devotion to duty on the part of the officers, privates, and civilians alike went far in making it possible for those primarily responsible for the success of the Ballistics Branch to carry what at times were heavy burdens. All contributed to the successes that it attained...
>
> —Forest Ray Moulton [1628, p. 6]

> Bliss arrived here a few days ago and seems to have taken hold well and to be enjoying the work very much. At the least I hope that he will pick up a lot of information that will be valuable to his work at the University. However, I desire very much that we could have his services for the period of the war.
>
> —letter from Oswald Veblen at Aberdeen Proving Ground to E. H. Moore, October 12, 1918

The first of the two quotes above comes from Moulton's *History of the Ballistics Branch*; it is an introductory remark stressing how much community spirit reigned in the group he headed. An entire page is given to a description of this camaraderie.

Moulton views the Branch as more than just a collection of individuals devoted to turning out results. The second quote is a mild sentiment of a similar nature, given by Veblen to Moore, regarding the ease with which Bliss had fit in to the other ballistics community; much longer and descriptive passages portraying Aberdeen as a smoothly working community, especially as seen by Norbert Wiener, are available elsewhere. In each quote, one sees effectively the existence of a mathematical research community and, in the second case, one which existed in parallel with a proving ground, as pointed out by Grier. In this section, an attempt is made to discuss the contents of the joint research efforts, contents which consist to a large extent of material related to differential variations. Though few new technical documents will be introduced, the task is to justify the assertion that each of these groups had as a major focus the material described in this study. An attempt is also made to describe how the arc of Bliss' mathematical work from his early studies with Bolza through that of his students Graves and Goldstine fits in with the conception of the University of Chicago as a school with a program in analysis, a topic explored in the recent book of Karen Parshall [1643].

In fact, in a chapter entitled "Surveying the 1920s Research Landscape," Parshall devotes a section to Research in Analysis, a section with several points of contact with the present study. The section begins by citing the two major centers of analysis in America, as seen by Veblen in a comprehensive survey: Harvard, with Osgood, Bôcher, and Birkhoff, and the University of Chicago, with Bolza and Bliss. It thus transpired that the tradition of analysis in America was "directly imported from Germany in the early 1890s, and it was carried into the twentieth century by members - like Birkhoff and Gilbert Biss - of America's second mathematical research generation" [1643, p. 47]. A result of the Harvard analysts, the 1922 paper of Birkhoff and Kellogg on invariant points [1371], of peripheral interest in the stream of the present work, is cited. Hildebrandt, a student of Moore's whose joint work with Lawrence Graves resulted in 1927s [1500], in this study given as the source of more general implicit function theorems in general analysis, is also cited. He is presented as an explorer of functional analysis at the University of Michigan who, along with another student of Moore, "gave the first general proof of the principle of uniform boundedness for what would come to be called Banach Spaces" [1643, pp. 50–51]. (Parshall notes that this principle was independently proved four years later by Banach and Steinhaus and gives references.) The work of Bliss and some of his students is given as well [pp. 52–54].

In a later section, "Analysis in the 1930s," an interesting work from the 1920s is highlighted [p. 280]. Leonida Tonelli, an Italian analyst, published in 1921 and 1923 a two-volume set in which he approached the calculus of variations from a point of view independent of the usual Weierstrassian development. (References can be found in [1643].) Bliss was so impressed with this approach that he translated a summary of Tonelli's from Italian into English and published it as [1701]. Readers of the present study will not be surprised by Bliss' interest in Tonelli's work on reading the latter's orienting remarks: "Having thus specified the scope of the proposed investigation a natural method presents itself, that of the functional calculus according to the ideals of Volterra. The integrals which are considered in the calculus of variations ...depend on the most part upon curves or surfaces and are real-valued functions of such curves and surfaces. These belong therefore in the domain of functional calculus..." [1701, p. 164]. Bliss' 1910s interest in Volterra's and Fréchet's work and his efforts to have his students LeStourgeon and Fischer push these in to the realm of the calculus of variations can be considered supplementary to the account of Tonelli's orientation. The work of McShane and Graves in the 1930s is also described in this section of Parshall's book [p. 282].

It is curious that with the attention paid to American work on functions of a line, the theory of functionals,[23] and functional analysis, the path leading from Volterra's remarks on an implicit function theorem to Graves' 1965 MAA entry [1499] on the general implicit function theorem, is not mentioned. Perhaps its close relation to differential and integral equation theory has sequestered it from the usual basic functional analytic theorems such as uniform boundedness and the open mapping theorem, which arose from Fourier series, among other places. One tends to find the general implicit function theorem mentioned in books on "applied mathematics" or differential equations. In could be included in a survey of mathematics in 1920s America. Also missing is the journey from Bliss' work on smoothness of solutions to differential equations to his incorporation of the Fréchet differential into their study. Again, these lines can be considered supplementary to Parshall's surveys.

Turning now to the issue of research communities,[24] that is, groups of mathematicians with a common research agenda, we ask if it possible to consider the joint Aberdeen/Washington efforts in this light. They have been thought of in this way by some writers on the history of the war; Grier, for example, states that "It [Aberdeen] became a center of mathematical research..." [1502, p. 923] and "The legacy of their work was the community it created and the discipline it instilled" [p. 930]. In the introductory paragraph of the March 12, 1924 letter from Philip Schwartz to Veblen in which the former decries the state of the post-War ballistics education in America, Schwartz writes: "...I am writing apropos of a conversation which I had with you just before I left Princeton in 1921. You remarked that you believed that many of the Aberdeen reports were worth publishing in order to have somewhere, a permanent record of the activities of the section during the war." It thus appears

[23] On p. 14 is reproduced a 1928 list by Veblen of America's research mathematicians in which E. H. Moore, Harvard's G. C. Evans, Hildebrandt and Graves are listed under "General Analysis and Theory of Functionals."

[24] In [1642, p. 384], Parshall argues that "In the United States Eliakim Hastings Moore and his students, Leonard Eugene Dickskon, Oswald Veblen, and Robert L. Moore at the University of Chicago, together with Edward V. Huntington at Harvard, responded particularly to Hilbert's call for foundational research and defined what would come to be known as the 'American school of postulate theory', perhaps the first geographically extended, thematically oriented, research level 'school' of mathematics to be supported within the context of the emergent American mathematical community." The material that follows is not meant to challenge this claim but simply to provide a similar example of an organized, single-themed group, differences notwithstanding.

that Veblen himself briefly considered the Aberdeen group to have achieved some print-worthy mathematics. One wonders what material he would have chosen.[25]

One might consider the numerical methods of integration as chief among the contributions of these groups and indeed as has been by some writers. But the reservations of the present author to this have been stated several times and given in detail in [1482]. Briefly, Moulton himself downplayed his role in this activity, stating that many people could have done what he did, though perhaps not many pure mathematicians; his chief pride was saved for his proof of convergence of his method, the result of his adherence to the Weierstrassian standards of rigor. The method was in use in French ballistics for decades prior to the War and in celestial mechanics frequently. In [1482], it is argued that almost simultaneously with the appearance of Moulton's method came alternatives suggested by Milne, Gronwall, Bennett, and others. Not only does this argue against the uniqueness of his contribution but, if discussed in print, would result in comparisons between methods which would be of limited mathematical interest, given the primitive stage of numerical analysis in America at this time. Indeed, in spite of the success of Moulton's method, it is also argued in [1482] that the result in the mathematical community could best be described as an "increase in the level of awareness" of the subject.

It seems that the subject of differential variations is a good candidate for a focus of these joint "schools." Recall the observation of earlier chapters that the twin topics of differential variations and numerical integration were often held up as the chief mathematical innovations in American ballistics coming out of the World War. There is, to begin with, evidence of a certain urgency to address this material, comparable to that associated with Moulton's mind with the great inadequacies of the Siacci theory. Yet again we recall his statements: "... the methods of determining the effects of such disturbing factors as winds, abnormal air densities, and the rotation of the earth were imperfect in the extreme and in some cases gave results of the wrong sign" [1628, p. 10]. "In my first paper, written shortly after I entered the service, I seized the problem of computing differential variations and pulled it out of the fogs in which it was involved and place it on a solid foundation" [1633, p. 321]. Here Moulton stresses that it was in his *first* paper that these problems were addressed. In spite of this attention, Bliss reported that on his arrival at Aberdeen "F. R. Moulton, Veblen, and their associates had devised effective methods of successive approximation for solving the differential equations of trajectories in all cases, but the theory and application of the so-called 'differential corrections' of trajectories was still very imperfect" [1400, p. 601]. In a letter to Veblen dated

[25] Veblen and Moulton would be the twin leaders of this group. In un undated letter from Veblen to E. H. Moore written from Aberdeen a weary Veblen remarks: "I expect to come in contact with Major Moulton in his new job. Our rôles will be reversed, for I am now the practical man who has spent his time with the guns, mules, and mud to the exclusion of reading and reflection, while Moulton will be the scholastic visionary who can tell me why things happened as I have seen them." [University of Chicago]

August 14, 1919, Gronwall[26] writes " Thus far, I have done next to nothing on the jump card question. As soon as I arrived here [Moulton's office] some problems in differential corrections were put up to me, and the result was a rather lengthy investigation of various aspects of the theory..."[Veblen Papers]

Among the researchers in this community, one could list F. R. Moulton, G. A. Bliss, T. H. Gronwall, J. F. Ritt, W. E. Milne, Dunham Jackson, and Philip Franklin. In his *History*, Moulton lists technical reports written by these men and distributed to interested parties. Five of the ten papers by Moulton himself were devoted at least partly to differential variations, two of the three by Milne were similarly concerned, and five papers by Ritt on the subject were listed. Moulton's work inspired a general variational technique which was explained in his textbook on differential equations. The reports of Bliss which were examined in detail earlier and formed the basis of FOLIB and DECAF as well as his articles for the JUSA/CAJ appear in the collection of Aberdeen documents. In particular, the supplement dealing with his antiaircraft work culminating in the three-page computation sheet exists only in the form of this technical report, having never achieved a proper published form. Gronwall's key contribution [1510], whose contents only received limited mention in 1944s [1397] and nowhere else, is part of the collection. Gronwall's published papers [1512] and [1513] are given in preliminary form as well in this collection. One can also include the papers [1506] by Gronwall and its predecessor, Ritt's note [1657], as "spinoff" items, the first containing Gronwall's Lemma. These papers are the core of the research effort contributed by the mathematicians. An interesting topic for further research would be the possible effects of this material on Bliss' future work in differential equations the calculus of variations.

The mathematicians were also involved to a limited extent with the practical side of the subject. Gronwall appears yet again as the author of [1511] a cartographical exercise for determining range and deflection of trial shots from observed data. He also constructed two nomograms for this purpose [1451, pp. 28–31].[27] Numerous examples of provisional range-firing tables with differential effects are to be found among the Aberdeen documents; production of such was the material goal of the group. Educational efforts include Dunham Jackson's widely used ordnance textbook and Philip Franklin's alternative derivation of the basic equation for variations, the derivation having been scooped up by Hoar and used in his book. Alexander's computational device for use by battery commanders may also be listed.[28]

---

[26] From the preceding paragraph in the letter, it appears that Gonwall had been asked to look into the problem of analyzing "jump cards," cardboard screens placed at intervals in front of the gun. As the projectile leaves the gun, it penetrates each card, leaving an image which shows the yaw.

[27] In another memorandum on range computation Philip Alger, the Navy ballistician, remarks that "This method has been widely used in engineering to expedite complicated numerical processes. For example, the French use a nomogram to determine air density from barometer and temperature readings, while some engineers use a nomogram to determine desirable sizes for steel bars in reinforced concrete columns" [1344, p. 5].

[28] Informal contacts among the mathematicians resulted in some professional interactions. In particular, in a letter [NARA ENIAC] of March 31, 1919, from Gronwall to the Dean of the Graduate School at Princeton, one has a recommendation of Philip Franklin for an applied-for

The last mathematician standing from the original group seems to have been Gronwall, who stayed with Moulton's people, now known as the Technical Staff, until 1922. During this time, much of his work was taken up with differential variations. In a list of "Mathematical Papers" (see [1451] for references), one finds papers 240–249 devoted to his output, including such matters as cross-wind factors, density tables, and differential variations in range to any point on a trajectory; the third was apparently of interest to others besides Moulton and Bliss. All are dated 1921 or 1922. Earlier papers on approximate formulas for variations due to rotation of the earth and changes in gravity are also available.

Taken as a whole, the above material suggests an enormous effort expended on this material, and in volume of product if not in length of existence, the joint arms of ballistics inquiry in the First World War justify the description of a research community devoted at least in part to one topic. How closely this community resembles other schools of mathematics in America as described by Parshall, for example, is a matter for further investigation.

## 5.7 Exterior Ballistics for The English-Speaking Public

> If it had not been for the World War, there would not have been any occasion, so early as 1922, for a Supplement to the Eleventh Edition of the *Encyclopaedia Britannica*, as published in 1911. But for the exceptional situation so created, the original intention not to take in hand anything equivalent to a Twelfth Edition until a much later date would undoubtedly have been maintained. So colossal a convulsion, however, as that of the war, with consequences shown in so many unexpected directions and radically changing the world-outlook under the new conditions, made the need for this prompt addition to universal history absolutely imperative, as a record and illumination of so peculiarly dark and complex a period. ... The reader has before him what may be described as an international stock-taking, by carefully selected authorities, of the march of events all over the world from 1909–1910 to 1920–1921 and of the nature and critical value of such advances as were made in the principal branches of knowledge during that period. In this respect, the New Volumes aim at giving a key to the problems of today, so far as these contemporary problems are bound up–as indeed they are to an unprecedented extent–with the new social and economic issues which only began to emerge in their present magnitude or to impress themselves on the public, as the result of the tremendous upheaval caused by the World War.
>
> —Editorial Preface, The Encyclopaedia Britannica, The New Volumes, The First of the New Volumes Volume XXX, 1922 [1695]

Although not much of the content of the War-developed mathematical ballistics was available to the English-speaking public (who in any event would likely have limited interest in it) at least for several years after the armistice, one could find

---

fellowship. Gronwall bases his recommendation on Franklin's work in ballistics as well as his solution of a problem in gamma functions suggested by the former. The solution was to appear in the Annals of Mathematics. Ritt also received assistance from Gronwall on his 1920 paper [1659], as acknowledged in a footnote.

a nontrivial description in the New Volumes of the Encyclopaedia Britiannica of 1922.

The editors had the difficult task of issuing a catch-up volume to the venerable series, trying to balance developments in all fields from 1911 to 1922 with reports on the War itself. The result is a fascinating mix of civilian topics with such Wartime subjects as, in volume 30, aeronautics, ammunition, artillery, many foldout battle maps, military bridging, military engineers, demobilization, and resettlement.

In this volume, an article extending from page 383 to page 394 has an attempt made to cover developments in ballistics. The article is split into two parts, the first of which is an account of interior ballistics by the British Lieutenant-Colonel Robert K. Hezlett, who also authored one of the earliest English-language books on nomography. The second, by the American Colonel W. H. Tschappat, Chief of Ordnance, deals with exterior ballistics. Here a worthy summary of the American methods, with bows to the French and British allies, is given, again falling roughly into the twin categories of numerical integration and differential variations. The article concludes with the latest British and American experiments accounting for rotation and shape of the projectile. The list of names (American) associated with this work consists of F. R. Moulton, G. A. Bliss, Dr. T. H. Gronwall, R. H. Kent, G. F. Hull, and L. J. Briggs; the last is a scientist whose previous posting was the Bureau of Standards. Five pages are given up to differential variations.

One senses the difficulty of relating mathematical wartime contributions to even the mathematically educated public. How much easier to celebrate the achievements of ballistics' celestial sister, as Charbonnier would have it, exemplified by the readily available schedules of the 1910 appearances of Halley's Comet mentioned earlier. One wonders at the reaction of the reader to the stream of prose in 2 mm high letters on onion skin paper, with accompanying charts, tables, diagrams, mathematical formulae, and graphs. But at least for the interested mathematically literate layman here was a "record and illumination" of one aspect of the contribution of mathematicians to the colossal convulsion. The material could be viewed from as many perspectives as there were readers of it, from dismissal of mathematics professors at their writing tables to admiration of the perceived solution to all ballistics problems.

## 5.8 Epilogue

The following letter was handwritten on US Army stationary.

> 420 Riverside Drive
> New York, N.Y.
> Aug. 6, 1945
> Dear Professor Veblen,
>
> Having arrived here, I tried to get in touch with you without success. If you get down this way I would very much like to discuss with you the general armament situation in the Air Forces both from the organizational point of view and the scientific as well. Recent developments as publicized by criticisms of present weapons on the one hand, and praise of

atomic bombs, on the other hand, as well as my probable assignment to Armament works at Wright Field[29] may offer an opportunity for you to put into effect some of your ideas.

Although I am traveling around considerably a phone message to me at Monument 2–4695 will reach me and permit me to get in touch with you when you visit this part of the country.

With best regards to Mrs. Veblen,
Sincerely,
P. Schwartz

[Veblen Papers]

## Appendix: Existence, Embedding, Implicit Function Theorems; The Problem Of Bolza

The first three theorems in the following are taken from [1399, pp. 269–283].

I. Existence Theorem for Implicit Functions

Let

$$f_i(x_1, \ldots, x_m; y_1, \ldots, y_n) = 0, \; i = 1, \ldots, n \tag{5.1}$$

be a given set of equations, we write for convenience $x = (x_1, \ldots, x_m)$, $y = (y_1, \ldots, y_n)$ so that the argument of the preceding functions can be written as $(x, y)$. Sets $(a, b)$ will be interpreted similarly. The neighborhood $(a, b)_\epsilon$ is the set of points $(x, y)$ for which

$$|x_r - a_r| \leq \epsilon, \; |y_i - b_i| \leq \epsilon, \; r = 1, \ldots, m; \; i = 1, \ldots, n \tag{5.2}$$

with $a_\epsilon$ and $b_\epsilon$ defined similarly. Then let $p = (a, b)$ be a point such that

(1) The functions $f_i(x, y)$ are continuous and have continuous derivatives $\frac{\partial f_i}{\partial y_k}$ in a neighborhood $N$ of $p$
(2) $f_i(a, b) = 0, \; i = 1, \ldots, n$
(3) $|\frac{\partial f_i}{\partial y_k}|(a, b) \neq 0$

Then there exists a set of functions $y_i(x_1, \ldots, x_m)$ single valued and continuous in a neighborhood $a_\delta$ and having the following properties:

(1) The points $(x, y(x_1, \ldots, x_m))$ which they define are in $N$ and satisfy the equations $f_i = 0$
(2) There exists a constant $\epsilon$ such that for each $x$ in $a_\delta$ the set $(x, y(x_1, \ldots, x_m))$ is the only solution $(x, y)$ of the equations $f_i = 0$ satisfying the inequalities

[29] Located near Riverside, Ohio.

$$y_i(x_1, \ldots, x_m) - \epsilon < y_i < y_i(x_1, \ldots, x_m) + \epsilon \tag{5.3}$$

(3) $y_i(a_1, \ldots, a_m) = b_i,\ i = 1, \ldots, n$

(4) In a sufficiently small neighborhood $a_\delta$ the functions $y_i(x_1, \ldots, x_m)$ have continuous partial derivatives of as many orders as are possessed by the functions $f_i$ in the neighborhood $N$.

II. Differential Equations: Existence of a Solution Through an Initial Point

We define, for $y = (y_1, \ldots, y_n)$, $\text{mod}(y) = (y_1^2 + \ldots + y_n^2)^{1/2}$.
Then let

$$\frac{dy_i}{dx} = f_i(x, y_1, \ldots, y_n),\ i = 1, \ldots, n \tag{5.4}$$

with initial point $(\xi, \eta) = (\xi, \eta_1, \ldots, \eta_n)$. The functions $f_i(x, y)$ satisfy:

(1) They are single valued and continuous in a neighborhood of $(\xi, \eta)_\epsilon$

(2) There is a constant $\kappa$ such that for every pair of points $(x, y)$ and $(x, Y)$ in $(\xi, \eta)_\epsilon$ one has

$$|f_i(x, y) - f_i(x, Y)| < \frac{\kappa}{n^{1/2}}\text{mod}(y - Y),\ i = 1, \ldots, n \tag{5.5}$$

Let $M = \max|f_i|,\ i = 1, \ldots, n$ in the neighborhood $(\xi, \eta)_\epsilon$. By a solution of the above equations is meant a set of functions $y_i(x)$ which, with their derivatives $y_i'(x)$, are continuous on an interval $[x_1, x_2]$ and satisfy the equations.

Then one has the following:

If the functions $f_i(x, y)$ have the properties above in a neighborhood $(\xi, \eta)_\epsilon$, then there exists in $(\xi, \eta)_\epsilon$ one and only one solution $y_i(x)$ of (5.4) through the initial point $(\xi, \eta)$. The solution $y_i(x)$ is defined and interior to $(\xi, \eta)_\epsilon$, at least on the interval $|x - \xi| \leq l$, where $l$ is the smaller of $\epsilon$ and $\frac{\epsilon}{M}$.

III. Differential Equations. The Imbedding Theorem

If in a neighborhood $R$ of the points $(x, y)$ on an arc $E$ having equations of the form

$$y_i = y_i(x),\ x_1 \leq x \leq x_2 \tag{5.6}$$

the functions $f_i(x, y)$ have the continuity properties 1) and 2) of II above, and if $E$ is a solution of the differential equations in II, then through every point $(\xi, \eta)$ sufficiently near to $E$ there passes one and but one solution

$$y_i = y_i(x, \xi, \eta) \tag{5.7}$$

of the differential equations. The functions $y_i(x, \xi, \eta)$ and their derivatives $y_{ix}(x, \xi, \eta)$ are continuous and satisfy the differential equations of II identically for all sets $(x, \xi, \eta)$ in a neighborhood $N$ of those belonging to E.

IV. Graves' General Implicit Function Theorem of 1927: Modern Wording

In what follows $(x)_\alpha$ is the set of all elements of a normed linear space which are at a distance strictly less than $\alpha$ from $x$.

Let $X$ be a normed linear space, $Y$ a Banach space, and $R_0$ a region in the Cartesian product space $X \times Y$. Let $G : R_0 \to Y$ and $(x_0, y_0)$ in $R_0$ satisfy

(H1) $G(x_0, y_0) = 0$
(H2) $G$ is of class $C^n$ on $R_0$
(H3) $d_y G(x_0, y_0; dy) = K_0(dy)$ has an inverse $L_0$

Then there exist positive constants $a$ and $b$ and a function $Y$ on $(x_0)_b$ to $(y_0)_a$ with the following properties:

(C1) the region $((x_0)_b, (y_0)_a)$ is contained in $R_0$
(C2) the point $(x, Y(x))$ is a solution of $G(x, y) = 0$ for every $x$ in $(x_0)_b$, and there is no other solution with the same $x$, having $y$ in $(y_0)_a$
(C3) the differential $d_y G(x, Y(x); dy)$ has an inverse for every $x$ in $(x_0)_b$
(C4) the function $Y$ is of class $C^n$ on $(x_0)_b$

V. The Problem of Bolza

The following is taken from [1487, p. 374].

Given the functional

$$U = \int_{t_0}^{t_1} f(y_1, \ldots, y_n, y_1', \ldots, y_n')\, dt + G(y_{10}, \ldots, y_{n0}; y_{11}, \ldots, y_{n1}) \tag{5.8}$$

to find a class of arcs satisfying $p$ differential and $q$ finite equations

$$\begin{aligned} &\phi_\alpha(y_1, \ldots, y_n, y_1', \ldots, y_n') = 0, \ \alpha = 1, 2, \ldots, p \\ &\psi_\beta(y_1, \ldots, y_n) = 0, \ \beta = 1, 2, \ldots, q \end{aligned} \tag{5.9}$$

as well as $r$ equations on the endpoints

$$\chi_\gamma(y_{10}, \ldots, y_{n0}; y_{11}, \ldots, y_{n1}) = 0, \ \gamma = 1, 2, \ldots, r \tag{5.10}$$

which minimizes $U$.

## Archival Materials (Abbreviations)

[Veblen Papers] Oswald Veblen Papers, Archives of the Library of Congress, Library of Congress, Washington, D.C.

[NARA Aberdeen] Record Group 156–866, National Archives and Records Administration, College Park, Maryland

[NARA Washington] Record Group 156.7.18, National Archives and Records Administration, College Park, Maryland

[NARA Ordnance School] Record Group 156–1512A, Box 2 , National Archives and Records Administration, Philadelphia, Pennsylvania
[NARA ENIAC] Record Group 156–1512A, Box 9, National Archives and Records Administration, Philadelphia, Pennsylvania
[Columbia University] University Archives and Columbiana Library, Columbia University in the City of New York
[Princeton University] University Archives Project, Seeley G. Mudd Manuscript Library, Princeton University, Princeton, New Jersey
[MIT] MIT Institute Archives and Special Collections, Massachusetts Institute of Technology, Cambridge, Massachusetts
[University of Chicago] Special Collections Research Center, University of Chicago Library, Chicago, Illinois
[Concord] Hoar Family Papers, Concord Free Public Library, Concord, Massachusetts
[Harvard] Office of Harvard War Records, Harvard University, Cambridge, Massachusetts

# References

## Preface


1. Chester, James, *The Theoretical Instruction of Gunners*, JUSA, **1**, No. 3, July 1892, pp. 171-206
2. Chester, James, *Whistler's Tables of Fire. Artillery Circular C*, Journal of the Military Service Institution of the United States, **13**, 1892, pp. 1028-1030
3. Chester, James, *Sea-Coast Artillery Instruction*, JUSA, **5**,1896, pp. 209-218
4. Hagood, Johnson, *Spotting for the Coast Artillery*, CAJ, **60**, No. 2, 1924, pp. 87-92
5. Hamilton, Alston, *Ballistics, Part II (Exterior Ballistics)*, Coast Artillery School Press, Fort Monroe, Virginia, 1908
6. Ingalls, James M., *Exterior Ballistics in the Plane of Fire*, D. Van Nostrand, New York, 1886
7. Schwartz, Phlilp, *The Variable Probable Error*, CAJ, **67**, 1927, pp. 405-413
8. Bennett, A. .A., The Problem of Exterior Ballistics, Army Ordnance, Vol. 3, No. 13, July-August 1922, pp. 38-40.
9. Schwartz, Philip, *More About Probable Error*, CAJ, **73**, 1930, pp. 111-120


## Introduction


10. Hamilton, Alston, *Ballistics, Part II (Exterior Ballistics)*, Coast Artillery School Press, Fort Monroe, Virginia, 1908
11. Stevenson, Craig, *Periodic Orbits of the Three Body Problem: F. R. Moulton's Quest for a New Lunar Theory*, Ph.D. Thesis, Open University, 2012
12. Stevenson, Craig, *Periodic Orbits: F. R. Moulton's Quest for a New Lunar Theory*, American Mathematical Society, 1921




A. Gluchoff, *From Fréchet Differentials to Firing Tables*,
https://doi.org/10.1007/978-3-031-68267-4

## Chapter 1

## Anonymous Editorials, Announcements, and Book Reviews in Military Journals


13. Announcement, JUSA **1**, 1892, pp. 3–4
14. Book Review, A Course of Instruction in Ordnance and Gunnery, Henry Metcalfe, JUSA, **1** 1892, pp. 74–76
15. Progress In Ballistics CAJ, **51**, 1919, pp. 69–71
16. Editorial, *Change Versus Progress*, JUSA, **56**, 1922, pp. 373–374
17. *FRANKFORD ARSENAL: Range, Deflection, and Wind Correction Computer*, Army Ordnance, March-April 1924, **4**, No. 23, pp. 339–341
18. Coast Artillerymen and the Army Ordnance School, CAJ, **2**, 1924, p.363
19. Ballistic Research at Marquette University, Army Ordnance, **7**, No.5, 1927, p. 399
20. Professor Moulton's Book, Army Ordnance, **8**, 1927, No. 1, pp. 12–15
21. *The Coast Artillery School, 1824–1927* CAJ, **66**, 1927, pp. 1–85
22. Book Review, New Methods in Exterior Ballistics, Forest Ray Moulton, CAJ, **66**, 1927, pp. 383–384
23. Book Review, Textbook of Ordnance and Gunnery, Earl McFarland, CAJ, **70**, 1929, pp. 533–534


## Manuscripts, Published Books and Articles


24. Alger, P. L., *Charts for the Calculation of the Effect of Small Changes in the Elements of Fire* JUSA, Vol. 51, 1919, pp. 585–603
25. Ames, J. S., *The Trained Man of Science in the War*, Science, New Series, 9, **480**, No. 1243, Oct. 25, 1918, pp. 401–410
26. Archibald, Thomas, Della Dumbaugh and Deborah Kent, *A Mobilized Community: Mathematicians in the United States during the First World War*, in *The War of Guns and Mathematics*, pp. 229–271
27. Askey, Richard, *Orthogonal Polynomials and Special Functions*, Society for Industrial and Applied Mathematics, Philadelphia, 1975
28. Aubin, David, and Catherine Goldstein, eds., *The War of Guns and Mathematics*, American Mathematical Society, 2014
29. Ball, H. P., *Of Responsible Command: A History of the U.S. Army War College*, Alumni Association of the United States Army War College, Carlisle Barracks, Pennsylvania, 1983
30. *Ballisticians in War and Peace, Volume I, A History of the United States Army Ballistic Research Laboratories 1914–1956*, Aberdeen Proving Ground, Maryland
31. Banach, S. *Théorie des Opérations Linéaires*, Second Edition, reprint of first edition of 1932, Chelsea Publishing Company, New York, 1955
32. Barrow-Green, June, *Poincaré and the Three Body Problem*, American Mathematical Society, 1997
33. Barrow-Green, June, *"Anti-aircraft guns all day long": Karl Pearson and computing for the Ministry of Munitions*, Revue d'histoire des mathématiques, **21**, fascicule 1, 2015, pp. 111–150
34. Beckman, H. S., *The Elements of Firing Table Construction*, Army Ordnance, **1**, No. 1, pp. 46–454

35. Bell, E. T., *The Development of Mathematics*, Macmillan and Co., 2nd edition, 1945
36. Bennett, A. A., *Introduction to Ballistics*, Engineer Reproduction Plant, Washington Barracks, D. C., 1921
37. Bennett, A. A., *The Problem of Exterior Ballistics*, Army Ordnance, **3**, July-August 1922, No. 13, pp. 38–40
38. Bennett, A. A., *Exterior Ballistic Tables Based on Numerical Integration*, United States Army, Ordnance Dept., Washington, Government Printing Office, 1924 (Document 1107), 1931, 1936
39. Bennett, A. A., *The Differential Analyzer*, Ballistic Research Laboratories, BRL Report No. 319, December, 1942
40. Benton, J. G., *A Course of Instruction in Ordnance and Gunnery, Third Edition*, D. Van Nostrand, New York, 1867
41. Berry, W. J., *Some Theorems Concerning Systems of Linear Partial Differential Equations*, Annals of Mathematics, **12**, No. 1, 1910, pp. 27–38
42. Birkhoff, Garrett, and Gian-Carlo Rota, *Ordinary Differential Equations*, Ginn and Co., Boston, 1962
43. Birkhoff, G., *Some Leaders In American Mathematics, 1891–1941*, in: Tarwater [1977], pp. 27–78
44. Birkhoff, G., and Erwin Kreyzig, *The Establishment of Functional Analysis*, Historia Mathematica, **11**, 1984 pp. 258–321
45. Birkhoff, G. D., *Boundary Value and Expansion Problems of Ordinary Linear Differential Equations*, Transactions of the American Mathematical Society, **9**, No. 4, 1908, pp. 373–395
46. Birkhoff, G. D., and O. D. Kellogg, *Invariant Points in Function Space*, Transactions of the American Mathematical Society, **23**, No. 1, January 1922, pp. 96–115
47. Bliss, Gilbert Ames, *An Existence Theorem for a Differential Equation of the Second Order, With an Application to the Calculus of Variations*, Trans. Am. Math. Soc., **5**, 1904, pp. 113–125
48. Bliss, Gilbert Ames, *The Solutions of Differential Equations of the First Order as Functions of Their Initial Conditions*, Ann. of Math., **6**, 1905, pp. 49–68
49. Bliss, Gilbert Ames, *A Generalization of the Notion of Angle*, Transactions of the American Mathematical Society, **7**, 1906, pp. 184–196
50. Bliss, Gilbert Ames, *The Scientific Work of Eliakim Hastings Moore*, Bull. Am. Math. Soc., **40**, 1910, pp. 207–213
51. Bliss, Gilbert Ames, *Integral Equations*, review of *An Introduction to the Study of Integral Equations* by Maxime Bôcher, *Bulletin of the American Mathematical Society*, **16**, No. 4, 1910, pp. 207–213
52. Bliss, Gilbert Ames, *A New Proof of the Implicit Function Theorem*, Bull. Am. Math. Soc., **18**, 1912, pp. 175–179
53. Bliss, Gilbert Ames, *Fundamental Existence Theorems*, The Princeton Colloquium, Part I, American Mathematical Society, 1913
54. Bliss, Gilbert Ames, *A Note on Functions of Lines*, Proc. Nat. Acad. Sci, U.S.A., **10**, 1915, pp. 173–175
55. Bliss, Gilbert Ames, review, V. Volterra, *Fonctions des Lignes*, Bull. Am. Math. Soc., **23**, 1915, pp. 344–355
56. Bliss, Gilbert Ames, *Solutions of Differential Equations as Functions of the Constants of Integration*, Bull. Amer. Math. Soc., **25**, 1918, pp. 15–26
57. Bliss, Gilbert Ames, *Differential Corrections*, NARA Aberdeen Ms A-29, No. 10, Nov. 1918
58. Bliss, Gilbert Ames, *The Problem of Mayer With Variable Endpoints*, Trans. Am. Math. Soc., **19**, 1918, pp. 305–316
59. Bliss, Gilbert Ames, *A Method of Computing Differential Corrections for a Trajectory*, CAJ/JUSA, **50**, 1918, pp. 455–460
60. Bliss, Gilbert Ames, *A Method of Computing Differential Corrections for a Trajectory*, CAJ/JUSA, **51**, 1919, pp. 445–449

61. Bliss, Gilbert Ames, *The Use of Adjoint Systems in the Problem of Differential Corrections for Trajectories*, JUSA, **51**, 1919, pp. 296–311
62. Bliss, Gilbert Ames, *Differential Equations Containing Arbitrary Functions*, Transactions of the American Mathematical Society, **210**, No. 9, 1920, pp. 79–92
63. Bliss, Gilbert Ames, *Functions of Lines in Ballistics*, ibid, pp. 93–106
64. Bliss, Gilbert Ames, *Calculus of Variations*, Mathematical Association of America, 1925
65. Bliss, Gilbert Ames, *A Boundary Value Problem for a System of Ordinary Differential Equations of the First Order*, Transactions of the American Mathematical Society, **28**, No. 4, 1926, pp. 561–584
66. Bliss, Gilbert Ames, *Mathematics*, Scientific Monthly, **24**, Issue 4, 1927, pp. 308–319
67. Bliss, Gilbert Ames, *Existence Theorems for Implicit Functions, Existence Theorems for Differential Equations*, University of Chicago, 1928
68. Bliss. Gilbert Ames, *The Problem of Lagrange in the Calculus of Variations*, Am. J. Math., **52**, No. 4, 1930, pp. 673–744
69. Bliss, Gilbert Ames, *Mathematical Interpretation of Geometrical And Physical Phenomena*, American Mathematical Monthly, **40**, 1933, pp. 472–480
70. Bliss, Gilbert Ames, *The Evolution of Problems of the Calculus of Variations*, The American Mathematical Monthly, **43**, No. 10, December 1936, pp. 598–609
71. Bliss, Gilbert Ames, *Mathematics for Exterior Ballistics*, John Wiley and Sons, New York, 1944
72. Bliss, Gilbert Ames, *Lectures on the Calculus of Variations*, The University of Chicago Press, Chicago, 1946
73. Bliss, Gilbert Ames, *Autobiographical Notes*, American Mathematical Monthly, **59**, November 1952, pp. 595–605
74. Bliss, Gilbert Ames, and Max Mason, *A Problem of the Calculus of Variations in Which the Integrand is Discontinuous*, Trans. Am. Math. Soc., **7**, No. 2, April 1906, pp. 325–336
75. Bliss, Gilbert Ames, and Max Mason, *The Properties of Curves in Space Which Minimize a Definite Integral*, Trans. Amer. Math. Soc., **9**, October 1908, pp. 440–466
76. Bliss, Gilbert Ames, and Max Mason, *Fields of Extremals in Space*, Transactions of the American Mathematical Society, **11**, 1910, pp. 325–340
77. Bôcher, M., *Green's Functions in Space of One Dimension*, Bulletin of the American Mathematical Society, **7**, 1901, pp. 297–299
78. Bôcher, M., *An Introduction to the Study of Integral Equations*, Cambridge Tracts in Mathematics and Mathematical Physics, No. 10, Cambridge University Press, 1909
79. Bôcher, M., *Boundary Problems and Green's Functions for Linear Differential and Difference Equations*, Annals of Mathematics, **13**, 1911–1912, pp. 71–88
80. Bôcher, M., *Boundary Problems in One Dimension*, in *Proceedings of the Fifth International Congress of Mathematicians, Cambridge, 22–28 August 1912*, Princeton University Press, 1913, pp. 163–195
81. Bôcher, M., *Applications and Generalizations of the Conception of Adjoint Systems*, Transactions of the American Mathematical Society, **14**, October 1913, pp. 403–420
82. Bolza, O., *Lectures on the Calculus of Variations*, University of Chicago Press, 1904, reprinted unabridged and corrected, Dover Publications, 1961
83. Bolza, Oskar, *Vorlesung Über Variationsrechnung*, Druck und Verlag von B.G. Teubner, 1909
84. Bolza, Oskar, *Integral Equations*, lecture notes by William Vernon Lovitt from a summer course given during summer 1913 at University of Chicago
85. Bruff, L. L., *A Textbook of Ordnance and Gunnery*, John Wiley and Sons, New York, 1896
86. Campbell, G. A., *Mathematics in Industrial Research*, Bell System Technical Journal, 1926, pp. 550–557
87. Carathéodory, C., review, Hadamard, *Leçons sur le calcul des variations*, **35**, Bulletin des Sciences mathématiques, pp. 123–141, 1911
88. Carll, L. B., *A Treatise On The Calculus of Variations*, MacMillan and Co., London, 1885

89. Carmichael, R. D., *The Provision Made by Mathematics for the Needs of Science*, Science, New Series, **45**, Issue 1168 (May 18, 1917), pp. 465–474
90. Case, Homer, *The Variable Probable Error and The Adjustment Problem*, CAJ, **68**, 1928, pp. 492–501
91. Charbonnier, P., *The Gâvre Commission: An Historical Sketch Of The Progress Of Exterior Ballistics* JUSA, **27**, 1907, pp. 238–262
92. Charbonnier, P., *The Gâvre Commission: An Historical Sketch Of The Progress Of Exterior Ballistics* JUSA, **28**, 1908, pp. 49–73
93. Charbonnier, P., *Sur L'État Actuel De La Balistique Extérieure Théorique*, Proceedings of the International Mathematical Congress held in Toronto, August 11–16, 1924, ed. Fields, John Charles, Nendeln, Kraus reprint, 1924 pp. 315–334
94. Coddington, E. A. and N. Levinson, *Theory of Ordinary Differential Equations*, McGraw-Hill Book Company, 1955
95. *The Confederate Field Manual*, Ritchie and Dunnavant, Richmond, 1862, reprinted 1984
96. Coxe, W. H. and Bugless. E., *Exterior Ballistics Charts*, E.I. du Pont de Nemours and Company, Inc., 1926
97. Craig, Thomas, *A Treatise on Linear Differential Equations*, New York, John Wiley and Sons, 1889
98. *The Applications of Elliptic Functions by Alfred George Greenhill, M.A., F.R.S., Professor of Mathematics in the Artillery College, Woolwich*, book review, Thomas Craig, JUSA, **2**, 1893, pp. 298–306
99. Cranz, C. and Becker, K., *Handbook of Ballistics, Volume I, Exterior Ballistics*, translated from the second German edition, H. M. Stationery Office, London, 1921
100. Darboux, G., *Sur les équations aux dérivés partielle*, Comptes Rendus, **96**, March 19, 1883, pp. 766–769
101. Darboux, G., *Leçons Sur La Théorie Générale Des Surfaces, Deuxieme Partie*, Gauthier-Villars et Fils, Paris, 1889
102. Darboux, G., *Leçons Sur La Théorie Générale Des Surfaces, Quatrième Partie*, Gauthier-Villars et Fils, Paris, 1896
103. Dederick, L. S., *Exterior Ballistics*, Bulletin of the American Mathematical Society, **34**, No. 5, 1928, p. 667
104. Dederick, L. S., *The Mathematics of Exterior Ballistics Computation*, American Mathematical Monthly, **47**, No. 9, November, 1940, pp. 628–634
105. Devorkin, D. H., *Henry Norris Russell, Dean of American Astronomers*, Princeton University Press, Princeton and Oxford, 2000
106. Dickson, L. E., *Mathematics in War Perspective*, Bulletin of the American Mathematical Society, **25**, No. 7, 1919, pp. 289–311
107. Dickson, T.C., *Course of Instruction Prescribed for Student Officers In the Ordnance School of Application in Differential Equations*, July 24, 1907, revised April 12, 1910, Government Printing Office, Washington
108. Dieudonné, J., *History of Functional Analysis*, Amsterdam: north Holland, 1981
109. Doerfler, R., Gluchoff,A., Guthery, S, and Paul Hamburg, *Calculating Curves*, Docent Press, 2012
110. Drach, Jules, *L'équation differentielle de la balistique extérieure et son integration par quadratures*, Annales Scientifiques de l'E.N.S., Serie 3, tome 37, 1920 pp. 1–94
111. Drenick, R., *The Perturbation Calculus in Missile Ballistics*, Journal of the Franklin Institute, **25**, 1951, pp. 423–436
112. Drenick, R., *Rudolph F. Drenick, Veteran*, interviewed by Mike Russert, Wayne Clark, November 18, 2003, South Setauket, New York, dmna.ny.gov/historic/veterans/transcripts/Drenick-Rudolf-F.pdf
113. du Bois-Reymond, P., *Über lineare partielle Differentialgleichungen zweiter Ordnung*, Journal für Mathematik, **104**, pp. 241–310
114. Evans, G. C., *Functionals and Their Applications*, American Mathematical Society Colloquium Lectures, **5**, Part I, 1918

115. Faulkner, F. D., *Optimum Interception of a Ballistic Missile at Intermediate Range*, Naval Postgraduate School, Monterey, California, October 1961
116. Faulkner, F. D., *A Comparison Between Some Methods For Computing Optimum Paths in the Problem of Bolza*, Research Paper No. 43, United States Naval Postgraduate School, Monterey, California, March 1964
117. Faulkner, F. D., *Newton's Method Applied To The Solution Of Classical Problems In Calculus Of Variations And Differential Games*, dissertation submitted in partial fulfillment of the requirements for the degree of Doctor of Philosophy in the University of Michigan, 1969
118. Feffer, Loren Butler, *Oswald Veblen and the Capitalization of American mathematics*, Isis, **89**, pp. 474–497
119. Fenster, Della, Kent, Deborah and Archibald, Thomas, *The First World War and Mathematics In the United States*, preprint
120. Fischer, C. A., *A Generalization of Volterra's Derivative of a Function of a Curve*, American Journal of Mathematics, **35**, No. 4, 1913, pp. 369–394
121. Fischer, C. A., *Note on the order of continuity of functions of lines*, Bulletin of the American Mathematical Society, **23**, (1916–1917), pp. 88–90
122. Flanagan, Francis, *Harmonic Functions*, Dover Press, 1972
123. Ford, L., *Instruction Memorandum No. 20, Subject: Elementary Mathematics For Field Artillery*, Headquarters, Field Artillery Central Officers Training School, Camp Zachary Taylor, Ky., 1919
124. Fowler, R. H., E. G. Gallop, C. N. H. Lock and H. W. Richmond, *The Aerodynamics of a Spinning Shell*, Philosophical Transactions of the Royal Society, A221, 1920, pp. 295–387
125. Franklin, Philip, *An Alternative Derivation of the Differential Equations Used in Computing Differential Corrections*, [NARA Aberdeen], No. 86, 1919
126. Franklin, Philip, *Exterior Ballistics, A Historical Survey*, Brown University Graduate School, 1943
127. *Philip Franklin of M.I.T. is Dead*, New York Times, 1/28/65
128. Fréchet, M. *Sur La Notion de la Differentielle D'Une Fonction de Ligne*, Transactions of the American Mathematical Society, **10**, 1913, pp. 135–161
129. Fuchs, l., Über Relationen, welche für die zwischen je zwei singularen Punkten erstrecken integrale der lösungen linearer differentialgleichungen stattfinden Journal für die reine und angewandte Mathematik, **76**, 1874, pp. 177–213
130. Galileo, *Two New Sciences*, A New Translation With Introduction and Notes, Stillman Drake, University of Wisconsin Press, 1974
131. Gasteyer, C.E., *Forest Ray Moulton, 1872–1952* , Biographical Memoirs of the National Academy of Sciences, **41**, pp. 340–355
132. Gateaux, R., *Fonctions d'une infinité de variables indépendantes*, Bulletin de la S. M. F., **47**, 1919, pp. 70–96
133. Glennon, J.H., *Accuracy and Probability of Fire*, Baltimore Press of Isaac Friedenwald, 1888
134. Gluchoff, Alan, *Pure Mathematics Applied in Early Twentieth-Century America: The case of T.H. Gronwall, Consulting Mathematician*, Historia Mathematica, **32**, 2005, pp. 312–357
135. Gluchoff, Alan, *Artillerymen and Mathematicians: Forest Ray Moulton and Changes in American Exterior Ballistics, 1885–1934*, Historia Mathematica, **38**, 2011, pp. 506–547
136. Goldstine, Herman H., *Conditions For A Minimum Of A Functional*, The University of Chicago Libraries, Chicago, Illinois, 1937
137. Goldstine, Herman. H., *A Multiplier Rule In Abstract Spaces*, Bulletin of the American Mathematical Society, **44**, 1938, pp. 388–394
138. Goldstine, Herman H., *The Computer from Pascal to von Neumann*, Princeton University Press, 1972
139. Oral History Transcript - Dr. Herman Goldstine, Interview with Dr. Herman Goldstine by Nancy Stern at Institute for Advanced Study, Princeton, March 14, 1977, http://www.aip.org/history/ohilist/4636.html
140. Goldstine, Herman H., *A History of the Calculus of Variations From the 17th Through the 19th Centuries*, Springer-Verlag, 1980

141. Goldstine, Herman H., interviewed in *The Princeton Mathematics Community in the 1930s*, Transcript Number 15, (PMC15), The Trustees of Princeton University, 1985
142. Goldstine, Herman, interview with author July 12, 2004
143. Goursat, Édouard, *Sur la Théorie des functions implicites*, Bull. de la S. M. F., **31**, 1903, pp. 184–192
144. Goursat, Édouard, *A Course In Mathematical Analysis*, Vol. II, Part II, translated by Hedrick, E.R., and Dunkel, Otto, Ginn and Co., 1917
145. Goursat, Édouard, *Cours d'Analyse mathématique, Tome III*, Troisième Edition, Gauthier-Villars, 1923
146. Graves, L., *Implicit Functions and Differential Equations in General Analysis*, Transactions of the American Mathematical Society, **29**, No. 3, July 1927, pp. 514–552
147. Graves, L., *Topics in the Functional Calculus*, Bulletin of the American Mathematical Society, **41**, 1935, pp. 641–662
148. Graves, L., *What Is A Functional?*, The American Mathematical Monthly, **55**, No. 8, October 1948, pp. 467–472
149. Graves, L. M., *Gilbert Ames Bliss, 1876–1951*, Bulletin of the American Mathematical Society, **58**, (1952), pp. 251–264
150. Graves, L., *The Theory of Functions of Real Variables*, $2^{nd}$ Edition, McGraw-Hill, 1956, reprinted by Dover Publications, 2009
151. Graves, L., *Nonlinear Mappings between Banach Spaces*, in *Studies in Real and Complex Analysis*, ed. I. I. Hirschman, The Mathematical Association of America, 1965, pp. 34–54
152. Graves, L., and T. H. Hildebrandt, *Implicit Functions and Their Differentials in General Analysis*, Transactions of the American Mathematical Society, **29**, No. 1, January 1927, pp. 127–153
153. Green, George, *An Essay on the Application of Mathematical Analysis to the Theories of Electricity and Magnetism*, Journal für Mathematik, **39**, 1850, pp. 73–89; **44**, 1852, pp. 356–374; **47**, 1854, pp. 161–221
154. Grier, David Alan, *Dr. Veblen Takes a Uniform: Mathematics in the First World War*, American Mathematical Monthly, **108**, December 2001, pp. 922–931
155. Grier, David Alan, *Dr. Veblen At Aberdeen: Mathematics, Military Applications and Mass Production*, in *Instrumental in War*, Boston, Brill, 2005, pp. 253–270
156. Grier, David Alan, *When Computers Were Human*, Princeton University Press, 2005
157. Gronwall, T. H., *Note on the Derivative with Respect to a Parameter of the Solutions of a System of Differential Equations*, Annals of Mathematics, Second Series, **20**, No. 4, July 1919, pp. 292–296
158. Gronwall, T. H., *On a System if Linear Partial Differential Equations of Hyperbolic Type*, Annals of Mathematics, 20, No. 4, July 1919, pp. 274–278
159. Gronwall, T. H., *On Differential Corrections and Weighting Factors*, NARA Washington
160. Gronwall, T.H., *Qualitative Properties of the Ballistic Trajectory*, Annals of Mathematics, Ser. II, Vol. 21, 1920, pp. 44–65
161. Gronwall, T. H., *Differential Variations In Ballistics With Applications to the Qualitative Properties of the Trajectory*, Transactions of the American Mathematical Society, **22**, No. 4, 1921, pp. 505–525
162. Guion, James L., review, *New Methods in Exterior Ballistics*, Army Ordnance, 1926, **7**, pp. 152–155
163. *Gunnery For Heavy Artillery*, Coast Artillery Corps, War Department, 1925
164. Hadamard, J., *Leçons sur la propagation des ondes*, Paris, Hermann, 1903
165. Hadamard, J., *Leçons sur le calcul des variations*, Hermann et fils, Paris, 1910
166. Hadamard, J., *Lectures on Cauchy's Problem in Linear Partial Differential Equations*, Yale University Press, 1923
167. Hadamard, J., *Le Développement et le role Scientifique Du Calcul Fonctionel*, Oevres de Jacques Hadamard, Tome I, Éditions du Centre National de la Recherche Scientifique, 1968, pp. 435–453

168. Hadamard, J., *Le Calcul Fonctionel*, Oevres de Jacques Hadamard, Tome I, Éditions du Centre National de la Recherche Scientifique 1968, pp. 2253–2266
169. Hagood, Johnson, *Spotting for the Coast Artillery*, CAJ, **60**, No. 2, 1924, pp. 87–92
170. Hamilton, Alston, *Ballistics, Part I*, Coast Artillery School Press, Fort Monroe, Virginia, 1908
171. Hamilton, Alston, *Ballistics, Part II (Exterior Ballistics)*, Coast Artillery School Press, Fort Monroe, Virginia, 1908
172. Hamilton, Alston, *A Practical Method of Computing High Angle Trajectories* NARA Aberdeen, No. 1, 1917
173. Hancock, Harris, *Lectures on the Calculus of Variations (The Weierstrassian Theory)*, University of Cincinnati Bulletin of Mathematics, 1904
174. Hancock, Harris, *Theory of Maxima and Minima*, Ginn and Co., 1917
175. Hardy, G. H., *A Mathematician's Apology*, Cambridge University Press, 1940
176. Hartman, Philip, *Ordinary Differential Equations*, John Wiley and Sons, 1964
177. Hatch, H.J., *A Reply to General Hagood*, CAJ, **60**, No. 3, 1924, pp. 171–176
178. Hedrick, H. B., review, Hadamard, J., *Leçons sur le calcul des variations*, Bull. Am. Math. Society, **21**, 1914, pp. 30–32
179. Hedrick, H. B., *Ballistic Effects Due To the Rotation of the Earth*, CAJ, **70**, No. 6, 1929, pp. 268–269
180. Hestenes, M., *The Calculus of Variations and Optimal Control Theory*, Wiley, New York, 1966
181. Hill, G. W., *Researches in the Lunar Theory*, American Journal of Mathematics, **1**, No. 1, 1876, pp. 5–21
182. Hitchcock, H. P., *Computation of Firing Tables for the U. S. Army*, Aberdeen Proving Ground, 1934
183. Hoar, Roger Sherman, *The Pickering Polaris Attachment*, JUSA, **50**, 1919, pp. 230–239
184. Hoar, Roger Sherman, *The New Ballistics*, JUSA, **51**, 1919, pp. 285–295
185. Hoar, Roger Sherman, *The New Artillery Grid Maps*, JUSA, **54**, 1921, 477–479
186. Hoar, Roger Sherman, *Problems for Solution* [2876–2882], American Mathematical Monthly, **28**, No. 2, Feb. 1921, pp. 89–90
187. Hoar, Roger Sherman, *A Course In Exterior Ballistics*, Government Printing Office, Washington, 1921
188. Hoar, Roger Sherman, *Discussions: On Proof by Mathematical Induction*, American Mathematical Monthly, **29**, No. 4, April 1922, pp. 162–169
189. Hoar, Roger Sherman, *The Rotation of the Earth*, ,**67**, No. 5 , 1927, pp. 377–388
190. Hoar, Roger Sherman, review, *New Methods in Exterior Ballistics* by Forest Ray Moulton, American Mathematical Monthly, **34**, No. 5, June-July 1927, pp. 325–326
191. Hoar, Roger Sherman, review, *New Methods In Exterior Ballistics* by Forest Ray Moulton, Army Ordnance, **7**, 1927, pp. 395–396
192. Hoar, Roger Sherman, *"New Methods in Exterior Ballistics"*, Army Ordnance, **9**, 1928, pp. 50–51
193. Hoar, Roger Sherman, *Curved versus Cartesian Coordinates in Ballistics*, Army Ordnance, **9**, No. 3, 1928, pp. 371–376
194. Hoar, Roger Sherman, entry in *National Cyclopedia of American Biography*, **50**, Ann Arbor, Michigan., p. 622
195. Hull, Gordon, *Some Aspects Of Physics In War And Peace*, Science, **51**, No. 1314, March 5, 1920, pp. 221–233
196. Hung, J. C., *Method of Adjoint Systems and its Engineering Applications*, The University of Tennessee Department of Electrical Engineering, Technical Note No. 1, Control Theory Group, Prepared for National Aeronautics and Space Administration, October 10, 1964
197. Hunter, Patti, *Drawing the Boundaries: Mathematical Statistics in 20th-Century America*, Historia Mathematica, **23**, 1996, pp. 7–30
198. Huntington, Edward V., *The Notion of Probable Error in Elementary Statistics*, Science, New Series, **66**, No. 1722, 1927, pp. 633–637

199. Ingalls, James M., *Exterior Ballistics in the Plane of Fire*, D. Van Nostrand, New York, 1886
200. Ingalls, James M., *Handbook of Problems in Direct Fire*, John Wiley and Sons, New York, 1890
201. Ingalls, James M., *Range Tables For The 12-Inch Cast-Iron B.L. Mortar*, JUSA **1**, 1892, pp. 55–61
202. Ingalls, James M., *General Ballistic Tables For Mortar Firing*, JUSA, **8**, No. 1, 1896, pp. 52–57
203. Ingalls, James M., *A New General Ballistic Table*, JUSA, **9**, No. 3, 1897, pp. 295–299
204. Ingalls, James M., *Ballistics Tables (Artillery Circular M)*, War Department Document 111, Government Printing Office, Washington, 1900
205. Irwin, F., *The Invariants of Linear Differential Expressions*, Proceedings of the American Mathematical Society, **44**, No. 1, 1908, pp. 3–60
206. Jackson, A. M., *Spotting*, CAJ, **61**, 1924, pp. 180–181
207. Jackson, D., *The Method of Numerical Integration in Exterior Ballistics*, Washington, Government Printing Office, 1921
208. Jackson, D., *The Human Significance of Mathematics*, American Mathematical Monthly, **35**, No. 8, October 1928, pp. 406–411
209. Jacobi, C. G. J., *Theoria Novi Multiplicatoris Systematiaequationum Differentialium Vulgarium Applicandi*, in *Gesammelte Werke*, Vierter Band, Cheslea Publishing Company, New York, N.Y., 1969, pp. 317–509
210. Jacobi, C. G. J., *Zur Theorie der Variations-Rechnung und der Differentialgleighungen*, Journal für Mathematik, **17**, 1837, p. 68 ff.
211. Johnson, J. J., *Computation of a Trajectory by Short Arcs*, CAJ, Vol. 55, 1921, pp. 50–67
212. Johnson, J. J., *An Introduction to the Construction of Nomographic or Alignment Charts*, Coast Artillery Journal, **59**, No. 1, July 1923, pp. 157–172
213. Kellogg, Oliver D., *Foundations of Potential Theory*, Springer, Berlin, 1929, reprinted by Dover Publications, 1953
214. Kent, R. H., *Projectile Design*, Army Ordnance, November-December 1921, **2**, No. 9, pp. 204–205
215. Kline, Morris, *Mathematical Thought from Ancient to Modern Times*, Oxford University Press, 1972
216. Kneser, Adolph, *Lehrbuch Der Variationsrechnung*, Friederich Vieweg Und Sohn, Braunschweig, 1900
217. Lagrange, Joseph Louis, *Essai D'une Nouvelle Méthode Pour Déterminer Les Maxima et Les Minima des Formules Integrales Indefinies*, Misc. Soc. Turin, **2**, 1762, pp. 173–195
218. Lagrange, Joseph Louis, *Solution de differens problemes de calcul integral*, Misc. Tour. **3**, 1762 1765, p. 179
219. Lamson, K., *A General Implicit Function Theorem With An Application To Problems Of Relative Minima*, American Journal of Mathematics, **41**, No. 4, October 1920, pp. 243–256
220. Lecat, Maurice, *Bibliographie du Calcul des Variations, 1850–1913*, Ghent and Paris, 1913
221. Legendre, A. L., *Mémoire sure la manière de distinguer les maximas des minimas dans le calcul des variations*, Mémoires de l'Académie des Sciences, 1786
222. Leitmann, G., ed., *Optimization Techniques*, Academic Press, New York, 1962
223. Le Stourgeon, E., *Minima of Functions of Lines*, Trans. Am. Math. Soc., **21**, No. 4, October 1920, pp. 357–383
224. Liouville, Joseph, *Second Mémoire Sur le developpment des functions ou parties de fonctions en series*, Journal des Mathématiques, **2**, 1837, pp. 16–55
225. Lissak, Ormand, *Ordnance and Gunnery*, John Wiley and Sons, New York, 1908
226. Lissak, Ormand, *Ordnance and Gunnery*, John Wiley and Sons, New York, 1915
227. Lorch, E. R., *Joseph Fels Ritt*, Bulletin of the American Mathematical Society, **57**, No. 4, 1951, pp. 307–318
228. Love, A. E. H., *A Treatise on the Mathematical Theory of Elasticity*, Fourth Edition, Cambridge University Press, 1927, reprinted by Dover Publications, New York, 1944

229. Lovitt, William Vernon, *Linear Integral Equations*, McGraw-Hill Book Company, 1924, reprinted by Dover Publications, 1950
230. Mason, Max, *Zur Theorie de Randwertaufgaben*, Mathematische Annalen, **63**, 1904, pp. 528–544
231. Mason, Max, *Green's Theorem and Green's Functions for Certain Systems of Differential Equations*, Transactions of the American Mathematical Society, **5**, 1904, pp. 220–225
232. Mason, Max, *On the boundary Value Problems of Linear Ordinary Differential Equations of the Second Order*, Transactions of the American Mathematical Society, **7**, 1906, pp. 340–344
233. Mason, Max, *On the Differential Equation of the Hyperbolic Type*, Mathematische Annalen, **65**, 1908, pp. 570–575
234. Mason, Max, *Selected Topics in the Theory of Boundary Value Problems of Differential Equations*, in The New Haven Mathematical Colloquium, Yale University Press, 1910, pp. 173–222
235. Mayer, A., *Über Das allgemeinste Problem der Variations-rechnung bei einer einzigen unabhangingen Variabeln*, Leip. Ber., **30**, 1878, pp. 16–32
236. Mayer, A., *Begründung der Lagrange'schen Multiplicatorenmethode in der Variationsrechnung*, Math. Ann., **26**, 1886, pp. 74–82
237. McCalla, T. R., *Application of the Adjoint System of Differential Equations in the Solution of the Bang-Bang Control Problem*, United States Naval Postgraduate School, Monterey, California, 1961
238. McFarland, Earl, *Textbook of Ordnance and Gunnery*, John Wiley and Sons, New York, 1929
239. McShane, E. J., *Differentials of Certain Functions in Exterior Ballistics*, Duke Mathematical Journal, **17**, 1950, pp. 115–134
240. McShane, E. J., *Gilbert Ames Bliss, 1876–1951 A Biographical Memoir by E.J. McShane*, National Academy of Sciences, 1958, pp. 32–53
241. McShane, E. J., *The Calculus of Variations from the Beginning through Optimal Control Theory*, SIAM Control and Optimization, **27**, No. 5, September 1989, pp. 916–939
242. McShane, Edward J., Kelley, John L., Reno, Franklin V., *Exterior Ballistics*, The University of Denver Press, Denver, Colorado, 1953
243. Meigs, J.F. and Ingersoll, R.R., *Exterior Ballistics*, Issac Friedenwald, Baltimore, 1885
244. Metcalfe, Henry, *A Course Of Instruction in Ordnance and Gunnery*, John Wiley and Sons, New York, 1891
245. Metcalfe, Henry, *Ordnance and Gunnery Part II: Cannon, Carriages, Small Arms, Ammunition and Accuracy*, West Point, United States Military Academy Press, 1891
246. Milne, W. E., *Equations of Differential Variations in Exterior Ballistics*, CAJ, **51**, 1919, pp. 154–159
247. Milne, W. E., review, *Mathematics For Exterior Ballistics*, Mathematics Reviews, MR 0010480; (6.24d) 71.0X
248. Moore, E. H., *On a Form of General Analysis With Applications to Linear Differential and Integral Equations*, Atti del IV Congresso Internazionale dei Matematici, **1**, Roma 1908, **2**, Roma 1909, pp. 98–114
249. Moore, E. H., E. J. Wilczynski and Max Mason, *The New Haven Mathematical Colloquium*, Yale University Press, New Haven, 1910
250. Moulton, Forest Ray, *On Certain Rigorous Methods of Treating Problems in Celestial Mechanics*, Decennial Publications of the University of Chicago, **8**, No. 191, (42), 1902
251. Moulton, Forest Ray, *The Influence of Astronomy on Mathematics*, Science, New Series, **33**, No. 835, March 10, 1911, pp. 357–364
252. Moulton, Forest Ray, *M. Henri Poincaré*, Popular Astronomy, **20**, No. 10, December 1912, pp. 620–634
253. Moulton, Forest Ray, *An Introduction To Celestial Mechanics*, Second Revised Edition, Macmillan Company, 1914, reprinted by Dover Publications, New York, 1970
254. Moulton, Forest Ray, *Lectures on the Modern Theory of Differential Equations*, University of Chicago, 1916–1917

255. Moulton, Forest Ray, *History of the Ballistics Branch of the Artillery Ammunition Section, Engineering Division of the Ordnance Department for the Period April 6, 1918 to April 2, 1919*, War Department, Office of the Chief of Ordnance, Washington, 1919
256. Moulton, Forest Ray, *Numerical Integration of Differential Equations*, CAJ, **51**, 1919, pp. 40–55
257. Mouton, Forest Ray, *Periodic Orbits*, Coast Artillery Journal, *57*, 1922, p. 587
258. Mouton, Forest Ray, *Numerical Solution of Differential Equations*, in *Smithsonian Mathematical Formulae and Tables of Elliptic Functions*, Smithsonian Miscellaneous Collections, **74**, No. 1, Publication 2672, Smithsonian Institution, 1922
259. Moulton, Forest Ray, *New Methods In Exterior Ballistics*, University of Chicago Press, 1926, reprinted by Dover Publications, 1962
260. Moulton, Forest Ray, discussion of review of *New Methods in Exterior Ballistics* by Roger Sherman Hoar, American Mathematical Monthly, **35**, No. 5, May 1928, pp. 246–250
261. Moulton, Forest Ray, reply to a review of *New Methods in Exterior Ballistics* by Roger Sherman Hoar, Army Ordnance, **8**, 1928, pp. 320–321
262. Moulton, Forest Ray, *Differential Equations*, MacMillan Company 1930, reprinted by Dover Publications, New York, 1958
263. Moulton, F. R., entry in *Notable Scientists: From 1900 to the Present*, Gale Group, 2001, Anonymous.
264. Nenninger, T.K., *The Leavenworth Schools and the Old Army: Education, Professionalism, and the Officer Corps of the United States Army, 1881–1918*, Greenwood Press, Westport, Connecticut, 1978
265. d'Ocagne, Maurice, *Traité de Nomographie, Deuxième Édition*, Gauthier Villars, Paris, 1921
266. Osgood, William Fogg, *Advanced Calculus*, The MacMillan Company, New York, 1925
267. Paaswell, G., *An Appeal to Producing Mathematicians*, Bulletin of the American Mathematical Society, **21**, No. 3, 1914, pp. 127–130
268. Pachpatte, B. G., *Inequalities for Differential and Integral Equations*, San Diego Press, 1998
269. Parshall, Karen Hunger, and Rowe, David E., *The Emergence of the American Mathematical Research Community, 1876–1900: J. J. Sylvester, Felix Klein, and E. H. Moore*, American Mathematical Society, 1994
270. Picard, É., *Mémoire sur la théorie des éqations aux dérivées partielles et la méthode des approximations successives*, Journal des Mathematiques, series 4, **6**, 1890, pp. 145–210
271. Picard, É. *Sur l'application des méthodes d'approximations successives à l'étude de certaines équations différentielles ordinaires*, Journal des Mathematiques series 4, **9**, 1893, pp. 217–271
272. Pincherle, S., *Sull'operazione aggiunta*, Rendiconto della R. Accademia delle Science dell'Instituto di Bologna, **2**, (1897 1898), pp. 130 139
273. Pincherle, S., *Sull'operazione agguinta di Lagrange*, Annali de metematica pur ed applicata, **21**, 1913 pp. 143–151
274. Poincaré, Henri, *Les Méthodes Nouvelles de la Mécanique Céleste , Tome 1*, Dover Publications, New York, reprint of original 1892 edition
275. *Lecture on The Probability and Dispersion of Fire*, translated at the Army War College From a French Document, Government Printing Office, 1917
276. *Provisional Range Tables for the British 75-mm. Gun, Model 1917*, Ordnance Department, U.S. Army, War Plans Division, Government Printing Office, 1918
277. Rademacher, Hans, *On the Accumulation of Errors In Numerical Integration on the ENIAC*, in *Theory and Techniques for Design of Electronic Digital Computers*, Lectures Given at the Moore School, 8 July 1946–31 August 1946, Volume II, Lecture 19, University of Pennsylvania, Moore School of Electrical Engineering, Philadelphia, Pennsylvania, November 1, 1947
278. *Range Tables For 75-mm. Model 1916 American and Model 1917 British Guns*, Prepared At Aberdeen Proving Ground, Engineer Reproduction Plant, U.S. Army, Washington Barracks, November 1921

279. Riemann, Bernhard, *Über die Fortpflanzung ebener Luftwellen von endlicher Schwingingungsweite*, Göttingen Abhandl., **8**, 1860, pp. 145–164
280. Ritt, J., *On Weighting Factor Curves For Low Elevation*, NARA Aberdeen, Technical Report, February 25, 1919
281. Ritt, J., *On the Differentiability of the Solution of a Differential Equation With Respect to a Parameter*, Annals of Mathematics, Second Series, **20**, No. 4, July 1919, pp. 289–291
282. Ritt, J., *On Weighting Factors in Flat Fire*, CAJ, **53**, 1920, pp. 404–410
283. Ritt, J., *On the Iteration of Rational Functions*, Transactions of the American Mathematical Society, **21**, 1920, pp. 313–320
284. Roberts, David L., *Albert Henry Wheeler (1873–1950): A Case Study in the Stratification of American Mathematical Activity*, Historia Mathematica, **23**, 1996, pp. 269–287
285. Roberts, David L., *Republic of Numbers*, Johns Hopkins University Press, 2019
286. Rosenstein, George M., *The Best Method: American Calculus Textbooks of the Nineteenth Century*, from *A Century of Mathematics In America, Part III*, ed. Duren, P., Askey, R., Merzbach, U., and Edwards, H., Providence, Rhode Island, American Mathematical Society, 1988–1989, pp. 77–109
287. Roth, William E., *The Exact Differential Equations Of A Trajectory In Terms Of The Hoar Curvilinear Coordinates*, CAJ, **68**, 1928, pp. 505–512
288. Rowe, J. E., *Moulton on Exterior Ballistics*, American Mathematical Monthly, **35**, No. 4, March-April 1928, pp. 229–232
289. Ruckman, John W., *The Effect Of Wind On The Motion Of A Projectile*, JUSA, **1**, 1892, pp. 5–24
290. Salmon, E. H., *Columns*, Oxford Technical Publications, Henry Frowde and Hodder and Stoughton, London, 1921
291. Sanchez, D., *Ordinary Differential Equations and Stability Theory*, Freeman and Company, 1968
292. Schwartz, Philip, *A Critical Survey of the Advances in Practical Ballistics in the Ordnance Department, U.S. Army Since 1917*, [NARA Abderdeen] A-I-31, 1920
293. Schwartz, Philip, *Measuring the Life of a Gun*, Army Ordnance, **5**, No. 2, 1924, pp. 111–112
294. Schwartz, Philip, *Accuracy Firing*, Army Ordnance, **5**, No. 3, 1924, pp. 521–522
295. Schwartz, Philip, *Experimental Investigation of The Effect of the Wind Upon the Motion of a Projectile*, CAJ, **60**, 1924, pp. 93–97
296. Schwartz, Philip, *Studies of Motion of Projectiles*, CAJ, **60**, 1924, pp. 430–431
297. Schwartz, Philip, *The English Long Range Gun*, CAJ, **61**, 1924, pp. 91–92
298. Schwartz, Philip, *Weighting Factors*, CAJ, **61**, 1924, pp. 371–372
299. Schwartz, Philip, *Remarks Concerning the Use of Firing Table Values*, CAJ, **3**, 1925, pp. 399–400
300. Schwartz, Phlilp, *The Variable Probable Error*, CAJ, **67**, 1927, pp. 405–413
301. Schwartz, Philip, *Much About Little* CAJ, **68**, 1928, pp. 291–297
302. Schwartz, Philip, *The Fork*, Field Artillery Journal, **18**, No. 3, 1929, pp. 604–609
303. Schwartz, Philip, *More About Probable Error*, CAJ, **73**, 1930, pp. 111–120
304. Siacci, F., *Balistique Extérieure*, Traduction Annoteé par P. Laurent, Berger-Levrault et Cic, Editours, Paris and Nancy, 1892
305. Siegmund-Schultze, R.,*Eliakim Hastings Moore's "General Analysis"*, Arch. Hist. Exact. Sci., **52**, 1998, pp. 51–89
306. Siegmund-Schultze, R., *The Late Arrival of Academic Applied Mathematics in the United States: a Paradox, Theses, and Literature*, NTM, **11**, 2002, pp. 116–12
307. Siegmund-Schultze, R., *The Ideology of Applied Mathematics Within Mathematics in Germany and The U. S. Until the End of World War II*, ILUIL, **27**, 2004, pp. 791–811
308. Sleight, E. R., *Mathematics Applied*, National Mathematics Magazine, **9**, No. 8, May 1935, pp. 219–223
309. Smith, Paul A., *Joseph Fels Ritt, 1893–1951*, United States National Academy of Sciences, Biographical Memoir, 1956, pp. 253–264

310. Spinrad, William, *Early History of The Ordnance School*, The Ordnance Sergeant **3**, No. 1, 1942. pp. 1–4
311. Stackel, P., *Abhandlungen Über Variations-Rechnung, Zweiter Theil*, Leipzig, Verlag Von Wilhelm Engelmann, 1894
312. Steele, Bret D., *Muskets and Pendulums: Benjamin Robins, Leonhard Euler, and the Ballistics Revolution*, Technology and Culture, **35**, No. 2, April 1994, pp. 348–382
313. Stevenson, Craig, *Periodic Orbits of the Three Body Problem: F. R. Moulton's Quest for a New Lunar Theory*, Ph.D. Thesis, Open University, 2012
314. Stevenson, Craig, *Periodic Orbits: F. R. Moulton's Quest for a New Lunar Theory*, American Mathematical Society, 2021
315. Tannehill, I. R., *The Ballistic Wind*, Journal of the United States Artillery, **52**, 1920, pp. 553–576
316. Tarwater, J. D., (Ed.), *The Bicentennial Tribute to American Mathematics, 1776–1976* Papers Presented to the Fifty-Ninth Annual Meeting of the Mathematical Association of America, Mathematical Association of America, Washington, D. C., 1977
317. Todhunter, I., *A Treatise on the Integral Calculus and its Applications*, MacMillan and Co., London, 1880
318. Tonelli, L., *Fondamenti di calcolo delle variazioni*, Nicola Zanichelli, I, 1921; II, 1923
319. Tropp, H. S., *Forest Ray Moulton*, Dictionary of Scientific Biography, **9**, pp. 552–553
320. Truesdell, Clifford, review of *Exterior Ballistics* by McSchane, Kelley and Reno, Scripta Mathematica, **20**, 1954, pp. 172–174
321. Tschappat, W.H., *Ballistics*, article in Encyclopedia Britannica, 1922, pp. 383–394
322. Tsien, H. S., *Engineering Cybernetics*, McGraw-Hill Book Co., 1954
323. Veblen, Oswald, *Ballistic Theories and the Preparation of Range Tables*, NARA Aberdeen, No. 54, 1921
324. Veblen, O., *Professor Veblen on Ballistics in the World War*, Army Ordnance, **4**, March-April 1924, pp. 337–338
325. Very, Edward W., *Computation of Range Tables*, Annapolis, Md., 1877
326. Volterra, V., *Leçons sur les Fonctions des Lignes*, Paris, Gauthier- Villars, 1913
327. Wertheimer, Albert, *A Method of Computing Differential Corrections For a Trajectory*, Journal of the Franklin Institute, **217**, No. 6, June 1934
328. Wiener, Norbert *Ex-Prodigy*, MIT Press, Cambridge, 1953
329. Williamson, Benjamin, *An Elementary Treatise on The Integral Calculus*, Longmans, Green and Co., London, 1916
330. Zeidler, E., *Applied Functional Analysis*, Springer-Verlag, 1995
331. www.math.colostate.edu/~pauld/M345/Some20theory.pdf

## Chapter 2

## Anonymous Editorials, Announcements, and Book Reviews in Military Journals

332. Announcement, JUSA **1**, 1892, pp. 3–4
333. Book Review, A Course of Instruction in Ordnance and Gunnery, Henry Metcalfe, JUSA, **1** 1892, pp. 74–76
334. Book Review, *Balistique Extérieure Rationelle (Problèmes balisiques secondaires)*, P. Carbonnier, JUSA, **27**, 1907, pp. 310–311
335. Progress In Ballistics CAJ, **51**, 1919, pp. 69–71
336. Editorial, *Change Versus Progress*, JUSA, **56**, 1922, pp. 373–374

337. *FRANKFORD ARSENAL: Range, Deflection, and Wind Correction Computer*, Army Ordnance, March-April 1924, **4**, No. 23, pp. 339–341
338. Coast Artillerymen and the Army Ordnance School, CAJ, **2**, 1924, p.363
339. Ballistic Research at Marquette University, Army Ordnance, **7**, No.5, 1927, p. 399
340. Professor Moulton's Book, Army Ordnance, **8**, 1927, No. 1, pp. 12–15
341. *The Coast Artillery School, 1824–1927* CAJ, **66**, 1927, pp. 1–85
342. Book Review, New Methods in Exterior Ballistics, Forest Ray Moulton, CAJ, **66**, 1927, pp. 383–384
343. Book Review, Textbook of Ordnance and Gunnery, Earl McFarland, CAJ, **70**, 1929, pp. 533–534

## Manuscripts, Published Books and Articles

344. Alger, P. L., *The Groundwork of Practical Naval Gunnery*, 2nd Edition, Annapolis, Md., The United States Naval Institute, 1917
345. Alger, P. L., *Charts for the Calculation of the Effect of Small Changes in the Elements of Fire* JUSA, Vol. 51, 1919, pp. 585–603
346. Ames, J. S., *The Trained Man of Science in the War*, Science, New Series, 9, **480**, No. 1243, Oct. 25, 1918, pp. 401–410
347. Archibald, Thomas, Della Dumbaugh and Deborah Kent, *A Mobilized Community: Mathematicians in the United States during the First World War*, in *The War of Guns and Mathematics*, pp. 229–271
348. Arnaud, J. J., *Under-Water Trajectories*, discussion, Journal of the U.S. Naval Institute, **47**, 1921, pp. 1417–1418
349. Askey, Richard, *Orthogonal Polynomials and Special Functions*, Society for Industrial and Applied Mathematics, Philadelphia, 1975
350. Aubin, David, and Catherine Goldstein, eds., *The War of Guns and Mathematics*, American Mathematical Society, 2014
351. Ball, H. P., *Of Responsible Command: A History of the U.S. Army War College*, Alumni Association of the United States Army War College, Carlisle Barracks, Pennsylvania, 1983
352. *Ballisticians in War and Peace, Volume I, A History of the United States Army Ballistic Research Laboratories 1914–1956*, Aberdeen Proving Ground, Maryland
353. Banach, S. *Théorie des Opérations Linéaires*, Second Edition, reprint of first edition of 1932, Chelsea Publishing Company, New York, 1955
354. Barrow-Green, June, *Poincaré and the Three Body Problem*, American Mathematical Society, 1997
355. Barrow-Green, June, *"Anti-aircraft guns all day long": Karl Pearson and computing for the Ministry of Munitions*, Revue d'histoire des mathématiques, **21**, fascicule 1, 2015, pp. 111–150
356. Beckman, H. S., *The Elements of Firing Table Construction*, Army Ordnance, **1**, No. 1, pp. 46–454
357. Bell, E. T., *The Development of Mathematics*, Macmillan and Co., 2nd edition, 1945
358. Bennett, A. A., *Physical Bases of Ballistic Table Computations, Ordnance Textbook*, Washington Government Printing Office, 1920
359. Bennett, A. A., *Introduction to Ballistics*, Engineer Reproduction Plant, Washington Barracks, D. C., 1921
360. Bennett, A. A., *The Problem of Exterior Ballistics*, Army Ordnance, **3**, July-August 1922, No. 13, pp. 38–40
361. Bennett, A. A., review, *Traité de Balistique Extérieure, Vol. II*, P. Charbonnier, Bulletin of the American Mathematical Society, **34**, No. 4, 1928, pp. 517–518

362. Bennett, A. A., *Exterior Ballistic Tables Based on Numerical Integration*, United States Army, Ordnance Dept., Washington, Government Printing Office, 1924 (Document 1107), 1931, 1936
363. Bennett, A. A., *The Differential Analyzer*, Ballistic Research Laboratories, BRL Report No. 319, December, 1942
364. Benton, J. G., *A Course of Instruction in Ordnance and Gunnery, Third Edition*, D. Van Nostrand, New York, 1867
365. Berry, W. J., *Some Theorems Concerning Systems of Linear Partial Differential Equations*, Annals of Mathematics, **12**, No. 1, 1910, pp. 27–38
366. Birkhoff, Garrett, and Gian-Carlo Rota, *Ordinary Differential Equations*, Ginn and Co., Boston, 1962
367. Birkhoff, G., *Some Leaders In American Mathematics, 1891–1941*, in: Tarwater [1977], pp. 27–78
368. Birkhoff, G., and Erwin Kreyzig, *The Establishment of Functional Analysis*, Historia Mathematica, **11**, 1984 pp. 258–321
369. Birkhoff, G. D., *Boundary Value and Expansion Problems of Ordinary Linear Differential Equations*, Transactions of the American Mathematical Society, **9**, No. 4, 1908, pp. 373–395
370. Birkhoff, G. D., and O. D. Kellogg, *Invariant Points in Function Space*, Transactions of the American Mathematical Society, **23**, No. 1, January 1922, pp. 96–115
371. Bliss, Gilbert Ames, *An Existence Theorem for a Differential Equation of the Second Order, With an Application to the Calculus of Variations*, Trans. Am. Math. Soc., **5**, 1904, pp. 113–125
372. Bliss, Gilbert Ames, *The Solutions of Differential Equations of the First Order as Functions of Their Initial Conditions*, Ann. of Math., **6**, 1905, pp. 49–68
373. Bliss, Gilbert Ames, *A Generalization of the Notion of Angle*, Transactions of the American Mathematical Society, **7**, 1906, pp. 184–196
374. Bliss, Gilbert Ames, *The Scientific Work of Eliakim Hastings Moore*, Bull. Am. Math. Soc., **40**, 1910, pp. 207–213
375. Bliss, Gilbert Ames, *Integral Equations*, review of *An Introduction to the Study of Integral Equations* by Maxime Bôcher, *Bulletin of the American Mathematical Society*, **16**, No. 4, 1910, pp. 207–213
376. Bliss, Gilbert Ames, *A New Proof of the Implicit Function Theorem*, Bull. Am. Math. Soc., **18**, 1912, pp. 175–179
377. Bliss, Gilbert Ames, *Fundamental Existence Theorems*, The Princeton Colloquium, Part I, American Mathematical Society, 1913
378. Bliss, Gilbert Ames, *A Note on Functions of Lines*, Proc. Nat. Acad. Sci, U.S.A., **10**, 1915, pp. 173 175
379. Bliss, Gilbert Ames, review, V. Volterra, *Fonctions des Lignes*, Bull. Am. Math. Soc., **23**, 1915, pp. 344–355
380. Bliss, Gilbert Ames, *Solutions of Differential Equations as Functions of the Constants of Integration*, Bull. Amer. Math. Soc., **25**, 1918, pp. 15–26
381. Bliss, Gilbert Ames, *Differential Corrections*, NARA Aberdeen Ms A-29, No. 10, Nov. 1918
382. Bliss, Gilbert Ames, *The Problem of Mayer With Variable Endpoints*, Trans. Am. Math. Soc., **19**, 1918, pp. 305–316
383. Bliss, Gilbert Ames, *A Method of Computing Differential Corrections for a Trajectory*, CAJ/JUSA, **50**, 1918, pp. 455–460
384. Bliss, Gilbert Ames, *A Method of Computing Differential Corrections for a Trajectory*, CAJ/JUSA, **51**, 1919, pp. 445–449
385. Bliss, Gilbert Ames, *The Use of Adjoint Systems in the Problem of Differential Corrections for Trajectories*, JUSA, **51**, 1919, pp. 296–311
386. Bliss, Gilbert Ames, *Differential Equations Containing Arbitrary Functions*, Transactions of the American Mathematical Society, **210**, No. 9, 1920, pp. 79–92
387. Bliss, Gilbert Ames, *Functions of Lines in Ballistics*, ibid, pp. 93–106
388. Bliss, Gilbert Ames, *Calculus of Variations*, Mathematical Association of America, 1925

389. Bliss, Gilbert Ames, *A Boundary Value Problem for a System of Ordinary Differential Equations of the First Order*, Transactions of the American Mathematical Society, **28**, No. 4, 1926, pp. 561–584
390. Bliss, Gilbert Ames, *Mathematics*, Scientific Monthly, **24**, Issue 4, 1927, pp. 308–319
391. Bliss, Gilbert Ames, *Existence Theorems for Implicit Functions, Existence Theorems for Differential Equations*, University of Chicago, 1928
392. Bliss. Gilbert Ames, *The Problem of Lagrange in the Calculus of Variations*, Am. J. Math., **52**, No. 4, 1930, pp. 673–744
393. Bliss, Gilbert Ames, *Mathematical Interpretation of Geometrical And Physical Phenomena*, American Mathematical Monthly, **40**, 1933, pp. 472–480
394. Bliss, Gilbert Ames, *The Evolution of Problems of the Calculus of Variations*, The American Mathematical Monthly, **43**, No. 10, December 1936, pp. 598–609
395. Bliss, Gilbert Ames, *Mathematics for Exterior Ballistics*, John Wiley and Sons, New York, 1944
396. Bliss, Gilbert Ames, *Lectures on the Calculus of Variations*, The University of Chicago Press, Chicago, 1946
397. Bliss, Gilbert Ames, *Autobiographical Notes*, American Mathematical Monthly, **59**, November 1952, pp. 595–605
398. Bliss, Gilbert Ames, and Max Mason, *A problem of the Calculus of Variations in Which the Integrand is Discontinuous*, Trans. Am. Math. Soc., **7**, No. 2, April 1906, pp. 325–336
399. Bliss, Gilbert Ames, and Max Mason, *The Properties of Curves in Space Which Minimize a Definite Integral*, Trans. Amer. Math. Soc., **9**, October 1908, pp. 440–466
400. Bliss, Gilbert Ames, and Max Mason, *Fields of Extremals in Space*, Transactions of the American Mathematical Society, **11**, 1910, pp. 325–340
401. Bôcher, M., *Green's Functions in Space of One Dimension*, Bulletin of the American Mathematical Society, **7**, 1901, pp. 297–299
402. Bôcher, M., *An Introduction to the Study of Integral Equations*, Cambridge Tracts in Mathematics and Mathematical Physics, No. 10, Cambridge University Press, 1909
403. Bôcher, M., *Boundary Problems and Green's Functions for Linear Differential and Difference Equations*, Annals of Mathematics, **13**, 1911–1912, pp. 71–88
404. Bôcher, M., *Boundary Problems in One Dimension*, in *Proceedings of the Fifth International Congress of Mathematicians, Cambridge, 22–28 August 1912*, Princeton University Press, 1913, pp. 163–195
405. Bôcher, M., *Applications and Generalizations of the Conception of Adjoint Systems*, Transactions of the American Mathematical Society, **14**, October 1913, pp. 403–420
406. Bolza, O., *Lectures on the Calculus of Variations*, University of Chicago Press, 1904, reprinted unabridged and corrected, Dover Publications, 1961
407. Bolza, Oskar, *Vorlesung Über Variationsrechnung*, Druck und Verlag von B.G. Teubner, 1909
408. Bolza, Oskar, *Integral Equations*, lecture notes by William Vernon Lovitt from a summer course given during summer 1913 at University of Chicago
409. Bruff, L. L., *A Textbook of Ordnance and Gunnery*, John Wiley and Sons, New York, 1896
410. Campbell, G. A., *Mathematics in Industrial Research*, Bell System Technical Journal, 1926, pp. 550–557
411. Carathéodory, C., review, Hadamard, *Leçons sur le calcul des variations*, **35**, Bulletin des Sciences mathématiques, pp. 123–141, 1911
412. Carll, L. B., *A Treatise On The Calculus of Variations*, MacMillan and Co., London, 1885
413. Carmichael, R. D., *The Provision Made by Mathematics for the Needs of Science*, Science, New Series, **45**, Issue 1168 (May 18, 1917), pp. 465–474
414. Case, Homer, *The Variable Probable Error and The Adjustment Problem*, CAJ, **68**, 1928, pp. 492–501
415. Charbonnier, P., *Balistique Extérieure Rationelle (Problèmes Balistique Secondaires)*, Encyclopèdie Scientifique de Méchanique Appliqué et Génie, M. D'Ocagne, directeur, Octave Doin, Paris, 1907

416. Charbonnier, P., *The Gâvre Commission: An Historical Sketch Of The Progress Of Exterior Ballistics* JUSA, **27**, 1907, pp. 238–262
417. Charbonnier, P., *The Gâvre Commission: An Historical Sketch Of The Progress Of Exterior Ballistics* JUSA, **28**, 1908, pp. 49–73
418. Charbonnier, P., *Sur L'État Actuel De La Balistique Extérieure Théorique*, Proceedings of the International Mathematical Congress held in Toronto, August 11–16, 1924, ed. Fields, John Charles, Nendeln, Kraus reprint, 1924 pp. 315–334
419. Charbonnier, P., *Traité de Balistique Extérieure, Tome I*, Gaston Doin, Gauthier-Villars; Paris, 1921
420. Charbonnier, P., *Traité de Balistique Extérieure, Tome II*, Gaston Doin, Gauthier-Villars; Paris, 1927
421. Coddington, E. A. and N. Levinson, *Theory of Ordinary Differential Equations*, McGraw-Hill Book Company, 1955
422. *The Confederate Field Manual*, Ritchie and Dunnavant, Richmond, 1862, reprinted 1984
423. Coxe, W. H. and Bugless. E., *Exterior Ballistics Charts*, E.I. du Pont de Nemours and Company, Inc., 1926
424. Craig, Thomas, *A Treatise on Linear Differential Equations*, New York, John Wiley and Sons, 1889
425. *The Applications of Elliptic Functions by Alfred George Greenhill, M.A., F.R.S., Professor of Mathematics in the Artillery College, Woolwich*, book review, Thomas Craig, JUSA, **2**, 1893, pp. 298–306
426. Cranz, C. and Becker, K., *Handbook of Ballistics, Volume I, Exterior Ballistics*, translated from the second German edition, H. M. Stationery Office, London, 1921
427. Darboux, G., *Sur les équations aux dérivés partielle*, Comptes Rendus, **96**, March 19, 1883, pp. 766–769
428. Darboux, G., *Leçons Sur La Théorie Générale Des Surfaces, Deuxieme Partie*, Gauthier-Villars et Fils, Paris, 1889
429. Darboux, G., *Leçons Sur La Théorie Générale Des Surfaces, Quatrième Partie*, Gauthier-Villars et Fils, Paris, 1896
430. Dederick, L. S., *Exterior Ballistics*, Bulletin of the American Mathematical Society, **34**, No. 5, 1928, p. 667
431. Dederick, L. S., *The Mathematics of Exterior Ballistics Computation*, American Mathematical Monthly, **47**, No. 9, November, 1940, pp. 628–634
432. Denslow, William R., *10,000 Famous Freemasons*, Missouri Lodge of Research, Independence, Mo., 1957
433. Devorkin, D. H., *Henry Norris Russell, Dean of American Astronomers*, Princeton University Press, Princeton and Oxford, 200
434. Dickson, L. E., *Mathematics in War Perspective*, Bulletin of the American Mathematical Society, **25**, No. 7, 1919, pp. 289–311
435. Dickson, T.C., *Course of Instruction Prescribed for Student Officers In the Ordnance School of Application in Differential Equations*, July 24, 1907, revised April 12, 1910, Government Printing Office, Washington
436. Dieudonné, J., *History of Functional Analysis*, Amsterdam: north Holland, 1981
437. Doerfler, R., Gluchoff, A.,Guthery, S., and Paul Hamburg, *Calculating Curves*, Docent Press, 2012
438. Drach, Jules, *L'équation differentielle de la balistique extérieure et son integration par quadratures*, Annales Scientifiques de l'E.N.S., Serie 3, tome 37, 1920 pp. 1–94
439. Drenick, R., *The Perturbation Calculus in Missile Ballistics*, Journal of the Franklin Institute, **25**, 1951, pp. 423–436
440. Drenick, R., *Rudolph F. Drenick, Veteran*, interviewed by Mike Russert, Wayne Clark, November 18, 2003, South Setauket, New York, dmna.ny.gov/historic/veterans/transcripts/Drenick-Rudolf-F.pdf
441. du Bois-Reymond, P., *Über lineare partielle Differentialgleichungen zweiter Ordnung*, Journal für Mathematik, **104**, pp. 241–310

442. Eggert, E. F., *Accurate Trajectories By Mechanical Integration*, Proceedings of the U.S. Naval Institute, **45** No.'s 1–6, January, 1919, pp. 71–84
443. Eggert, E. F., *Trajectories and Their Corrections*, discussion, Proceedings of the U.S. Naval Institute, **45** No.'s 7–12, 1919, pp. 1761–1765
444. Evans, G. C., *Functionals and Their Applications*, American Mathematical Society Colloquium Lectures, **5**, Part I, 1918
445. Faulkner, F. D., *Optimum Interception of a Ballistic Missile at Intermediate Range*, Naval Postgraduate School, Monterey, California, October 1961
446. Faulkner, F. D., *A Comparison Between Some Methods For Computing Optimum Paths in the Problem of Bolza*, Research Paper No. 43, United States Naval Postgraduate School, Monterey, California, March 1964
447. Faulkner, F. D., *Newton's Method Applied To The Solution Of Classical Problems In Calculus Of Variations And Differential Games*, dissertation submitted in partial fulfillment of the requirements for the degree of Doctor of Philosophy in the University of Michigan, 1969
448. Feffer, Loren Butler, *Oswald Veblen and the Capitalization of American mathematics*, Isis, **89**, pp. 474–497
449. Fenster, Della, Kent, Deborah and Archibald, Thomas, *The First World War and Mathematics In the United States*, preprint
450. Fischer, C. A., *A Generalization of Volterra's Derivative of a Function of a Curve*, American Journal of Mathematics, **35**, No. 4, 1913, pp. 369–394
451. Fischer, C. A., *Note on the order of continuity of functions of lines*, Bulletin of the American Mathematical Society, **23**, (1916–1917), pp. 88–90
452. Flanagan, Francis, *Harmonic Functions*, Dover Press, 1972
453. Ford, L., *Instruction Memorandum No. 20, Subject: Elementary Mathematics For Field Artillery*, Headquarters, Field Artillery Central Officers Training School, Camp Zachary Taylor, Ky., 1919
454. Fowler, R. H., E. G. Gallop, C. N. H. Lock and H. W. Richmond, *The Aerodynamics of a Spinning Shell*, Philosophical Transactions of the Royal Society, A221, 1920, pp. 295–387
455. Franklin, Philip, *An Alternative Derivation of the Differential Equations Used in Computing Differential Corrections*, [NARA Aberdeen], No. 86, 1919
456. Franklin, Philip, *Exterior Ballistics, A Historical Survey*, Brown University Graduate School, 1943
457. *Philip Franklin of M.I.T. is Dead*, New York Times, 1/28/65
458. Fréchet, M. *Sur La Notion de la Differentielle D'Une Fonction de Ligne*, Transactions of the American Mathematical Society, **10**, 1913, pp. 135–161
459. Fuchs, I., Über Relationen, welche für die zwischen je zwei singularen Punkten erstrecken integrale der lösungen linearer differentialgleichungen stattfinden Journal für die reine und angewandte Mathematik, **76**, 1874, pp. 177–213
460. Galileo, *Two New Sciences*, A New Translation With Introduction and Notes, Stillman Drake, University of Wisconsin Press, 1974
461. Gasteyer, C.E., *Forest Ray Moulton, 1872–1952* , Biographical Memoirs of the National Academy of Sciences, **41**, pp. 340–355
462. Gateaux, R., *Fonctions d'une infinité de variables indépendantes*, Bulletin de la S. M. F., **47**, 1919, pp. 70–96
463. Glennon, J.H., *Accuracy and Probability of Fire*, Baltimore Press of Isaac Friedenwald, 1888
464. Gluchoff, Alan, *Pure Mathematics Applied in Early Twentieth-Century America: The case of T.H. Gronwall, Consulting Mathematician*, Historia Mathematica, **32**, 2005, pp. 312–357
465. Gluchoff, Alan, *Artillerymen and Mathematicians: Forest Ray Moulton and Changes in American Exterior Ballistics, 1885–1934*, Historia Mathematica, **38**, 2011, pp. 506–547
466. Goldstine, Herman H., *Conditions For A Minimum Of A Functional*, The University of Chicago Libraries, Chicago, Illinois, 1937
467. Goldstine, Herman. H., *A Multiplier Rule In Abstract Spaces*, Bulletin of the American Mathematical Society, **44**, 1938, pp. 388–394

468. Goldstine, Herman H., *The Computer from Pascal to von Neumann*, Princeton University Press, 1972
469. Oral History Transcript - Dr. Herman Goldstine, Interview with Dr. Herman Goldstine by Nancy Stern at Institute for Advanced Study, Princeton, March 14, 1977, http://www.aip.org/history/ohilist/4636.html
470. Goldstine, Herman H., *A History of the Calculus of Variations From the 17th Through the 19th Centuries*, Springer-Verlag, 1980
471. Goldstine, Herman H., interviewed in *The Princeton Mathematics Community in the 1930s*, Transcript Number 15, (PMC15), The Trustees of Princeton University, 1985
472. Goldstine, Herman, interview with author July 12, 2004
473. Goursat, Édouard, *Sur la Théorie des functions implicites*, Bull. de la S. M. F., **31**, 1903, pp. 184–192
474. Goursat, Édouard, *A Course In Mathematical Analysis*, Vol. II, Part II, translated by Hedrick, E.R., and Dunkel, Otto, Ginn and Co., 1917
475. Goursat, Édouard, *Cours d'Analyse mathématique, Tome III*, Troisième Edition, Gauthier-Villars, 1923
476. Graves, L., *Implicit Functions and Differential Equations in General Analysis*, Transactions of the American Mathematical Society, **29**, No. 3, July 1927, pp. 514–552
477. Graves, L., *Topics in the Functional Calculus*, Bulletin of the American Mathematical Society, **41**, 1935, pp. 641–662
478. Graves, L., *What Is A Functional?*, The American Mathematical Monthly, **55**, No. 8, October 1948, pp. 467–472
479. Graves, L. M., *Gilbert Ames Bliss, 1876–1951*, Bulletin of the American Mathematical Society, **58**, (1952), pp. 251–264
480. Graves, L., *The Theory of Functions of Real Variables*, $2^{nd}$ Edition, McGraw-Hill, 1956, reprinted by Dover Publications, 2009
481. Graves, L., *Nonlinear Mappings between Banach Spaces*, in *Studies in Real and Complex Analysis*, ed. I. I. Hirschman, The Mathematical Association of America, 1965, pp. 34–54
482. Graves, L., and T. H. Hildebrandt, *Implicit Functions and Their Differentials in General Analysis*, Transactions of the American Mathematical Society, **29**, No. 1, January 1927, pp. 127–153
483. Green, George, *An Essay on the Application of Mathematical Analysis to the Theories of Electricity and Magnetism*, Journal für Mathematik, **39**, 1850, pp. 73–89; **44**, 1852, pp. 356–374; **47**, 1854, pp. 161–221
484. Grier, David Alan, *Dr. Veblen Takes a Uniform: Mathematics in the First World War*, American Mathematical Monthly, **108**, December 2001, pp. 922–931
485. Grier, David Alan, *Dr. Veblen At Aberdeen: Mathematics, Military Applications and Mass Production*, in *Instrumental in War*, Boston, Brill, 2005, pp. 253–270
486. Grier, David Alan, *When Computers Were Human*, Princeton University Press, 2005
487. Gronwall, T. H., *Note on the Derivative with Respect to a Parameter of the Solutions of a System of Differential Equations*, Annals of Mathematics, Second Series, **20**, No. 4, July 1919, pp. 292–296
488. Gronwall, T. H., *On a System if Linear Partial Differential Equations of Hyperbolic Type*, Annals of Mathematics, **20**, No. 4, July 1919, pp. 274–278
489. Gronwall, T. H., *On Differential Corrections and Weighting Factors*, NARA Washington 4, July 1919, pp. 274–278
490. Gronwall, T.H., *Qualitative Properties of the Ballistic Trajectory*, Annals of Mathematics, Ser. II, Vol. 21, 1920, pp. 44–65
491. Gronwall, T. H., *Differential Variations In Ballistics With Applications to the Qualitative Properties of the Trajectory*, Transactions of the American Mathematical Society, **22**, No. 4, 1921, pp. 505–525
492. Gronwall, T. H., review, *Traité de Balistique Extérieure, Vol. I*, P. Charbonnier, Bulletin of the American Mathematical Society, **33**, No. 1, 1927, p. 120

493. Guion, James L., review, *New Methods in Exterior Ballistics*, Army Ordnance, 1926, **7**, pp. 152–155
494. *Gunnery For Heavy Artillery*, Coast Artillery Corps, War Department, 1925
495. Hadamard, J., *Leçons sur la propagation des ondes*, Paris, Hermann, 1903
496. Hadamard, J., *Leçons sur le calcul des variations*, Hermann et fils, Paris, 1910
497. Hadamard, J., *Rapport sur les examinés par la Commission de Balistique pendant la durée de la guerre*, Comptes Rendus de l'Académie des Sciences, **170**, No, 8, 1920, pp. 436–445
498. Hadamard, J., *Lectures on Cauchy's Problem in Linear Partial Differential Equations*, Yale University Press, 1923
499. Hadamard, J., *Le Développement et le role Scientifique Du Calcul Fonctionel*, Oevres de Jacques Hadamard, Tome I, Éditions du Centre National de la Recherche Scientifique, 1968, pp. 435–453
500. Hadamard, J., *Le Calcul Fonctionel*, Oevres de Jacques Hadamard, Tome I, Éditions du Centre National de la Recherche Scientifique 1968, pp. 2253–2266
501. Hagood, Johnson, *Spotting for the Coast Artillery*, CAJ, **60**, No. 2, 1924, pp. 87–92
502. Hamilton, Alston, *Ballistics, Part I*, Coast Artillery School Press, Fort Monroe, Virginia, 1908
503. Hamilton, Alston, *Ballistics, Part II (Exterior Ballistics)*, Coast Artillery School Press, Fort Monroe, Virginia, 1908
504. Hamilton, Alston, *A Practical Method of Computing High Angle Trajectories* NARA Aberdeen, No. 1, 1917
505. Hancock, Harris, *Lectures on the Calculus of Variations (The Weierstrassian Theory)*, University of Cincinnati Bulletin of Mathematics, 1904
506. Hancock, Harris, *Theory of Maxima and Minima*, Ginn and Co., 1917
507. Hardy, G. H., *A Mathematician's Apology*, Cambridge University Press, 1940
508. Hartman, Philip, *Ordinary Differential Equations*, John Wiley and Sons, 1964
509. Hawkesworth, A. S., *Under-Water Trajectories*, Proceedings of the U.S.Naval Institute, **47**, February 1921, pp. 207–213
510. Hawkesworth, A. S., *New Ballistic Corrections*, Proceedings of the U.S. Naval Institute, **47**, 1921, pp. 1233–1249
511. Hatch, H.J., *A Reply to General Hagood*, CAJ, **60**, No. 3, 1924, pp. 171–176
512. Hedrick, H. B., review, Hadamard, J., *Leçons sur le calcul des variations*, Bull. Am. Math. Society, **21**, 1914, pp. 30–32
513. Hedrick, H. B., *Ballistic Effects Due To the Rotation of the Earth*, CAJ, **70**, No. 6, 1929, pp. 268–269
514. Hermann, Ernest E., *Exterior Ballistics 1935*, U. S. Naval Institute, Annapolis, Maryland, 1935
515. Hestenes, M., *The Calculus of Variations and Optimal Control Theory*, Wiley, New York, 1966
516. Hill, G. W., *Researches in the Lunar Theory*, American Journal of Mathematics, **1**, No. 1, 1876, pp. 5–21
517. Hitchcock, H. P., *Computation of Firing Tables for the U. S. Army*, Aberdeen Proving Ground, 1934
518. Hoar, Roger Sherman, *The Pickering Polaris Attachment*, JUSA, **50**, 1919, pp. 230–239
519. Hoar, Roger Sherman, *The New Ballistics*, JUSA, **51**, 1919, pp. 285–295
520. Hoar, Roger Sherman, *The New Artillery Grid Maps*, JUSA, **54**, 1921, 477–479
521. Hoar, Roger Sherman, *Problems for Solution* [2876–2882], American Mathematical Monthly, **28**, No. 2, Feb. 1921, pp. 89–90
522. Hoar, Roger Sherman, *A Course In Exterior Ballistics*, Government Printing Office, Washington, 1921
523. Hoar, Roger Sherman, *Discussions: On Proof by Mathematical Induction*, American Mathematical Monthly, **29**, No. 4, April 1922, pp. 162–169
524. Hoar, Roger Sherman, *The Rotation of the Earth*, ,**67**, No. 5 , 1927, pp. 377–388

525. Hoar, Roger Sherman, review, *New Methods in Exterior Ballistics* by Forest Ray Moulton, American Mathematical Monthly, **34**, No. 5, June-July 1927, pp. 325–326
526. Hoar, Roger Sherman, review, *New Methods In Exterior Ballistics* by Forest Ray Moulton, Army Ordnance, **7**, 1927, pp. 395–396
527. Hoar, Roger Sherman, *"New Methods in Exterior Ballistics"*, Army Ordnance, **9**, 1928, pp. 50–51
528. Hoar, Roger Sherman, *Curved versus Cartesian Coordinates in Ballistics*, Army Ordnance, **9**, No. 3, 1928, pp. 371–376
529. Hoar, Roger Sherman, entry in *National Cyclopedia of American Biography*, **50**, Ann Arbor, Michigan., p. 622
530. Hughes, Thomas P., *American Genesis*, Viking, 1989
531. Hull, Gordon, *Some Aspects Of Physics In War And Peace*, Science, **51**, No. 1314, March 5, 1920, pp. 221–233
532. Hung, J. C., *Method of Adjoint Systems and its Engineering Applications*, The University of Tennessee Department of Electrical Engineering, Technical Note No. 1, Control Theory Group, Prepared for National Aeronautics and Space Administration, October 10, 1964
533. Hunter, Patti, *Drawing the Boundaries: Mathematical Statistics in 20th-Century America*, Historia Mathematica, **23**, 1996, pp. 7–30
534. Huntington, Edward V., *The Notion of Probable Error in Elementary Statistics*, Science, New Series, **66**, No. 1722, 1927, pp. 633–637
535. Ingalls, James M., *Exterior Ballistics in the Plane of Fire*, D. Van Nostrand, New York, 1886
536. Ingalls, James M., *Handbook of Problems in Direct Fire*, John Wiley and Sons, New York, 1890
537. Ingalls, James M., *Range Tables For The 12-Inch Cast-Iron B.L. Mortar*, JUSA **1**, 1892, pp. 55–61
538. Ingalls, James M., *General Ballistic Tables For Mortar Firing*, JUSA, **8**, No. 1, 1896, pp. 52–57
539. Ingalls, James M., *A New General Ballistic Table*, JUSA, **9**, No. 3, 1897, pp. 295–299
540. Ingalls, James M., *Ballistics Tables (Artillery Circular M)*, War Department Document 111, Government Printing Office, Washington, 1900
541. Irwin, F., *The Invariants of Linear Differential Expressions*, Proceedings of the American Mathematical Society, **44**, No. 1, 1908, pp. 3–60
542. Jackson, A. M., *Spotting*, CAJ, **61**, 1924, pp. 180–181
543. Jackson, D., *The Method of Numerical Integration in Exterior Ballistics*, Washington, Government Printing Office, 1921
544. Jackson, D., *The Human Significance of Mathematics*, American Mathematical Monthly, **35**, No. 8, October 1928, pp. 406–411
545. Jacobi, C. G. J., *Theoria Novi Multiplicatoris Systematiaequationum Differentialium Vulgarium Applicandi*, in *Gesammelte Werke*, Vierter Band, Cheslea Publishing Company, New York, N.Y., 1969, pp. 317–509
546. Jacobi, C. G. J., *Zur Theorie der Variations-Rechnung und der Differentialgleighungen*, Journal für Mathematik, **17**, 1837, p. 68 ff.
547. Johnson, J. J., *Computation of a Trajectory by Short Arcs*, CAJ, Vol. 55, 1921, pp. 50–67
548. Johnson, J. J., *An Introduction to the Construction of Nomographic or Alignment Charts*, Coast Artillery Journal, **59**, No. 1, July 1923, pp. 157–172
549. Kellogg, Oliver D., *Foundations of Potential Theory*, Springer, Berlin, 1929, reprinted by Dover Publications, 1953
550. Kent, R. H., *Projectile Design*, Army Ordnance, November-December 1921, **2**, No. 9, pp. 204–205
551. Kirk, A. G., *Trajectories and Their Corrections*, Proceedings of the U.S. Naval Institute, **45** No.s 5–8, 1919, pp. 1375–1395
552. Kirk, A. G., *New Methods of Exterior Ballistics*, Proceedings of the U.S. Naval Institute, **46**, 1920 pp. 1776–1784
553. Meigs, J.F. and Ingersoll, R.R., *Exterior Ballistics*, Issac Friedenwald, Baltimore, 1885

554. Kline, Morris, *Mathematical Thought from Ancient to Modern Times*, Oxford University Press, 1972
555. Kneser, Adolph, *Lehrbuch Der Variationsrechnung*, Friederich Vieweg Und Sohn, Braunschweig, 1900
556. Lagrange, Joseph Louis, *Essai D'une Nouvelle Méthode Pour Déterminer Les Maxima et Les Minima des Formules Integrales Indefinies*, Misc. Soc. Turin, **2**, 1762, pp. 173–195
557. Lagrange, Joseph Louis, *Solution de differens problemes de calcul integral*, Misc. Tour. **3**, 1762–1765, p. 179
558. Lamson, K., *A General Implicit Function Theorem With An Application To Problems Of Relative Minima*, American Journal of Mathematics, **41**, No. 4, October 1920, pp. 243–256
559. Lecat, Maurice, *Bibliographie du Calcul des Variations, 1850–1913*, Ghent and Paris, 1913
560. Legendre, A. L., *Mémoire sure la manière de distinguer les maximas des minimas dans le calcul des variations*, Mémoires de l'Académie des Sciences, 1786
561. Leitmann, G., ed., *Optimization Techniques*, Academic Press, New York, 1962
562. Le Stourgeon, E., *Minima of Functions of Lines*, Trans. Am. Math. Soc., **21**, No. 4, October 1920, pp. 357–383
563. Liouville, Joseph, *Second Mémoire Sur le developpment des functions ou parties de fonctions en series*, Journal des Mathématiques, **2**, 1837, pp. 16–55
564. Lissak, Ormand, *Ordnance and Gunnery*, John Wiley and Sons, New York, 1908
565. Lissak, Ormand, *Ordnance and Gunnery*, John Wiley and Sons, New York, 1915
566. Lorch, E. R., *Joseph Fels Ritt*, Bulletin of the American Mathematical Society, **57**, No. 4, 1951, pp. 307–318
567. Love, A. E. H., *A Treatise on the Mathematical Theory of Elasticity*, Fourth Edition, Cambridge University Press, 1927, reprinted by Dover Publications, New York, 1944
568. Lovitt, William Vernon, *Linear Integral Equations*, McGraw-Hill Book Company, 1924, reprinted by Dover Publications, 1950
569. Mason, Max, *Zur Theorie de Randwertaufgaben*, Mathematische Annalen, **63**, 1904, pp. 528–544
570. Mason, Max, *Green's Theorem and Green's Functions for Certain Systems of Differential Equations*, Transactions of the American Mathematical Society, **5**, 1904, pp. 220–225
571. Mason, Max, *On the boundary Value Problems of Linear Ordinary Differential Equations of the Second Order*, Transactions of the American Mathematical Society, **7**, 1906, pp. 340–344
572. Mason, Max, *On the Differential Equation of the Hyperbolic Type*, Mathematische Annalen, **65**, 1908, pp. 570–575
573. Mason, Max, *Selected Topics in the Theory of Boundary Value Problems of Differential Equations*, in The New Haven Mathematical Colloquium, Yale University Press, 1910, pp. 173–222
574. Mayer, A., *Über Das allgemeinste Problem der Variations-rechnung bei einer einzigen unabhangingen Variabeln*, Leip. Ber., **30**, 1878, pp. 16–32
575. Mayer, A., *Begründung der Lagrange'schen Multiplicatorenmethode in der Variationsrechnung*, Math. Ann., **26**, 1886, pp. 74–82
576. McCalla, T. R., *Application of the Adjoint System of Differential Equations in the Solution of the Bang-Bang Control Problem*, United States Naval Postgraduate School, Monterey, California, 1961
577. McFarland, Earl, *Textbook of Ordnance and Gunnery*, John Wiley and Sons, New York, 1929
578. McShane, E. J., *Differentials of Certain Functions in Exterior Ballistics*, Duke Mathematical Journal, **17**, 1950, pp. 115–134
579. McShane, E. J., *Gilbert Ames Bliss, 1876–1951 A Biographical Memoir by E.J. McShane*, National Academy of Sciences, 1958, pp. 32–53
580. McShane, E. J., *The Calculus of Variations from the Beginning through Optimal Control Theory*, SIAM Control and Optimization, **27**, No. 5, September 1989, pp. 916–939
581. McShane, Edward J., Kelley, John L., Reno, Franklin V., *Exterior Ballistics*, The University of Denver Press, Denver, Colorado, 1953

582. Meigs, J.F. and Ingersoll, R.R., *Exterior Ballistics*, Issac Friedenwald, Baltimore, 1885
583. Metcalfe, Henry, *A Course Of Instruction in Ordnance and Gunnery*, John Wiley and Sons, New York, 1891
584. Metcalfe, Henry, *Ordnance and Gunnery Part II: Cannon, Carriages, Small Arms, Ammunition and Accuracy*, West Point, United States Military Academy Press, 1891
585. Milne, W. E., *Equations of Differential Variations in Exterior Ballistics*, CAJ, **51**, 1919, pp. 154–159
586. Milne, W. E., review, *Mathematics For Exterior Ballistics*, Mathematics Reviews, MR 0010480; (6.24d) 71.0X
587. Moore, E. H., *On a Form of General Analysis With Applications to Linear Differential and Integral Equations*, Atti del IV Congresso Internazionale dei Matematici, **1**, Roma 1908, **2**, Roma 1909, pp. 98–114
588. Moore, E. H., E. J. Wilczynski and Max Mason, *The New Haven Mathematical Colloquium*, Yale University Press, New Haven, 1910
589. Moulton, Forest Ray, *On Certain Rigorous Methods of Treating Problems in Celestial Mechanics*, Decennial Publications of the University of Chicago, **8**, No. 191, (42), 1902
590. Moulton, Forest Ray, *The Influence of Astronomy on Mathematics*, Science, New Series, **33**, No. 835, March 10, 1911, pp. 357–364
591. Moulton, Forest Ray, *M. Henri Poincaré*, Popular Astronomy, **20**, No. 10, December 1912, pp. 620–634
592. Moulton, Forest Ray, *An Introduction To Celestial Mechanics*, Second Revised Edition, Macmillan Company, 1914, reprinted by Dover Publications, New York, 1970
593. Moulton, Forest Ray, *Lectures on the Modern Theory of Differential Equations*, University of Chicago, 1916–1917
594. Moulton, Forest Ray, *History of the Ballistics Branch of the Artillery Ammunition Section, Engineering Division of the Ordnance Department for the Period April 6, 1918 to April 2, 1919*, War Department, Office of the Chief of Ordnance, Washington, 1919
595. Moulton, Forest Ray, *Numerical Integration of Differential Equations*, CAJ, **51**, 1919, pp. 40–55
596. Mouton, Forest Ray, *Numerical Solution of Differential Equations*, in *Smithsonian Mathematical Formulae and Tables of Elliptic Functions*, Smithsonian Miscellaneous Collections, **74**, No. 1, Publication 2672, Smithsonian Institution, 1922
597. Moulton, Forest Ray, *New Methods In Exterior Ballistics*, University of Chicago Press, 1926, reprinted by Dover Publications, 1962
598. Moulton, Forest Ray, discussion of review of *New Methods in Exterior Ballistics* by Roger Sherman Hoar, American Mathematical Monthly, **35**, No. 5, May 1928, pp. 246–250
599. Moulton, Forest Ray, reply to a review of *New Methods in Exterior Ballistics* by Roger Sherman Hoar, Army Ordnance, **8**, 1928, pp. 320–321
600. Moulton, Forest Ray, *Differential Equations*, MacMillan Company 1930, reprinted by Dover Publications, New York, 1958
601. Moulton, F. R., entry in *Notable Scientists: From 1900 to the Present*, Gale Group, 2001, Anonymous.
602. Nenninger, T.K., *The Leavenworth Schools and the Old Army: Education, Professionalism, and the Officer Corps of the United States Army, 1881–1918*, Greenwood Press, Westport, Connecticut, 1978
603. d'Ocagne, Maurice, *Traité de Nomographie, Deuxième Édition*, Gauthier Villars, Paris, 1921
604. Osgood, William Fogg, *Advanced Calculus*, The MacMillan Company, New York, 1925
605. Paaswell, G., *An Appeal to Producing Mathematicians*, Bulletin of the American Mathematical Society, **21**, No. 3, 1914, pp. 127–130
606. Pachpatte, B. G., *Inequalities for Differential and Integral Equations*, San Diego Press, 1998
607. Parshall, Karen Hunger, and Rowe, David E., *The Emergence of the American Mathematical Research Community, 1876–1900: J. J. Sylvester, Felix Klein, and E. H. Moore*, American Mathematical Society, 1994

608. Picard, É., *Mémoire sur la théorie des éqations aux dérivées partielles et la méthode des approximations successives*, Journal des Mathematiques, series 4, **6**, 1890, pp. 145–210
609. Picard, É. *Sur l'application des méthodes d'approximations successives à l'étude de certaines équations différentielles ordinaires*, Journal des Mathematiques series 4, **9**, 1893, pp. 217–271
610. Pincherle, S., *Sull'operazione aggiunta*, Rendiconto della R. Accademia delle Science dell'Instituto di Bologna, **2**, (1897–1898), pp. 130–139
611. Pincherle, S., *Sull'operazione agguinta di Lagrange*, Annali de metematica pur ed applicata, **21**, 1913 pp. 143–151
612. Poincaré, Henri, *Les Méthodes Nouvelles de la Mécanique Céleste , Tome 1*, Dover Publications, New York, reprint of original 1892 edition
613. *Lecture on The Probability and Dispersion of Fire*, translated at the Army War College From a French Document, Government Printing Office, 1917
614. *Provisional Range Tables for the British 75-mm. Gun, Model 1917*, Ordnance Department, U.S. Army, War Plans Division, Government Printing Office, 1918
615. Rademacher, Hans, *On the Accumulation of Errors In Numerical Integration on the ENIAC*, in *Theory and Techniques for Design of Electronic Digital Computers*, Lectures Given at the Moore School, 8 July 1946–31 August 1946, Volume II, Lecture 19, University of Pennsylvania, Moore School of Electrical Engineering, Philadelphia, Pennsylvania, November 1, 1947
616. *Range Tables For 75-mm. Model 1916 American and Model 1917 British Guns*, Prepared At Aberdeen Proving Ground, Engineer Reproduction Plant, U.S. Army, Washington Barracks, November 1921
617. Riemann, Bernhard, *Über die Fortpflanzung ebener Luftwellen von endlicher Schwingingungsweite*, Gottingen Abhandl., **8**, 1860, pp. 145–164
618. Ritt, J., *On Weighting Factor Curves For Low Elevation*, NARA Aberdeen, Technical Report, February 25, 1919
619. Ritt, J., *On the Differentiability of the Solution of a Differential Equation With Respect to a Parameter*, Annals of Mathematics, Second Series, **20**, No. 4, July 1919, pp. 289–291
620. Ritt, J., *On Weighting Factors in Flat Fire*, CAJ, **53**, 1920, pp. 404–410
621. Ritt, J., *On the Iteration of Rational Functions*, Transactions of the American Mathematical Society, **21**, 1920, pp. 313–320
622. Roberts, David L., *Albert Henry Wheeler (1873–1950): A Case Study in the Stratification of American Mathematical Activity*, Historia Mathematica, **23**, 1996, pp. 269–287
623. Roberts, David L., *Republic of Numbers*, Johns Hopkins University Press, 2019
624. Rosenstein, George M., *The Best Method: American Calculus Textbooks of the Nineteenth Century*, from *A Century of Mathematics In America, Part III*, ed. Duren, P., Askey, R., Merzbach, U., and Edwards, H., Providence, Rhode Island, American Mathematical Society, 1988–1989, pp. 77–109
625. Roth, William E., *The Exact Differential Equations Of A Trajectory In Terms Of The Hoar Curvilinear Coordinates*, CAJ, **68**, 1928, pp. 505–512
626. Rowe, J. E., *Moulton on Exterior Ballistics*, American Mathematical Monthly, **35**, No. 4, March-April 1928, pp. 229–232
627. Ruckman, John W., *The Effect Of Wind On The Motion Of A Projectile*, JUSA, **1**, 1892, pp. 5–24
628. Salmon, E. H., *Columns*, Oxford Technical Publications, Henry Frowde and Hodder and Stoughton, London, 1921
629. Sanchez, D., *Ordinary Differential Equations and Stability Theory*, Freeman and Company, 1968
630. Schwartz, Philip, *A Critical Survey of the Advances in Practical Ballistics in the Ordnance Department, U.S. Army Since 1917*, [NARA Abderdeen] A-I-31, 1920
631. Schwartz, Philip, *Measuring the Life of a Gun*, Army Ordnance, **5**, No. 2, 1924, pp. 111–112
632. Schwartz, Philip, *Accuracy Firing*, Army Ordnance, **5**, No. 3, 1924, pp. 521–522

633. Schwartz, Philip, *Experimental Investigation of The Effect of the Wind Upon the Motion of a Projectile*, CAJ, **60**, 1924, pp. 93–97
634. Schwartz, Philip, *Studies of Motion of Projectiles*, CAJ, **60**, 1924, pp. 430–431
635. Schwartz, Philip, *The English Long Range Gun*, CAJ, **61**, 1924, pp. 91–92
636. Schwartz, Philip, *Weighting Factors*, CAJ, **61**, 1924, pp. 371–372
637. Schwartz, Philip, *Remarks Concerning the Use of Firing Table Values*, CAJ, **3**, 1925, pp. 399–400
638. Schwartz, Phlilp, *The Variable Probable Error*, CAJ, **67**, 1927, pp. 405–413
639. Schwartz, Philip, *Much About Little* CAJ, **68**, 1928, pp. 291–297
640. Schwartz, Philip, *The Fork*, Field Artillery Journal, **18**, No. 3, 1929, pp. 604–609
641. Schwartz, Philip, *More About Probable Error*, CAJ, **73**, 1930, pp. 111–120
642. Siacci, F., *Balistique Extérieure*, Traduction Annoteé par P. Laurent, Berger-Levrault et Cic, Editours, Paris and Nancy, 1892
643. Siegmund-Schultze, R.,*Eliakim Hastings Moore's "General Analysis"*, Arch. Hist. Exact. Sci., **52**, 1998, pp. 51–89
644. Siegmund-Schultze, R., *The Late Arrival of Academic Applied Mathematics in the United States: a Paradox, Theses, and Literature*, NTM, **11**, 2002, pp. 116–12
645. Siegmund-Schultze, R., *The Ideology of Applied Mathematics Within Mathematics in Germany and The U. S. Until the End of World War II*, ILUIL, **27**, 2004, pp. 791–811
646. Sleight, E. R., *Mathematics Applied*, National Mathematics Magazine, **9**, No. 8, May 1935, pp. 219–223
647. Smith, Paul A., *Joseph Fels Ritt, 1893–1951*, United States National Academy of Sciences, Biographical Memoir, 1956, pp. 253–264
648. Spinrad, William, *Early History of The Ordnance School*, The Ordnance Sergeant **3**, No. 1, 1942. pp. 1–4
649. Stackel, P., *Abhandlungen Über Variations-Rechnung, Zweiter Theil*, Leipzig, Verlag Von Wilhelm Engelmann, 1894
650. Steele, Bret D., *Muskets and Pendulums: Benjamin Robins, Leonhard Euler, and the Ballistics Revolution*, Technology and Culture, **35**, No. 2, April 1994, pp. 348–382
651. Stevenson, Craig, *Periodic Orbits of the Three Body Problem: F. R. Moulton's Quest for a New Lunar Theory*, PhD Thesis, Open University, 2012
652. Tannehill, I. R., *The Ballistic Wind*, Journal of the United States Artillery, **52**, 1920, pp. 553–576
653. Tarwater, J. D., (Ed.), *The Bicentennial Tribute to American Mathematics, 1776–1976* Papers Presented to the Fifty-Ninth Annual Meeting of the Mathematical Association of America, Mathematical Association of America, Washington, D. C., 1977
654. Todhunter, I., *A Treatise on the Integral Calculus and its Applications*, MacMillan and Co., London, 1880
655. Tonelli, L., *Fondamenti di calcolo delle variazioni*, Nicola Zanichelli, I, 1921; II, 1923
656. Tropp, H. S., *Forest Ray Moulton*, Dictionary of Scientific Biography, **9**, pp. 552–553
657. Truesdell, Clifford, review of *Exterior Ballistics* by McSchane, Kelley and Reno, Scripta Mathematica, **20**, 1954, pp. 172–174
658. Tschappat, W.H., *Ballistics*, article in Encyclopedia Britannica, 1922, pp. 383–394
659. Tsien, H. S., *Engineering Cybernetics*, McGraw-Hill Book Co., 1954
660. Veblen, Oswald, *Ballistic Theories and the Preparation of Range Tables*, NARA Aberdeen, No. 54, 1921
661. Veblen, O., *Professor Veblen on Ballistics in the World War*, Army Ordnance, **4**, March-April 1924, pp. 337–338
662. Very, Edward W., *Computation of Range Tables*, Annapolis, Md., 1877
663. Volterra, V., *Leçons sur les Fonctions des Lignes*, Paris, Gauthier- Villars, 1913
664. Wertheimer, Albert, *A Method of Computing Differential Corrections For a Trajectory*, Journal of the Franklin Institute, **217**, No. 6, June 1934
665. Wiener, Norbert *Ex-Prodigy*, MIT Press, Cambridge, 1953

666. Williamson, Benjamin, *An Elementary Treatise on The Integral Calculus*, Longmans, Green and Co., London, 1916
667. Zeidler, E., *Applied Functional Analysis*, Springer-Verlag, 1995
668. www.math.colostate.edu/~pauld/M345/Some20theory.pdf

# Chapter 3

## Anonymous Editorials, Announcements, and Book Reviews in Military Journals

669. Announcement, JUSA **1**, 1892, pp. 3–4
670. Book Review, A Course of Instruction in Ordnance and Gunnery, Henry Metcalfe, JUSA, **1** 1892, pp. 74–76
671. Progress In Ballistics CAJ, **51**, 1919, pp. 69–71
672. Editorial, *Change Versus Progress*, JUSA, **56**, 1922, pp. 373–374
673. *FRANKFORD ARSENAL: Range, Deflection, and Wind Correction Computer*, Army Ordnance, March-April 1924, **4**, No. 23, pp. 339–341
674. Coast Artillerymen and the Army Ordnance School, CAJ, **2**, 1924, p.363
675. Ballistic Research at Marquette University, Army Ordnance, **7**, No.5, 1927, p. 399
676. Professor Moulton's Book, Army Ordnance, **8**, 1927, No. 1, pp. 12–15
677. *The Coast Artillery School, 1824–1927* CAJ, **66**, 1927, pp. 1–85
678. Book Review, New Methods in Exterior Ballistics, Forest Ray Moulton, CAJ, **66**, 1927, pp. 383–384
679. Book Review, Textbook of Ordnance and Gunnery, Earl McFarland, CAJ, **70**, 1929, pp. 533–534

## Manuscripts, Published Books and Articles

680. Alger, P. L., *Charts for the Calculation of the Effect of Small Changes in the Elements of Fire* JUSA, Vol. 51, 1919, pp. 585–603
681. Ames, J. S., *The Trained Man of Science in the War*, Science, New Series, 9, **480**, No. 1243, Oct. 25, 1918, pp. 401–410
682. Archibald, Thomas, Della Dumbaugh and Deborah Kent, *A Mobilized Community: Mathematicians in the United States during the First World War*, in *The War of Guns and Mathematics*, pp. 229–271
683. Askey, Richard, *Orthogonal Polynomials and Special Functions*, Society for Industrial and Applied Mathematics, Philadelphia, 1975
684. Aubin, David, and Catherine Goldstein, eds., *The War of Guns and Mathematics*, American Mathematical Society, 2014
685. Ball, H. P., *Of Responsible Command: A History of the U.S. Army War College*, Alumni Association of the United States Army War College, Carlisle Barracks, Pennsylvania, 1983
686. *Ballisticians in War and Peace, Volume I, A History of the United States Army Ballistic Research Laboratories 1914–1956*, Aberdeen Proving Ground, Maryland
687. Banach, S. *Théorie des Opérations Linéaires*, Second Edition, reprint of first edition of 1932, Chelsea Publishing Company, New York, 1955
688. Barrow-Green, June, *Poincaré and the Three Body Problem*, American Mathematical Society, 1997

689. Barrow-Green, June, *"Anti-aircraft guns all day long": Karl Pearson and computing for the Ministry of Munitions*, Revue d'histoire des mathématiques, **21**, fascicule 1, 2015, pp. 111–150
690. Beckman, H. S., *The Elements of Firing Table Construction*, Army Ordnance, **1**, No. 1, pp. 46–454
691. Bell, E. T., *The Development of Mathematics*, Macmillan and Co., 2nd edition, 1945
692. Bennett, A. A., *Introduction to Ballistics*, Engineer Reproduction Plant, Washington Barracks, D. C., 1921
693. Bennett, A. A., *The Problem of Exterior Ballistics*, Army Ordnance, **3**, July-August 1922, No. 13, pp. 38–40
694. Bennett, A. A., *Exterior Ballistic Tables Based on Numerical Integration*, United States Army, Ordnance Dept., Washington, Government Printing Office, 1924 (Document 1107), 1931, 1936
695. Bennett, A. A., *The Differential Analyzer*, Ballistic Research Laboratories, BRL Report No. 319, December, 1942
696. Benton, J. G., *A Course of Instruction in Ordnance and Gunnery, Third Edition*, D. Van Nostrand, New York, 1867
697. Berry, W. J., *Some Theorems Concerning Systems of Linear Partial Differential Equations*, Annals of Mathematics, **12**, No. 1, 1910, pp. 27–38
698. Birkhoff, Garrett, and Gian-Carlo Rota, *Ordinary Differential Equations*, Ginn and Co., Boston, 1962
699. Birkhoff, G., *Some Leaders In American Mathematics, 1891–1941*, in: Tarwater [1977], pp. 27–78
700. Birkhoff, G., and Erwin Kreyzig, *The Establishment of Functional Analysis*, Historia Mathematica, **11**, 1984 pp. 258–321
701. Birkhoff, G. D., *Boundary Value and Expansion Problems of Ordinary Linear Differential Equations*, Transactions of the American Mathematical Society, **9**, No. 4, 1908, pp. 373–395
702. Birkhoff, G. D., and O. D. Kellogg, *Invariant Points in Function Space*, Transactions of the American Mathematical Society, **23**, No. 1, January 1922, pp. 96–115
703. Bliss, Gilbert Ames, *An Existence Theorem for a Differential Equation of the Second Order, With an Application to the Calculus of Variations*, Trans. Am. Math. Soc., **5**, 1904, pp. 113–125
704. Bliss, Gilbert Ames, *The Solutions of Differential Equations of the First Order as Functions of Their Initial Conditions*, Ann. of Math., **6**, 1905, pp. 49–68
705. Bliss, Gilbert Ames, *A Generalization of the Notion of Angle*, Transactions of the American Mathematical Society, **7**, 1906, pp. 184–196
706. Bliss, Gilbert Ames, *The Scientific Work of Eliakim Hastings Moore*, Bull. Am. Math. Soc., **40**, 1910, pp. 207–213
707. Bliss, Gilbert Ames, *Integral Equations*, review of *An Introduction to the Study of Integral Equations* by Maxime Bôcher, *Bulletin of the American Mathematical Society*, **16**, No. 4, 1910, pp. 207–213
708. Bliss, Gilbert Ames, *A New Proof of the Implicit Function Theorem*, Bull. Am. Math. Soc., **18**, 1912, pp. 175–179
709. Bliss, Gilbert Ames, *Fundamental Existence Theorems*, The Princeton Colloquium, Part I, American Mathematical Society, 1913
710. Bliss, Gilbert Ames, *A Note on Functions of Lines*, Proc. Nat. Acad. Sci, U.S.A., **10**, 1915, pp. 173–175
711. Bliss, Gilbert Ames, review, V. Volterra, *Fonctions des Lignes*, Bull. Am. Math. Soc., **23**, 1915, pp. 344–355
712. Bliss, Gilbert Ames, *Solutions of Differential Equations as Functions of the Constants of Integration*, Bull. Amer. Math. Soc., **25**, 1918, pp. 15–26
713. Bliss, Gilbert Ames, *Differential Corrections*, NARA Aberdeen Ms A-29, No. 10, Nov. 1918
714. Bliss, Gilbert Ames, *The Problem of Mayer With Variable Endpoints*, Trans. Am. Math. Soc., **19**, 1918, pp. 305–316

715. Bliss, Gilbert Ames, *A Method of Computing Differential Corrections for a Trajectory*, CAJ/JUSA, **50**, 1918, pp. 455–460
716. Bliss, Gilbert Ames, *A Method of Computing Differential Corrections for a Trajectory*, CAJ/JUSA, **51**, 1919, pp. 445–449
717. Bliss, Gilbert Ames, *The Use of Adjoint Systems in the Problem of Differential Corrections for Trajectories*, JUSA, **51**, 1919, pp. 296–311
718. Bliss, Gilbert Ames, *Differential Equations Containing Arbitrary Functions*, Transactions of the American Mathematical Society, **210**, No. 9, 1920, pp. 79–92
719. Bliss, Gilbert Ames, *Functions of Lines in Ballistics*, ibid, pp. 93–106
720. Bliss, Gilbert Ames, *Calculus of Variations*, Mathematical Association of America, 1925
721. Bliss, Gilbert Ames, *A Boundary Value Problem for a System of Ordinary Differential Equations of the First Order*, Transactions of the American Mathematical Society, **28**, No. 4, 1926, pp. 561–584
722. Bliss, Gilbert Ames, *Mathematics*, Scientific Monthly, **24**, Issue 4, 1927, pp. 308–319
723. Bliss, Gilbert Ames, *Existence Theorems for Implicit Functions, Existence Theorems for Differential Equations*, University of Chicago, 1928
724. Bliss. Gilbert Ames, *The Problem of Lagrange in the Calculus of Variations*, Am. J. Math., **52**, No. 4, 1930, pp. 673–744
725. Bliss, Gilbert Ames, *Mathematical Interpretation of Geometrical And Physical Phenomena*, American Mathematical Monthly, **40**, 1933, pp. 472–480
726. Bliss, Gilbert Ames, *The Evolution of Problems of the Calculus of Variations*, The American Mathematical Monthly, **43**, No. 10, December 1936, pp. 598–609
727. Bliss, Gilbert Ames, *Mathematics for Exterior Ballistics*, John Wiley and Sons, New York, 1944
728. Bliss, Gilbert Ames, *Lectures on the Calculus of Variations*, The University of Chicago Press, Chicago, 1946
729. Bliss, Gilbert Ames, *Autobiographical Notes*, American Mathematical Monthly, **59**, November 1952, pp. 595–605
730. Bliss, Gilbert Ames, and Max Mason, *A problem of the Calculus of Variations in Which the Integrand is Discontinuous*, Trans. Am. Math. Soc., **7**, No. 2, April 1906, pp. 325–336
731. Bliss, Gilbert Ames, and Max Mason, *The Properties of Curves in Space Which Minimize a Definite Integral*, Trans. Amer. Math. Soc., **9**, October 1908, pp. 440–466
732. Bliss, Gilbert Ames, and Max Mason, *Fields of Extremals in Space*, Transactions of the American Mathematical Society, **11**, 1910, pp. 325–340
733. Bôcher, M., *Green's Functions in Space of One Dimension*, Bulletin of the American Mathematical Society, **7**, 1901, pp. 297–299
734. Bôcher, M., *An Introduction to the Study of Integral Equations*, Cambridge Tracts in Mathematics and Mathematical Physics, No. 10, Cambridge University Press, 1909
735. Bôcher, M., *Boundary Problems and Green's Functions for Linear Differential and Difference Equations*, Annals of Mathematics, **13**, 1911–1912, pp. 71–88
736. Bôcher, M., *Boundary Problems in One Dimension*, in *Proceedings of the Fifth International Congress of Mathematicians, Cambridge, 22–28 August 1912*, Princeton University Press, 1913, pp. 163–195
737. Bôcher, M., *Applications and Generalizations of the Conception of Adjoint Systems*, Transactions of the American Mathematical Society, **14**, October 1913, pp. 403–420
738. Bolza, O., *Lectures on the Calculus of Variations*, University of Chicago Press, 1904, reprinted unabridged and corrected, Dover Publications, 1961
739. Bolza, Oskar, *Vorlesung Über Variationsrechnung*, Druck und Verlag von B.G. Teubner, 1909
740. Bolza, Oskar, *Integral Equations*, lecture notes by William Vernon Lovitt from a summer course given during summer 1913 at University of Chicago
741. Bruff, L. L., *A Textbook of Ordnance and Gunnery*, John Wiley and Sons, New York, 1896
742. Campbell, G. A., *Mathematics in Industrial Research*, Bell System Technical Journal, 1926, pp. 550–557

743. Carathéodory, C., review, Hadamard, *Leçons sur le calcul des variations*, **35**, Bulletin des Sciences mathématiques, pp. 123–141, 1911
744. Carll, L. B., *A Treatise On The Calculus of Variations*, MacMillan and Co., London, 1885
745. Carmichael, R. D., *The Provision Made by Mathematics for the Needs of Science*, Science, New Series, **45**, Issue 1168 (May 18, 1917), pp. 465–474
746. Case, Homer, *The Variable Probable Error and The Adjustment Problem*, CAJ, **68**, 1928, pp. 492–501
747. Charbonnier, P., *The Gâvre Commission: An Historical Sketch Of The Progress Of Exterior Ballistics* JUSA, **27**, 1907, pp. 238–262
748. Charbonnier, P., *The Gâvre Commission: An Historical Sketch Of The Progress Of Exterior Ballistics* JUSA, **28**, 1908, pp. 49–73
749. Charbonnier, P., *Sur L'État Actuel De La Balistique Extérieure Théorique*, Proceedings of the International Mathematical Congress held in Toronto, August 11–16, 1924, ed. Fields, John Charles, Nendeln, Kraus reprint, 1924 pp. 315–334
750. Coddington, E. A. and N. Levinson, *Theory of Ordinary Differential Equations*, McGraw-Hill Book Company, 1955
751. *The Confederate Field Manual*, Ritchie and Dunnavant, Richmond, 1862, reprinted 1984
752. Coxe, W. H. and Bugless. E., *Exterior Ballistics Charts*, E.I. du Pont de Nemours and Company, Inc., 1926
753. Craig, Thomas, *A Treatise on Linear Differential Equations*, New York, John Wiley and Sons, 1889
754. *The Applications of Elliptic Functions by Alfred George Greenhill, M.A., F.R.S., Professor of Mathematics in the Artillery College, Woolwich*, book review, Thomas Craig, JUSA, **2**, 1893, pp. 298–306
755. Cranz, C. and Becker, K., *Handbook of Ballistics, Volume I, Exterior Ballistics*, translated from the second German edition, H. M. Stationery Office, London, 1921
756. Darboux, G., *Sur les équations aux dérivés partielle*, Comptes Rendus, **96**, March 19, 1883, pp. 766–769
757. Darboux, G., *Leçons Sur La Théorie Générale Des Surfaces, Deuxieme Partie*, Gauthier-Villars et Fils, Paris, 1889
758. Darboux, G., *Leçons Sur La Théorie Générale Des Surfaces, Quatrième Partie*, Gauthier-Villars et Fils, Paris, 1896
759. Dederick, L. S., *Exterior Ballistics*, Bulletin of the American Mathematical Society, **34**, No. 5, 1928, p. 667
760. Dederick, L. S., *The Mathematics of Exterior Ballistics Computation*, American Mathematical Monthly, **47**, No. 9, November, 1940, pp. 628–634
761. Devorkin, D. H., *Henry Norris Russell, Dean of American Astronomers*, Princeton University Press, Princeton and Oxford, 200
762. Dickson, L. E., *Mathematics in War Perspective*, Bulletin of the American Mathematical Society, **25**, No. 7, 1919, pp. 289–311
763. Dickson, T.C., *Course of Instruction Prescribed for Student Officers In the Ordnance School of Application in Differential Equations*, July 24, 1907, revised April 12, 1910, Government Printing Office, Washington
764. Dieudonné, J., *History of Functional Analysis*, Amsterdam: north Holland, 1981
765. Doerfler, R., Gluchoff, A.,Guthery, S., and Paul Hamburg, *Calculating Curves*, Docent Press, 2012
766. Drach, Jules, *L'équation differentielle de la balistique extérieure et son integration par quadratures*, Annales Scientifiques de l'E.N.S., Serie 3, tome 37, 1920 pp. 1–94
767. Drenick, R., *The Perturbation Calculus in Missile Ballistics*, Journal of the Franklin Institute, **25**, 1951, pp. 423–436
768. Drenick, R., *Rudolph F. Drenick, Veteran*, interviewed by Mike Russert, Wayne Clark, November 18, 2003, South Setauket, New York, dmna.ny.gov/historic/veterans/transcripts/Drenick-Rudolf-F.pdf

769. du Bois-Reymond, P., *Über lineare partielle Differentialgleichungen zweiter Ordnung*, Journal für Mathematik, **104**, pp. 241–310
770. Evans, G. C., *Functionals and Their Applications*, American Mathematical Society Colloquium Lectures, **5**, Part I, 1918
771. Faulkner, F. D., *Optimum Interception of a Ballistic Missile at Intermediate Range*, Naval Postgraduate School, Monterey, California, October 1961
772. Faulkner, F. D., *A Comparison Between Some Methods For Computing Optimum Paths in the Problem of Bolza*, Research Paper No. 43, United States Naval Postgraduate School, Monterey, California, March 1964
773. Faulkner, F. D., *Newton's Method Applied To The Solution Of Classical Problems In Calculus Of Variations And Differential Games*, dissertation submitted in partial fulfillment of the requirements for the degree of Doctor of Philosophy in the University of Michigan, 1969
774. Feffer, Loren Butler, *Oswald Veblen and the Capitalization of American mathematics*, Isis, **89**, pp. 474–497
775. Fenster, Della, Kent, Deborah and Archibald, Thomas, *The First World War and Mathematics In the United States*, preprint
776. Fischer, C. A., *A Generalization of Volterra's Derivative of a Function of a Curve*, American Journal of Mathematics, **35**, No. 4, 1913, pp. 369–394
777. Fischer, C. A., *Note on the order of continuity of functions of lines*, Bulletin of the American Mathematical Society, **23**, (1916–1917), pp. 88–90
778. Flanagan, Francis, *Harmonic Functions*, Dover Press, 1972
779. Ford, L., *Instruction Memorandum No. 20, Subject: Elementary Mathematics For Field Artillery*, Headquarters, Field Artillery Central Officers Training School, Camp Zachary Taylor, Ky., 1919
780. Fowler, R. H., E. G. Gallop, C. N. H. Lock and H. W. Richmond, *The Aerodynamics of a Spinning Shell*, Philosophical Transactions of the Royal Society, A221, 1920, pp. 295–387
781. Franklin, Philip, *An Alternative Derivation of the Differential Equations Used in Computing Differential Corrections*, [NARA Aberdeen], No. 86, 1919
782. Franklin, Philip, *Exterior Ballistics, A Historical Survey*, Brown University Graduate School, 1943
783. *Philip Franklin of M.I.T. is Dead*, New York Times, 1/28/65
784. Fréchet, M. *Sur La Notion de la Differentielle D'Une Fonction de Ligne*, Transactions of the American Mathematical Society, **10**, 1913, pp. 135–161
785. Fuchs, l., Über Relationen, welche für die zwischen je zwei singularen Punkten erstrecken integrale der lösungen linearer differentialgleichungen stattfinden Journal für die reine und angewandte Mathematik, **76**, 1874, pp. 177–213
786. Galileo, *Two New Sciences*, A New Translation With Introduction and Notes, Stillman Drake, University of Wisconsin Press, 1974
787. Gasteyer, C.E., *Forest Ray Moulton, 1872–1952* , Biographical Memoirs of the National Academy of Sciences, **41**, pp. 340–355
788. Gateaux, R., *Fonctions d'une infinité de variables indépendantes*, Bulletin de la S. M. F., **47**, 1919, pp. 70–96
789. Glennon, J.H., *Accuracy and Probability of Fire*, Baltimore Press of Isaac Friedenwald, 1888
790. Gluchoff, Alan, *Pure Mathematics Applied in Early Twentieth-Century America: The case of T.H. Gronwall, Consulting Mathematician*, Historia Mathematica, **32**, 2005, pp. 312–357
791. Gluchoff, Alan, *Artillerymen and Mathematicians: Forest Ray Moulton and Changes in American Exterior Ballistics, 1885–1934*, Historia Mathematica, **38**, 2011, pp. 506–547
792. Goldstine, Herman H., *Conditions For A Minimum Of A Functional*, The University of Chicago Libraries, Chicago, Illinois, 1937
793. Goldstine, Herman. H., *A Multiplier Rule In Abstract Spaces*, Bulletin of the American Mathematical Society, **44**, 1938, pp. 388–394
794. Goldstine, Herman H., *The Computer from Pascal to von Neumann*, Princeton University Press, 1972

795. Oral History Transcript - Dr. Herman Goldstine, Interview with Dr. Herman Goldstine by Nancy Stern at Institute for Advanced Study, Princeton, March 14, 1977, http://www.aip.org/history/ohilist/4636.html
796. Goldstine, Herman H., *A History of the Calculus of Variations From the 17th Through the 19th Centuries*, Springer-Verlag, 1980
797. Goldstine, Herman H., interviewed in *The Princeton Mathematics Community in the 1930s*, Transcript Number 15, (PMC15), The Trustees of Princeton University, 1985
798. Goldstine, Herman, interview with author July 12, 2004
799. Goursat, Édouard, *Sur la Théorie des functions implicites*, Bull. de la S. M. F., **31**, 1903, pp. 184–192
800. Goursat, Édouard, *A Course In Mathematical Analysis*, Vol. II, Part II, translated by Hedrick, E.R., and Dunkel, Otto, Ginn and Co., 1917
801. Goursat, Édouard, *Cours d'Analyse mathématique, Tome III*, Troisième Edition, Gauthier-Villars, 1923
802. Graves, L., *Implicit Functions and Differential Equations in General Analysis*, Transactions of the American Mathematical Society, **29**, No. 3, July 1927, pp. 514–552
803. Graves, L., *Topics in the Functional Calculus*, Bulletin of the American Mathematical Society, **41**, 1935, pp. 641–662
804. Graves, L., *What Is A Functional?*, The American Mathematical Monthly, **55**, No. 8, October 1948, pp. 467–472
805. Graves, L. M., *Gilbert Ames Bliss, 1876–1951*, Bulletin of the American Mathematical Society, **58**, (1952), pp. 251–264
806. Graves, L., *The Theory of Functions of Real Variables*, $2^{nd}$ Edition, McGraw-Hill, 1956, reprinted by Dover Publications, 2009
807. Graves, L., *Nonlinear Mappings between Banach Spaces*, in *Studies in Real and Complex Analysis*, ed. I. I. Hirschman, The Mathematical Association of America, 1965, pp. 34–54
808. Graves, L., and T. H. Hildebrandt, *Implicit Functions and Their Differentials in General Analysis*, Transactions of the American Mathematical Society, **29**, No. 1, January 1927, pp. 127–153
809. Green, George, *An Essay on the Application of Mathematical Analysis to the Theories of Electricity and Magnetism*, Journal für Mathematik, **39**, 1850, pp. 73–89; **44**, 1852, pp. 356–374; **47**, 1854, pp. 161–221
810. Grier, David Alan, *Dr. Veblen Takes a Uniform: Mathematics in the First World War*, American Mathematical Monthly, **108**, December 2001, pp. 922–931
811. Grier, David Alan, *Dr. Veblen At Aberdeen: Mathematics, Military Applications and Mass Production*, in *Instrumental in War*, Boston, Brill, 2005, pp. 253–270
812. Grier, David Alan, *When Computers Were Human*, Princeton University Press, 2005
813. Gronwall, T. H., *Note on the Derivative with Respect to a Parameter of the Solutions of a System of Differential Equations*, Annals of Mathematics, Second Series, **20**, No. 4, July 1919, pp. 292–296
814. Gronwall, T. H., *On a System if Linear Partial Differential Equations of Hyperbolic Type*, Annals of Mathematics, **20**, No. 4, July 1919, pp. 274–278
815. Gronwall, T. H., *On Differential Corrections and Weighting Factors*, NARA Washington 4, July 1919, pp. 274–278
816. Gronwall, T.H., *Qualitative Properties of the Ballistic Trajectory*, Annals of Mathematics, Ser. II, Vol. 21, 1920, pp. 44–65
817. Gronwall, T. H., *Differential Variations In Ballistics With Applications to the Qualitative Properties of the Trajectory*, Transactions of the American Mathematical Society, **22**, No. 4, 1921, pp. 505–525
818. Guion, James L., review, *New Methods in Exterior Ballistics*, Army Ordnance, 1926, **7**, pp. 152–155
819. *Gunnery For Heavy Artillery*, Coast Artillery Corps, War Department, 1925
820. Hadamard, J., *Leçons sur la propagation des ondes*, Paris, Hermann, 1903
821. Hadamard, J., *Leçons sur le calcul des variations*, Hermann et fils, Paris, 1910

822. Hadamard, J., *Lectures on Cauchy's Problem in Linear Partial Differential Equations*, Yale University Press, 1923
823. Hadamard, J., *Le Développement et le role Scientifique Du Calcul Fonctionel*, Oevres de Jacques Hadamard, Tome I, Éditions du Centre National de la Recherche Scientifique, 1968, pp. 435–453
824. Hadamard, J., *Le Calcul Fonctionel*, Oevres de Jacques Hadamard, Tome I, Éditions du Centre National de la Recherche Scientifique 1968, pp. 2253–2266
825. Hagood, Johnson, *Spotting for the Coast Artillery*, CAJ, **60**, No. 2, 1924, pp. 87–92
826. Hamilton, Alston, *Ballistics, Part I*, Coast Artillery School Press, Fort Monroe, Virginia, 1908
827. Hamilton, Alston, *Ballistics, Part II (Exterior Ballistics)*, Coast Artillery School Press, Fort Monroe, Virginia, 1908
828. Hamilton, Alston, *A Practical Method of Computing High Angle Trajectories* NARA Aberdeen, No. 1, 1917
829. Hancock, Harris, *Lectures on the Calculus of Variations (The Weierstrassian Theory)*, University of Cincinnati Bulletin of Mathematics, 1904
830. Hancock, Harris, *Theory of Maxima and Minima*, Ginn and Co., 1917
831. Hardy, G. H., *A Mathematician's Apology*, Cambridge University Press, 1940
832. Hartman, Philip, *Ordinary Differential Equations*, John Wiley and Sons, 1964
833. Hatch, H.J., *A Reply to General Hagood*, CAJ, **60**, No. 3, 1924, pp. 171–176
834. Hedrick, H. B., review, Hadamard, J., *Leçons sur le calcul des variations*, Bull. Am. Math. Society, **21**, 1914, pp. 30–32
835. Hedrick, H. B., *Ballistic Effects Due To the Rotation of the Earth*, CAJ, **70**, No. 6, 1929, pp. 268–269
836. Hestenes, M., *The Calculus of Variations and Optimal Control Theory*, Wiley, New York, 1966
837. Hill, G. W., *Researches in the Lunar Theory*, American Journal of Mathematics, **1**, No. 1, 1876, pp. 5–21
838. Hitchcock, H. P., *Computation of Firing Tables for the U. S. Army*, Aberdeen Proving Ground, 1934
839. Hoar, Roger Sherman, *The Pickering Polaris Attachment*, JUSA, **50**, 1919, pp. 230–239
840. Hoar, Roger Sherman, *The New Ballistics*, JUSA, **51**, 1919, pp. 285–295
841. Hoar, Roger Sherman, *The New Artillery Grid Maps*, JUSA, **54**, 1921, 477–479
842. Hoar, Roger Sherman, *Problems for Solution* [2876–2882], American Mathematical Monthly, **28**, No. 2, Feb. 1921, pp. 89–90
843. Hoar, Roger Sherman, *A Course In Exterior Ballistics*, Government Printing Office, Washington, 1921
844. Hoar, Roger Sherman, *Discussions: On Proof by Mathematical Induction*, American Mathematical Monthly, **29**, No. 4, April 1922, pp. 162–169
845. Hoar, Roger Sherman, *The Rotation of the Earth*, **67**, No. 5, 1927, pp. 377–388
846. Hoar, Roger Sherman, review, *New Methods in Exterior Ballistics* by Forest Ray Moulton, American Mathematical Monthly, **34**, No. 5, June-July 1927, pp. 325–326
847. Hoar, Roger Sherman, review, *New Methods In Exterior Ballistics* by Forest Ray Moulton, Army Ordnance, **7**, 1927, pp. 395–396
848. Hoar, Roger Sherman, *"New Methods in Exterior Ballistics"*, Army Ordnance, **9**, 1928, pp. 50–51
849. Hoar, Roger Sherman, *Curved versus Cartesian Coordinates in Ballistics*, Army Ordnance, **9**, No. 3, 1928, pp. 371–376
850. Hoar, Roger Sherman, entry in *National Cyclopedia of American Biography*, **50**, Ann Arbor, Michigan., p. 622
851. Hull, Gordon, *Some Aspects Of Physics In War And Peace*, Science, **51**, No. 1314, March 5, 1920, pp. 221–233
852. Hung, J. C., *Method of Adjoint Systems and its Engineering Applications*, The University of Tennessee Department of Electrical Engineering, Technical Note No. 1, Control Theory Group, Prepared for National Aeronautics and Space Administration, October 10, 1964

853. Hunter, Patti, *Drawing the Boundaries: Mathematical Statistics in 20th-Century America*, Historia Mathematica, **23**, 1996, pp. 7–30
854. Huntington, Edward V., *The Notion of Probable Error in Elementary Statistics*, Science, New Series, **66**, No. 1722, 1927, pp. 633–637
855. Ingalls, James M., *Exterior Ballistics in the Plane of Fire*, D. Van Nostrand, New York, 1886
856. Ingalls, James M., *Handbook of Problems in Direct Fire*, John Wiley and Sons, New York, 1890
857. Ingalls, James M., *Range Tables For The 12-Inch Cast-Iron B.L. Mortar*, JUSA **1**, 1892, pp. 55–61
858. Ingalls, James M., *General Ballistic Tables For Mortar Firing*, JUSA, **8**, No. 1, 1896, pp. 52–57
859. Ingalls, James M., *A New General Ballistic Table*, JUSA, **9**, No. 3, 1897, pp. 295–299
860. Ingalls, James M., *Ballistics Tables (Artillery Circular M)*, War Department Document 111, Government Printing Office, Washington, 1900
861. Irwin, F., *The Invariants of Linear Differential Expressions*, Proceedings of the American Mathematical Society, **44**, No. 1, 1908, pp. 3–60
862. Jackson, A. M., *Spotting*, CAJ, **61**, 1924, pp. 180–181
863. Jackson, D., *The Method of Numerical Integration in Exterior Ballistics*, Washington, Government Printing Office, 1921
864. Jackson, D., *The Human Significance of Mathematics*, American Mathematical Monthly, **35**, No. 8, October 1928, pp. 406–411
865. Jacobi, C. G. J., *Theoria Novi Multiplicatoris Systematiaequationum Differentialium Vulgarium Applicandi*, in *Gesammelte Werke*, Vierter Band, Cheslea Publishing Company, New York, N.Y., 1969, pp. 317–509
866. Jacobi, C. G. J., *Zur Theorie der Variations-Rechnung und der Differentialgleighungen*, Journal für Mathematik, **17**, 1837, p. 68 ff.
867. Johnson, J. J., *Computation of a Trajectory by Short Arcs*, CAJ, Vol. 55, 1921, pp. 50–67
868. Johnson, J. J., *An Introduction to the Construction of Nomographic or Alignment Charts*, Coast Artillery Journal, **59**, No. 1, July 1923, pp. 157–172
869. Kellogg, Oliver D., *Foundations of Potential Theory*, Springer, Berlin, 1929, reprinted by Dover Publications, 1953
870. Kent, R. H., *Projectile Design*, Army Ordnance, November-December 1921, **2**, No. 9, pp. 204–205
871. Kline, Morris, *Mathematical Thought from Ancient to Modern Times*, Oxford University Press, 1972
872. Kneser, Adolph, *Lehrbuch Der Variationsrechnung*, Friederich Vieweg Und Sohn, Braunschweig, 1900
873. Lagrange, Joseph Louis, *Essai D'une Nouvelle Méthode Pour Déterminer Les Maxima et Les Minima des Formules Integrales Indefinies*, Misc. Soc. Turin, **2**, 1762, pp. 173–195
874. Lagrange, Joseph Louis, *Solution de differens problemes de calcul integral*, Misc. Tour. **3**, 1762–1765, p. 179
875. Lamson, K., *A General Implicit Function Theorem With An Application To Problems Of Relative Minima*, American Journal of Mathematics, **41**, No. 4, October 1920, pp. 243–256
876. Lecat, Maurice, *Bibliographie du Calcul des Variations, 1850–1913*, Ghent and Paris, 1913
877. Legendre, A. L., *Mémoire sure la manière de distinguer les maximas des minimas dans le calcul des variations*, Mémoires de l'Académie des Sciences, 1786
878. Leitmann, G., ed., *Optimization Techniques*, Academic Press, New York, 1962
879. Le Stourgeon, E., *Minima of Functions of Lines*, Trans. Am. Math. Soc., **21**, No. 4, October 1920, pp. 357–383
880. Liouville, Joseph, *Second Mémoire Sur le developpment des functions ou parties de fonctions en series*, Journal des Mathématiques, **2**, 1837, pp. 16–55
881. Lissak, Ormand, *Ordnance and Gunnery*, John Wiley and Sons, New York, 1908
882. Lissak, Ormand, *Ordnance and Gunnery*, John Wiley and Sons, New York, 1915
883. Lorch, E. R., *Joseph Fels Ritt*, Bulletin of the American Mathematical Society, **57**, No. 4, 1951, pp. 307–318

884. Love, A. E. H., *A Treatise on the Mathematical Theory of Elasticity*, Fourth Edition, Cambridge University Press, 1927, reprinted by Dover Publications, New York, 1944
885. Lovitt, William Vernon, *Linear Integral Equations*, McGraw-Hill Book Company, 1924, reprinted by Dover Publications, 1950
886. Mason, Max, *Zur Theorie de Randwertaufgaben*, Mathematische Annalen, **63**, 1904, pp. 528–544
887. Mason, Max, *Green's Theorem and Green's Functions for Certain Systems of Differential Equations*, Transactions of the American Mathematical Society, **5**, 1904, pp. 220–225
888. Mason, Max, *On the boundary Value Problems of Linear Ordinary Differential Equations of the Second Order*, Transactions of the American Mathematical Society, **7**, 1906, pp. 340–344
889. Mason, Max, *On the Differential Equation of the Hyperbolic Type*, Mathematische Annalen, **65**, 1908, pp. 570–575
890. Mason, Max, *Selected Topics in the Theory of Boundary Value Problems of Differential Equations*, in The New Haven Mathematical Colloquium, Yale University Press, 1910, pp. 173–222
891. Mayer, A., *Über Das allgemeinste Problem der Variations-rechnung bei einer einzigen unabhangingen Variabeln*, Leip. Ber., **30**, 1878, pp. 16–32
892. Mayer, A., *Begründung der Lagrange'schen Multiplicatorenmethode in der Variationsrechnung*, Math. Ann., **26**, 1886, pp. 74–82
893. McCalla, T. R., *Application of the Adjoint System of Differential Equations in the Solution of the Bang-Bang Control Problem*, United States Naval Postgraduate School, Monterey, California, 1961
894. McFarland, Earl, *Textbook of Ordnance and Gunnery*, John Wiley and Sons, New York, 1929
895. McShane, E. J., *Differentials of Certain Functions in Exterior Ballistics*, Duke Mathematical Journal, **17**, 1950, pp. 115–134
896. McShane, E. J., *Gilbert Ames Bliss, 1876–1951 A Biographical Memoir by E.J. McShane*, National Academy of Sciences, 1958, pp. 32–53
897. McShane, E. J., *The Calculus of Variations from the Beginning through Optimal Control Theory*, SIAM Control and Optimization, **27**, No. 5, September 1989, pp. 916–939
898. McShane, Edward J., Kelley, John L., Reno, Franklin V., *Exterior Ballistics*, The University of Denver Press, Denver, Colorado, 1953
899. Meigs, J.F. and Ingersoll, R.R., *Exterior Ballistics*, Issac Friedenwald, Baltimore, 1885
900. Metcalfe, Henry, *A Course Of Instruction in Ordnance and Gunnery*, John Wiley and Sons, New York, 1891
901. Metcalfe, Henry, *Ordnance and Gunnery Part II: Cannon, Carriages, Small Arms, Ammunition and Accuracy*, West Point, United States Military Academy Press, 1891
902. Milne, W. E., *Equations of Differential Variations in Exterior Ballistics*, CAJ, **51**, 1919, pp. 154–159
903. Milne, W. E., review, *Mathematics For Exterior Ballistics*, Mathematics Reviews, MR 0010480; (6.24d) 71.0X
904. Moore, E. H., *On a Form of General Analysis With Applications to Linear Differential and Integral Equations*, Atti del IV Congresso Internazionale dei Matematici, **1**, Roma 1908, **2**, Roma 1909, pp. 98–114
905. Moore, E. H., E. J. Wilczynski and Max Mason, *The New Haven Mathematical Colloquium*, Yale University Press, New Haven, 1910
906. Moulton, Forest Ray, *On Certain Rigorous Methods of Treating Problems in Celestial Mechanics*, Decennial Publications of the University of Chicago, **8**, No. 191, (42), 1902
907. Moulton, Forest Ray, *The Influence of Astronomy on Mathematics*, Science, New Series, **33**, No. 835, March 10, 1911, pp. 357–364
908. Moulton, Forest Ray, *M. Henri Poincaré*, Popular Astronomy, **20**, No. 10, December 1912, pp. 620–634
909. Moulton, Forest Ray, *An Introduction To Celestial Mechanics*, Second Revised Edition, Macmillan Company, 1914, reprinted by Dover Publications, New York, 1970

910. Moulton, Forest Ray, *Lectures on the Modern Theory of Differential Equations*, University of Chicago, 1916–1917
911. Moulton, Forest Ray, *History of the Ballistics Branch of the Artillery Ammunition Section, Engineering Division of the Ordnance Department for the Period April 6, 1918 to April 2, 1919*, War Department, Office of the Chief of Ordnance, Washington, 1919
912. Moulton, Forest Ray, *Numerical Integration of Differential Equations*, CAJ, **51**, 1919, pp. 40–55
913. Mouton, Forest Ray, *Numerical Solution of Differential Equations*, in *Smithsonian Mathematical Formulae and Tables of Elliptic Functions*, Smithsonian Miscellaneous Collections, **74**, No. 1, Publication 2672, Smithsonian Institution, 1922
914. Moulton, Forest Ray, *New Methods In Exterior Ballistics*, University of Chicago Press, 1926, reprinted by Dover Publications, 1962
915. Moulton, Forest Ray, discussion of review of *New Methods in Exterior Ballistics* by Roger Sherman Hoar, American Mathematical Monthly, **35**, No. 5, May 1928, pp. 246–250
916. Moulton, Forest Ray, reply to a review of *New Methods in Exterior Ballistics* by Roger Sherman Hoar, Army Ordnance, **8**, 1928, pp. 320–321
917. Moulton, Forest Ray, *Differential Equations*, MacMillan Company 1930, reprinted by Dover Publications, New York, 1958
918. Moulton, F. R., entry in *Notable Scientists: From 1900 to the Present*, Gale Group, 2001, Anonymous.
919. Nenninger, T.K., *The Leavenworth Schools and the Old Army: Education, Professionalism, and the Officer Corps of the United States Army, 1881–1918*, Greenwood Press, Westport, Connecticut, 1978
920. d'Ocagne, Maurice, *Traité de Nomographie, Deuxième Édition*, Gauthier Villars, Paris, 1921
921. Osgood, William Fogg, *Advanced Calculus*, The MacMillan Company, New York, 1925
922. Paaswell, G., *An Appeal to Producing Mathematicians*, Bulletin of the American Mathematical Society, **21**, No. 3, 1914, pp. 127–130
923. Pachpatte, B. G., *Inequalities for Differential and Integral Equations*, San Diego Press, 1998
924. Parshall, Karen Hunger, and Rowe, David E., *The Emergence of the American Mathematical Research Community, 1876–1900: J. J. Sylvester, Felix Klein, and E. H. Moore*, American Mathematical Society, 1994
925. Picard, É., *Mémoire sur la théorie des éqations aux dérivées partielles et la méthode des approximations successives*, Journal des Mathematiques, series 4, **6**, 1890, pp. 145–210
926. Picard, É. *Sur l'application des méthodes d'approximations successives à l'étude de certaines équations différentielles ordinaires*, Journal des Mathematiques series 4, **9**, 1893, pp. 217–271
927. Pincherle, S., *Sull'operazione aggiunta*, Rendiconto della R. Accademia delle Science dell'Instituto di Bologna, **2**, (1897–1898), pp. 130–139
928. Pincherle, S., *Sull'operazione agguinta di Lagrange*, Annali de metematica pur ed applicata, **21**, 1913 pp. 143–151
929. Poincaré, Henri, *Les Méthodes Nouvelles de la Mécanique Céleste, Tome 1*, Dover Publications, New York, reprint of original 1892 edition
930. *Lecture on The Probability and Dispersion of Fire*, translated at the Army War College From a French Document, Government Printing Office, 1917
931. *Provisional Range Tables for the British 75-mm. Gun, Model 1917*, Ordnance Department, U.S. Army, War Plans Division, Government Printing Office, 1918
932. Rademacher, Hans, *On the Accumulation of Errors In Numerical Integration on the ENIAC*, in *Theory and Techniques for Design of Electronic Digital Computers*, Lectures Given at the Moore School, 8 July 1946–31 August 1946, Volume II, Lecture 19, University of Pennsylvania, Moore School of Electrical Engineering, Philadelphia, Pennsylvania, November 1, 1947
933. *Range Tables For 75-mm. Model 1916 American and Model 1917 British Guns*, Prepared At Aberdeen Proving Ground, Engineer Reproduction Plant, U.S. Army, Washington Barracks, November 1921

934. Riemann, Bernhard, *Über die Fortpflanzung ebener Luftwellen von endlicher Schwingingungsweite*, Gottingen Abhandl., **8**, 1860, pp. 145–164
935. Ritt, J., *On Weighting Factor Curves For Low Elevation*, NARA Aberdeen, Technical Report, February 25, 1919
936. Ritt, J., *On the Differentiability of the Solution of a Differential Equation With Respect to a Parameter*, Annals of Mathematics, Second Series, **20**, No. 4, July 1919, pp. 289–291
937. Ritt, J., *On Weighting Factors in Flat Fire*, CAJ, **53**, 1920, pp. 404–410
938. Ritt, J., *On the Iteration of Rational Functions*, Transactions of the American Mathematical Society, **21**, 1920, pp. 313–320
939. Roberts, David L., *Albert Henry Wheeler (1873–1950): A Case Study in the Stratification of American Mathematical Activity*, Historia Mathematica, **23**, 1996, pp. 269–287
940. Roberts, David L., *Republic of Numbers*, Johns Hopkins University Press, 2019
941. Rosenstein, George M., *The Best Method: American Calculus Textbooks of the Nineteenth Century*, from *A Century of Mathematics In America, Part III*, ed. Duren, P., Askey, R., Merzbach, U., and Edwards, H., Providence, Rhode Island, American Mathematical Society, 1988–1989, pp. 77–109
942. Roth, William E., *The Exact Differential Equations Of A Trajectory In Terms Of The Hoar Curvilinear Coordinates*, CAJ, **68**, 1928, pp. 505–512
943. Rowe, J. E., *Moulton on Exterior Ballistics*, American Mathematical Monthly, **35**, No. 4, March-April 1928, pp. 229–232
944. Ruckman, John W., *The Effect Of Wind On The Motion Of A Projectile*, JUSA, **1**, 1892, pp. 5–24
945. Salmon, E. H., *Columns*, Oxford Technical Publications, Henry Frowde and Hodder and Stoughton, London, 1921
946. Sanchez, D., *Ordinary Differential Equations and Stability Theory*, Freeman and Company, 1968
947. Schwartz, Philip, *A Critical Survey of the Advances in Practical Ballistics in the Ordnance Department, U.S. Army Since 1917*, [NARA Abderdeen] A-I-31, 1920
948. Schwartz, Philip, *Measuring the Life of a Gun*, Army Ordnance, **5**, No. 2, 1924, pp. 111–112
949. Schwartz, Philip, *Accuracy Firing*, Army Ordnance, **5**, No. 3, 1924, pp. 521–522
950. Schwartz, Philip, *Experimental Investigation of The Effect of the Wind Upon the Motion of a Projectile*, CAJ, **60**, 1924, pp. 93–97
951. Schwartz, Philip, *Studies of Motion of Projectiles*, CAJ, **60**, 1924, pp. 430–431
952. Schwartz, Philip, *The English Long Range Gun*, CAJ, **61**, 1924, pp. 91–92
953. Schwartz, Philip, *Weighting Factors*, CAJ, **61**, 1924, pp. 371–372
954. Schwartz, Philip, *Remarks Concerning the Use of Firing Table Values*, CAJ, **3**, 1925, pp. 399–400
955. Schwartz, Phlilp, *The Variable Probable Error*, CAJ, **67**, 1927, pp. 405–413
956. Schwartz, Philip, *Much About Little* CAJ, **68**, 1928, pp. 291–297
957. Schwartz, Philip, *The Fork*, Field Artillery Journal, **18**, No. 3, 1929, pp. 604–609
958. Schwartz, Philip, *More About Probable Error*, CAJ, **73**, 1930, pp. 111–120
959. Siacci, F., *Balistique Extérieure*, Traduction Annoteé par P. Laurent, Berger-Levrault et Cic, Editours, Paris and Nancy, 1892
960. Siegmund-Schultze, R.,*Eliakim Hastings Moore's "General Analysis"*, Arch. Hist. Exact. Sci., **52**, 1998, pp. 51–89
961. Siegmund-Schultze, R., *The Late Arrival of Academic Applied Mathematics in the United States: a Paradox, Theses, and Literature*, NTM, **11**, 2002, pp. 116–12
962. Siegmund-Schultze, R., *The Ideology of Applied Mathematics Within Mathematics in Germany and The U. S. Until the End of World War II*, ILUIL, **27**, 2004, pp. 791–811
963. Sleight, E. R., *Mathematics Applied*, National Mathematics Magazine, **9**, No. 8, May 1935, pp. 219–223
964. Smith, Paul A., *Joseph Fels Ritt, 1893–1951*, United States National Academy of Sciences, Biographical Memoir, 1956, pp. 253–264
965. Spinrad, William, *Early History of The Ordnance School*, The Ordnance Sergeant **3**, No. 1, 1942. pp. 1–4

966. Stackel, P., *Abhandlungen Über Variations-Rechnung, Zweiter Theil*, Leipzig, Verlag Von Wilhelm Engelmann, 1894
967. Steele, Bret D., *Muskets and Pendulums: Benjamin Robins, Leonhard Euler, and the Ballistics Revolution*, Technology and Culture, **35**, No. 2, April 1994, pp. 348–382
968. Stevenson, Craig, *Periodic Orbits of the Three Body Problem: F. R. Moulton's Quest for a New Lunar Theory*, Ph.D. Thesis, Open University, 2012
969. Tannehill, I. R., *The Ballistic Wind*, Journal of the United States Artillery, **52**, 1920, pp. 553–576
970. Tarwater, J. D., (Ed.), *The Bicentennial Tribute to American Mathematics, 1776–1976* Papers Presented to the Fifty-Ninth Annual Meeting of the Mathematical Association of America, Mathematical Association of America, Washington, D. C., 1977
971. Todhunter, I., *A Treatise on the Integral Calculus and its Applications*, MacMillan and Co., London, 1880
972. Tonelli, L., *Fondamenti di calcolo delle variazioni*, Nicola Zanichelli, I, 1921; II, 1923
973. Tropp, H. S., *Forest Ray Moulton*, Dictionary of Scientific Biography, **9**, pp. 552–553
974. Truesdell, Clifford, review of *Exterior Ballistics* by McSchane, Kelley and Reno, Scripta Mathematica, **20**, 1954, pp. 172–174
975. Tschappat, W.H., *Ballistics*, article in Encyclopedia Britannica, 1922, pp. 383–394
976. Tsien, H. S., *Engineering Cybernetics*, McGraw-Hill Book Co., 1954
977. Veblen, Oswald, *Ballistic Theories and the Preparation of Range Tables*, NARA Aberdeen, No. 54, 1921
978. Veblen, O., *Professor Veblen on Ballistics in the World War*, Army Ordnance, **4**, March-April 1924, pp. 337–338
979. Very, Edward W., *Computation of Range Tables*, Annapolis, Md., 1877
980. Volterra, V., *Leçons sur les Fonctions des Lignes*, Paris, Gauthier- Villars, 1913
981. Wertheimer, Albert, *A Method of Computing Differential Corrections For a Trajectory*, Journal of the Franklin Institute, **217**, No. 6, June 1934
982. Wiener, Norbert *Ex-Prodigy*, MIT Press, Cambridge, 1953
983. Williamson, Benjamin, *An Elementary Treatise on The Integral Calculus*, Longmans, Green and Co., London, 1916
984. Zeidler, E., *Applied Functional Analysis*, Springer-Verlag, 1995
985. www.math.colostate.edu/~pauld/M345/Some20theory.pdf

## Chapter 4

## Anonymous Editorials, Announcements, and Book Reviews in Military Journals

## JUSA = Journal of The United States Artillery, CAJ = Coast Artillery Journal

986. Announcement, JUSA ( = Journal of the United States Artillery) **1**, 1892, pp. 3–4
987. Book Review, A Course of Instruction in Ordnance and Gunnery, Henry Metcalfe, JUSA, **1** 1892, pp. 74–76
988. Book Review, *Balistique Extérieure Rationelle (Problèmes balisiques secondaires)*, P. Carbonnier, JUSA, **27**, 1907, pp. 310–311
989. Range Table For The 10-Inch B. L. Rifle, Steel JUSA, **5**, 1896, p. 218
990. Progress In Ballistics CAJ (= Coast Artillery Journal), **51**, 1919, pp. 69–71

991. Editorial, *Change Versus Progress*, JUSA, **56**, 1922, pp. 373–374
992. *FRANKFORD ARSENAL: Range, Deflection, and Wind Correction Computer*, Army Ordnance, March-April 1924, **4**, No. 23, pp. 339–341
993. Coast Artillerymen and the Army Ordnance School, CAJ, **2**, 1924, p.363
994. Ballistic Research at Marquette University, Army Ordnance, **7**, No.5, 1927, p. 399
995. Professor Moulton's Book, Army Ordnance, **8**, 1927, No. 1, pp. 12–15
996. *The Coast Artillery School, 1824–1927* CAJ, **66**, 1927, pp. 1–85
997. Book Review, New Methods in Exterior Ballistics, Forest Ray Moulton, CAJ, **66**, 1927, pp. 383–384
998. Book Review, Textbook of Ordnance and Gunnery, Earl McFarland, CAJ, **70**, 1929, pp. 533–534

## Manuscripts, Published Books and Articles

999. Alger, P. L., *Charts for the Calculation of the Effect of Small Changes in the Elements of Fire* JUSA, Vol. 51, 1919, pp. 585–603
1000. Ames, J. S., *The Trained Man of Science in the War*, Science, New Series, 9, **480**, No. 1243, Oct. 25, 1918, pp. 401–410
1001. Archibald, Thomas, Della Dumbaugh and Deborah Kent, *A Mobilized Community: Mathematicians in the United States during the First World War*, in *The War of Guns and Mathematics*, pp. 229–271
1002. Askey, Richard, *Orthogonal Polynomials and Special Functions*, Society for Industrial and Applied Mathematics, Philadelphia, 1975
1003. Aubin, David, and Catherine Goldstein, eds., *The War of Guns and Mathematics*, American Mathematical Society, 2014
1004. Ball, H. P., *Of Responsible Command: A History of the U.S. Army War College*, Alumni Association of the United States Army War College, Carlisle Barracks, Pennsylvania, 1983
1005. *Ballisticians in War and Peace, Volume I, A History of the United States Army Ballistic Research Laboratories 1914–1956*, Aberdeen Proving Ground, Maryland
1006. Banach, S. *Théorie des Opérations Linéaires*, Second Edition, reprint of first edition of 1932, Chelsea Publishing Company, New York, 1955
1007. Barrow-Green, June, *Poincaré and the Three Body Problem*, American Mathematical Society, 1997
1008. Barrow-Green, June, *"Anti-aircraft guns all day long": Karl Pearson and computing for the Ministry of Munitions*, Revue d'histoire des mathématiques, **21**, fascicule 1, 2015, pp. 111–150
1009. Beckman, H. S., *The Elements of Firing Table Construction*, Army Ordnance, **1**, No. 1, pp. 46–454
1010. Bell, E. T., *The Development of Mathematics*, Macmillan and Co., 2nd edition, 1945
1011. Bennett, A. A., *Physical Bases of Ballistic Table Computations, Ordnance Textbook*, Washington Government Printing Office, 1920
1012. Bennett, A. A., *Introduction to Ballistics*, Engineer Reproduction Plant, Washington Barracks, D. C., 1921
1013. Bennett, A. A., *The Problem of Exterior Ballistics*, Army Ordnance, **3**, July-August 1922, No. 13, pp. 38–40
1014. Bennett, A. A., review, *Traité de Balistique Extérieure, Vol. II*, P. Charbonnier, Bulletin of the American Mathematical Society, **34**, No. 4, 1928, pp. 517–518
1015. Bennett, A. A., *Exterior Ballistic Tables Based on Numerical Integration*, United States Army, Ordnance Dept., Washington, Government Printing Office, 1924 (Document 1107), 1931, 1936

1016. Bennett, A. A., *The Differential Analyzer*, Ballistic Research Laboratories, BRL Report No. 319, December, 1942
1017. Benton, J. G., *A Course of Instruction in Ordnance and Gunnery, Third Edition*, D. Van Nostrand, New York, 1867
1018. Berry, W. J., *Some Theorems Concerning Systems of Linear Partial Differential Equations*, Annals of Mathematics, **12**, No. 1, 1910, pp. 27–38
1019. Birkhoff, Garrett, and Gian-Carlo Rota, *Ordinary Differential Equations*, Ginn and Co., Boston, 1962
1020. Birkhoff, G., *Some Leaders In American Mathematics, 1891–1941*, in: Tarwater [1977], pp. 27–78
1021. Birkhoff, G., and Erwin Kreyzig, *The Establishment of Functional Analysis*, Historia Mathematica, **11**, 1984 pp. 258–321
1022. Birkhoff, G. D., *Boundary Value and Expansion Problems of Ordinary Linear Differential Equations*, Transactions of the American Mathematical Society, **9**, No. 4, 1908, pp. 373–395
1023. Birkhoff, G. D., and O. D. Kellogg, *Invariant Points in Function Space*, Transactions of the American Mathematical Society, **23**, No. 1, January 1922, pp. 96–115
1024. Bliss, Gilbert Ames, *An Existence Theorem for a Differential Equation of the Second Order, With an Application to the Calculus of Variations*, Trans. Am. Math. Soc., **5**, 1904, pp. 113–125
1025. Bliss, Gilbert Ames, *The Solutions of Differential Equations of the First Order as Functions of Their Initial Conditions*, Ann. of Math., **6**, 1905, pp. 49–68
1026. Bliss, Gilbert Ames, *A Generalization of the Notion of Angle*, Transactions of the American Mathematical Society, **7**, 1906, pp. 184–196
1027. Bliss, Gilbert Ames, *The Scientific Work of Eliakim Hastings Moore*, Bull. Am. Math. Soc., **40**, 1910, pp. 207–213
1028. Bliss, Gilbert Ames, *Integral Equations*, review of *An Introduction to the Study of Integral Equations* by Maxime Bôcher, *Bulletin of the American Mathematical Society*, **16**, No. 4, 1910, pp. 207–213
1029. Bliss, Gilbert Ames, *A New Proof of the Implicit Function Theorem*, Bull. Am. Math. Soc., **18**, 1912, pp. 175–179
1030. Bliss, Gilbert Ames, *Fundamental Existence Theorems*, The Princeton Colloquium, Part I, American Mathematical Society, 1913
1031. Bliss, Gilbert Ames, *A Note on Functions of Lines*, Proc. Nat. Acad. Sci, U.S.A., **10**, 1915, pp. 173–175
1032. Bliss, Gilbert Ames, review, V. Volterra, *Fonctions des Lignes*, Bull. Am. Math. Soc., **23**, 1915, pp. 344–355
1033. Bliss, Gilbert Ames, *Solutions of Differential Equations as Functions of the Constants of Integration*, Bull. Amer. Math. Soc., **25**, 1918, pp. 15–26
1034. Bliss, Gilbert Ames, *Differential Corrections*, NARA Aberdeen Ms A-29, No. 10, Nov. 1918
1035. Bliss, Gilbert Ames, *The Problem of Mayer With Variable Endpoints*, Trans. Am. Math. Soc., **19**, 1918, pp. 305–316
1036. Bliss, Gilbert Ames, *A Method of Computing Differential Corrections for a Trajectory*, CAJ/JUSA, **50**, 1918, pp. 455–460
1037. Bliss, Gilbert Ames, *A Method of Computing Differential Corrections for a Trajectory*, CAJ/JUSA, **51**, 1919, pp. 445–449
1038. Bliss, Gilbert Ames, *The Use of Adjoint Systems in the Problem of Differential Corrections for Trajectories*, JUSA, **51**, 1919, pp. 296–311
1039. Bliss, Gilbert Ames, *Differential Equations Containing Arbitrary Functions*, Transactions of the American Mathematical Society, **210**, No. 9, 1920, pp. 79–92
1040. Bliss, Gilbert Ames, *Functions of Lines in Ballistics*, ibid, pp. 93–106
1041. Bliss, Gilbert Ames, *Calculus of Variations*, Mathematical Association of America, 1925
1042. Bliss, Gilbert Ames, *A Boundary Value Problem for a System of Ordinary Differential Equations of the First Order*, Transactions of the American Mathematical Society, **28**, No. 4, 1926, pp. 561–584

1043. Bliss, Gilbert Ames, *Mathematics*, Scientific Monthly, **24**, Issue 4, 1927, pp. 308–319

1044. Bliss, Gilbert Ames, *Existence Theorems for Implicit Functions, Existence Theorems for Differential Equations*, University of Chicago, 1928

1045. Bliss. Gilbert Ames, *The Problem of Lagrange in the Calculus of Variations*, Am. J. Math., **52**, No. 4, 1930, pp. 673–744

1046. Bliss, Gilbert Ames, *Mathematical Interpretation of Geometrical And Physical Phenomena*, American Mathematical Monthly, **40**, 1933, pp. 472–480

1047. Bliss, Gilbert Ames, *The Evolution of Problems of the Calculus of Variations*, The American Mathematical Monthly, **43**, No. 10, December 1936, pp. 598–609

1048. Bliss, Gilbert Ames, *Mathematics for Exterior Ballistics*, John Wiley and Sons, New York, 1944

1049. Bliss, Gilbert Ames, *Lectures on the Calculus of Variations*, The University of Chicago Press, Chicago, 1946

1050. Bliss, Gilbert Ames, *Autobiographical Notes*, American Mathematical Monthly, **59**, November 1952, pp. 595–605

1051. Bliss, Gilbert Ames, and Max Mason, *A problem of the Calculus of Variations in Which the Integrand is Discontinuous*, Trans. Am. Math. Soc., **7**, No. 2, April 1906, pp. 325–336

1052. Bliss, Gilbert Ames, and Max Mason, *The Properties of Curves in Space Which Minimize a Definite Integral*, Trans. Amer. Math. Soc., **9**, October 1908, pp. 440–466

1053. Bliss, Gilbert Ames, and Max Mason, *Fields of Extremals in Space*, Transactions of the American Mathematical Society, **11**, 1910, pp. 325–340

1054. Bôcher, M., *Green's Functions in Space of One Dimension*, Bulletin of the American Mathematical Society, **7**, 1901, pp. 297–299

1055. Bôcher, M., *An Introduction to the Study of Integral Equations*, Cambridge Tracts in Mathematics and Mathematical Physics, No. 10, Cambridge University Press, 1909

1056. Bôcher, M., *Boundary Problems and Green's Functions for Linear Differential and Difference Equations*, Annals of Mathematics, **13**, 1911–1912, pp. 71–88

1057. Bôcher, M., *Boundary Problems in One Dimension*, in *Proceedings of the Fifth International Congress of Mathematicians, Cambridge, 22–28 August 1912*, Princeton University Press, 1913, pp. 163–195

1058. Bôcher, M., *Applications and Generalizations of the Conception of Adjoint Systems*, Transactions of the American Mathematical Society, **14**, October 1913, pp. 403–420

1059. Bolza, O., *Lectures on the Calculus of Variations*, University of Chicago Press, 1904, reprinted unabridged and corrected, Dover Publications, 1961

1060. Bolza, Oskar, *Vorlesung Über Variationsrechnung*, Druck und Verlag von B.G. Teubner, 1909

1061. Bolza, Oskar, *Integral Equations*, lecture notes by William Vernon Lovitt from a summer course given during summer 1913 at University of Chicago

1062. Bruff, L. L., *A Textbook of Ordnance and Gunnery*, John Wiley and Sons, New York, 1896

1063. Campbell, G. A., *Mathematics in Industrial Research*, Bell System Technical Journal, 1926, pp. 550–557

1064. Carathéodory, C., review, Hadamard, *Leçons sur le calcul des variations*, **35**, Bulletin des Sciences mathématiques, pp. 123–141, 1911

1065. Carll, L. B., *A Treatise On The Calculus of Variations*, MacMillan and Co., London, 1885

1066. Carmichael, R. D., *The Provision Made by Mathematics for the Needs of Science*, Science, New Series, **45**, Issue 1168 (May 18, 1917), pp. 465–474

1067. Case, Homer, *The Variable Probable Error and The Adjustment Problem*, CAJ, **68**, 1928, pp. 492–501

1068. Chang, Iris, *Thread of the Silkworm*, BasicBooks, A Division of HarperCollins Publishers, Inc., 1995

1069. Charbonnier, P., *Balistique Extérieure Rationelle (Problèmes Balistique Secondaires)*, Encyclopèdie Scientifique de Méchanique Appliqué et Génie, M. D'Ocagne, directeur, Octave Doin, Paris, 1907

1070. Charbonnier, P., *The Gâvre Commission: An Historical Sketch Of The Progress Of Exterior Ballistics* JUSA, **27**, 1907, pp. 238–262

1071. Charbonnier, P., *The Gâvre Commission: An Historical Sketch Of The Progress Of Exterior Ballistics* JUSA, **28**, 1908, pp. 49–73
1072. Charbonnier, P., *Sur L'État Actuel De La Balistique Extérieure Théorique*, Proceedings of the International Mathematical Congress held in Toronto, August 11–16, 1924, ed. Fields, John Charles, Nendeln, Kraus reprint, 1924 pp. 315–334
1073. Charbonnier, P., *Traité de Balistique Extérieure, Tome I*, Gaston Doin, Gauthier-Villars; Paris, 1921
1074. Charbonnier, P., *Traité de Balistique Extérieure, Tome II*, Gaston Doin, Gauthier-Villars; Paris, 1927
1075. Chester, James, *The Theoretical Instruction of Gunners*, JUSA, **1**, No. 3, July 1892, pp. 171–206
1076. Chester, James, *Whistler's Tables of Fire. Artillery Circular C*, Journal of the Military Service Institution of the United States, **13**, 1892, pp. 1028–1030
1077. Chester, James, *Sea-Coast Artillery Instruction*, JUSA, **5**,1896, pp. 209–218
1078. Coddington, E. A. and N. Levinson, *Theory of Ordinary Differential Equations*, McGraw-Hill Book Company, 1955
1079. *The Confederate Field Manual*, Ritchie and Dunnavant, Richmond, 1862, reprinted 1984
1080. Coxe, W. H. and Bugless. E., *Exterior Ballistics Charts*, E.I. du Pont de Nemours and Company, Inc., 1926
1081. Craig, Thomas, *A Treatise on Linear Differential Equations*, New York, John Wiley and Sons, 1889
1082. *The Applications of Elliptic Functions by Alfred George Greenhill, M.A., F.R.S., Professor of Mathematics in the Artillery College, Woolwich*, book review, Thomas Craig, JUSA, **2**, 1893, pp. 298–306
1083. Cranz, C. and Becker, K., *Handbook of Ballistics, Volume I, Exterior Ballistics*, translated from the second German edition, H. M. Stationery Office, London, 1921
1084. Darboux, G., *Sur les équations aux dérivés partielle*, Comptes Rendus, **96**, March 19, 1883, pp. 766–769
1085. Darboux, G., *Leçons sur la théorie générale des surfaces, deuxieme partie*, Gauthier-Villars et Fils, Paris, 1889
1086. Darboux, G., *Leçons sur la théorie générale des surfaces, quatrième partie*, Gauthier-Villars et Fils, Paris, 1896
1087. Dederick, L. S., *Exterior Ballistics*, Bulletin of the American Mathematical Society, **34**, No. 5, 1928, p. 667
1088. Dederick, L. S., *The Mathematics of Exterior Ballistics Computation*, American Mathematical Monthly, **47**, No. 9, November, 1940, pp. 628–634
1089. Devorkin, D. H., *Henry Norris Russell, Dean of American Astronomers*, Princeton University Press, Princeton and Oxford, p. 200
1090. Dickson, L. E., *Mathematics in War Perspective*, Bulletin of the American Mathematical Society, **25**, No. 7, 1919, pp. 289–311
1091. Dickson, T.C., *Course of Instruction Prescribed for Student Officers In the Ordnance School of Application in Differential Equations*, July 24, 1907, revised April 12, 1910, Government Printing Office, Washington
1092. Dieudonné, J., *History of Functional Analysis*, Amsterdam: north Holland, 1981
1093. Doerfler, R., Gluchoff, A.,Guthery, S., and Paul Hamburg, *Calculating Curves*, Docent Press, 2012
1094. Drach, Jules, *L'équation differentielle de la balistique extérieure et son integration par quadratures*, Annales Scientifiques de l'E.N.S., Serie 3, tome 37, 1920 pp. 1–94
1095. Drenick, R., *The Perturbation Calculus in Missile Ballistics*, Journal of the Franklin Institute, **25**, 1951, pp. 423–436
1096. Drenick, R., *Rudolph F. Drenick, Veteran*, interviewed by Mike Russert, Wayne Clark, November 18, 2003, South Setauket, New York, dmna.ny.gov/historic/veterans/transcripts/Drenick-Rudolf-F.pdf

1097. du Bois-Reymond, P., *Über lineare partielle Differentialgleichungen zweiter Ordnung*, Journal für Mathematik, **104**, pp. 241–310
1098. Evans, G. C., *Functionals and Their Applications*, American Mathematical Society Colloquium Lectures, **5**, Part I, 1918
1099. Faulkner, F. D., *Optimum Interception of a Ballistic Missile at Intermediate Range*, Naval Postgraduate School, Monterey, California, October 1961
1100. Faulkner, F. D., *A Comparison Between Some Methods For Computing Optimum Paths in the Problem of Bolza*, Research Paper No. 43, United States Naval Postgraduate School, Monterey, California, March 1964
1101. Faulkner, F. D., *Newton's Method Applied To The Solution Of Classical Problems In Calculus Of Variations And Differential Games*, dissertation submitted in partial fulfillment of the requirements for the degree of Doctor of Philosophy in the University of Michigan, 1969
1102. Feffer, Loren Butler, *Oswald Veblen and the Capitalization of American mathematics*, Isis, **89**, pp. 474–497
1103. Fenster, Della, Kent, Deborah and Archibald, Thomas, *The First World War and Mathematics In the United States*, preprint
1104. Fischer, C. A., *A Generalization of Volterra's Derivative of a Function of a Curve*, American Journal of Mathematics, **35**, No. 4, 1913, pp. 369–394
1105. Fischer, C. A., *Note on the order of continuity of functions of lines*, Bulletin of the American Mathematical Society, **23**, (1916–1917), pp. 88–90
1106. Flanagan, Francis, *Harmonic Functions*, Dover Press, 1972
1107. Ford, L., *Instruction Memorandum No. 20, Subject: Elementary Mathematics For Field Artillery*, Headquarters, Field Artillery Central Officers Training School, Camp Zachary Taylor, Ky., 1919
1108. Fowler, R. H., E. G. Gallop, C. N. H. Lock and H. W. Richmond, *The Aerodynamics of a Spinning Shell*, Philosophical Transactions of the Royal Society, A221, 1920, pp. 295–387
1109. Franklin, Philip, *An Alternative Derivation of the Differential Equations Used in Computing Differential Corrections*, [NARA Aberdeen], No. 86, 1919
1110. Franklin, Philip, *Exterior Ballistics, A Historical Survey*, Brown University Graduate School, 1943
1111. *Philip Franklin of M.I.T. is Dead*, New York Times, 1/28/65
1112. Fréchet, M. *Sur La Notion de la Differentielle D'Une Fonction de Ligne*, Transactions of the American Mathematical Society, **10**, 1913, pp. 135–161
1113. Friedlander, Alan, *A Study of Guidance Sensitivity for Various Low-THrust Transfers from Earth to Mars*, Technical Note D-1183, Lewis Research Center, Clevelandm Ohio, February 1962
1114. Fuchs, L., Über Relationen, welche für die zwischen je zwei singulären Punkten erstreckten integrale der lösungen linearer differentialgleichungen stattfinden, Journal für die reine und angewandte Mathematik, **76**, 1874, pp. 177–213
1115. Galileo, *Two New Sciences*, A New Translation With Introduction and Notes, Stillman Drake, University of Wisconsin Press, 1974
1116. Gasteyer, C.E., *Forest Ray Moulton, 1872–1952* , Biographical Memoirs of the National Academy of Sciences, **41**, pp. 340–355
1117. Gateaux, R., *Fonctions d'une infinité de variables indépendantes*, Bulletin de la S. M. F., **47**, 1919, pp. 70–96
1118. Glennon, J.H., *Accuracy and Probability of Fire*, Baltimore Press of Isaac Friedenwald, 1888
1119. Gluchoff, Alan, *Pure Mathematics Applied in Early Twentieth-Century America: The case of T.H. Gronwall, Consulting Mathematician*, Historia Mathematica, **32**, 2005, pp. 312–357
1120. Gluchoff, Alan, *Artillerymen and Mathematicians: Forest Ray Moulton and Changes in American Exterior Ballistics, 1885–1934*, Historia Mathematica, **38**, 2011, pp. 506–547
1121. Goldstine, Herman H., *Conditions For A Minimum Of A Functional*, The University of Chicago Libraries, Chicago, Illinois, 1937

1122. Goldstine, Herman. H., *A Multiplier Rule In Abstract Spaces*, Bulletin of the American Mathematical Society, **44**, 1938, pp. 388–394
1123. Goldstine, Herman H., *The Computer from Pascal to von Neumann*, Princeton University Press, 1972
1124. Oral History Transcript - Dr. Herman Goldstine, Interview with Dr. Herman Goldstine by Nancy Stern at Institute for Advanced Study, Princeton, March 14, 1977, http://www.aip.org/history/ohilist/4636.html
1125. Goldstine, Herman H., *A History of the Calculus of Variations From the 17th Through the 19th Centuries*, Springer-Verlag, 1980
1126. Goldstine, Herman H., interviewed in *The Princeton Mathematics Community in the 1930s*, Transcript Number 15, (PMC15), The Trustees of Princeton University, 1985
1127. Goldstine, Herman, interview with author July 12, 2004
1128. Goodman, T. R., G. N. Lance, *The Numerical Integration of Two-Point Boundary Value Problems*, Mathematical Tables and Other Aids to Computation, April 1956, **10**, No. 54, pp. 82–86
1129. Goursat, Édouard, *Sur la Théorie des functions implicites*, Bull. de la S. M. F., **31**, 1903, pp. 184–192
1130. Goursat, Édouard, *A Course In Mathematical Analysis*, Vol. II, Part II, translated by Hedrick, E.R., and Dunkel, Otto, Ginn and Co., 1917
1131. Goursat, Édouard, *Cours d'Analyse mathématique, Tome III*, Troisième Edition, Gauthier-Villars, 1923
1132. Graves, L., *Implicit Functions and Differential Equations in General Analysis*, Transactions of the American Mathematical Society, **29**, No. 3, July 1927, pp. 514–552
1133. Graves, L., *Topics in the Functional Calculus*, Bulletin of the American Mathematical Society, **41**, 1935, pp. 641–662
1134. Graves, L., *What Is A Functional?*, The American Mathematical Monthly, **55**, No. 8, October 1948, pp. 467–472
1135. Graves, L. M., *Gilbert Ames Bliss, 1876–1951*, Bulletin of the American Mathematical Society, **58**, (1952), pp. 251–264
1136. Graves, L., *The Theory of Functions of Real Variables*, $2^{nd}$ Edition, McGraw-Hill, 1956, reprinted by Dover Publications, 2009
1137. Graves, L., *Nonlinear Mappings between Banach Spaces*, in *Studies in Real and Complex Analysis*, ed. I. I. Hirschman, The Mathematical Association of America, 1965, pp. 34–54
1138. Graves, L., and T. H. Hildebrandt, *Implicit Functions and Their Differentials in General Analysis*, Transactions of the American Mathematical Society, **29**, No. 1, January 1927, pp. 127–153
1139. Green, George, *An Essay on the Application of Mathematical Analysis to the Theories of Electricity and Magnetism*, Journal für Mathematik, **39**, 1850, pp. 73–89; **44**, 1852, pp. 356–374; **47**, 1854, pp. 161–221
1140. Grier, David Alan, *Dr. Veblen Takes a Uniform: Mathematics in the First World War*, American Mathematical Monthly, **108**, December 2001, pp. 922–931
1141. Grier, David Alan, *Dr. Veblen At Aberdeen: Mathematics, Military Applications and Mass Production*, in *Instrumental in War*, Steven A. Walton, Ed., Boston, Brill, 2005, pp. 253–270
1142. Grier, David Alan, *When Computers Were Human*, Princeton University Press, 2005
1143. Gronwall, T. H., *Note on the Derivative with Respect to a Parameter of the Solutions of a System of Differential Equations*, Annals of Mathematics, Second Series, **20**, No. 4, July 1919, pp. 292–296
1144. Gronwall, T. H., *On a System if Linear Partial Differential Equations of Hyperbolic Type*, Annals of Mathematics, **20**, No. 4, July 1919, pp. 274–278
1145. Gronwall, T. H., *On Differential Corrections and Weighting Factors*, NARA Washington 4, July 1919, pp. 274–278
1146. Gronwall, T.H., *Qualitative Properties of the Ballistic Trajectory*, Annals of Mathematics, Ser. II, Vol. 21, 1920, pp. 44–65

1147. Gronwall, T. H., *Differential Variations In Ballistics With Applications to the Qualitative Properties of the Trajectory*, Transactions of the American Mathematical Society, **22**, No. 4, 1921, pp. 505–525
1148. Gronwall, T. H., review, *Traité de Balistique Extérieure, Vol. I*, P. Charbonnier, Bulletin of the American Mathematical Society, **33**, No. 1, 1927, p. 120
1149. Guion, James L., review, *New Methods in Exterior Ballistics*, Army Ordnance, 1926, **7**, pp. 152–155
1150. *Gunnery For Heavy Artillery*, Coast Artillery Corps, War Department, 1925
1151. Hadamard, J., *Leçons sur la propagation des ondes*, Paris, Hermann, 1903
1152. Hadamard, J., *Leçons sur le calcul des variations*, Hermann et fils, Paris, 1910
1153. Hadamard, J., *Rapport sur les examinés par la Commission de Balistique pendant la durée de la guerre*, Comptes Rendus de l'Académie des Sciences, **170**, No, 8, 1920, pp. 436–445
1154. Hadamard, J., *Lectures on Cauchy's Problem in Linear Partial Differential Equations*, Yale University Press, 1923
1155. Hadamard, J., *Le Développement et le role Scientifique Du Calcul Fonctionel*, Oevres de Jacques Hadamard, Tome I, Éditions du Centre National de la Recherche Scientifique, 1968, pp. 435–453
1156. Hadamard, J., *Le Calcul Fonctionel*, Oevres de Jacques Hadamard, Tome I, Éditions du Centre National de la Recherche Scientifique 1968, pp. 2253–2266
1157. Hagood, Johnson, *Spotting for the Coast Artillery*, CAJ, **60**, No. 2, 1924, pp. 87–92
1158. Hamilton, Alston, *Ballistics, Part I*, Coast Artillery School Press, Fort Monroe, Virginia, 1908
1159. Hamilton, Alston, *Ballistics, Part II (Exterior Ballistics)*, Coast Artillery School Press, Fort Monroe, Virginia, 1908
1160. Hamilton, Alston, *A Practical Method of Computing High Angle Trajectories* NARA Aberdeen, No. 1, 1917
1161. Hancock, Harris, *Lectures on the Calculus of Variations (The Weierstrassian Theory)*, University of Cincinnati Bulletin of Mathematics, 1904
1162. Hancock, Harris, *Theory of Maxima and Minima*, Ginn and Co., 1917
1163. Hardy, G. H., *A Mathematician's Apology*, Cambridge University Press, 1940
1164. Hartman, Philip, *Ordinary Differential Equations*, John Wiley and Sons, 1964
1165. Hatch, H.J., *A Reply to General Hagood*, CAJ, **60**, No. 3, 1924, pp. 171–176
1166. Hayes, Thomas J., *Elements of Ordnance*, John Wiley and Sons, Inc., New York, 1938
1167. Hedrick, H. B., review, Hadamard, J., *Leçons sur le calcul des variations*, Bull. Am. Math. Society, **21**, 1914, pp. 30–32
1168. Hedrick, H. B., *Ballistic Effects Due To the Rotation of the Earth*, CAJ, **70**, No. 6, 1929, pp. 268–269
1169. Hestenes, M., *The Calculus of Variations and Optimal Control Theory*, Wiley, New York, 1966
1170. Hill, G. W., *Researches in the Lunar Theory*, American Journal of Mathematics, **1**, No. 1, 1876, pp. 5–21
1171. Hitchcock, H. P., *Computation of Firing Tables for the U. S. Army*, Aberdeen Proving Ground, 1934
1172. Hoar, Roger Sherman, *The Pickering Polaris Attachment*, JUSA, **50**, 1919, pp. 230–239
1173. Hoar, Roger Sherman, *The New Ballistics*, JUSA, **51**, 1919, pp. 285–295
1174. Hoar, Roger Sherman, *The New Artillery Grid Maps*, JUSA, **54**, 1921, 477–479
1175. Hoar, Roger Sherman, *Problems for Solution* [2876–2882], American Mathematical Monthly, **28**, No. 2, Feb. 1921, pp. 89–90
1176. Hoar, Roger Sherman, *A Course In Exterior Ballistics*, Government Printing Office, Washington, 1921
1177. Hoar, Roger Sherman, *Discussions: On Proof by Mathematical Induction*, American Mathematical Monthly, **29**, No. 4, April 1922, pp. 162–169
1178. Hoar, Roger Sherman, *The Rotation of the Earth*, ,**67**, No. 5 , 1927, pp. 377–388

1179. Hoar, Roger Sherman, review, *New Methods in Exterior Ballistics* by Forest Ray Moulton, American Mathematical Monthly, **34**, No. 5, June-July 1927, pp. 325–326
1180. Hoar, Roger Sherman, review, *New Methods In Exterior Ballistics* by Forest Ray Moulton, Army Ordnance, **7**, 1927, pp. 395–396
1181. Hoar, Roger Sherman, *"New Methods in Exterior Ballistics"*, Army Ordnance, **9**, 1928, pp. 50–51
1182. Hoar, Roger Sherman, *Curved versus Cartesian Coordinates in Ballistics*, Army Ordnance, **9**, No. 3, 1928, pp. 371–376
1183. Hoar, Roger Sherman, entry in *National Cyclopedia of American Biography*, **50**, Ann Arbor, Michigan., p. 622
1184. Hull, Gordon, *Some Aspects Of Physics In War And Peace*, Science, **51**, No. 1314, March 5, 1920, pp. 221–233
1185. Hung, J. C., *Method of Adjoint Systems and its Engineering Applications*, The University of Tennessee Department of Electrical Engineering, Technical Note No. 1, Control Theory Group, Prepared for National Aeronautics and Space Administration, October 10, 1964
1186. Hunter, Patti, *Drawing the Boundaries: Mathematical Statistics in 20th-Century America*, Historia Mathematica, **23**, 1996, pp. 7–30
1187. Huntington, Edward V., *The Notion of Probable Error in Elementary Statistics*, Science, New Series, **66**, No. 1722, 1927, pp. 633–637
1188. Ingalls, James M., *Exterior Ballistics in the Plane of Fire*, D. Van Nostrand, New York, 1886
1189. Ingalls, James M., *Handbook of Problems in Direct Fire*, John Wiley and Sons, New York, 1890
1190. Ingalls, James M., *Range Tables For The 12-Inch Cast-Iron B.L. Mortar*, JUSA **1**, 1892, pp. 55–61
1191. Ingalls, James M., *General Ballistic Tables For Mortar Firing*, JUSA, **8**, No. 1, 1896, pp. 52–57
1192. Ingalls, James M., *A New General Ballistic Table*, JUSA, **9**, No. 3, 1897, pp. 295–299
1193. Ingalls, James M., *Ballistics Tables (Artillery Circular M)*, War Department Document 111, Government Printing Office, Washington, 1900
1194. Irwin, F., *The Invariants of Linear Differential Expressions*, Proceedings of the American Mathematical Society, **44**, No. 1, 1908, pp. 3–60
1195. Jackson, A. M., *Spotting*, CAJ, **61**, 1924, pp. 180–181
1196. Jackson, D., *The Method of Numerical Integration in Exterior Ballistics*, Washington, Government Printing Office, 1921
1197. Jackson, D., *The Human Significance of Mathematics*, American Mathematical Monthly, **35**, No. 8, October 1928, pp. 406–411
1198. Jacobi, C. G. J., *Theoria Novi Multiplicatoris Systemati Aequationum Differentialium Vulgarium Applicandi*, in *Gesammelte Werke*, Vierter Band, Cheslea Publishing Company, New York, N.Y., 1969, pp. 317–509
1199. Jacobi, C. G. J., *Zur Theorie der Variations-Rechnung und der Differentialgleighungen*, Journal für Mathematik, **17**, 1837, p. 68 ff.
1200. Johnson, J. J., *Computation of a Trajectory by Short Arcs*, CAJ, Vol. 55, 1921, pp. 50–67
1201. Johnson, J. J., *An Introduction to the Construction of Nomographic or Alignment Charts*, Coast Artillery Journal, **59**, No. 1, July 1923, pp. 157–172
1202. Kellogg, Oliver D., *Foundations of Potential Theory*, Springer, Berlin, 1929, reprinted by Dover Publications, 1953
1203. Kent, R. H., *Projectile Design*, Army Ordnance, November-December 1921, **2**, No. 9, pp. 204–205
1204. Kent, R. H., *Professor Moulton's "New Methods in Exterior Ballistics" Arouses Comment*, Army ORdnance, **7**. No. 41, March 1927, p. 395
1205. Kline, Morris, *Mathematical Thought from Ancient to Modern Times*, Oxford University Press, 1972
1206. Kneser, Adolph, *Lehrbuch Der Variationsrechnung*, Friederich Vieweg Und Sohn, Braunschweig, 1900

1207. Lagrange, Joseph Louis, *Essai d'une nouvelle méthode pour déterminer les maxima et les minima des formules integrales indefinies*, Misc. Soc. Turin, **2**, 1762, pp. 173–195
1208. Lagrange, Joseph Louis, *Solution de differens problemes de calcul integral*, Misc. Tour. **3**, 1762–1765, p. 179
1209. Lamson, K., *A General Implicit Function Theorem With An Application To Problems Of Relative Minima*, American Journal of Mathematics, **41**, No. 4, October 1920, pp. 243–256
1210. Lecat, Maurice, *Bibliographie du Calcul des Variations, 1850–1913*, Ghent and Paris, 1913
1211. Legendre, A. L., *Mémoire sure la manière de distinguer les maximas des minimas dans le calcul des variations*, Mémoires de l'Académie des Sciences, 1786
1212. Leitmann, G., ed., *Optimization Techniques*, Academic Press, New York, 1962
1213. Le Stourgeon, E., *Minima of Functions of Lines*, Trans. Am. Math. Soc., **21**, No. 4, October 1920, pp. 357–383
1214. Liouville, Joseph, *Second Mémoire Sur le developpment des functions ou parties de fonctions en series*, Journal des Mathématiques, **2**, 1837, pp. 16–55
1215. Lissak, Ormand, *Ordnance and Gunnery*, John Wiley and Sons, New York, 1908
1216. Lissak, Ormand, *Ordnance and Gunnery*, John Wiley and Sons, New York, 1915
1217. Lorch, E. R., *Joseph Fels Ritt*, Bulletin of the American Mathematical Society, **57**, No. 4, 1951, pp. 307–318
1218. Love, A. E. H., *A Treatise on the Mathematical Theory of Elasticity*, Fourth Edition, Cambridge University Press, 1927, reprinted by Dover Publications, New York, 1944
1219. Lovitt, William Vernon, *Linear Integral Equations*, McGraw-Hill Book Company, 1924, reprinted by Dover Publications, 1950
1220. Lowan, Arnold N. *On The Bliss Algorithm and its application to the Treatment of Many-point Boundary Problems*, University of California Radiation Laboratory, Livermore, California, June 1958
1221. Mason, Max, *Zur Theorie de Randwertaufgaben*, Mathematische Annalen, **63**, 1904, pp. 528–544
1222. Mason, Max, *Green's Theorem and Green's Functions for Certain Systems of Differential Equations*, Transactions of the American Mathematical Society, **5**, 1904, pp. 220–225
1223. Mason, Max, *On the boundary Value Problems of Linear Ordinary Differential Equations of the Second Order*, Transactions of the American Mathematical Society, **7**, 1906, pp. 340–344
1224. Mason, Max, *On the Differential Equation of the Hyperbolic Type*, Mathematische Annalen, **65**, 1908, pp. 570–575
1225. Mason, Max, *Selected Topics in the Theory of Boundary Value Problems of Differential Equations*, in The New Haven Mathematical Colloquium, Yale University Press, 1910, pp. 173–222
1226. Mayer, A., *Über Das allgemeinste Problem der Variationsrechnung bei einer einzigen unabhangingen Variabeln*, Leip. Ber., **30**, 1878, pp. 16–32
1227. Mayer, A., *Begründung der Lagrange'schen Multiplicatorenmethode in der Variationsrechnung*, Math. Ann., **26**, 1886, pp. 74–82
1228. McCalla, T. R., *Application of the Adjoint System of Differential Equations in the Solution of the Bang-Bang Control Problem*, United States Naval Postgraduate School, Monterey, California, 1961
1229. McFarland, Earl, *Textbook of Ordnance and Gunnery*, John Wiley and Sons, New York, 1929
1230. McShane, E. J., *Differentials of Certain Functions in Exterior Ballistics*, Duke Mathematical Journal, **17**, 1950, pp. 115–134
1231. McShane, E. J., *Gilbert Ames Bliss, 1876–1951 A Biographical Memoir by E.J. McShane*, National Academy of Sciences, 1958, pp. 32–53
1232. McShane, E. J., *The Calculus of Variations from the Beginning through Optimal Control Theory*, SIAM Control and Optimization, **27**, No. 5, September 1989, pp. 916–939
1233. McShane, Edward J., Kelley, John L., Reno, Franklin V., *Exterior Ballistics*, The University of Denver Press, Denver, Colorado, 1953

1234. Meigs, J.F. and Ingersoll, R.R., *Exterior Ballistics*, Issac Friedenwald, Baltimore, 1885
1235. Metcalfe, Henry, *A Course Of Instruction in Ordnance and Gunnery*, John Wiley and Sons, New York, 1891
1236. Metcalfe, Henry, *Ordnance and Gunnery Part II: Cannon, Carriages, Small Arms, Ammunition and Accuracy*, West Point, United States Military Academy Press, 1891
1237. Milne, W. E., *Equations of Differential Variations in Exterior Ballistics*, CAJ, **51**, 1919, pp. 154–159
1238. Milne, W. E., review, *Mathematics For Exterior Ballistics*, Mathematics Reviews, MR 0010480; (6.24d) 71.0X
1239. Moore, E. H., *On a Form of General Analysis With Applications to Linear Differential and Integral Equations*, Atti del IV Congresso Internazionale dei Matematici, **1**, Roma 1908, **2**, Roma 1909, pp. 98–114
1240. Moore, E. H., E. J. Wilczynski and Max Mason, *The New Haven Mathematical Colloquium*, Yale University Press, New Haven, 1910
1241. Moulton, Forest Ray, *On Certain Rigorous Methods of Treating Problems in Celestial Mechanics*, Decennial Publications of the University of Chicago, **8**, No. 191, (42), 1902
1242. Moulton, Forest Ray, *The Influence of Astronomy on Mathematics*, Science, New Series, **33**, No. 835, March 10, 1911, pp. 357–364
1243. Moulton, Forest Ray, *M. Henri Poincaré*, Popular Astronomy, **20**, No. 10, December 1912, pp. 620–634
1244. Moulton, Forest Ray, *An Introduction To Celestial Mechanics*, Second Revised Edition, Macmillan Company, 1914, reprinted by Dover Publications, New York, 1970
1245. Moulton, Forest Ray, *Lectures on the Modern Theory of Differential Equations*, University of Chicago, 1916–1917
1246. Moulton, Forest Ray, *History of the Ballistics Branch of the Artillery Ammunition Section, Engineering Division of the Ordnance Department for the Period April 6, 1918 to April 2, 1919*, War Department, Office of the Chief of Ordnance, Washington, 1919
1247. Moulton, Forest Ray, *Numerical Integration of Differential Equations*, CAJ, **51**, 1919, pp. 40–55
1248. Mouton, Forest Ray, *Numerical Solution of Differential Equations*, in *Smithsonian Mathematical Formulae and Tables of Elliptic Functions*, Smithsonian Miscellaneous Collections, **74**, No. 1, Publication 2672, Smithsonian Institution, 1922
1249. Moulton, Forest Ray, *New Methods In Exterior Ballistics*, University of Chicago Press, 1926, reprinted by Dover Publications, 1962
1250. Moulton, Forest Ray, discussion of review of *New Methods in Exterior Ballistics* by Roger Sherman Hoar, American Mathematical Monthly, **35**, No. 5, May 1928, pp. 246–250
1251. Moulton, Forest Ray, reply to a review of *New Methods in Exterior Ballistics* by Roger Sherman Hoar, Army Ordnance, **8**, 1928, pp. 320–321
1252. Moulton, Forest Ray, *Differential Equations*, MacMillan Company 1930, reprinted by Dover Publications, New York, 1958
1253. Moulton, F. R., entry in *Notable Scientists: From 1900 to the Present*, Gale Group, 2001, Anonymous.
1254. Nenninger, T.K., *The Leavenworth Schools and the Old Army: Education, Professionalism, and the Officer Corps of the United States Army, 1881–1918*, Greenwood Press, Westport, Connecticut, 1978
1255. d'Ocagne, Maurice, *Traité de Nomographie, Deuxième Édition*, Gauthier Villars, Paris, 1921
1256. Osgood, William Fogg, *Advanced Calculus*, The MacMillan Company, New York, 1925
1257. Paaswell, G., *An Appeal to Producing Mathematicians*, Bulletin of the American Mathematical Society, **21**, No. 3, 1914, pp. 127–130
1258. Pachpatte, B. G., *Inequalities for Differential and Integral Equations*, San Diego Press, 1998
1259. Parshall, Karen Hunger, and Rowe, David E., *The Emergence of the American Mathematical Research Community, 1876–1900: J. J. Sylvester, Felix Klein, and E. H. Moore*, American Mathematical Society, 1994

1260. Picard, É., *Mémoire sur la théorie des éqations aux dérivées partielles et la méthode des approximations successives*, Journal des Mathematiques, series 4, **6**, 1890, pp. 145–210
1261. Picard, É. *Sur l'application des méthodes d'approximations successives à l'étude de certaines équations différentielles ordinaires*, Journal des Mathematiques series 4, **9**, 1893, pp. 217–271
1262. Pincherle, S., *Sull'operazione aggiunta*, Rendiconto della R. Accademia delle Science dell'Instituto di Bologna, **2**, (1897–1898), pp. 130–139
1263. Pincherle, S., *Sull'operazione agguinta di Lagrange*, Annali de metematica pur ed applicata, **21**, 1913 pp. 143–151
1264. Poincaré, Henri, *Les Méthodes Nouvelles de la Mécanique Céleste , Tome 1*, Dover Publications, New York, reprint of original 1892 edition
1265. Polachek, Harry, *The Men Behind The Men Behind The Guns*, Scripta Mathematica, **9**, March 1943, pp. 27–32
1266. Polachek, Harry, *Solution of the Differential Equations of Motion of a Projectile in a Medium of Quasi-Newtonian Resistance*, Quarterly Applied Math. **7**, 1949, pp. 275–291
1267. *Lecture on The Probability and Dispersion of Fire*, author unknown, translated at the Army War College From a French Document, Government Printing Office, 1917
1268. *Provisional Range Tables for the British 75-mm. Gun, Model 1917*, Ordnance Department, U.S. Army, War Plans Division, Government Printing Office, 1918
1269. Rademacher, Hans, *On the Accumulation of Errors In Numerical Integration on the ENIAC*, in *Theory and Techniques for Design of Electronic Digital Computers*, Lectures Given at the Moore School, 8 July 1946–31 August 1946, Volume II, Lecture 19, University of Pennsylvania, Moore School of Electrical Engineering, Philadelphia, Pennsylvania, November 1, 1947
1270. *Range Tables For 75-mm. Model 1916 American and Model 1917 British Guns*, Prepared At Aberdeen Proving Ground, Engineer Reproduction Plant, U.S. Army, Washington Barracks, November 1921
1271. Riemann, Bernhard, *Über die Fortpflanzung ebener Luftwellen von endlicher Schwingingungsweite*, Gottingen Abhandl., **8**, 1860, pp. 145–164
1272. Ritt, J., *On Weighting Factor Curves For Low Elevation*, NARA Aberdeen, Technical Report, February 25, 1919
1273. Ritt, J., *On the Differentiability of the Solution of a Differential Equation With Respect to a Parameter*, Annals of Mathematics, Second Series, **20**, No. 4, July 1919, pp. 289–291
1274. Ritt, J., *On Weighting Factors in Flat Fire*, CAJ, **53**, 1920, pp. 404–410
1275. Ritt, J., *On the Iteration of Rational Functions*, Transactions of the American Mathematical Society, **21**, 1920, pp. 313–320
1276. Roberts, David L., *Albert Henry Wheeler (1873–1950): A Case Study in the Stratification of American Mathematical Activity*, Historia Mathematica, **23**, 1996, pp. 269–287
1277. Roberts, David L., *Republic of Numbers*, Johns Hopkins University Press, 2019
1278. Rosenstein, George M., *The Best Method: American Calculus Textbooks of the Nineteenth Century*, from *A Century of Mathematics In America, Part III*, ed. Duren, P., Askey, R., Merzbach, U., and Edwards, H., Providence, Rhode Island, American Mathematical Society, 1988–1989, pp. 77–109
1279. Roth, William E., *The Exact Differential Equations Of A Trajectory In Terms Of The Hoar Curvilinear Coordinates*, CAJ, **68**, 1928, pp. 505–512
1280. Rowe, J. E., *A Course In Exterior Ballistics*, Bulletin of the American Mathematical Society, **29** , No. 6, , June 1923, pp. 277–279
1281. Rowe, J. E., *Moulton on Exterior Ballistics*, American Mathematical Monthly, **35**, No. 4, March-April 1928, pp. 229–232
1282. Ruckman, John W., *The Effect Of Wind On The Motion Of A Projectile*, JUSA, **1**, 1892, pp. 5–24
1283. Salmon, E. H., *Columns*, Oxford Technical Publications, Henry Frowde and Hodder and Stoughton, London, 1921

1284. Sanchez, D., *Ordinary Differential Equations and Stability Theory*, Freeman and Company, 1968
1285. Schwartz, Philip, *A Critical Survey of the Advances in Practical Ballistics in the Ordnance Department, U.S. Army Since 1917*, [NARA Abderdeen] A-I-31, 1920
1286. Schwartz, Philip, *Measuring the Life of a Gun*, Army Ordnance, **5**, No. 2, 1924, pp. 111–112
1287. Schwartz, Philip, *Accuracy Firing*, Army Ordnance, **5**, No. 3, 1924, pp. 521–522
1288. Schwartz, Philip, *Experimental Investigation of The Effect of the Wind Upon the Motion of a Projectile*, CAJ, **60**, 1924, pp. 93–97
1289. Schwartz, Philip, *Studies of Motion of Projectiles*, CAJ, **60**, 1924, pp. 430–431
1290. Schwartz, Philip, *The English Long Range Gun*, CAJ, **61**, 1924, pp. 91–92
1291. Schwartz, Philip, *Weighting Factors*, CAJ, **61**, 1924, pp. 371–372
1292. Schwartz, Philip, *Remarks Concerning the Use of Firing Table Values*, CAJ, **3**, 1925, pp. 399–400
1293. Schwartz, Phlilp, *The Variable Probable Error*, CAJ, **67**, 1927, pp. 405–413
1294. Schwartz, Philip, *Much About Little* CAJ, **68**, 1928, pp. 291–297
1295. Schwartz, Philip, *The Fork*, Field Artillery Journal, **18**, No. 3, 1929, pp. 604–609
1296. Schwartz, Philip, *More About Probable Error*, CAJ, **73**, 1930, pp. 111–120
1297. Siacci, F., *Balistique Extérieure*, Traduction Annoteé par P. Laurent, Berger-Levrault et Cic, Editours, Paris and Nancy, 1892
1298. Siegmund-Schultze, R.,*Eliakim Hastings Moore's "General Analysis"*, Arch. Hist. Exact. Sci., **52**, 1998, pp. 51–89
1299. Siegmund-Schultze, R., *The Late Arrival of Academic Applied Mathematics in the United States: a Paradox, Theses, and Literature*, NTM, **11**, 2002, pp. 116–12
1300. Siegmund-Schultze, R., *The Ideology of Applied Mathematics Within Mathematics in Germany and The U. S. Until the End of World War II*, ILUIL, **27**, 2004, pp. 791–811
1301. Sleight, E. R., *Mathematics Applied*, National Mathematics Magazine, **9**, No. 8, May 1935, pp. 219–223
1302. Smith, Paul A., *Joseph Fels Ritt, 1893–1951*, United States National Academy of Sciences, Biographical Memoir, 1956, pp. 253–264
1303. Spinrad, William, *Early History of The Ordnance School*, The Ordnance Sergeant **3**, No. 1, 1942. pp. 1–4
1304. Stäckel, P., *Abhandlungen Über Variations-Rechnung, Zweiter Theil*, Leipzig, Verlag Von Wilhelm Engelmann, 1894
1305. Steele, Bret D., *Muskets and Pendulums: Benjamin Robins, Leonhard Euler, and the Ballistics Revolution*, Technology and Culture, **35**, No. 2, April 1994, pp. 348–382
1306. Stevenson, Craig, *Periodic Orbits of the Three Body Problem: F. R. Moulton's Quest for a New Lunar Theory*, PhD Thesis, Open University, 2012
1307. Stevenson, Craig, *Periodic Orbits: F. R. Moulton's quest for a New Lunar Theory*, American Mathematical Society, 1921
1308. Tannehill, I. R., *The Ballistic Wind*, Journal of the United States Artillery, **52**, 1920, pp. 553–576
1309. Tarwater, J. D., (Ed.), *The Bicentennial Tribute to American Mathematics, 1776–1976* Papers Presented to the Fifty-Ninth Annual Meeting of the Mathematical Association of America, Mathematical Association of America, Washington, D. C., 1977
1310. Todhunter, I., *A Treatise on the Integral Calculus and its Applications*, MacMillan and Co., London, 1880
1311. Tonelli, L., *Fondamenti di calcolo delle variazioni*, Nicola Zanichelli, I, 1921; II, 1923
1312. Tropp, H. S., *Forest Ray Moulton*, Dictionary of Scientific Biography, **9**, pp. 552–553
1313. Truesdell, Clifford, review of *Exterior Ballistics* by McShane, Kelley and Reno, Scripta Mathematica, **20**, 1954, pp. 172–174
1314. Tschappat, W.H., *Ballistics*, article in Encyclopedia Britannica, 1922, pp. 383–394
1315. Tsien, H. S., Adamson, T. C., E. L. Knuth, *Automatic Navigation of a Long Range Rocket Vehicle*, Journal of the American Rocket Society, July-August, 1952

1316. Tsien, H. S., *Engineering Cybernetics*, McGraw-Hill Book Co., 1954
1317. Veblen, Oswald, *Ballistic Theories and the Preparation of Range Tables*, NARA Aberdeen, No. 54, 1921
1318. Veblen, O., *Professor Veblen on Ballistics in the World War*, Army Ordnance, **4**, March-April 1924, pp. 337–338
1319. Very, Edward W., *Computation of Range Tables*, Annapolis, Md., 1877
1320. Volterra, V., *Leçons sur les Fonctions des Lignes*, Paris, Gauthier- Villars, 1913
1321. Wertheimer, Albert, *A Method of Computing Differential Corrections For a Trajectory*, Journal of the Franklin Institute, **217**, No. 6, June 1934
1322. Wiener, Norbert *Ex-Prodigy*, MIT Press, Cambridge, 1953
1323. Williamson, Benjamin, *An Elementary Treatise on The Integral Calculus*, Longmans, Green and Co., London, 1916
1324. Wilson, Edwin B., Advanced Calculus, Dover reprint of the original 1912 edition, 1958
1325. Wilson, Edwin B., *The Mathematics of Aerodynamics*, Science, **25**, No. 7, September 1918, pp. 191–197
1326. Zeidler, E., *Applied Functional Analysis*, Springer-Verlag, 1995
1327. www.math.colostate.edu/~pauld/M345/Some20theory.pdf

# Chapter 5

## Anonymous Editorials, Announcements, And Book Reviews In Military

## Journals JUSA = Journal of The United States Artillery, CAJ = Coast Artillery Journal

1328. Announcement, JUSA ( = Journal of the United States Artillery) **1**, 1892, pp. 3–4
1329. Book Review, A Course of Instruction in Ordnance and Gunnery, Henry Metcalfe, JUSA, **1** 1892, pp. 74–76
1330. Technical Publications, “An Investigation of the Strength of Columns Leading to some New Formulas,” *American Machinist*, July 28, 1898, p. 565
1331. Book Review, *Balistique Extérieure Rationelle (Problèmes balisiques secondaires)*, P. Carbonnier, JUSA, **27**, 1907, pp. 310–311
1332. Range Table For The 10-Inch B. L. Rifle, Steel JUSA, **5**, 1896, p. 218
1333. Progress In Ballistics CAJ (= Coast Artillery Journal), **51**, 1919, pp. 69–71
1334. Editorial, *Change Versus Progress*, JUSA, **56**, 1922, pp. 373–374
1335. *FRANKFORD ARSENAL: Range, Deflection, and Wind Correction Computer*, Army Ordnance, March-April 1924, **4**, No. 23, pp. 339–341
1336. Coast Artillerymen and the Army Ordnance School, CAJ, **2**, 1924, p.363
1337. Ballistic Research at Marquette University, Army Ordnance, **7**, No.5, 1927, p. 399
1338. Professor Moulton’s Book, Army Ordnance, **8**, 1927, No. 1, pp. 12–15
1339. *The Coast Artillery School, 1824–1927* CAJ, **66**, 1927, pp. 1–85
1340. Book Review, New Methods in Exterior Ballistics, Forest Ray Moulton, CAJ, **66**, 1927, pp. 383–384
1341. Book Review, Textbook of Ordnance and Gunnery, Earl McFarland, CAJ, **70**, 1929, pp. 533–534

## Manuscripts, Published Books and Articles

1342. Alger, P. L., *The Groundwork of Practical Naval Gunnery*, 2nd Edition, Annapolis, Md., The United States Naval Institute, 1917
1343. Alger, P. L., *Charts for the Calculation of the Effect of Small Changes in the Elements of Fire* JUSA, Vol. 51, 1919, pp. 585–603
1344. Alger, P. L., *Memorandum On Range Computations*, Aberdeen Proving Ground, November 28, 1919, [Aberdeen]
1345. Ames, J. S., *The Trained Man of Science in the War*, Science, New Series, 9, 480, No. 1243, Oct. 25, 1918, pp. 401–410
1346. Archibald, Thomas, Della Dumbaugh and Deborah Kent, *A Mobilized Community: Mathematicians in the United States during the First World War*, in *The War of Guns and Mathematics*, pp. 229–271
1347. Arnaud, J. J., *Under-Water Trajectories*, discussion, Journal of the U.S. Naval Institute, 47, 1921, pp. 1417–1418
1348. Askey, Richard, *Orthogonal Polynomials and Special Functions*, Society for Industrial and Applied Mathematics, Philadelphia, 1975
1349. Aubin, David, and Catherine Goldstein, eds., *The War of Guns and Mathematics*, American Mathematical Society, 2014
1350. Aubin, David, *'I'm Just a Mathematician': Why and How Mathematicians Collaborated with Military Ballisticians at Gavre*, in Aubin, David, and Catherine Goldstein, eds., *The War of Guns and Mathematics*, American Mathematical Society, 2014
1351. Ball, H. P., *Of Responsible Command: A History of the U.S. Army War College*, Alumni Association of the United States Army War College, Carlisle Barracks, Pennsylvania, 1983
1352. *Ballisticians in War and Peace, Volume I, A History of the United States Army Ballistic Research Laboratories 1914–1956*, Aberdeen Proving Ground, Maryland
1353. Banach, S. *Théorie des Opérations Linéaires*, Second Edition, reprint of first edition of 1932, Chelsea Publishing Company, New York, 1955
1354. Barrow-Green, June, *Poincaré and the Three Body Problem*, American Mathematical Society, 1997
1355. Barrow-Green, June, *"Anti-aircraft guns all day long": Karl Pearson and computing for the Ministry of Munitions*, Revue d'histoire des mathématiques, **21**, fascicule 1, 2015, pp. 111–150
1356. Barth, Carl G., *An Investigation of the Strength of Columns Leading to Some New Formulas*, Journal of the Association of Engineering Societies, **20**, No. 4, April 1898, pp. 239–256
1357. Beckman, H. S., *The Elements of Firing Table Construction*, Army Ordnance, **1**, No. 1, pp. 46–454
1358. Bell, E. T., *The Development of Mathematics*, Macmillan and Co., 2nd edition, 1945
1359. Bennett, A. A., *Physical Bases of Ballistic Table Computations, Ordnance Textbook*, Washington Government Printing Office, 1920
1360. Bennett, A. A., *Introduction to Ballistics*, Engineer Reproduction Plant, Washington Barracks, D. C., 1921
1361. Bennett, A. A., *The Problem of Exterior Ballistics*, Army Ordnance, **3**, July-August 1922, No. 13, pp. 38–40
1362. Bennett, A. A., review, *Traité de Balistique Extérieure, Vol. II*, P. Charbonnier, Bulletin of the American Mathematical Society, **34**, No. 4, 1928, pp. 517–518
1363. Bennett, A. A., *Exterior Ballistic Tables Based on Numerical Integration*, United States Army, Ordnance Dept., Washington, Government Printing Office, 1924 (Document 1107), 1931, 1936
1364. Bennett, A. A., *The Differential Analyzer*, Ballistic Research Laboratories, BRL Report No. 319, December, 1942
1365. Benton, J. G., *A Course of Instruction in Ordnance and Gunnery, Third Edition*, D. Van Nostrand, New York, 1867

1366. Berry, W. J., *Some Theorems Concerning Systems of Linear Partial Differential Equations*, Annals of Mathematics, **12**, No. 1, 1910, pp. 27–38
1367. Birkhoff, Garrett, and Gian-Carlo Rota, *Ordinary Differential Equations*, Ginn and Co., Boston, 1962
1368. Birkhoff, G., *Some Leaders In American Mathematics, 1891–1941*, in: Tarwater [1977], pp. 27–78
1369. Birkhoff, G., and Erwin Kreyzig, *The Establishment of Functional Analysis*, Historia Mathematica, **11**, 1984 pp. 258–321
1370. Birkhoff, G. D., *Boundary Value and Expansion Problems of Ordinary Linear Differential Equations*, Transactions of the American Mathematical Society, **9**, No. 4, 1908, pp. 373–395
1371. Birkhoff, G. D., and O. D. Kellogg, *Invariant Points in Function Space*, Transactions of the American Mathematical Society, **23**, No. 1, January 1922, pp. 96–115
1372. Bliss, Gilbert Ames, *An Existence Theorem for a Differential Equation of the Second Order, With an Application to the Calculus of Variations*, Trans. Am. Math. Soc., **5**, 1904, pp. 113–125
1373. Bliss, Gilbert Ames, *The Solutions of Differential Equations of the First Order as Functions of Their Initial Conditions*, Ann. of Math., **6**, 1905, pp. 49–68
1374. Bliss, Gilbert Ames, *A Generalization of the Notion of Angle*, Transactions of the American Mathematical Society, **7**, 1906, pp. 184–196
1375. Bliss, Gilbert Ames, *The Scientific Work of Eliakim Hastings Moore*, Bull. Am. Math. Soc., 40, 1910, pp. 207–213
1376. Bliss, Gilbert Ames, *Integral Equations*, review of *An Introduction to the Study of Integral Equations* by Maxime Bôcher, *Bulletin of the American Mathematical Society*, **16**, No. 4, 1910, pp. 207–213
1377. Bliss, Gilbert Ames, *A New Proof of the Implicit Function Theorem*, Bull. Am. Math. Soc., **18**, 1912, pp. 175–179
1378. Bliss, Gilbert Ames, *Fundamental Existence Theorems*, The Princeton Colloquium, Part I, American Mathematical Society, 1913
1379. Bliss, Gilbert Ames, *A Note on Functions of Lines*, Proc. Nat. Acad. Sci, U.S.A., **10**, 1915, pp. 173–175
1380. Bliss, Gilbert Ames, review, V. Volterra, *Fonctions des Lignes*, Bull. Am. Math. Soc., **23**, 1915, pp. 344–355
1381. Bliss, Gilbert Ames, *Solutions of Differential Equations as Functions of the Constants of Integration*, Bull. Amer. Math. Soc., **25**, 1918, pp. 15–26
1382. Bliss, Gilbert Ames, *Differential Corrections*, NARA Aberdeen Ms A-29, No. 10, Nov. 1918
1383. Bliss, Gilbert Ames, *The Problem of Mayer With Variable Endpoints*, Trans. Am. Math. Soc., **19**, 1918, pp. 305–316
1384. Bliss, Gilbert Ames, *A Method of Computing Differential Corrections for a Trajectory*, CAJ/JUSA, 50, 1918, pp. 455–460
1385. Bliss, Gilbert Ames, *A Method of Computing Differential Corrections for a Trajectory*, CAJ/JUSA, **51**, 1919, pp. 445–449
1386. Bliss, Gilbert Ames, *The Use of Adjoint Systems in the Problem of Differential Corrections for Trajectories*, JUSA, **51**, 1919, pp. 296–311
1387. Bliss, Gilbert Ames, *Differential Equations Containing Arbitrary Functions*, Transactions of the American Mathematical Society, **210**, No. 9, 1920, pp. 79–92
1388. Bliss, Gilbert Ames, *Functions of Lines in Ballistics*, ibid, pp. 93–106
1389. Bliss, Gilbert Ames, *Calculus of Variations*, Mathematical Association of America, 1925
1390. Bliss, Gilbert Ames, *The Calculus of Variations*, Bulletin of the American Mathematical Society, 31, March-April 1925, pp. 163–172
1391. Bliss, Gilbert Ames, *A Boundary Value Problem for a System of Ordinary Differential Equations of the First Order*, Transactions of the American Mathematical Society, **28**, No. 4, 1926, pp. 561–584
1392. Bliss, Gilbert Ames, *Mathematics*, Scientific Monthly, **24**, Issue 4, 1927, pp. 308–319

1393. Bliss, Gilbert Ames, *Existence Theorems for Implicit Functions, Existence Theorems for Differential Equations*, University of Chicago, 1928
1394. Bliss. Gilbert Ames, *The Problem of Lagrange in the Calculus of Variations*, Am. J. Math., **52**, No. 4, 1930, pp. 673–744
1395. Bliss, Gilbert Ames, *Mathematical Interpretation of Geometrical And Physical Phenomena*, American Mathematical Monthly, **40**, 1933, pp. 472–480
1396. Bliss, Gilbert Ames, *The Evolution of Problems of the Calculus of Variations*, The American Mathematical Monthly, **43**, No. 10, December 1936, pp. 598–609
1397. Bliss, Gilbert Ames, *Mathematics for Exterior Ballistics*, John Wiley and Sons, New York, 1944
1398. Bliss, Gilbert Ames, *Oskar Bolza - In Memoriam*, Bulletin of the American Mathematical Society, **50**, 1944. pp 478–489
1399. Bliss, Gilbert Ames, *Lectures on the Calculus of Variations*, The University of Chicago Press, Chicago, 1946
1400. Bliss, Gilbert Ames, *Autobiographical Notes*, American Mathematical Monthly, **59**, November 1952, pp. 595–605
1401. Bliss, Gilbert Ames, and Max Mason, *A problem of the Calculus of Variations in Which the Integrand is Discontinuous*, Trans. Am. Math. Soc., **7**, No. 2, April 1906, pp. 325–336
1402. Bliss, Gilbert Ames, and Max Mason, *The Properties of Curves in Space Which Minimize a Definite Integral*, Trans. Amer. Math. Soc., **9**, October 1908, pp. 440–466
1403. Bliss, Gilbert Ames, and Max Mason, *Fields of Extremals in Space*, Transactions of the American Mathematical Society, **11**, 1910, pp. 325–340
1404. Bôcher, M., *Green's Functions in Space of One Dimension*, Bulletin of the American Mathematical Society, **7**, 1901, pp. 297–299
1405. Bôcher, M., *An Introduction to the Study of Integral Equations*, Cambridge Tracts in Mathematics and Mathematical Physics, No. 10, Cambridge University Press, 1909
1406. Bôcher, M., *Boundary Problems and Green's Functions for Linear Differential and Difference Equations*, Annals of Mathematics, **13**, 1911–1912, pp. 71–88
1407. Bôcher, M., *Boundary Problems in One Dimension*, in *Proceedings of the Fifth International Congress of Mathematicians, Cambridge, 22–28 August 1912*, Princeton University Press, 1913, pp. 163–195
1408. Bôcher, M., *Applications and Generalizations of the Conception of Adjoint Systems*, Transactions of the American Mathematical Society, **14**, October 1913, pp. 403–420
1409. Bolza, O., *Lectures on the Calculus of Variations*, University of Chicago Press, 1904, reprinted unabridged and corrected, Dover Publications, 1961
1410. Bolza, Oskar, *Vorlesung Über Variationsrechnung*, Druck und Verlag von B.G. Teubner, 1909
1411. Bolza, Oskar, *Integral Equations*, lecture notes by William Vernon Lovitt from a summer course given during summer 1913 at University of Chicago
1412. Booss-Bavnbek, Bernhelm, and Jens Hoyrup, *Mathematics and War: An Invitation To Revisit*, The Mathematical Intelligencer, **25**, No. 3, 2003, pp. 12–25
1413. Bruff, L. L., *A Textbook of Ordnance and Gunnery*, John Wiley and Sons, New York, 1896
1414. Burr, William H., *Note on Gordon's Formula for Long Columns*, Engineering News, **8**, 1881, p. 307
1415. Burr, William H., *The Elasticity and Resistance Of The Materials of Engineering*, Fifth Edition, John Wiley and Sons, New York and London, Chapman and Hall, Ltd., 1897
1416. Cain, William, *Theory of the Ideal Column*, Transactions of the American Society of Civil Engineers, **39**, June 1898, pp. 96–120
1417. Cain, William, **A Brief Course in the Calculus**, 3rd Edition, revised, D. van Nostrand Company,New York, 1905, 1907
1418. Calvert, Monte A., *The Mechanical Engineer in America, 1830–1910*, The Johns Hopkins Press, Baltimore, 1967
1419. Campbell, G. A., *Mathematics in Industrial Research*, Bell System Technical Journal, 1926, pp. 550–557

1420. Carathéodory, C., review, Hadamard, *Leçons sur le calcul des variations*, **35**, Bulletin des Sciences mathématiques, pp. 123–141, 1911
1421. Carll, L. B., *A Treatise On The Calculus of Variations*, MacMillan and Co., London, 1885
1422. Carmichael, R. D., *The Provision Made by Mathematics for the Needs of Science*, Science, New Series, **45**, Issue 1168 (May 18, 1917), pp. 465–474
1423. Case, Homer, *The Variable Probable Error and The Adjustment Problem*, CAJ, **68**, 1928, pp. 492–501
1424. Chang, Iris, *Thread of the Silkworm*, BasicBooks, A Division of HarperCollins Publishers, Inc., 1995
1425. Charbonnier, P., *Balistique Extérieure Rationelle (Problèmes Balistique Secondaires)*, Encyclopèdie Scientifique de Méchanique Appliqué et Génie, M. D'Ocagne, directeur, Octave Doin, Paris, 1907
1426. Charbonnier, P., *The Gâvre Commission: An Historical Sketch Of The Progress Of Exterior Ballistics* JUSA, **27**, 1907, pp. 238–262
1427. Charbonnier, P., *The Gâvre Commission: An Historical Sketch Of The Progress Of Exterior Ballistics* JUSA, **28**, 1908, pp. 49–73
1428. Charbonnier, P., *Sur L'État Actuel De La Balistique Extérieure Théorique*, Proceedings of the International Mathematical Congress held in Toronto, August 11–16, 1924, ed. Fields, John Charles, Nendeln, Kraus reprint, 1924 pp. 315–334
1429. Charbonnier, P., *Traité de Balistique Extérieure, Tome I*, Gaston Doin, Gauthier-Villars; Paris, 1921
1430. Charbonnier, P., *Traité de Balistique Extérieure, Tome II*, Gaston Doin, Gauthier-Villars; Paris, 1927
1431. Chester, James, *The Theoretical Instruction of Gunners*, JUSA, **1**, No. 3, July 1892, pp. 171–206
1432. Chester, James, *Whistler's Tables of Fire. Artillery Circular C*, Journal of the Military Service Institution of the United States, **13**, 1892, pp. 1028–1030
1433. Chester, James, *Sea-Coast Artillery Instruction*, JUSA, **5**,1896, pp. 209–218
1434. Coddington, E. A. and N. Levinson, *Theory of Ordinary Differential Equations*, McGraw-Hill Book Company, 1955
1435. *The Confederate Field Manual*, Ritchie and Dunnavant, Richmond, 1862, reprinted 1984
1436. Cousins, Robert H., *Strength of Beams and Columns*, E. and F. N. Spon, New York, London, 1889
1437. Coxe, W. H. and Bugless. E., *Exterior Ballistics Charts*, E.I. du Pont de Nemours and Company, Inc., 1926
1438. Craig, Thomas, *A Treatise on Linear Differential Equations*, New York, John Wiley and Sons, 1889
1439. *The Applications of Elliptic Functions by Alfred George Greenhill, M.A., F.R.S., Professor of Mathematics in the Artillery College, Woolwich*, book review, Thomas Craig, JUSA, **2**, 1893, pp. 298–306
1440. Cranz, C. and Becker, K., *Handbook of Ballistics, Volume I, Exterior Ballistics*, translated from the second German edition, H. M. Stationery Office, London, 1921
1441. Darboux, G., *Sur les équations aux dérivés partielle*, Comptes Rendus, **96**, March 19, 1883, pp. 766–769
1442. Darboux, G., *Leçons sur la théorie générale des surfaces, deuxieme partie*, Gauthier-Villars et Fils, Paris, 1889
1443. Darboux, G., *Leçons sur la théorie générale des surfaces, quatrième partie*, Gauthier-Villars et Fils, Paris, 1896
1444. Dederick, L. S., *Exterior Ballistics*, Bulletin of the American Mathematical Society, **34**, No. 5, 1928, p. 667
1445. Dederick, L. S., *The Mathematics of Exterior Ballistics Computation*, American Mathematical Monthly, **47**, No. 9, November, 1940, pp. 628–634
1446. Denslow, William R., *10,000 Famous Freemasons*, Missouri Lodge of Research, Independence, Mo., 1957

1447. Devorkin, D. H., *Henry Norris Russell, Dean of American Astronomers*, Princeton University Press, Princeton and Oxford, p. 200
1448. Dickson, L. E., *Mathematics in War Perspective*, Bulletin of the American Mathematical Society, **25**, No. 7, 1919, pp. 289–311
1449. Dickson, T.C., *Course of Instruction Prescribed for Student Officers In the Ordnance School of Application in Differential Equations*, July 24, 1907, revised April 12, 1910, Government Printing Office, Washington
1450. Dieudonné, J., *History of Functional Analysis*, Amsterdam: north Holland, 1981
1451. Doerfler, R., Gluchoff, A.,Guthery, S., and Paul Hamburg, *Calculating Curves*, Docent Press, 2012
1452. Drach, Jules, *L'équation differentielle de la balistique extérieure et son integration par quadratures*, Annales Scientifiques de l'E.N.S., Serie 3, tome 37, 1920 pp. 1–94
1453. Drenick, R., *The Perturbation Calculus in Missile Ballistics*, Journal of the Franklin Institute, **25**, 1951, pp. 423–436
1454. Drenick, R., *Rudolph F. Drenick, Veteran*, interviewed by Mike Russert, Wayne Clark, November 18, 2003, South Setauket, New York, dmna.ny.gov/historic/veterans/transcripts/-Drenick-Rudolf-F.pdf
1455. du Bois-Reymond, P., *Über lineare partielle Differentialgleichungen zweiter Ordnung*, Journal für Mathematik, **104**, pp. 241–310
1456. Eddy, H. T., *The Most Economical Cut-Off for Steam Compression*, Power, **17** August 1897, p.12
1457. Eggert, E. F., *Accurate Trajectories By Mechanical Integration*, Proceedings of the U.S. Naval Institute, **45** No.'s 1–6, January, 1919, pp. 71–84
1458. Eggert, E. F., *Trajectories and Their Corrections*, discussion, Proceedings of the U.S. Naval Institute, **45** No.'s 7–12, 1919, pp. 1761–1765
1459. Evans, G. C., *Functionals and Their Applications*, American Mathematical Society Colloquium Lectures, **5**, Part I, 1918
1460. Faulkner, F. D., *Optimum Interception of a Ballistic Missile at Intermediate Range*, Naval Postgraduate School, Monterey, California, October 1961
1461. Faulkner, F. D., *A Comparison Between Some Methods For Computing Optimum Paths in the Problem of Bolza*, Research Paper No. 43, United States Naval Postgraduate School, Monterey, California, March 1964
1462. Faulkner, F. D., *Newton's Method Applied To The Solution Of Classical Problems In Calculus Of Variations And Differential Games*, dissertation submitted in partial fulfillment of the requirements for the degree of Doctor of Philosophy in the University of Michigan, 1969
1463. Feffer, Loren Butler, *Oswald Veblen and the Capitalization of American mathematics*, Isis, 89, pp. 474–497
1464. Fenster, Della, Kent, Deborah and Archibald, Thomas, *The First World War and Mathematics In the United States*, preprint
1465. Fischer, C. A., *A Generalization of Volterra's Derivative of a Function of a Curve*, American Journal of Mathematics, **35**, No. 4, 1913, pp. 369–394
1466. Fischer, C. A., *Note on the order of continuity of functions of lines*, Bulletin of the American Mathematical Society, **23**, (1916–1917), pp. 88–90
1467. Flanagan, Francis, *Harmonic Functions*, Dover Press, 1972
1468. Ford, L., *Instruction Memorandum No. 20, Subject: Elementary Mathematics For Field Artillery*, Headquarters, Field Artillery Central Officers Training School, Camp Zachary Taylor, Ky., 1919
1469. Fowler, R. H., E. G. Gallop, C. N. H. Lock and H. W. Richmond, *The Aerodynamics of a Spinning Shell*, Philosophical Transactions of the Royal Society, A221, 1920, pp. 295–387
1470. Franklin, Philip, *An Alternative Derivation of the Differential Equations Used in Computing Differential Corrections*, [NARA Aberdeen], No. 86, 1919
1471. Franklin, Philip, *Exterior Ballistics, A Historical Survey*, Brown University Graduate School, 1943

1472. *Philip Franklin of M.I.T. is Dead*, New York Times, 1/28/65
1473. Fréchet, M. *Sur La Notion de la Differentielle D'Une Fonction de Ligne*, Transactions of the American Mathematical Society, **10**, 1913, pp. 135–161
1474. Friedlander, Alan, *A Study of Guidance Sensitivity for Various Low-THrust Transfers from Earth to Mars*, Technical Note D-1183, Lewis Research Center, Clevelandm Ohio, February 1962
1475. Fuchs, L., Über Relationen, welche für die zwischen je zwei singulären Punkten erstreckten integrale der lösungen linearer differentialgleichungen stattfinden, Journal für die reine und angewandte Mathematik, **76**, 1874, pp. 177–213
1476. Galileo, *Two New Sciences*, A New Translation With Introduction and Notes, Stillman Drake, University of Wisconsin Press, 1974
1477. Gasteyer, C.E., *Forest Ray Moulton, 1872–1952* , Biographical Memoirs of the National Academy of Sciences, **41**, pp. 340–355
1478. Gateaux, R., *Fonctions d'une infinité de variables indépendantes*, Bulletin de la S. M. F., **47**, 1919, pp. 70–96
1479. Gilman, F., *The Path of a Projectile*, Journal of the Franklin Institute, **132**, October 1891, pp. 267–280
1480. Glennon, J.H., *Accuracy and Probability of Fire*, Baltimore Press of Isaac Friedenwald, 1888
1481. Gluchoff, Alan, *Pure Mathematics Applied in Early Twentieth-Century America: The case of T.H. Gronwall, Consulting Mathematician*, Historia Mathematica, **32**, 2005, pp. 312–357
1482. Gluchoff, Alan, *Artillerymen and Mathematicians: Forest Ray Moulton and Changes in American Exterior Ballistics, 1885–1934*, Historia Mathematica, **38**, 2011, pp. 506–547
1483. Goldstine, Herman H., *Conditions For A Minimum Of A Functional*, The University of Chicago Libraries, Chicago, Illinois, 1937
1484. Goldstine, Herman. H., *A Multiplier Rule In Abstract Spaces*, Bulletin of the American Mathematical Society, **44**, 1938, pp. 388–394
1485. Goldstine, Herman H., *The Computer from Pascal to von Neumann*, Princeton University Press, 1972
1486. Oral History Transcript - Dr. Herman Goldstine, Interview with Dr. Herman Goldstine by Nancy Stern at Institute for Advanced Study, Princeton, March 14, 1977, http://www.aip.org/history/ohilist/4636.html
1487. Goldstine, Herman H., *A History of the Calculus of Variations From the 17th Through the 19th Centuries*, Springer-Verlag, 1980
1488. Goldstine, Herman H., interviewed in *The Princeton Mathematics Community in the 1930s*, Transcript Number 15, (PMC15), The Trustees of Princeton University, 1985
1489. Goldstine, Herman, interview with author July 12, 2004
1490. Goodman, T. R., and G. N Lance, *The Numerical Integration of Two-Point Boundary Value Problems*, Math. Tables Aids Computation, **10**, No. 82, 1956
1491. Goursat, Édouard, *Sur la Théorie des functions implicites*, Bull. de la S. M. F., **31**, 1903, pp. 184–192
1492. Goursat, Édouard, *A Course In Mathematical Analysis*, Vol. II, Part II, translated by Hedrick, E.R., and Dunkel, Otto, Ginn and Co., 1917
1493. Goursat, Édouard, *Cours d'Analyse mathématique, Tome III*, Troisième Edition, Gauthier-Villars, 1923
1494. Graves, L., *Implicit Functions and Differential Equations in General Analysis*, Transactions of the American Mathematical Society, **29**, No. 3, July 1927, pp. 514–552
1495. Graves, L., *Topics in the Functional Calculus*, Bulletin of the American Mathematical Society, **41**, 1935, pp. 641–662
1496. Graves, L., *What Is A Functional?*, The American Mathematical Monthly, **55**, No. 8, October 1948, pp. 467–472
1497. Graves, L. M., *Gilbert Ames Bliss, 1876–1951*, Bulletin of the American Mathematical Society, **58**, (1952), pp. 251–264
1498. Graves, L., *The Theory of Functions of Real Variables*, $2^{nd}$ Edition, McGraw-Hill, 1956, reprinted by Dover Publications, 2009

1499. Graves, L., *Nonlinear Mappings between Banach Spaces*, in *Studies in Real and Complex Analysis*, ed. I. I. Hirschman, The Mathematical Association of America, 1965, pp. 34–54
1500. Graves, L., and T. H. Hildebrandt, *Implicit Functions and Their Differentials in General Analysis*, Transactions of the American Mathematical Society, **29**, No. 1, January 1927, pp. 127–153
1501. Green, George, *An Essay on the Application of Mathematical Analysis to the Theories of Electricity and Magnetism*, Journal für Mathematik, **39**, 1850, pp. 73–89; 44, 1852, pp. 356–374; 47, 1854, pp. 161–221
1502. Grier, David Alan, *Dr. Veblen Takes a Uniform: Mathematics in the First World War*, American Mathematical Monthly, **108**, December 2001, pp. 922–931
1503. Grier, David Alan, *Dr. Veblen At Aberdeen: Mathematics, Military Applications and Mass Production*, in *Instrumental in War*, Steven A. Walton, Ed., Boston, Brill, 2005, pp. 253–270
1504. Grier, David Alan, *When Computers Were Human*, Princeton University Press, 2005
1505. Grodzovskii, G. L., Yu. N. Ivanov, V. Y. Tokarev, *Mechanics of Low Thrust Spaceflight*,1966, translated from Russian, Published for the National Aeronautics and Space Administration and the National Science Foundation by the Israeli Program for Scientific Translation, 1969
1506. Gronwall, T. H., *Note on the Derivative with Respect to a Parameter of the Solutions of a System of Differential Equations*, Annals of Mathematics, Second Series, 20, No. 4, July 1919, pp. 292–296
1507. Gronwall, T. H., *Elastic Stresses In An Infinite Solid With Spherical Cavity*, Annals of Mathematics, Second Series, **19**, No. 4, June 1918, pp. 295–296
1508. Gronwall, T. H., *On the Influence of Keyways on the Stress Distribution in Cylindrical Shafts*, Trans. Am. Math. Soc. **20** pp. 234–244
1509. Gronwall, T. H., *On a System of Linear Partial Differential Equations of Hyperbolic Type*, Annals of Mathematics, 20, No. 4, July 1919, pp. 274–278
1510. Gronwall, T. H., *On Differential Corrections and Weighting Factors*, NARA Washington 4, July 1919, pp. 274–278
1511. Gronwall, T. H., *On the Construction of Maps for the Determination of Range and Deflection From Observed Azimuth* Aberdeen E-III-7, No. 81, July 1919
1512. Gronwall, T.H., *Qualitative Properties of the Ballistic Trajectory*, Annals of Mathematics, Ser. II, Vol. 21, 1920, pp. 44–65
1513. Gronwall, T. H., *Differential Variations In Ballistics With Applications to the Qualitative Properties of the Trajectory*, Transactions of the American Mathematical Society, 22, No. 4, 1921, pp. 505–525
1514. Gronwall, T. H., review, *Traité de Balistique Extérieure, Vol. I*, P. Charbonnier, Bulletin of the American Mathematical Society, **33**, No. 1, 1927, p. 120
1515. Guion, James L., review, *New Methods in Exterior Ballistics*, Army Ordnance, 1926, **7**, pp. 152–155
1516. *Gunnery For Heavy Artillery*, Coast Artillery Corps, War Department, 1925
1517. Hadamard, J., *Leçons sur la propagation des ondes*, Paris, Hermann, 1903
1518. Hadamard, J., *Leçons sur le calcul des variations*, Hermann et fils, Paris, 1910
1519. Hadamard, J., *Rapport sur les examinés par la Commission de Balistique pendant la durée de la guerre*, Comptes Rendus de l'Académie des Sciences, **170**, No, 8, 1920, pp. 436–445
1520. Hadamard, J., *Lectures on Cauchy's Problem in Linear Partial Differential Equations*, Yale University Press, 1923
1521. Hadamard, J., *Le Développement et le role Scientifique Du Calcul Fonctionel*, Oevres de Jacques Hadamard, Tome I, Éditions du Centre National de la Recherche Scientifique, 1968, pp. 435–453
1522. Hadamard, J., *Le Calcul Fonctionel*, Oevres de Jacques Hadamard, Tome I, Éditions du Centre National de la Recherche Scientifique 1968, pp. 2253–2266
1523. Hagood, Johnson, *Spotting for the Coast Artillery*, CAJ, 60, No. 2, 1924, pp. 87–92
1524. Hamilton, Alston, *Ballistics, Part I*, Coast Artillery School Press, Fort Monroe, Virginia, 1908
1525. Hamilton, Alston, *Ballistics, Part II (Exterior Ballistics)*, Coast Artillery School Press, Fort Monroe, Virginia, 1908

1526. Hamilton, Alston, *A Practical Method of Computing High Angle Trajectories* NARA Aberdeen, No. 1, 1917
1527. Hancock, Harris, *Lectures on the Calculus of Variations (The Weierstrassian Theory)*, University of Cincinnati Bulletin of Mathematics, 1904
1528. Hancock, Harris, *Theory of Maxima and Minima*, Ginn and Co., 1917
1529. Hardy, G. H., *A Mathematician's Apology*, Cambridge University Press, 1940
1530. Hartman, Philip, *Ordinary Differential Equations*, John Wiley and Sons, 1964
1531. Hatch, H.J., *A Reply to General Hagood*, CAJ, **60**, No. 3, 1924, pp. 171–176
1532. Hawkesworth, A. S., *Under-Water Trajectories*, Proceedings of the U.S.Naval Institute, **47**, February 1921, pp. 207–213
1533. Hawkesworth, A. S., *New Ballistic Corrections*, Proceedings of the U.S. Naval Institute, **47**, 1921, pp. 1233–1249
1534. Hayes, Thomas J., *Elements of Ordnance*, John Wiley and Sons, Inc., New York, 1938
1535. Hedrick, H. B., review, Hadamard, J., *Leçons sur le calcul des variations*, Bull. Am. Math. Society, **21**, 1914, pp. 30–32
1536. Hedrick, H. B., *Ballistic Effects Due To the Rotation of the Earth*, CAJ, **70**, No. 6, 1929, pp. 268–269
1537. Hermann, Ernest E., *Exterior Ballistics 1935*, U. S. Naval Institute, Annapolis, Maryland, 1935
1538. Hestenes, M., *The Calculus of Variations and Optimal Control Theory*, Wiley, New York, 1966
1539. Hill, G. W., *Researches in the Lunar Theory*, American Journal of Mathematics, **1**, No. 1, 1876, pp. 5–21
1540. Hitchcock, H. P., *Computation of Firing Tables for the U. S. Army*, Aberdeen Proving Ground, 1934
1541. Hoar, Roger Sherman, *The Pickering Polaris Attachment*, JUSA, **50**, 1919, pp. 230–239
1542. Hoar, Roger Sherman, *The New Ballistics*, JUSA, **51**, 1919, pp. 285–295
1543. Hoar, Roger Sherman, *The New Artillery Grid Maps*, JUSA, **54**, 1921, 477–479
1544. Hoar, Roger Sherman, *Problems for Solution* [2876–2882], American Mathematical Monthly, **28**, No. 2, Feb. 1921, pp. 89–90
1545. Hoar, Roger Sherman, *A Course In Exterior Ballistics*, Government Printing Office, Washington, 1921
1546. Hoar, Roger Sherman, *Discussions: On Proof by Mathematical Induction*, American Mathematical Monthly, **29**, No. 4, April 1922, pp. 162–169
1547. Hoar, Roger Sherman, *The Rotation of the Earth*, ,**67**, No. 5 , 1927, pp. 377–388
1548. Hoar, Roger Sherman, review, *New Methods in Exterior Ballistics* by Forest Ray Moulton, American Mathematical Monthly, **34**, No. 5, June-July 1927, pp. 325–326
1549. Hoar, Roger Sherman, review, *New Methods In Exterior Ballistics* by Forest Ray Moulton, Army Ordnance, **7**, 1927, pp. 395–396
1550. Hoar, Roger Sherman, *"New Methods in Exterior Ballistics"*, Army Ordnance, **9**, 1928, pp. 50–51
1551. Hoar, Roger Sherman, *Curved versus Cartesian Coordinates in Ballistics*, Army Ordnance, **9**, No. 3, 1928, pp. 371–376
1552. Hoar, Roger Sherman, entry in *National Cyclopedia of American Biography*, **50**, Ann Arbor, Michigan., p. 622
1553. Hughes, Thomas P., *American Genesis*, Viking, 1989
1554. Hull, Gordon, *Some Aspects Of Physics In War And Peace*, Science, **51**, No. 1314, March 5, 1920, pp. 221–233
1555. Hull. Gordon, *Some Applications of Physics to Ordnance Problems*, Journal of the Franklin Institute, **192**, 1921, pp. 327–347
1556. Hung, J. C., *Method of Adjoint Systems and its Engineering Applications*, The University of Tennessee Department of Electrical Engineering, Technical Note No. 1, Control Theory Group, Prepared for National Aeronautics and Space Administration, October 10, 1964

1557. Hunter, Patti, *Drawing the Boundaries: Mathematical Statistics in 20th-Century America*, Historia Mathematica, **23**, 1996, pp. 7–30
1558. Huntington, Edward V., *The Notion of Probable Error in Elementary Statistics*, Science, New Series, **66**, No. 1722, 1927, pp. 633–637
1559. Ingalls, James M., *Exterior Ballistics in the Plane of Fire*, D. Van Nostrand, New York, 1886
1560. Ingalls, James M., *Handbook of Problems in Direct Fire*, John Wiley and Sons, New York, 1890
1561. Ingalls, James M., *Range Tables For The 12-Inch Cast-Iron B.L. Mortar*, JUSA **1**, 1892, pp. 55–61
1562. Ingalls, James M., *General Ballistic Tables For Mortar Firing*, JUSA, **8**, No. 1, 1896, pp. 52–57
1563. Ingalls, James M., *A New General Ballistic Table*, JUSA, **9**, No. 3, 1897, pp. 295–299
1564. Ingalls, James M., *Ballistics Tables (Artillery Circular M)*, War Department Document 111, Government Printing Office, Washington, 1900
1565. Irwin, F., *The Invariants of Linear Differential Expressions*, Proceedings of the American Mathematical Society, **44**, No. 1, 1908, pp. 3–60
1566. Jackson, A. M., *Spotting*, CAJ, **61**, 1924, pp. 180–181
1567. Jackson, D., *The Method of Numerical Integration in Exterior Ballistics*, Washington, Government Printing Office, 1921
1568. Jackson, D., *The Human Significance of Mathematics*, American Mathematical Monthly, **35**, No. 8, October 1928, pp. 406–411
1569. Jacobi, C. G. J., *Theoria Novi Multiplicatoris Systemati Aequationum Differentialium Vulgarium Applicandi*, in *Gesammelte Werke*, Vierter Band, Cheslea Publishing Company, New York, N.Y., 1969, pp. 317–509
1570. Jacobi, C. G. J., *Zur Theorie der Variations-Rechnung und der Differentialgleighungen*, Journal für Mathematik, **17**, 1837, p. 68 ff.
1571. Johnson, J. B., Bryan, C. W., and F. E. Turneaure, *The Theory and Practice of Modern Framed Structures*, Second Edition, John Wiley and Sons, New York, 1894
1572. Johnson, J. J., *Computation of a Trajectory by Short Arcs*, CAJ, Vol. 55, 1921, pp. 50–67
1573. Johnson, J. J., *An Introduction to the Construction of Nomographic or Alignment Charts*, Coast Artillery Journal, **59**, No. 1, July 1923, pp. 157–172
1574. Johnson, Thomas H., *On the Strength of Columns: Discussing the Experiments Which have Been Accumulated, and Proposing New Formulas*, Transactions of the American Society of Civil Engineers, **15**, July 1886, pp. 518–550
1575. Kanigel, Robert, *The One Best Way*, Penguin Books, New York, 1997
1576. Kellogg, O. D., *A Decade of American Mathematics*, Science, New Series, **53**, No. 1381, June 17, 1921, pp. 541–548
1577. Kellogg, Oliver D., *Foundations of Potential Theory*, Springer, Berlin, 1929, reprinted by Dover Publications, 1953
1578. Kent, R. H., *Projectile Design*, Army Ordnance, November-December 1921, **2**, No. 9, pp. 204–205
1579. Kent, R. H., *Professor Moulton's "New Methods in Exterior Ballistics" Arouses Comment*, Army ORdnance, **7**. No. 41, March 1927, p. 395
1580. Kirk, A. G., *Trajectories and Their Corrections*, Proceedings of the U.S. Naval Institute, **45** No.s 5–8, 1919, pp. 1375–1395
1581. Kirk, A. G., *New Methods of Exterior Ballistics*, Proceedings of the U.S. Naval Institute, **46**, 1920 pp. 1776–1784
1582. Kline, Morris, *Mathematical Thought from Ancient to Modern Times*, Oxford University Press, 1972
1583. Kneser, Adolph, *Lehrbuch Der Variationsrechnung*, Friederich Vieweg Und Sohn, Braunschweig, 1900
1584. Lagrange, Joseph Louis, *Essai d'une nouvelle méthode pour déterminer les maxima et les minima des formules integrales indefinies*, Misc. Soc. Turin, **2**, 1762, pp. 173–195

1585. Lagrange, Joseph Louis, *Solution de differens problemes de calcul integral*, Misc. Tour. **3**, 1762–1765, p. 179
1586. Lamson, K., *A General Implicit Function Theorem With An Application To Problems Of Relative Minima*, American Journal of Mathematics, **41**, No. 4, October 1920, pp. 243–256
1587. Lanza, Gaetano, *Applied Mechanics*, eigth Edition, John Wiley and Sons, New York, 1904
1588. Lecat, Maurice, *Bibliographie du Calcul des Variations, 1850–1913*, Ghent and Paris, 1913
1589. Legendre, A. L., *Mémoire sure la manière de distinguer les maximas des minimas dans le calcul des variations*, Mémoires de l'Académie des Sciences, 1786
1590. Leitmann, G., ed., *Optimization Techniques*, Academic Press, New York, 1962
1591. Le Stourgeon, E., *Minima of Functions of Lines*, Trans. Am. Math. Soc., 21, No. 4, October 1920, pp. 357–383
1592. Liouville, Joseph, *Second Mémoire Sur le developpment des functions ou parties de fonctions en series*, Journal des Mathématiques, **2**, 1837, pp. 16–55
1593. Lissak, Ormand, *Ordnance and Gunnery*, John Wiley and Sons, New York, 1908
1594. Lissak, Ormand, *Ordnance and Gunnery*, John Wiley and Sons, New York, 1915
1595. Lorch, E. R., *Joseph Fels Ritt*, Bulletin of the American Mathematical Society, **57**, No. 4, 1951, pp. 307–318
1596. Love, A. E. H., *A Treatise on the Mathematical Theory of Elasticity*, Fourth Edition, Cambridge University Press, 1927, reprinted by Dover Publications, New York, 1944
1597. Lovitt, William Vernon, *Linear Integral Equations*, McGraw-Hill Book Company, 1924, reprinted by Dover Publications, 1950
1598. Lowan, Arnold N. *On The Bliss Algorithm and its application to the Treatment of Many-point Boundary Problems*, University of California Radiation Laboratory, Livermore, California, June 1958
1599. Manning, Florence M., *Carl G. Barth: a Sketch*, Thesis, A. B. 1926, Submitted in partial satisfaction of the requirements for the degree of MASTER OF ARTS in Economics, in the Graduate Division, of the University of California
1600. Mason, Max, *Zur Theorie de Randwertaufgaben*, Mathematische Annalen, **63**, 1904, pp. 528–544
1601. Mason, Max, *Green's Theorem and Green's Functions for Certain Systems of Differential Equations*, Transactions of the American Mathematical Society, **5**, 1904, pp. 220–225
1602. Mason, Max, *On the boundary Value Problems of Linear Ordinary Differential Equations of the Second Order*, Transactions of the American Mathematical Society, **7**, 1906, pp. 340–344
1603. Mason, Max, *On the Differential Equation of the Hyperbolic Type*, Mathematische Annalen, **65**, 1908, pp. 570–575
1604. Mason, Max, *Selected Topics in the Theory of Boundary Value Problems of Differential Equations*, in The New Haven Mathematical Colloquium, Yale University Press, 1910, pp. 173–222
1605. Mayer, A., *Über Das allgemeinste Problem der Variationsrechnung bei einer einzigen unabhangingen Variabeln*, Leip. Ber., 30, 1878, pp. 16–32
1606. Mayer, A., *Begründung der Lagrange'schen Multiplicatorenmethode in der Variationsrechnung*, Math. Ann., **26**, 1886, pp. 74–82
1607. McCalla, T. R., *Application of the Adjoint System of Differential Equations in the Solution of the Bang-Bang Control Problem*, United States Naval Postgraduate School, Monterey, California, 1961
1608. McFarland, Earl, *Textbook of Ordnance and Gunnery*, John Wiley and Sons, New York, 1929
1609. McShane, E. J., *Differentials of Certain Functions in Exterior Ballistics*, Duke Mathematical Journal, 17, 1950, pp. 115–134
1610. McShane, E. J., *Gilbert Ames Bliss, 1876–1951 A Biographical Memoir by E.J. McShane*, National Academy of Sciences, 1958, pp. 32–53
1611. McShane, E. J., *The Calculus of Variations from the Beginning through Optimal Control Theory*, SIAM Control and Optimization, **27**, No. 5, September 1989, pp. 916–939

1612. McShane, Edward J., Kelley, John L., Reno, Franklin V., *Exterior Ballistics*, The University of Denver Press, Denver, Colorado, 1953
1613. Meigs, J.F. and Ingersoll, R.R., *Exterior Ballistics*, Issac Friedenwald, Baltimore, 1885
1614. Merriman, Mansfield, *On the Constants In Gordon's Formula For The Strength of Columns*, Journal of the Franklin Institute, January 1882, pp. 58–74
1615. Metcalfe, Henry, *A Course Of Instruction in Ordnance and Gunnery*, John Wiley and Sons, New York, 1891
1616. Metcalfe, Henry, *Ordnance and Gunnery Part II: Cannon, Carriages, Small Arms, Ammunition and Accuracy*, West Point, United States Military Academy Press, 1891
1617. Milne, W. E., *Equations of Differential Variations in Exterior Ballistics*, CAJ, **51**, 1919, pp. 154–159
1618. Milne, W. E., review, *Mathematics For Exterior Ballistics*, Mathematics Reviews, MR 0010480; (6.24d) 71.0X
1619. Moncrieff, J. M.,*The Practical Column Under Central or Eccentric Loads*, Transactions of the American Society of Civil Engineers, **45**, June 1901, pp. 334–460
1620. Moore, E. H., *On a Form of General Analysis With Applications to Linear Differential and Integral Equations*, Atti del IV Congresso Internazionale dei Matematici, **1**, Roma 1908, **2**, Roma 1909, pp. 98–114
1621. Moore, E. H., E. J. Wilczynski and Max Mason, *The New Haven Mathematical Colloquium*, Yale University Press, New Haven, 1910
1622. Moore, Herbert F., *The Effects of Keyways on the Strength of Shafts*, University of Illinois Bulletin, No. 42, 1909
1623. Moulton, Forest Ray, *On Certain Rigorous Methods of Treating Problems in Celestial Mechanics*, Decennial Publications of the University of Chicago, **8**, No. 191, (42), 1902
1624. Moulton, Forest Ray, *The Influence of Astronomy on Mathematics*, Science, New Series, **33**, No. 835, March 10, 1911, pp. 357–364
1625. Moulton, Forest Ray, *M. Henri Poincaré*, Popular Astronomy, **20**, No. 10, December 1912, pp. 620–634
1626. Moulton, Forest Ray, *An Introduction To Celestial Mechanics*, Second Revised Edition, Macmillan Company, 1914, reprinted by Dover Publications, New York, 1970
1627. Moulton, Forest Ray, *Lectures on the Modern Theory of Differential Equations*, University of Chicago, 1916–1917
1628. Moulton, Forest Ray, *History of the Ballistics Branch of the Artillery Ammunition Section, Engineering Division of the Ordnance Department for the Period April 6, 1918 to April 2, 1919*, War Department, Office of the Chief of Ordnance, Washington, 1919
1629. Moulton, Forest Ray, *Numerical Integration of Differential Equations*, CAJ, **51**, 1919, pp. 40–55
1630. Mouton, Forest Ray, *Numerical Solution of Differential Equations*, in *Smithsonian Mathematical Formulae and Tables of Elliptic Functions*, Smithsonian Miscellaneous Collections, **74**, No. 1, Publication 2672, Smithsonian Institution, 1922
1631. Moulton, Forest Ray, *New Methods In Exterior Ballistics*, University of Chicago Press, 1926, reprinted by Dover Publications, 1962
1632. Moulton, Forest Ray, discussion of review of *New Methods in Exterior Ballistics* by Roger Sherman Hoar, American Mathematical Monthly, **35**, No. 5, May 1928, pp. 246–250
1633. Moulton, Forest Ray, reply to a review of *New Methods in Exterior Ballistics* by Roger Sherman Hoar, Army Ordnance, **8**, 1928, pp. 320–321
1634. Moulton, Forest Ray, *Differential Equations*, MacMillan Company 1930, reprinted by Dover Publications, New York, 1958
1635. Moulton, F. R., entry in *Notable Scientists: From 1900 to the Present*, Gale Group, 2001, Anonymous.
1636. Nenninger, T.K., *The Leavenworth Schools and the Old Army: Education, Professionalism, and the Officer Corps of the United States Army, 1881–1918*, Greenwood Press, Westport, Connecticut, 1978
1637. d'Ocagne, Maurice, *Traité de Nomographie, Deuxième Édition*, Gauthier Villars, Paris, 1921

1638. Osgood, William Fogg, *Advanced Calculus*, The MacMillan Company, New York, 1925
1639. Paaswell, G., *An Appeal to Producing Mathematicians*, Bulletin of the American Mathematical Society, **21**, No. 3, 1914, pp. 127–130
1640. Pachpatte, B. G., *Inequalities for Differential and Integral Equations*, San Diego Press, 1998
1641. Parshall, Karen Hunger, and Rowe, David E., *The Emergence of the American Mathematical Research Community, 1876–1900: J. J. Sylvester, Felix Klein, and E. H. Moore*, American Mathematical Society, 1994
1642. Parshall, Karen Hunger, *Perspectives on American Mathematics*, Bulletin of the American Mathematical Society,**37**, No.4, pp 381–405
1643. Parshall, Karen Hunger, The New Era in American Mathematics, 1920–1950, Princeton University Press, 2020
1644. Picard, É., *Mémoire sur la théorie des éqations aux dérivées partielles et la méthode des approximations successives*, Journal des Mathematiques, series 4, **6**, 1890, pp. 145–210
1645. Picard, É. *Sur l'application des méthodes d'approximations successives à l'étude de certaines équations différentielles ordinaires*, Journal des Mathematiques series 4, **9**, 1893, pp. 217–271
1646. Pincherle, S., *Sull'operazione aggiunta*, Rendiconto della R. Accademia delle Science dell'Instituto di Bologna, **2**, (1897–1898), pp. 130–139
1647. Pincherle, S., *Sull'operazione agguinta di Lagrange*, Annali de metematica pur ed applicata, **21**, 1913 pp. 143–151
1648. Poincaré, Henri, *Les Méthodes Nouvelles de la Mécanique Céleste , Tome 1*, Dover Publications, New York, reprint of original 1892 edition
1649. Polachek, Harry, *The Men Behind The Men Behind The Guns*, Scripta Mathematica, **9**, March 1943, pp. 27–32
1650. Polachek, Harry, *Solution of the Differential Equations of Motion of a Projectile in a Medium of Quasi-Newtonian Resistance*, Quarterly Applied Math. **7**, 1949, pp. 275–291
1651. *Lecture on The Probability and Dispersion of Fire*, author unknown, translated at the Army War College From a French Document, Government Printing Office, 1917
1652. *Provisional Range Tables for the British 75-mm. Gun, Model 1917*, Ordnance Department, U.S. Army, War Plans Division, Government Printing Office, 1918
1653. Rademacher, Hans, *On the Accumulation of Errors In Numerical Integration on the ENIAC*, in *Theory and Techniques for Design of Electronic Digital Computers*, Lectures Given at the Moore School, 8 July 1946–31 August 1946, Volume II, Lecture 19, University of Pennsylvania, Moore School of Electrical Engineering, Philadelphia, Pennsylvania, November 1, 1947
1654. *Range Tables For 75-mm. Model 1916 American and Model 1917 British Guns*, Prepared At Aberdeen Proving Ground, Engineer Reproduction Plant, U.S. Army, Washington Barracks, November 1921
1655. Riemann, Bernhard, *Über die Fortpflanzung ebener Luftwellen von endlicher Schwingingungsweite*, Gottingen Abhandl., **8**, 1860, pp. 145–164
1656. Ritt, J., *On Weighting Factor Curves For Low Elevation*, NARA Aberdeen, Technical Report, February 25, 1919
1657. Ritt, J., *On the Differentiability of the Solution of a Differential Equation With Respect to a Parameter*, Annals of Mathematics, Second Series, **20**, No. 4, July 1919, pp. 289–291
1658. Ritt, J., *On Weighting Factors in Flat Fire*, CAJ, **53**, 1920, pp. 404–410
1659. Ritt, J., *On the Iteration of Rational Functions*, Transactions of the American Mathematical Society, **21**, 1920, pp. 313–320
1660. Roberts, David L., *Albert Henry Wheeler (1873–1950): A Case Study in the Stratification of American Mathematical Activity*, Historia Mathematica, **23**, 1996, pp. 269–287
1661. Roberts, David L., *Republic of Numbers*, Johns Hopkins University Press, 2019
1662. Rosenstein, George M., *The Best Method: American Calculus Textbooks of the Nineteenth Century*, from *A Century of Mathematics In America, Part III*, ed. Duren, P., Askey, R., Merzbach, U., and Edwards, H., Providence, Rhode Island, American Mathematical Society, 1988–1989, pp. 77–109

1663. Roth, William E., *The Exact Differential Equations Of A Trajectory In Terms Of The Hoar Curvilinear Coordinates*, CAJ, **68**, 1928, pp. 505–512
1664. Rowe, J. E., *A Course In Exterior Ballistics*, Bulletin of the American Mathematical Society, **29** , No. 6, , June 1923, pp. 277–279
1665. Rowe, J. E., *Moulton on Exterior Ballistics*, American Mathematical Monthly, **35**, No. 4, March-April 1928, pp. 229–232
1666. Ruckman, John W., *Teh Wind Problem in Gunnery*, School of Mines Quarterly, **11**, 1889/90, pp. 306–320
1667. Ruckman, John W., *The Effect Of Wind On The Motion Of A Projectile*, JUSA, **1**, 1892, pp. 5–24
1668. Salmon, E. H., *Columns*, Oxford Technical Publications, Henry Frowde and Hodder and Stoughton, London, 1921
1669. Sanchez, D., *Ordinary Differential Equations and Stability Theory*, Freeman and Company, 1968
1670. Schwartz, Philip, *A Critical Survey of the Advances in Practical Ballistics in the Ordnance Department, U.S. Army Since 1917*, [NARA Abderdeen] A-I-31, 1920
1671. Schwartz, Philip, *Measuring the Life of a Gun*, Army Ordnance, **5**, No. 2, 1924, pp. 111–112
1672. Schwartz, Philip, *Accuracy Firing*, Army Ordnance, **5**, No. 3, 1924, pp. 521–522
1673. Schwartz, Philip, *Experimental Investigation of The Effect of the Wind Upon the Motion of a Projectile*, CAJ, 60, 1924, pp. 93–97
1674. Schwartz, Philip, *Studies of Motion of Projectiles*, CAJ, **60**, 1924, pp. 430–431
1675. Schwartz, Philip, *The English Long Range Gun*, CAJ, **61**, 1924, pp. 91–92
1676. Schwartz, Philip, *Weighting Factors*, CAJ, **61**, 1924, pp. 371–372
1677. Schwartz, Philip, *A Critical Survey of the Advances in Practical Ballistics in the Ordnance Department, U. S. Army, Since 1917*, Coast ArtilleryJournal, **62**, January-June, 1925, pp. 54–66
1678. Schwartz, Philip, *Remarks Concerning the Use of Firing Table Values*, CAJ, **63**, 1925, pp. 399–400
1679. Schwartz, Phlilp, *The Variable Probable Error*, CAJ, **67**, 1927, pp. 405–413
1680. Schwartz, Philip, *Much About Little* CAJ, **68**, 1928, pp. 291–297
1681. Schwartz, Philip, *The Fork*, Field Artillery Journal, **18**, No. 3, 1929, pp. 604–609
1682. Schwartz, Philip, *More About Probable Error*, CAJ, **73**, 1930, pp. 111–120
1683. Siacci, F., *Balistique Extérieure*, Traduction Annoteé par P. Laurent, Berger-Levrault et Cic, Editours, Paris and Nancy, 1892
1684. Siegmund-Schultze, R.,*Eliakim Hastings Moore's "General Analysis"*, Arch. Hist. Exact. Sci., **52**, 1998, pp. 51–89
1685. Siegmund-Schultze, R., *The Late Arrival of Academic Applied Mathematics in the United States: a Paradox, Theses, and Literature*, NTM, **11**, 2002, pp. 116–12
1686. Siegmund-Schultze, R., *Military Work in Mathematics 1914–1945, An Attempt at an International Perspective*, in *Mathematics and War*, Birkhauser, Basel, 2003, pp. 23–82
1687. Siegmund-Schultze, R., *The Ideology of Applied Mathematics Within Mathematics in Germany and The U. S. Until the End of World War II*, ILUIL, 27, 2004, pp. 791–811
1688. Sleight, E. R., *Mathematics Applied*, National Mathematics Magazine, **9**, No. 8, May 1935, pp. 219–223
1689. Smith, Paul A., *Joseph Fels Ritt, 1893–1951*, United States National Academy of Sciences, Biographical Memoir, 1956, pp. 253–264
1690. Spinrad, William, *Early History of The Ordnance School*, The Ordnance Sergeant **3**, No. 1, 1942. pp. 1–4
1691. Stäckel, P., *Abhandlungen Über Variations-Rechnung, Zweiter Theil*, Leipzig, Verlag Von Wilhelm Engelmann, 1894
1692. Steele, Bret D., *Muskets and Pendulums: Benjamin Robins, Leonhard Euler, and the Ballistics Revolution*, Technology and Culture, **35**, No. 2, April 1994, pp. 348–382
1693. Stevenson, Craig, *Periodic Orbits of the Three Body Problem: F. R. Moulton's Quest for a New Lunar Theory*, Ph.D. Thesis, Open University, 2012

1694. Stevenson, Craig, *Periodic Orbits: F. R. Moulton's quest for a New Lunar Theory*, American Mathematical Society, 1921
1695. Stockley, Ernest Norman, *Military Bridging*, The Encyclopaedia Britannica, The New Volumes, 30, London, New York, 1922, pp. 500–504
1696. *Syllabus of Mathematics, A Symposium Compiled by The Committee on the Teaching of Mathematics to Students of Engineering*, The New Era Printing Company, Lancaster Pa., 1914
1697. Tannehill, I. R., *The Ballistic Wind*, Journal of the United States Artillery, **52**, 1920, pp. 553–576
1698. Tarwater, J. D., (Ed.), *The Bicentennial Tribute to American Mathematics, 1776–1976* Papers Presented to the Fifty-Ninth Annual Meeting of the Mathematical Association of America, Mathematical Association of America, Washington, D. C., 1977
1699. Todd, David, Halley's Comet, American Book Company, New York, Cincinnati, Chicago, 1910
1700. Todhunter, I., *A Treatise on the Integral Calculus and its Applications*, MacMillan and Co., London, 1880
1701. Tonelli, L., *Fondamenti di calcolo delle variazioni*, Nicola Zanichelli, I, 1921; II, 1923
1702. Tropp, H. S., *Forest Ray Moulton*, Dictionary of Scientific Biography, **9**, pp. 552–553
1703. Truesdell, Clifford, review of *Exterior Ballistics* by McShane, Kelley and Reno, Scripta Mathematica, **20**, 1954, pp. 172–174
1704. Tschappat, W.H., *Ballistics*, article in Encyclopedia Britannica, 1922, pp. 383–394
1705. Tsien, H. S., Adamson, T. C., E. L. Knuth, *Automatic Navigation of a Long Range Rocket Vehicle*, Journal of the American Rocket Society, July-August, 1952
1706. Tsien, H. S., *Engineering Cybernetics*, McGraw-Hill Book Co., 1954
1707. Veblen, Oswald, *Ballistic Theories and the Preparation of Range Tables*, NARA Aberdeen, No. 54, 1921
1708. Veblen, O., *Professor Veblen on Ballistics in the World War*, Army Ordnance, **4**, March-April 1924, pp. 337–338
1709. Very, Edward W., *Computation of Range Tables*, Annapolis, Md., 1877
1710. Volterra, V., *Leçons sur les Fonctions des Lignes*, Paris, Gauthier- Villars, 1913
1711. Wertheimer, Albert, *A Method of Computing Differential Corrections For a Trajectory*, Journal of the Franklin Institute, **217**, No. 6, June 1934
1712. Wiener, Norbert *Ex-Prodigy*, MIT Press, Cambridge, 1953
1713. Wilhelm, Glenn P., *Long Range Small Arms Firing, Part II*, Army Ordnance, **1**, pp. 273–285
1714. Williamson, Benjamin, *An Elementary Treatise on The Integral Calculus*, Longmans, Green and Co., London, 1916
1715. Wilson, Edwin B., Advanced Calculus, Dover reprint of the original 1912 edition, 1958
1716. Wlison, Edwin B., *The Mathematics of Aerodynamics*, Science, **25**, No. 7, September 1918, pp. 191–197
1717. Wilson, Edwin B., *Empiricism and Rationalism*, Science, **64**, No. 1646, July 16, 1926, pp. 47–57
1718. Worcester, J. R., *Safe Stresses in Steel Columns*, Transactions of the American Society of Civil Engineers, **61**, pp. 158–215, 1908
1719. Zeidler, E., *Applied Functional Analysis*, Springer-Verlag, 1995
1720. www.math.colostate.edu/~pauld/M345/Some20theory.pdf

# Index




A. Gluchoff, *From Fréchet Differentials to Firing Tables*,
https://doi.org/10.1007/978-3-031-68267-4

The manufacturer's authorised representative in the EU is Springer Nature Customer Service Centre GmbH, Europaplatz 3, 69115 Heidelberg, Germany. If you have any concerns regarding our products, please contact ProductSafety@springernature.com

Printed and bound by CPI Group (UK) Ltd, Croydon, CR0 4YY
30/09/2026
02225600-0014